Receptors and Recognition

Series A

Published

Volume 1 (1976)
M.F. Greaves (London), Cell Surface Receptors: A Biological Perspective
F. Macfarlane Burnet (Melbourne), The Evolution of Receptors and Recognition in the Immune System.
K. Resch (Heidelberg), Membrane Associated Events in Lymphocyte Activation
K.N. Brown (London), Specificity in Host—Parasite Interaction

Volume 2 (1976)
D. Givol (Jerusalem), A Structural Basis for Molecular Recognition: The Antibody Case
B.D. Gomperts (London), Calcium and Cell Activation
M.A.B. de Sousa (New York), Cell Traffic
D. Lewis (London), Incompatibility in Flowering Plants
A. Levitski (Jerusalem), Catecholamine Receptors

Volume 3 (1977)
J. Lindstrom (Salk, California), Antibodies to Receptors for Acetylcholine and other Hormones
M. Crandall (Kentucky), Mating-type Interaction in Micro-organisms
H. Furthmayr (New Haven), Erythrocyte Membrane Proteins
M. Silverman (Toronto), Specificity of Membrane Transport

Volume 4 (1977)
M. Sonenberg and A.S. Schneider (New York), Hormone Action at the Plasma Membrane: Biophysical Approaches
H. Metzger (NIH, Bethesda), The Cellular Receptor for IgE
T.P. Stossel (Boston), Endocytosis
A. Meager (Warwick), and R.C. Hughes (London), Virus Receptors
M.E. Eldefrawi and A.T. Eldefrawi (Baltimore), Acetylcholine Receptors

In preparation

Volume 5 (1978)
K. Weber (Gottingen), Actin
P.A. Lehman (Mexico), Stereoselectivity in Receptor Recognition
A.G. Lee (Southampton), Fluorescence and NMR Studies of Membranes
L. Kohn (NIH, Bethesda), Relationships in the Structure and Function of Receptors for Glycoprotein Hormones, Bacterial Toxins and Interferon

Series B

Published

The Specificity and Action of Animal, Bacterial and Plant Toxins (B1)
edited by P. Cuatrecasas (Burroughs Wellcome, North Carolina)

In preparation

Intercellular Junctions and Synapses (B2)
edited by J. Feldman (University of London), N.B. Gilula (Rockefeller
   University, New York) and J.D. Pitts (University of Glasgow)

Specificity of Embryological Interactions (B4)
edited by D. Garrod (University of Southampton)

Taxis and Behaviour (B5)
edited by G. Hazelbauer (Wallenberg Laboratory, Uppsala)

Virus Receptors (B6)
edited by L. Phillipson and L. Randall (University of Uppsala) and
K. Lonberg-Holm (Du Pont, Delaware)

Receptors and Recognition

General Editors: P. Cuatrecasas and M.F. Greaves

About the series

Cellular recognition — the process by which cells interact with, and respond to, molecular signals in their environment — plays a crucial role in virtually all important biological functions. These encompass fertilization, infectious interactions, embryonic development, the activity of the nervous system, the regulation of growth and metabolism by hormones and the immune response to foreign antigens. Although our knowledge of these systems has grown rapidly in recent years, it is clear that a full understanding of cellular recognition phenomena will require an integrated and multidisciplinary approach.

This series aims to expedite such an understanding by bringing together accounts by leading researchers of all biochemical, cellular and evolutionary aspects of recognition systems. The series will contain volumes of two types. First, there will be volumes containing about five reviews from different areas of the general subject written at a level suitable for all biologically oriented scientists (Receptors and Recognition, series A). Secondly, there will be more specialized volumes, (Receptors and Recognition, series B), each of which will be devoted to just one particularly important area.

Advisory Editorial Board

A.C. Allison, Clinical Research Centre, London, U.K.
E.A. Boyse, Memorial Sloan-Kettering Cancer Center, New York, U.S.A.
F.M. Burnet, University of Melbourne, Australia.
G. Gerisch, Biozentrum der Universität Basel, Switzerland.
D. Lewis, University College London, U.K.
V.T. Marchesi, Yale University, New Haven, U.S.A.
A.A. Moscona, University of Chicago, U.S.A.
G.L. Nicolson, University of California, Irvine, U.S.A.
L. Philipson, University of Uppsala, Sweden.
G.K. Radda, University of Oxford, U.K.
M. Raff, University College London, U.K.
H.P. Rang, St. George's Hospital Medical School, London, U.K.
M. Rodbell, National Institutes of Health, Bethesda, U.S.A.
M. Sela, The Weizmann Institute of Science, Israel.
L. Thomas, Memorial Sloan-Kettering Cancer Center, New York U.S.A.
D.F.H. Wallach, Tufts University School of Medicine, Boston, U.S.A.
L. Wolpert, The Middlesex Hospital Medical School, London, U.K.

Receptors and Recognition

Series B Volume 3

Microbial Interactions

Edited by
J.L. Reissig

*Department of Biology, C.W. Post College,
Long Island University, New York*

LONDON
CHAPMAN AND HALL

A Halsted Press Book
John Wiley & Sons, New York

First published 1977
by Chapman and Hall Ltd.,
11 New Fetter Lane, London EC4P 4EE

© 1977 Chapman and Hall

Typeset by C. Josée Utteridge-Faivre of Red Lion Setters
and printed in Great Britain
at the University Printing House, Cambridge

ISBN 0 412 14830 7

All rights reserved. No part of
this book may be reprinted, or reproduced
or utilized in any form or by any electronic,
mechanical or other means, now known or hereafter
invented, including photocopying and recording,
or in any information storage and retrieval
system, without permission in writing
from the Publisher.

Distributed in the U.S.A. by Halsted Press,
a Division of John Wiley & Sons, Inc., New York

Library of Congress Cataloging in Publication Data

Main entry under title:

Microbial interactions.

(Receptors and recognition: Series B; v.3
Includes index.
1. Micro-organisms – Physiology. 2, Cell interaction.
I. Reissig, Jose Luis. II. Series.
QR96.5.M5   576'.11   77–22325
ISBN 0–470–99226–3

Contents

		page	
	Contributors		viii
	Preface		ix
1	Aggregation and Cell Surface Receptors in Cellular Slime Molds Peter C. Newell		1
2	Bacterial Chemotaxis Gerald L. Hazelbauer and John S. Parkinson		59
3	Bacterial Receptors for Phages and Colicins as Constituents of Specific Transport Systems V. Braun and K. Hantke		99
4	The Attachment of Bacteria to the Surfaces of Animal Cells Garth W. Jones		139
5	Binding and Entry of DNA in Bacterial Transformation Sanford A. Lacks		177
6	A Redefinition of the Mating Phenomenon in Bacteria Mark Achtman and Ron Skurray		233
7	Cell–Cell Interactions during Mating in *Saccharomyces cerevisiae* Thomas R. Manney and James H. Meade		281
8	Mating Interactions in *Chlamydomonas* Ursula W. Goodenough		323
9	Cell–Cell Interactions in Ciliates: Evolutionary and Genetic Constraints D.L. Nanney		351
10	An Overview Jose L. Reissig		399
	Thesaurus of Microbial Interactions		417
	Index		427

Contributors

Mark Achtman, Max-Planck Institut für molekulare Genetik, Berlin, Germany.

Volkmar Braun, Lehrstuhl Mikrobiologie II, Universität Tübingen, Tübingen, West Germany.

Ursula W. Goodenough, The Biological Laboratories, Harvard University, Cambridge, Massachusetts, U.S.A.

K. Hantke, Lehrstuhl Mikrobiologie II, Universität Tübingen, Tübingen, West Germany.

Gerald L. Hazelbauer, Wallenberg Laboratory, University of Uppsala, Uppsala, Sweden.

Garth W. Jones, Department of Microbiology, University of Michigan, Ann Arbor, Michigan, U.S.A.

Sanford Lacks, Department of Biology, Brookhaven National Laboratory, Upton New York, U.S.A.

Thomas R. Manney, Department of Physics, Kansas State University, Manhattan, Kansas, U.S.A.

James H. Meade, Department of Physics, Kansas State University, Manhattan, Kansas, U.S.A.

David L. Nanney, Department of Genetics and Development, University of Illinois, Urbana, Illinois, U.S.A.

Peter C. Newell, Department of Biochemistry, University of Oxford, Oxford, U.K.

John S. Parkinson, Department of Biology, University of Utah, Salt Lake City, Utah, U.S.A.

Ron Skurray, Department of Molecular Biology, University of California, Berkeley, California, U.S.A.

Preface

Microbiology has undergone a number of metamorphoses in its relatively brief existence. It has been in approximate succession, morphology, epidemiology, biochemistry, genetics, and molecular biology. It is also becoming a significant parcel of cell surface studies. The one embodiment which has remained elusive — particularly for bacteriology — is the taxonomic one. This may have been a blessing in disguise because it encouraged microbiologists to deal with the general rather than the particular; promoting a search for unitary explanations, in the manner of Kluyver and van Niel, long before anyone knew about the universality of the genetic code, or could trace the genealogy of enzymes from the study of amino acid substitutions.

This volume is predicated on the idea that deep analogies underly the mechanisms of cellular interaction, and therefore belongs in the unitary tradition of microbiology. It occupies itself with a wide variety of micro-organisms, considering them from vantage points of considerable diversity, ranging from taxonomic irreverence to keen evolutionary awareness, and is concerned with areas which have developed independently of each other.

A perceptive librarian might place this book in a section of the stacks labeled 'Mystery Stories'. It contains nine pieces (Chapters 1 to 9) differing widely in degree of scrutability. The chapters on slime molds (Newell), bacterial chemotaxis (Hazelbauer and Parkinson) and transformation (Lacks) rank high in that regard: the plot has progressed towards the final stages, and there can be little doubt about the identity of some of the culprits. At the other end of the scale lies the review on the mode of attachment of microbes to animal cells (Jones), covering a story which has barely advanced beyond a partial presentation of the cast and the sketch of possible scenarios. Three chapters are at that very stage, all too familiar to detective story aficionados, at which new developments demand a reassessment of the whole situation. Such is the case of the stories on mating in bacteria (Achtman and Skurray), yeast (Manney and Meade) and algae (Goodenough). The two remaining chapters, concerned with receptors located on the outer membrane of bacteria (Braun and Hantke) and with mating in ciliates (Nanney), contain not one but a whole network of plots, ranging from the clear-cut to the downright baffling. Nanney's chapter would also be at home in the 'History' section of the library stacks, as it contributes a precious evolutionary perspective to our collection.

While the chapter by Braun and Hantke, as well as that of Lacks, refer to specific aspects of transport systems, no attempt has been made to deal here with the overall problem of biological transport. This is regrettable insofar as transport studies have played, and will continue to play, a pivotal role in our understanding of membrane function. However the field has become too formidable to attempt to deal with it in an isolated chapter, and there is no dearth of treatises on the subject.

The temptation to draw on these nine chapters in order to search for basic mechanisms of interaction, common threads and alternative evolutionary strategies, is not only hard to resist, but also deserves encouragement. Such encouragement is provided in two sections of the book: an Overview (Chapter 10) and what has been called a Thesaurus (Appendix). An alphabetical index is also provided. The Thesaurus is an index arranged conceptually. Its main purpose is to lay bare the contents of the book in a compact format, thus stimulating the reader to discover relationships which might otherwise remain hidden within the text. Inevitably, the construction of the Thesaurus required the selection of a specific conceptual framework around which it could be organized, but this framework should be viewed as nothing more than an aid for the orderly retrieval of information.

March, 1977   *Jose L. Reissig*

1 Aggregation and Cell Surface Receptors in Cellular Slime Molds

PETER C. NEWELL

1.1	Introduction	page	3
1.2	The aggregation mechanism		7
	1.2.1 Pulsed signal generation		7
	1.2.2 Signal reception		12
	1.2.3 Signal destruction		15
	1.2.4 Signal relay		17
	1.2.5 The chemotactic response		22
	1.2.6 The developmental initiation response		23
	1.2.7 The final docking manoeuvre		24
1.3	Cell surface cAMP receptors		25
1.4	Slime mold lectins and their receptors		31
	1.4.1 Discovery of discoidins and pallidins		31
	1.4.2 Evidence for the role of slime mold lectins as recognition and adhesion proteins		31
	1.4.3 Properties of slime mold lectins		34
	1.4.4 Membrane receptors for slime mold lectins		35
	1.4.5 Cell-binding lectins in other systems		35
1.5	Contact sites A		36
1.6	Other surface components implicated in cellular interaction		38
	1.6.1 Divalent metal ion receptors		38
	1.6.2 Glycoproteins binding concanavalin A		38
	1.6.3 Glycosphingolipids		41
	1.6.4 Purified plasma membrane components		42
	1.6.5 ^{125}Iodine surface-labelled components		43
	1.6.6 Plasma membrane particles		44
1.7	Further studies		46
1.8	Summary		48
	References		49

Acknowledgements

I would like to acknowledge the help of my many friends and colleagues in providing me with reprints and preprints of their publications and in particular I wish to thank Dr David Ratner for his critical reading of the manuscripts and Mr Frank Caddick for drawing the figures.

Microbial Interactions
(*Receptors and Recognition,* Series B, Volume 3)
Edited by J.L. Reissig
Published in 1977 by Chapman and Hall, 11 New Fetter Lane, London EC4P 4EE
© Chapman and Hall

1.1 INTRODUCTION

The cellular slime molds would, without much doubt, have been generally ignored had they not exhibited a very peculiar pattern of behaviour. Most living cells are either single independent entities or are dependent members of a cell community that functions as an organism. Slime molds, however, have achieved the art of existing in either state as conditions seem to warrant.

The visually remarkable phase is the transition period between the two forms of existence during which a population of amoebae which previously was unobtrusively feeding on bacteria suddenly moves towards a collecting point and in a few hours transforms itself into a co-ordinated moving slug-like being. This aggregation phase produces radiating streams or whorls of cells sometimes resembling supernovae (Fig. 1.1) or, in dense populations covering an agar surface, concentric or spiral rings can be seen that slowly move outward like waves from the aggregation center (Fig. 1.2). This phase is the overture to a process of development that ends with the production of a droplet of spores held aloft by a slender column of stalk cells.

During aggregation as many as 100 000 amoebae mass together in one aggregate from distances of over 20 mm. To enable such a process to occur, a long-range communication system must clearly be operating between the separate moving cells. Such a system is far more open to investigation than is the 'whispering' between cells that is all that passes for communication in developing systems where all the cells lie adjacent to each other. It is this feature that attracts so much current interest and prompts a detailed study of the aggregation mechanism and of the cell surface receptors that are involved. Various aspects of the aggregation phase and of the cell surface components that are involved have been reviewed before, (Bonner, 1967, 1971, 1973; Konijn, 1972a; Newell, 1971, 1975; Gerisch, 1968, Gerisch *et al.*, 1974b; Robertson and Grutsch, 1974; Loomis, 1975; Frazier, 1976.)

Our current concept of the process of aggregation in the species *Dictyostelium discoideum* (as revealed by the work of many laboratories – see Section 1.2 for details and references) may be briefly summarised as follows: The initial stimulus needed to start aggregation is starvation. This induces in some cells the ability to produce slow rhythmic pulses of the attractant cyclic adenosine monophosphate (cAMP) with an initial frequency of roughly one pulse every ten minutes. Meanwhile, the rest of the starving population produce cAMP receptors on their cell surface that enable them to perceive the pulsed signal. The cAMP signal does not diffuse very far from the centers of its production but is destroyed within 57 μm by phosphodiesterase enzymes (both free and membrane-bound) that are also produced by the starving amoebae. Two responses to this signal are noticeable. Firstly, the amoebae soon begin to move in the general direction of the signal source (and continue to move for 100 seconds covering a distance of about 20 μm) and secondly, about 12 seconds

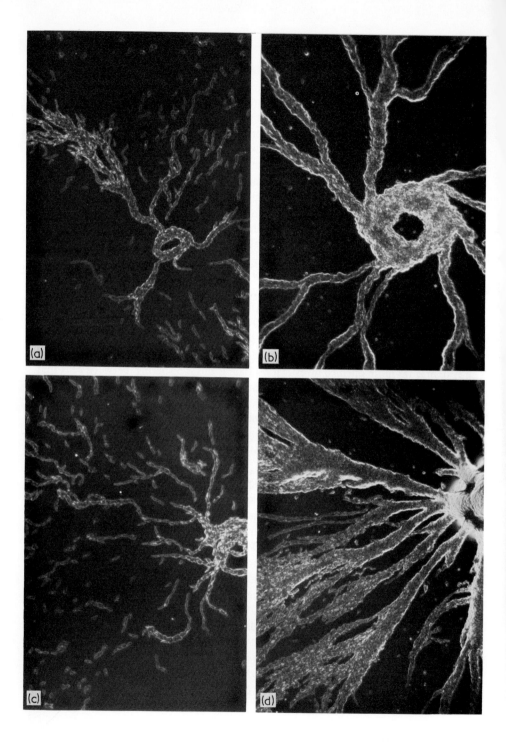

after receiving the signal, the amoebae themselves emit a pulse of cAMP. This pulse then diffuses away and reaches amoebae further out from the center. The amoebae closer to the center are prevented from relaying the signal inwards by their being refractory to further relay stimulation for several minutes after producing a pulse. (The relay refractory period). By this system of relay, a series of waves of cAMP production, destruction and response move outward from the center as the amoebae move inwards (Fig. 1.2). Such waves can be seen with the naked eye probably because of a difference in the shape of moving amoebae which tend to be elongated compared with still amoebae which tend to be more rounded and the waves are particularly clear when viewed as time-lapse movie films. Due to the relay of the pulse by the responding amoebae, their radial motion is unstable and amoebae tend to gather into streams which increasingly act as strong local sources of attraction. Amoebae can, at times, even be seen to move outwards from the center for a while in order to join a stream that happens to curve around behind them. Eventually, the amoebae in these moving streams all reach the aggregation center and a compact aggregate of cells is formed that then secretes a slime sheath over itself. The aggregation phase is now over and the morphogenetic phase of development begins. By rising vertically off the substratum, the aggregate converts itself into a finger-like body which flops over on to the substratum and migrates away as a 'slug' which eventually develops into the mature fruiting structure.

Not all genera of the cellular slime molds, however, use cAMP as the chemotactic agent (the general name of 'acrasins' being given to such agents). Recent work by Wurster *et al.* (1976) has shown that cells of the species *Polysphondylium violaceum* employ an acrasin (probably a peptide) of less than 1500 daltons. Like cAMP, this acrasin is specifically destroyed by an enzyme produced by the cells. *P. violaceum* is not responsive to cAMP nor is *Dictyostelium discoideum* responsive to the *P. violaceum* acrasin (Table 1.1). A curious fact that may be noticed in Table 1.1 is that while *P. violaceum* is unresponsive to cAMP as an acrasin, it actually does secrete it and also produces a phosphodiesterase. This finding suggests that cAMP with its phosphodiesterase play a role in slime mold aggregation or development that is additional to its role as an acrasin in *D. discoideum.*

This hypothesis is supported by a number of reports stemming from that of Bonner (1970) that indicated that high concentrations of cAMP could induce small

Fig. 1.1 Amoebae of *Dictysostelium discoideum* aggregating on thin layers of agar. Dilute suspensions of starving amoebae were plated in the absence of bacteria on to buffered agar and incubated for 16 hours at 7°C followed by incubation at 22°C for 2 hours. The pattern of aggrgation was followed using dark field microscopy. These pictures (taken by the author) show (a) the start of an open spiral aggregation center after a few minutes of incubation at 22°C and (b) the same aggregate 1.5 hours later. (c) and (d) show concentric aggregation patterns taken at similar times. The individual amoebae are approximately 10 μm in diameter.

Fig. 1.2 Concentric and spiral wave patterns seen in densely populated agar plates of starving *D. discoideum* amoebae, using dark field illumination. The light bands are bands of moving amoebae. The long side of the picture represents a distance of approximately 50 mm. (Photograph by courtesy of Drs Fernanda Alcântara and Marylin Monk.)

groups of starving *D. discoideum* amoebae to differentiate into stalk cells (Chia, 1975; Hamilton and Chia, 1975; Town *et al.*, 1976) and it is interesting that Francis (1975) has reported that this is also true for *P. pallidum*. A role for cAMP in determining which cell type (stalk cells or spores) the aggregated amoebae will finally differentiate into seems therefore possible. Such a scheme would not be inconsistent with the observation that the aggregation center forms the front tip of the slug, and it is this tip region that gives rise to the cells that finally form the stalk (Bonner, 1967). It has not yet been definitely shown, however, that the slug tip contains more cAMP than posterior regions. There is also evidence that, in *D. discoideum*, the cAMP pulses (but not a continuous gradient) have a role in initially switching on the post-aggregational program of morphogenetic development (Gerisch *et al.*, 1975a; Darmon *et al.*, 1975) (This is further described in Section 1.2.6).

The role of cAMP in development of the cellular slime molds therefore seems likely to be complex, but it would be surprising from the evidence so far collected if its several actions were completely unrelated. If its role during aggregation can be elucidated, then its effects in less amenable situations in lower and higher cells may be better understood.

Table 1.1 The secretion of cAMP and phosphodiesterase and the attraction to cAMP and the *P. violaceum* acrasin in various species of cellular slime mold. The number of +'s indicates the relative amounts of cAMP or phosphodiesterase secreted. 0 indicates no secretion or response (After Bonner *et al.*, 1972 and Wurster *et al.*, 1976).

Species	Extracellular cAMP	Extracellular phosphodiesterase	Attraction to cAMP	Attraction to *P. violaceum* acrasin
Dictyostelium discoideum	+ +	+ +	+ + + +	0
Dictyostelium purpureum	+ + +	+ +	+ + + +	0
Dictyostelium mucoroides	+ + +	+ +	+ + + +	0
Dictyostelium rosareum	+ + +	+	+ + + +	0
Dictyostelium minutum	+	0	0	0
Polysphondylium pallidum	+	0	0	+ + + +
Polysphondylium violaceum	+	+ +	0	+ + + +

1.2 THE AGGREGATION MECHANISM

This can be conveniently divided into seven steps: (1) Pulsed signal generation, (2) Signal reception, (3) Signal destruction, (4) Signal relay, (5) Chemotactic response, (6) Developmental initiation response, (7) cell recognition and adhesion (the final docking manoeuvre) (Fig. 1.3). These steps will be considered separately.

1.2.1 Pulsed signal generation

The simplest mechanism that might have accounted for a centripetal movement of amoebae would have been the establishment of a chemical gradient of attractant with its highest concentration at the center. The amoebae, however, seem to have had other ideas and although they can under certain conditions respond to continuous gradients of attractant, they seem to use a discontinuous signal in normal aggregation that is produced by a rhythmically pulsating attraction center.

The notion that the emitted signal is pulsatile rather than continuous was based initially on observations, made over a number of years, of waves of concentric or spiral rings emanating from the aggregation center as the amoebae move inwards

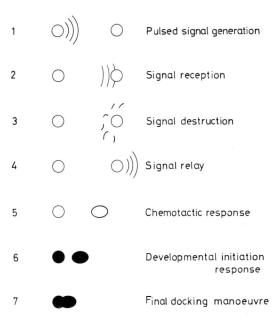

Fig. 1.3 Seven acts of aggregation. The left-hand circle schematically represents a signalling amoeba in the aggregation center (the three curved lines representing a single signal pulse being emitted) and the right-hand circle a responding amoeba. Steps 1–5 may be repeated several times before steps 6 and 7 are taken by the amoebae.

(Arndt, 1937; Bonner, 1944; Shaffer, 1957; Gerisch, 1968). These waves are best seen after incubation of starving wild-type *D. discoideum* amoebae at 7°C followed by incubation at 22°C and observation using dark field optics (Alcântara and Monk, 1974; Gross et al., 1976) (Fig. 1.2). The dark and light bands, which are thought to be caused by small differences in optical properties between moving and still amoebae, can be seen to move outwards at roughly 300 μm min^{-1} (at a cell density of 1×10^5 amoebae per cm^2).

Such visible bands do not of themselves, however, provide conclusive evidence for pulsatile signalling. The phenomenon could equally well be due to the amoebae sensing a continuous signal but being periodically refractory to stimulation. Other evidence in favour of pulsatile signalling comes from the work of Shaffer (1962) who noticed that amoebae responded alternately to two aggregation centers that happened to lie on either side of them, indicating a fluctuating signal rather than a fluctuating response to a static gradient.

Further evidence comes from the work of Gerisch and Hess (1974) who suspended washed *D. discoideum* amoebae at a cell density of 2×10^8 ml^{-1} in buffer in optical cuvettes supplied with bubbled oxygen. They found that amoebae from the vegetative phase showed no change in their optical density measured at 405 nm under such

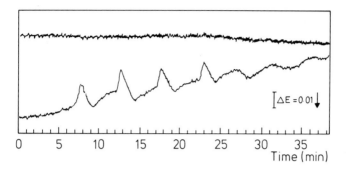

Fig. 1.4 Spontaneous oscillations of amoebae in suspension culture seen by changes in the optical density at 405 nm. No oscillations are seen in amoebae harvested from the vegetative or early aggregation phases (upper trace). Synchronous oscillation by cohesive amoebae (after approximately 9 hours of starvation) is shown in the lower trace. (From Gerisch and Hess, 1974).

conditions. However, amoebae harvested after a period of six hours of starvation and then resuspended, immediately produced oscillations in the optical density trace at 405 nm. Generally, an initial series of spikes was recorded followed by sinusoidal oscillations (Fig. 1.4). The spike amplitude was approximately 2% of the background density. Because the amplitude decreased steadily with increasing wave length between 380 nm and 600 nm (which is typical of light-scattering changes) the signals probably represent changes of cell shape. Significantly the oscillations were affected by added pulses of exogeneous cAMP (the chemotactic attractant for *D. discoideum*; Konijn, 1972b) to the extent that:

(1) a cAMP pulse given in the first half period after the spike caused a phase delay in the oscillations,
(2) a cAMP pulse given after the middle of the period caused the next spike to be precocious with subsequent spikes in phase with it, so that a phase change was produced,
(3) a cAMP pulse given just before the regular appearance of a spike caused the following spike to be inhibited but with no observable phase shift,
(4) a continuous addition of cAMP at 3–50 n mol min^{-1} l^{-1} suppressed the autonomous oscillations, and
(5) with cells harvested and resuspended at 2.5 to 4 hours after the start of development (a stage at which amoebae do not normally produce spontaneous oscillations) the optical density could be induced to oscillate by several added pulses of cAMP given at approximately 8 minute intervals. After this time the cells spontaneously produced oscillations. (The resulting premature appearance of other developmental events induced by the cAMP pulses will be further discussed in Section 1.2.6).

The most direct and convincing evidence for pulsatile signalling comes from the exciting work of Gerisch and Wick (1975) who showed that amoebae harvested after

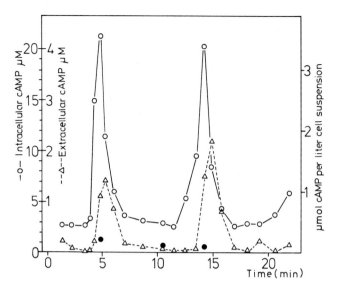

Fig. 1.5 Spontaneous oscillations of cAMP concentration in suspensions of cohesive amoebae. Left ordinate: intra- and extracellular concentrations of cAMP in either the intracellular or the extracellular compartment. The difference between the left- and right-hand ordinates results from the different volumes of cell mass and extracellular space. Filled circles represent the intracellular cAMP after incubation of the extracts with phosphodiesterase. (From Gerisch and Wick, 1975).

nine hours of development and used in suspension culture with bubbled oxygen would spontaneously synthesize and release pulses of cAMP (Fig. 1.5). This would seem to be excellent evidence for the existence of a system with an autonomous signal generation system. The measurement of such changes in both intra- and extracellular cAMP concentrations using a binding assay required the use of an ingenious silicone layer rapid extraction system to separate the cells from the supernatant and to prevent loss of signal by the ever present phosphodiesterase. With this system Gerisch and Wick showed that the periodicity of the intracellular spikes of cAMP were in phase with the changes in optical density at 405 nm seen by Gerisch and Hess. The intracellular cAMP concentration rose sharply in each period of synthesis and reached a peak within 1–2.5 minutes, the extracellular cAMP reaching its peak 30–40 seconds later.

In view of such evidence, there can be little doubt that the cAMP signal is indeed produced and released periodically by cells in suspension and it seems highly likely that this periodic signal corresponds to the periodic signalling by aggregation centres during aggregation on agar surfaces.

The origin of the autonomous oscillation in the pulsing amoebae is possibly the fluctuating redox state of the cytochrome chain of the mitochondria. This fluctuation is seen to have a periodicity that is similar to and in phase with the periodic changes of shape observed with amoebae in suspension (Gersich and Hess, 1974). The connection between such an internal oscillator and the pulsed production of the cAMP signal in the aggregation center is not yet known. Evidence that the internal oscillator is not directly coupled to cAMP production comes from the work of Gerisch and Hess (1974). These authors observed that an externally applied continuous cAMP signal that blocked the autonomous oscillatory shape changes of developing amoebae in suspension culture did not effect the oscillation of the cytochrome chain. Furthermore, induction of a phase change in the cAMP oscillators by external cAMP pulses did not induce a phase change in the cytochrome chain oscillations. A secondary oscillator may therefore be present, tuned to oscillate in harmony with the cytochrome system but responsive to external cAMP oscillations by a feedback mechanism. A possible candidate for such a secondary oscillator is the enzyme system of Rossomando and Sussman (1973). These authors observed that the enzyme ATP-pyrophosphohydrolase which splits ATP to AMP and pyrophosphate specifically, requires cAMP for activity. The adenylate cyclase enzyme that makes cAMP from ATP and which co-purifies with the pyrophosphohydrolase on agarose columns was found to be activated by 5'AMP. Because of the dual relationship between the activators and products of these two enzymes, Rossomando and Sussman postulated that they might be capable of a pulsatile synthesis of cAMP. The theoretical aspects of the interaction were investigated by Goldbeter (1975) who calculated that such a feedback system could indeed lead to an oscillatory synthesis of cAMP with a pulse interval consistent with the pulse frequency observed in aggregation centers. However the concentrations of cAMP and 5'AMP that were needed to activate the enzymes (about 2mM) are much higher than the concentrations recorded for these metabolites *in vivo*. (The concentration of cAMP, for example, is thought to fluctuate between 0 and 1 μM extracellularly and between 0 and 20 μM intracellularly (Gerisch and Wick, 1975). If such a system operated, it would, moreover, have to be intracellular rather than being on the external surface of the cell since Gerisch *et al.* (1975b) found that 5'AMP did not affect the spontaneous oscillations of cells in suspension culture nor could a binding site for 5'AMP be detected on the surface of aggregating amoebae (Malchow and Gerisch, 1974). The physiological significance of this particular oscillation scheme is therefore in some doubt, although an enzyme system of this general type would seem plausible as a mechanism for a secondary oscillator.

A recent study by Wurster (1976a) and by Nanjundiah *et al.* (1976) has revealed that the frequency of oscillations (but not the amplitude) seen in cells in suspension culture is dependent on the temperature. At lower temperatures, the oscillation frequency is decreased. The activation energy calculated from the data ($\Delta E = 15.8$ kcal mol^{-1}) was noted by Nanjundiah *et al.* to be closer to the activation energies of enzyme-catalyzed reactions than to the much smaller activation energies of known specific transport processes.

The frequency of the pulsed signal, even at a fixed temperature, is not a constant throughout development. By analysis of time-lapse films of concentric wave propagation of cells on agar, it has been found that at the start of signalling the period between signals is about 10 minutes but decreases rapidly during the early stages of aggregation to 5 minutes. This is then followed by a slower change to about 2.5 minutes later in aggregation (Gerisch, 1965; Durston, 1974a,b; Alcãntara and Monk, 1974). Durston (1974a) observed a bimodal distribution of the intervals between waves from individual pulsing centers (or 'pacemakers') with amplitudes of 5 minutes and 10 minutes. He suggested that the 5 minute period might correspond to the period of internal autonomous oscillation and that the frequency of the pulse initiation is simply limited by the relay refractory period of the amoebae (the period, after being stimulated to produce a cAMP pulse, during which they cannot be stimulated to produce another new pulse). The initial relay refractory period, which is thought from other evidence (see Section 1.2.4) to be longer than 5 minutes, means that the amoebae in the pulsing center would ignore every second autonomous oscillation and produce signal pulses initially every 10 minutes leading to the observed waves initially produced at this frequency. As the refractory period decreased during development it would soon get to the critical value of 5 minutes, at which point every internal oscillation at the center would stimulate an external pulse of cAMP and hence a wave every five minutes. Although the subsequent decrease in the signal period to about 2.5 minutes later in development is more difficult to explain, it might possibly be due to the establishment of a continuous signal source by the newly formed aggregate tip (a nipple-like structure on the top of the aggregate). Such a continuous source (evidence for which is as yet inconclusive) could elicit a propagated wave with the wave period being as short as the current relay refractory period of the aggregating amoebae i.e. the measured refractory period of 2.5 minutes late in aggregation would give a wave period of about 2.5 minutes as actually observed.

1.2.2 Signal reception

The pulsed signal of cAMP is received and temporarily bound by receptor molecules on the surface of responding amoebae (Malchow and Gerisch, 1974; Green and Newell, 1975; Henderson, 1975; Mato and Konijn, 1975a). The evidence for the existence of such receptors and their properties will be dealt with in Section 1.3.

The specificity of the receptor has been investigated by Konijn and Jastorff (1973) who tested the efficacy of a range of synthetic cAMP analogues with the chemotactic assay of Konijn. In this assay, small drops of salt solution containing about 500 amoebae were placed on hydrophobic agar and 0.1 μl drops of the test substance placed alongside them at a distance of about 0.3 mm. After intervals of five minutes and ten minutes further applications of the test substance were made. The test was scored positive if at least twice as many amoebae were pressed against the side of the drop closest to the cAMP or analogue as against the opposite side. After making and testing cAMP analogues substituted in the phosphate ring, these authors concluded that nearly all such substitutions greatly reduced potency of the molecules in the

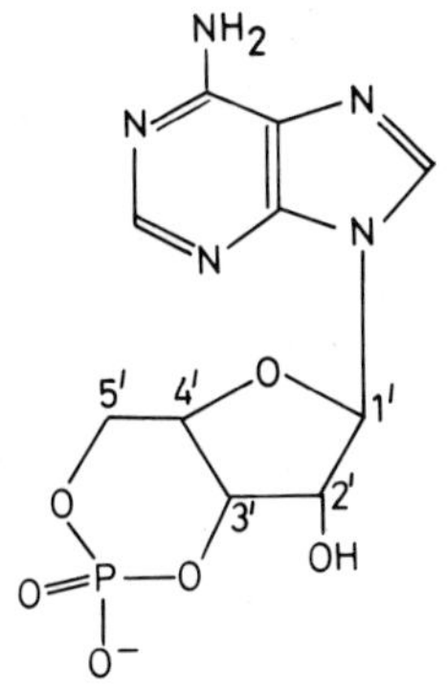

Fig. 1.6 The cAMP molecule. Substitution of the oxygen atom attached to the 5′ carbon atom by a methylene or an amido group has no, or only a slight, effect on the chemotactic response of amoebae. Other types of substitution drastically reduce activity whether in the phosphate ring or in the purine moiety (From Konijn and Jastorff, 1973).

chemotactic test, usually by several orders of magnitude. Surprisingly, the oxygen atom between the 5′ carbon of the ribose and the phosphorus atom could be replaced by a methylene group or an amido group without significant loss of chemotactic activity (See Fig. 1.6 for the numbering of the atoms) (This result differs from studies of mammalian cAMP receptors which show greatly reduced activity with such compounds). No replacements were found that increased chemotactic potency, including those that increased the lipophilic character of the cAMP molecule. This suggests that, unlike mammalian systems, the cAMP molecule fulfils its role without the need to cross a lipid bilayer.

Replacements of atoms of the adenine base also lead to reduced potency. While some replacements, such as adding a benzoyl group in place of a hydrogen atom on the C-6 amino group of the adenine, gave only small (about ten-fold) reduction in potency, replacement of most other components gave reductions of between 100 and 10 000 fold. Typically, cGMP was active only at concentrations roughly 1 000 times higher than cAMP (Konijn, 1973).

The mechanism of perception of the cAMP signal after its reception by the amoebal receptors is an area of considerable current interest. For example, it has been proposed that the cAMP receptor is a membrane-bound phosphodiesterase (mPD) that uses the free energy change of hydrolysis to work some intracellular endergonic reaction connected to cell movement. The mPD molecules would effectively count the relative distribution of cAMP molecules around the cell and translate this information into a directional awareness (Malchow et al., 1973). Strong evidence against such a dual role for the cAMP receptor and the mPD molecules however, is that the two activities clearly show different substrate specificities (Malchow et al., 1975).

Generally, the models proposed to explain perception of cAMP are of two types:

(1) The cell receptors monitor the difference in concentration of cAMP from the front to the back of the cell, and this is translated into a directional response by some unknown mechanism. The perception of a spatial gradient in this way is the oldest idea around (and may even be true!) but the tasks demanded of the cell receptors in such a model would be formidable. Not only would a difference in the concentration have to be detectable at two points on the cell surface which are, at most, one cell diameter (about 10 μm) apart in a concentration gradient of as low as 4×10^{-9} M mm^{-1}, but the cell would also have to take a 'chemotactic compass bearing' from the results in order to move off in the right direction! Recently, Mato et al., (1975) have calculated that if there are 10^6 receptors per cell (which may be a considerable overestimate) with a dissociation constant of 10 nM, then in a physiologically detectable gradient of about 4×10^{-9} M mm^{-1} a difference of 40 pM would be expected between the two ends of the amoebae. This is translatable (given the possibly questionable assumption of a cAMP receptor sensing period of 1 ms) as a difference of only 12 occupied receptors.

(2) The cell receptors monitor changes in cAMP concentration with time. A pulse of cAMP 'hitting' an amoeba could be registered on one side of the cell before the other. Since the amplitude of such pulses (0–2 μM observed in suspension culture by Gerisch and Wick, 1975) is far higher than the small differences calculable for spatial gradients, the detection of such a pulse would not require the comparison of small differences required in the first model described above. If the rest of the cell became refractory very quickly after the pulse reached the leading edge of the cell, then directionality of response could be achieved. With the finding of pulsed signals of cAMP this model clearly seems attractive. As pointed out before, however, (Gerisch et al., 1975b; Newell, 1975) the presence of a pulsatile signal source is not essential for aggregation as amoebae move quite effectively towards single drops of cAMP in the Konijn assay. (The gradients produced by such drops on agar are not in reality static, nevertheless, but vary continuously with time due to diffusion from the drop). Gerisch et al., (1975b) have proposed that static gradients could be sensed by filopodia that are constantly being flicked out in all directions. This rapid movement of the filopodia would give a temporal change in concentration in a static gradient. The idea is made more attractive by the finding of Kobilinsky and co-workers (1976) that cAMP can induce the formation of such filopodia (see Section 1.2.7). A possible difficulty with this model, however, is that if the calculations of Mato et al. of a difference of only 12 occupied receptors between the extremeties of the whole cell area are correct, then less than one occupied receptor difference would be expected along a filopod.

Mato et al. attempted to settle the question of which model is correct by linear regression analysis of a double logarithmic plot of distance against cAMP concentration detected by the amoebae. The slope of the plot, however, was 1/4.25 which lies between that of 1/4 expected for model 1 and 1/5 expected for model 2. The mechanism of signal perception therefore remains an open question.

1.2.3 Signal destruction

Destruction of the cAMP signal in *D. discoideum* amoebae occurs by means of phosphodiesterase enzymes. The existence of at least two basic types of this enzyme has been clearly demonstrated in this species: the membrane-bound form (mPD) and the extracellular form (ePD). The relationship of the soluble intracellular phosphodiesterase activity to these two activities is not totally clear.

The ePD enzyme appears to be a protein with a molecular weight of about 60 000 daltons (Chassy, 1972; Reidel *et al.*, 1972; Pannbacker and Bravard, 1972) and a K_m reported by Riedel *et al.* (1972) to be about 4 μM. The enzyme was reported by Chassy (1972) to be slowly converted after its release from the cell to a less active form with a K_m of 2 mM and a molecular weight of about 130 000 daltons (presumably by dimerisation) and it was in such a low activity form that the ePD was initially reported by Chang in 1968.

The ePD differs from the mPD in showing normal Michaelian kinetics, whereas the kinetics of the mPD are non-linear, and ePD also differs in binding to a protein inhibitor which is produced by amoebae early in aggregation (Riedel and Gerisch, 1971; Riedel *et al.* 1972; Malchow *et al.* 1975). The timing and extent of formation of inhibitor differs markedly between strains. (Fig. 1.7). The NC4-derived strains only produce inhibitor in excess during later development while V12-derived strains form excess inhibitor very much earlier. The curious result of this inhibitor production in V12 strains is that although ePD is produced and is active during the growth phase, its activity is greatly inhibited just at the time during aggregation when one would expect it to have been operative. As a consequence, Gerisch (1976) has concluded that the ePD can have no indispensible hydrolytic function during aggregation. This conclusion is consistent with the theoretical calculation of Nanjundiah and Malchow (1976) that suggested that the ePD could scarcely affect the concentration of cAMP under the normal laboratory conditions of aggregation and it is almost certainly the mPD with its lower K_m that rapidly destroys the cAMP pulses. (A role for the ePD in the regulation of the early events of development has been suggested recently, however; see Section 1.2.6.)

As mentioned in the introduction, it is noteworthy that other species and genera of cellular slime mold besides *D. discoideum* secrete ePD (see Table 1.1). It is interesting and possibly significant that the list of ePD secretors (and cAMP producers) includes *Polysphondylium violaceum* which does not use cAMP as the chemotactic attractant (Gerisch *et al.*, 1972; Bonner *et al.*, 1972).

The mPD enzyme was first detected by Pannbacker in 1972, and has since been found to be an interesting enzyme with an important role in aggregation. The enzyme is located on the cell surface. Evidence for this comes from the finding that the hydrolysis products of cAMP appeared extremely rapidly, i.e. in less than 5 seconds after adding cAMP to cohesive cells (Malchow and Gerisch, 1974). It was unlikely, therefore, that the cAMP entered the cells before being hydrolysed. In strong support of this idea is the cytochemical localization of mPD on the outer

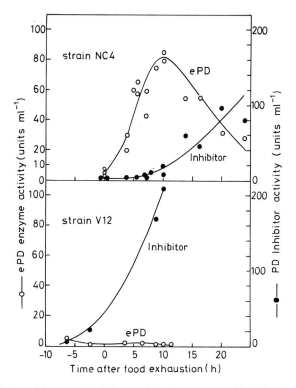

Fig. 1.7 Inhibitor (●) and extracellular phosphodiesterase (ePD) (○) activities in the agar layer of *D. discoideum* strain NC4 (top curves) and strain V12/M1 (bottom curves). The curves show the differences in ePD activity during aggregation of these strains that result from the different times of inhibitor production (After Gerisch, 1976).

plasma membrane surface by Farnham (1975) and also the presence of approximately 30% of the membrane-bound phosphodiesterase activity in purified plasma membrane fractions of cohesive amoebae (Green, 1976).

Malchow *et al.*, (1972) isolated the protein from crude whole cell membrane fractions and, after solubilization using lithium 3, 5-diiodosalicylate, they purified the mPD activity by ammonium sulfate fractionation and agarose gel filtration. The enzyme was eluted as a broad peak corresponding to approximately 500 000 daltons, this being, almost certainly, in a polymeric state. More recently, Huesgen and Gerisch (1975) solubilized the mPD using deoxycholate and determined a molecular weight of 200 000 by elution from Sephadex G-200.

It is of interest that the ePD inhibitor protein does not inhibit the mPD in whole cells or insoluble membrane fragments. The speculative possibility that the mPD is a multimeric form of the ePD is supported by the finding of Malchow *et al.* (1975)

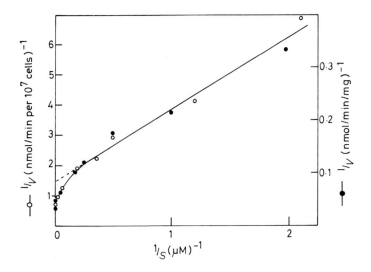

Fig. 1.8 Double reciprocal plots of cAMP and cGMP hydrolysis by membrane-bound phosphodiesterase (mPD) showing non-linear kinetics. The data of Malchow *et al.* (1975) using living whole cells with cGMP as substrate (○) is plotted together with the data of Green and Newell (1975) using cell membrane fragments with cAMP as substrate (●).

that after solubilization (which partially dissociates the enzyme) the mPD is to some extent inhibited by the inhibitor protein, with some (lower molecular weight) fractions being more sensitive than others. It is therefore an interesting possibility that the inhibitor site is present on the mPD but is masked by its multimeric state or by the presence of other membrane components.

Using Lineweaver–Burke double reciprocal plots, all preparations of the mPD were found to show non-linear kinetics (Malchow *et al.*, 1975; Green and Newell, 1975) with a pronounced downward curvature at higher substrate concentrations (Fig. 1.8). The significance of such non-linear kinetic properties has been suggested by Malchow *et al.* to be related to the large range of cAMP concentrations experienced by cells having to receive either weak or strong signals from sources far or near. The non-linear kinetics which effectively produce an affinity (apparent K_m) for cAMP varying from 0.2 to 30 μM would enable the enzyme to hydrolyse a signal of cAMP down to the 'noise' level in a much smaller time period than could an enzyme with conventional Michaelian linear kinetics.

1.2.4 Signal relay

Evidence that the signal emitted by the aggregation center is relayed outwards rather than simply diffusing outwards over long distances comes initially from the observations of Bonner (1949) that amoebae tended to aggregate into streams rather than in a direct radial path toward the center. Sometimes amoebae can even be observed to move outwards from the center temporarily to get into a nearby stream. Bonner concluded that the streams of amoebae as well as the center must be capable of releasing the attractive chemical signal.

Shaffer, in 1957 and 1958, proposed that the large aggregation territories of certain slime molds (including *D. discoideum*) were achieved by a relay system in which the signal initially produced by the aggregation center not only attracted amoebae towards the center, but also induced the amoebae to secrete the signal themselves. This idea fitted well with the observations of Arndt (1937), Bonner (1944) and Shaffer (1957) showing that waves could be seen to move slowly outward from the center of aggregating cultures of amoebae on agar plates at the same time as the amoebae moved inwards.

The idea of a relay was studied in more detail by Gerisch (1965, 1968) who measured the rate at which the visible waves moved outwards on agar plate cultures of densely plated *D. discoideum* amoebae. He found that under these conditions the rate of signal propagation was 43 μm min^{-1}. With signals being generated from the center at the rate of one every five minutes, this allowed as many as 14 concentric waves to follow each other simultaneously over a very large aggregation field.

Using the data of Gerisch, it was calculated by Cohen and Robertson (1971) that, after being signalled, a cell took 15 seconds before emitting its own signal (the relay delay time). For this calculation they assumed an arbitrary value of one cell length for the range of the relayed signal and a negligible time taken for diffusion of the signal over this range.

Subsequent careful measurements of the range of the relayed signal from time lapse movie film by Alcântara and Monk (1974), using monolayers of amoebae on agar, have indicated that the relay range is fairly constant at 57 μm, with a relay delay time of 12 seconds under these conditions. In other words the signal is relayed in a series of steps every 12 seconds, rather than continuously, with the radial distance of each step being 57 μm (see Fig. 1.9). The amoebae within each step appear to be affected simultaneously by the relayed signal and they move inwards roughly in unison. This movement period then lasts for 100 seconds before the amoebae stop again and wait for the next relay front to reach them. Alcântara and Monk also found that the velocity of signal propagation varied with the population density, being faster at lower densities. Values of approximately 360 μm min^{-1} were recorded at densities of 5×10^4 amoebae cm^{-2}.

To ensure that a relayed signal is only propagated in one direction (outwards), Shaffer (1962) suggested that amoebae had to become refractory to further activation of signal relay after being signalled. This 'relay refractory period' (which should not be confused with the refractory period for chemotactic movement) has been investigated by Durston (1973) using the microelectrode system devised by Robertson *et al.* (1972). With this device a small electrical current flow at intervals produces a miniature pulse of cAMP from a hollow needle onto an agar plate spread

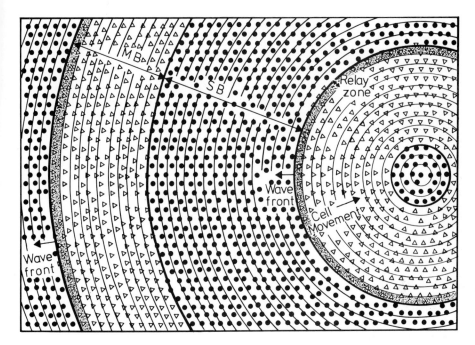

Fig. 1.9 Diagrammatic representation of the aggregation patterns seen in Fig. 1.2. The field of amoebae is divided into bands of centripetally moving cells (▷) and still cells (●) around the centre of attraction on the right of the diagram (MB = movement band; SB = still band). Two wave fronts are shown moving to the left in the diagram and represent the leading edges of the signals emitted by the center. The concentric lines drawn within the moving and still bands represent the successive 57 μm steps through which the wave front has passed in successive 12-second intervals starting from the release of the pulse from the center. The current relay zones are shown stippled. The width of the movement band is determined by the duration of movement of the amoebae after receiving a cAMP signal (movement is usually for about 100 seconds). The width of the still band is determined by the frequency of signal emission by the center: the higher the frequency of pulsation, the narrower the band (After Alcântara and Monk, 1974.)

with amoebae. After several hours of starvation, the amoebae are able to move towards the needle in response to the pulsed signal. Using a three minute pulse period, Durston found that whereas amoebae after 6 hours of development respond to every third pulse, 7-hour amoebae respond to every second pulse and 9-hour amoebae to every pulse. He concluded that the refractory period decreases during aggregation from greater than 6 minutes before 6 hours of development to less than 3 minutes after 9 hours.

Direct evidence that aggregating amoebae possess the machinery for relaying a

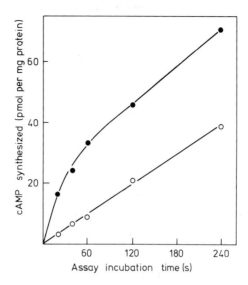

Fig. 1.10 Decay of the activated state of the adenylate cyclase during assay of extracts of cAMP-pulsed whole cells. Amoebae in washed suspension culture after 4 hours of starvation were pulsed with 200 nM cAMP (●) and transferred after 5 seconds of sonication into the incubation mixture within 15 seconds for the assay of adenylate cyclase. In the estimation of basal activity the cAMP was replaced by 200 nM 5′AMP (○). (After Roos and Gerisch, 1976.)

cAMP signal has come recently from the work of Shaffer (1975) and Roos *et al.* (1975) who elicited pulsed emissions of the cAMP signal from developing (8–12 hour) amoebae by supplying small pulses of cAMP of around 10^{-7} M. Such responses could not be elicited from vegetative amoebae. (These studies differ from those of Gerisch and Wick discussed in Section 1.2.1 above, in that the latter were observing spontaneous pulses rather than those induced by externally applied cAMP as discussed here.) To show the responses the amoebae were pre-labelled with tritiated adenine and were pulsed with unlabelled cAMP. Aliquots of extracellular medium were then tested for radioactive cAMP after its chromatographic separation. Both groups found that the amplification of the signal was somewhere between 10 and 40 times the strength of the input signal. The amplitude of the amplified signal under the conditions of Roos *et al.* was found to be about 0.5 μM which corresponds well with the amplitude found for spontaneous cAMP emissions of 1 μM by Gerisch and Wick (1975) at cell concentrations of 2×10^8 amoebae ml^{-1}. If the destruction by phosphodiesterase was taken into account, Roos *et al.* calculated that about 1.5×10^5 molecules of cAMP were released per cell in each pulse.

Very recently, Roos and Gerisch (1976) used a highly sensitive assay for the adenylate cyclase enzyme and showed that adding pulses of cAMP to whole cells

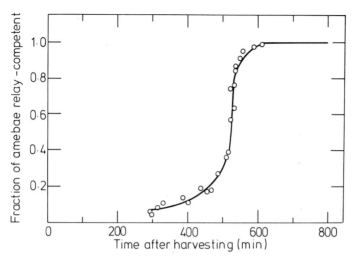

Fig. 1.11 The appearance of relay-competant amoebae during aggregation of *D. discoideum* on agar. (After Gingle and Robertson, 1976).

induced a very rapid rise in the activity of the enzyme measured by cAMP formed *in vivo* or in rapidly made sonic extracts. This enhanced activity then slowly decayed to the basal level over a period of about 60 seconds (Fig. 1.10). The effect was only observed if whole cells were pulsed, the extracts themselves being insensitive to added cAMP pulses. This work is a very beautiful demonstration that the enzyme for synthesizing cAMP is indeed stimulated by an incoming cAMP pulse. The fact that the increased activity is not immediately lost after sonication of the cells makes this system very amenable to further exciting exploration.

The time at which amoebae in a developing population become relay-competent has been determined in a series of experiments by Gingle and Robertson (1976). Using the nomenclature of these authors, the fraction (X_2) of relaying competent amoebae in a population of cell density N was determined as a function of time (Fig. 1.11). The ingenious method used to determine this value involved first finding the critical density of competent amoebae that must be present together on an agar surface in order to elicit a field-wide relay response from a cAMP-emitting microelectrode. This value (2.5×10^4 cells cm^{-2}; Gingle, 1976) is not in fact a constant but varies slightly with total cell density (N) and this is later allowed for. At the developmental time at which the critical number is reached the number of amoebae present that can relay ($X_2 N$) equals 2.5×10^4 cm^{-2}. By taking amoebae at different times of development and placing them on agar at a series of different densities and then testing their relay ability with the pulsing microelectrode, the lowest density of amoebae needed to ensure the presence of the critical number of relay-competent cells was determined. Dividing 2.5×10^4 cm^{-2} by this cell density gave the value of X_2 at each time point so analysed. The results (Fig. 1.11) indicate a sharp rise in the number of relay-competent amoebae. Under the conditions used by Gingle and Robertson, this occurred at 8–9 hours after the start of development.

1.2.5 The chemotactic response

Chemotaxis, strictly taken here to mean the movement response by the amoebae to the source of attraction, lasts for approximately 100 seconds, during which time the amoebae move approximately 20 μm inwards (Alcântara and Monk, 1974). After this the amoebae stop and await the next relay front moving outwards towards them (Fig. 1.9). The period of movement is not necessarily at a constant speed; Alcântara and Monk found that there is commonly a fast component to the movement response of about 20 seconds, followed by a progressively slower movement until the cell comes to rest. The details of the actual movement are not precisely understood but they can be envisaged as a sequence of actions including prolonged and orientated extension of pseudopodia and contraction of the rest of the cell. This could possibly be mediated by the triggered release of Ca^{2+} ions from vesicles into the cytoplasmic space (Gerisch et al., 1976). (Support for such a notion comes from the work of Mockrin and Spudich (1976) showing that the actin-activated myosin ATPase in *Dictyostelium* amoebae is indeed controlled by Ca^{2+} ions). The direction of movement appears to be perpendicular to the threshold concentration contour that first strikes the cell membrane. Usually, only one pseudopodium is protuded in response to receipt of a chemotactic stimulus and this soon becomes the new 'front' of the amoeba. When the movement ceases all pseudopodia are normally withdrawn and the amoeba takes on a somewhat more rounded shape. It is this difference in shape that Alcântara and Monk consider to be the basis of the difference in appearance of dark and light bands seen in aggregating amoebae on agar plates.

A curious problem that has been around for a long time but is still apparently unsolved (Shaffer, 1975) is that of why amoebae move consistently towards the signalling source after being stimulated, rather than moving outward again immediately a new signal pulse is generated just behind them. In the theoretical treatment of Cohen and Robertson (1971) this problem was obviated by assuming the signal to be of negligible duration (less than 0.2 seconds when it reached the cell) and the amoebae to possess a lengthy refractory period before being capable of re-orientated movement (commonly not distinguished from the refractory period for signal relay of 2.5–7 minutes). However, neither of these assumptions seems to be borne out by the experimental observations. Gerisch and Wick (1975), Shaffer (1975) and Roos *et al.* (1975) measured the duration of the cAMP signal released by cells in suspension and found it to be in excess of 60 seconds. The measurements of Alcântara and Monk (1974) of the refractory period for movement showed it to be equal to or somewhat less than 12 seconds. Moreover Gerisch *et al.* (1975c) found that localized cAMP stimuli applied alternately to either side of an amoeba by microcapillaries at intervals of less than 10 seconds could cause new pseudopodia to be formed within 5 seconds after each stimulus – even before the former pseudopodia pointing in the opposite direction had been retracted. The refractory period for chemotactic movement must therefore be very short and is clearly a different phenomenon from the refractory period for signal relay. However, as such a refractory period for movement would not (theoretically) support unidirectional movement in an aggregating field unless it was in *excess* of 12 seconds (the relay delay time) there remains something of a riddle to be solved.

1.2.6 The developmental initiation response

The initial trigger that ultimately brings about aggregation and subsequent morphogenetic development in the cellular slime molds is clearly starvation. There is ample evidence, however, that this condition cannot by itself be a direct regulator of the developmental process. For example, starvation of mutants that are blocked at some point in aggregation does not induce (without additional outside interference) any observable later events that are characteristic of either spore or stalk cell differentiation. Development generally has the characteristics of a sequential process, the initial trigger starting a chain reaction with each step dependent upon some earlier one. One may ask, therefore, what triggers this developmental chain reaction to be initiated? Is it linking up of cells at the end of the aggregation phase that triggers the start of morphogenetic development and the cellular differentiation of spores and stalk cells or is it some event in the prior aggregation phase? That cell differentiation is dependent upon proper cell contacts being established and maintained has been shown by Newell *et al.* (1971), Gregg (1971) and Yu and Gregg (1975), as indeed it has been shown for glutamine synthetase induction in embryonic chick retina by Morris and Moscona (1970, 1971). This cell contact cannot however be the first trigger for morphogenetic development because the receptors that are used during cell docking are themselves synthesized in response to some earlier stimulus and are already present when the events of cell-cell contact are occurring.

Using the synthesis of one such cell contact protein (contact site A, see Section 1.4.2) as a measure of the extent of initiation of development, Gerisch *et al.* (1975a) and Darmon *et al.* (1975) have shown that artificial pulses of cAMP at 5–7 minute intervals induce precocious formation of contact sites A and hence cell-cell adhesion. Chains of cells linked by EDTA-resistant cell contacts were formed in pulsed suspension cultures 2–3 hours before the unpulsed controls. Moreover, if some aggregation-deficient mutants were similarly pulsed, Darmon *et al.* found that they could overcome their deficiency and aggregate. Continuous addition of cAMP did not produce these effects. Klein and Darmon (1975) have since shown that ePD is also precociously formed in cAMP-pulsed cultures. The inference from these studies is that the pulse relay system is not only used in *D. discoideum* for causing the aggregation of amoebae, but is also an early trigger that induces other developmental events such as ePD production and formation of contact sites A. Whether it also directly induces still later events or whether these are all induced by the cell docking system remains to be established.

An enzyme whose activity is possibly related to the effect of cAMP pulses on development is the extracellular phosphodiesterase. This enzyme (or a molecule remarkably similar to it) seems to be able to induce the information of cohesiveness precociously in amoebae and this precocity seems to be additional to that induced by cAMP (Alcântara and Bazill, 1976; Klein and Darmon, 1976). These workers noticed that amoebae at low population density on agar took longer to aggregate than those at higher density and that a factor secreted by the amoebae at high density could speed up the aggregation of low density amoebae. A careful and thorough study by Alcântara and Bazill showed that there were strong grounds for believing that the aggregation-stimulating factor (called AF) was in fact the ePD.

Although the studies showing regulational effects on development are clearly at an early stage, the finding that development is in some way controlled by pulses of cAMP and possibly ePD could be the key to understanding the genetic control system that is operative. The connection between such external cAMP pulses and gene regulation in the nucleus could be cAMP-activated protein kinases that could move into the nucleus and phosphorylate non-histone proteins. Two such cAMP-activated kinases have recently been found by J. Sampson to be formed during the first few hours of starvation (personal communication).

1.2.7 The final docking manoeuvre

The final phase of aggregation involving cell recognition and specific cell adhesion which I refer to as cell 'docking' in this review (by analogy to astronautical docking manoeuvres) depends upon the prior synthesis of specific cell surface components. Two such types of component are known, namely contact sites A and slime mold lectins (such as discoidins and pallidins) and their membrane receptors, and these will be dealt with in detail in Section 1.4 and 1.5.

A somewhat speculative, but possibly very important, piece of work that should not be overlooked is that of Kobilinsky et al. (1976) who have recently suggested a role for filopodia at the stage at which cells approach each other. Using transmission electron microscopy, they showed, that axenically growing (or stationary phase) AX2 amoebae occasionally possessed small filiform projections of the cell surface of the type called filopodia (previously observed by Rossomando et al. (1974) in scanning electron micrographs) (see Fig. 1.12). They could also be seen protuding from amoebae under dark field microscopy, but generally less than 10% of the amoebae possessed such structures. However, treatment of logarithmically growing amoebae with 10^{-5} M cAMP provoked all the cells to form filopodia within 10 minutes and the average length of these structures was increased, often reaching out by the equivalent of the cell diameter. After the filopodia had been formed the cells tended to move towards each other and clump and the filopodia simultaneously retracted. Other nucleotides such as AMP, ADP or cGMP had no effects in similar amounts. It was also noticeable that cells possessing filopodia were much more easily agglutinated by the plant lectin concanavalin A (see Section 1.5.2) than were untreated cells without filopodia. The authors suggest that this filopodial response may be used during cAMP-induced

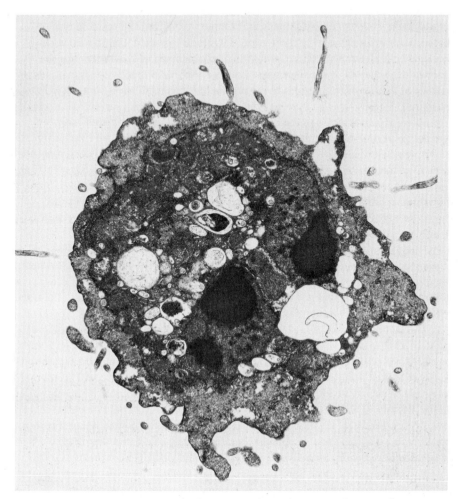

Fig. 1.12 Electron micrograph of a section through a vegetative amoeba showing the presence of filopodia. (By courtesy of Dr Lawrence Kobilinsky).

aggregation, and that the first contact between cells may be mediated through the intertwining of their filopodia. Retraction could then lead to the final approach, with the cells ultimately held firmly together with the slime mold lectins or contact sites A.

1.3 CELL SURFACE cAMP RECEPTORS

If aggregating amoebae can detect cAMP, it is reasonable to suppose that they possess receptor molecules capable of binding cAMP. Such receptors have been found by a number of groups (Malchow and Gerisch, 1974; Green and Newell, 1975; Henderson,

1975; Mato and Konijn, 1975a). Because of the impermeability of the plasma membrane of slime molds to cAMP (Moens and Konijn, 1974) and the rapid demise of extracellular cAMP signals (Malchow et al., 1972) the expected location for such receptors would be on the cell surface. Evidence in support of this location is the demonstration that all of the membrane-bound cAMP binding activity from cohesive amoebae of *D. discoideum* resides in the purified plasma membrane fractions (Green and Newell, 1975).

It is incidentally interesting that, in addition to the cell surface cAMP receptors, there are also cAMP-binding proteins in the soluble fraction of the cell (Malkinson and Ashworth, 1973) that probably play no direct part in cAMP signal detection and whose function can only as yet be guessed at. These proteins are distinguishable from the cell surface cAMP-binding proteins not only in their intracellular position, but also in their affinity for cAMP and cGMP. In the case of the soluble protein the affinity for cAMP and cGMP is similar, whereas for the cell surface cAMP receptors the affinity for cAMP is several thousand-fold greater than for cGMP.

The major problem that had to be overcome before the binding of cAMP to cell surface receptors could be determined was the high level of cAMP phosphodiesterase present in the developing slime mold cells. Unlike the phosphodiesterase of other cells, the slime mold enzyme was not markedly inhibited by theophylline or other commonly used phosphodiesterase inhibitors. (Chang, 1968).

Malchow and Gerisch (1973, 1974) solved this problem by measuring binding of tritiated cAMP to cell suspensions in the presence of a vast excess of cGMP (0.5 mM). The cGMP had a similar affinity compared with cAMP for the phosphodiesterase but had about at least a thousand-fold lower affinity than cAMP for the cAMP-binding protein, so the cGMP acted as a transient phosphodiesterase inhibitor. Using this system Malchow and Gerisch showed that no binding of cAMP could be detected in growth phase amoebae, whereas in amoebae tested after a few hours of development strong binding was apparent. From Scatchard plots they calculated that there were 5×10^5 cAMP receptors per cell with an apparent dissociation constant of 100 to 200 nM. The Scatchard plots were analysed using a linear interpretation but a degree of curvilinearity was noticeable at low cAMP concentrations. This was more pronounced in later work using a phosphodiesterase-minus mutant (*Wag*-6) in the absence of cGMP (Gerisch et al., 1975c).

In a study by Green and Newell (1975) the problem of the phosphodiesterase was overcome by the addition of 2 mM dithiothreitol to the cAMP-binding assay rather than excess cGMP. Binding was assayed directly by a membrane filtration system which allowed measurment of binding over a longer time period and produced a sensitive assay. It was found that Scatchard plots of the binding activity of developing cells (Fig. 1.13) were strongly curvilinear and were concave upwards. Such curvilinear plots can be due to a number of causes, the best-known being (a) the presence of multiple classes of binding sites that have different but constant affinities or (b) the existence of site-site interactions of the type defined as negatively co-operative. Such curvilinear plots are not uncommon in the binding of hormones (such as insulin) to target tissue

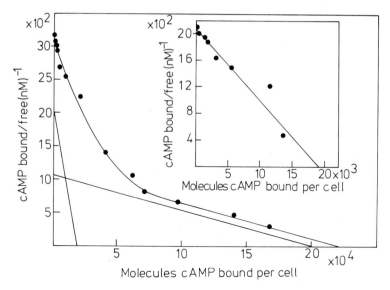

Fig. 1.13 Scatchard plot of cAMP-binding activity of AX3 cells at the 9.5 hour stage of development. The graph in the inset is the Scatchard plot which results when the experimental points obtained at cAMP concentrations lower than 30 nM are corrected for the contribution by the low affinity Type 1 binding activity. (After Green and Newell, 1975).

membrane receptors although some hormones such as growth hormone, calcitonin and prolactin show linear Scatchard plots (De Meyts et al., 1976). When the binding of cAMP to cohesive slime mold amoebae was analysed in terms of two sites of constant affinity, it was found that the sites with lower affinity (called class I sites) resembled those reported by Malchow and Gerisch, there being about 2×10^5 sites per cell with dissociation constants of about 150 to 200 nM. In contrast, the higher affinity sites (called Class II sites) were present in much smaller numbers − about 1 to 2×10^4 per cell − but showed a dissociation constant of about 10 nM (Fig. 1.13).

Henderson (1975) also independently assayed cAMP binding using an indirect method based on that of Malchow and Gerisch but using dithiothreitol to inhibit the phosphodiesterase. Using double reciprocal plots of the binding data, she found that there were about 10^5 sites per cell with a dissociation constant of about 36 nM. This value, may, however, be an average value of multiple classes of sites, as in some experiments the plots were seen to be non-linear. Heterogeneity of sites was also suggested by experiments in which partial rather than complete competition was observed between fluorescent and photolysable cAMP derivatives and tritiated cAMP.

Using a completely different approach, Mato and Konijn (1975a) minimized the activity of the phosphodiesterase by using low numbers of cells (1×10^5) on filters saturated with acetate buffer pH 4.6 (a pH at which the phosphodiesterase is much less active). Binding of tritiated cAMP was measured directly by finding the counts

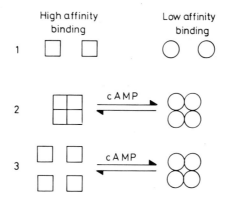

Fig. 1.14 Possible mechanisms to explain the appearance of curvilinear Scatchard plots of cAMP binding to cohesive cells of *D. discoideum*.
(1) Two distinct types of protein with different and fixed affinities.
(2) A negative co-operative interaction of identical subunits (arbitrarily shown as four subunits). A progressive shift from the high affinity form to the form with lower affinity (probably via a series of intermediate complexes) is induced by the binding of cAMP. (Note: Models involving negative co-operativity by slow transitions in *monomeric* proteins (Neet and Ainslie, 1976) are not considered to be likely here because the curvilinear plots are observed using ligand binding data. This would not be expected from slow transitions).
(3) A negative co-operative interaction involving aggregation of subunits.

bound to the amoebae after periods of incubation ranging from 1–11 hours. From control experiments (done under somewhat different conditions), it was deduced that the phosphodiesterase could not have significantly hydrolysed the cAMP on the filters over the incubation period. The ability of unlabelled cAMP to compete with the observed binding of tritiated cAMP was not reported but internalization of the label was shown to be negligible by the ease of removal of the tritium label from the cells by washing prior to measurement of radioactivity. Mato and Konijn found, with this assay, that binding activity increased during development in *D. discoideum*, *D. purpureum* and *D. mucoroides* amoebae but not in *P. pallidum*. By using double reciprocal plots of data from assays they calculated that *D. discoideum* had 1.2×10^6 receptors per cell with a dissociation constant of 6 nM. The restricted range of concentration used (2–10 nM) precluded measurement of any lower affinity (Class I) binding, so the single high affinity dissociation constant found is fully compatible with data from other workers. The calculation of 1.2×10^6 such sites being present per cell, however, is between one and two orders of magnitude greater than the numbers reported by other workers, and this remains as yet unexplained. Mato and Konijn (1975b) also reported that the presence of exogenous (1 mM) ATP stimulated the binding of cAMP to their cells. The physiological significance of effects of exogenous ATP at this concentration, however, are not altogether clear.

The appearance of two classes of binding sites with low and high affinity reported by Green and Newell may arise by several mechanisms of which three will be considered. (Fig. 1.14). Firstly, one possibility is that two distinct macromolecular species exist in the membrane with fixed but different affinities for cAMP. The Class I and Class II affinities then refer to the affinities of the two macromolecular species. In favour of this mechanism is the finding that the two classes can appear at slightly different times during development. This, unfortunately, is not a strong argument, not only because this difference is not always apparent, but also because even a co-operative subunit interaction scheme (see below) might also be capable of eliciting one form of affinity before the other.

Secondly, the binding protein may be composed of identical subunits that show co-operativity of cAMP binding between them. Fig. 1.14 shows the protein to be tetrameric, although this need not of necessity be the case. If the subunit interactions take the form of negative co-operativity (i.e. the affinity is highest when few sites are filled) then the observed Class II and Class I affinities would represent the affinities of the non-co-operating sites (at low cAMP concentrations) and the fully co-operating sites (at high cAMP concentrations) respectively.

The third possible mechanism is that the subunits are again negatively co-operative but by means of association of the subunits in the membrane. This type of reversible aggregation or dissociation of subunits through ligand-induced movements in the fluid membrane has been suggested by Singer and Nicholson (1972) to confer co-operative properties on membrane subunits.

Equilibrium studies of cAMP binding cannot of themselves distinguish between these possible mechanisms. This must await kinetic studies in which the effect of high unlabelled cAMP concentrations on the dissociation rate of low concentrations of bound cAMP is found. It is interesting to note that the use of kinetic analysis has recently suggested a negative co-operative mechanism to explain the curvilinear Scatchard plots of insulin binding to target tissue membranes (DeMeyts *et al.*, 1976) and for catecholamine binding to β-adrenergic receptors (Limbird *et al.*, 1975).

The third proposal gains some support from the effects of concanavalin A (Con A) on development. As described in Section 1.6.2, Con A has been found by Darmon and Klein (1976) to stimulate aggregation at concentrations up to those allowing saturation of the Con A receptor sites. At higher concentrations, Con A inhibits development until it is removed from the cell surface. That such action could be mediated by inhibiting the association of the cAMP receptor subunits in *D. discoideum*, is made plausible by the finding that Con A can indeed inhibit cluster formation induced in membranes by ligands (Edelman, 1974) and also that it inhibits the site—site interaction of insulin receptors in lymphocytes without binding to the biological receptor sites (DeMeyts *et al.*, 1974).

One possible reason for the existence of more than one class of binding activity was suggested by Green and Newell (1975) to be that the responses of chemotaxis and of signal relay might use different receptor systems. A mutant search was therefore conducted on the basis that any mutant strain of *D. discoideum* that was unable to

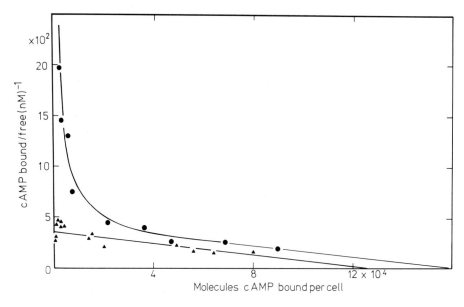

Fig. 1.15 Scatchard plot of cAMP binding activity of the mutant NP114 (▲), showing lack of the high binding affinity; and its parental strain X22 (●), showing high affinity binding.

relay the pulsed cAMP signal would be unable to form the large aggregation patterns and streams seen in the wild type. Mutagenized amoebae were screened by Newell and Green (unpublished work) and a mutant with the desired phenotype was isolated. This mutant (called NP114) aggregated with normal timing and produced small but normally proportioned fruiting bodies, but it only aggregated into small aggregates (seen particularly well in starving monolayers on thin agar plates) and did not produce the normal large streams in the process, choosing rather to form aggregates by localized, mainly radial, aggregation. It was of some interest to find that this mutant was capable of Class I cAMP binding but almost completely lacked the Class II (high affinity) binding (Fig. 1.15). A possible (but at this stage speculative) inference from this result is that the high-affinity Class II binding is required for signal relay. The finding of this mutant does not, however, greatly help to decide between the mechanisms of generation of the different cAMP binding affinities (as the mutant could be explained using any of the three models described above) and this must await the outcome of the kinetic experiments.

1.4 SLIME MOLD LECTINS AND THEIR RECEPTORS

1.4.1 Discovery of discoidins and pallidins

In 1973, Rosen *et al.* discovered that a protein that caused agglutination of formalin-treated sheep erythrocytes could be demonstrated in crude extracts of whole cells of *D. discoideum.* This protein (later called discoidin) was found to increase in quantity or activity over 400-fold during the period 3–12 hours of development when cells become progressively more cohesive. The important observation was also made that this erythrocyte agglutination activity was inhibited by the sugars N-acetyl-D-galactosamine, D-galactose and L-fucose. Using its galactose-binding ability, Rosen *et al.* (1973) purified the protein using a column of the galactose-containing polymer Sepharose 4-B as an affinity adsorbant. The purified protein was later shown by Simpson *et al.* (1974) and Frazier *et al.* (1975) to consist of two proteins, discoidin I and discoidin II. Following this work, a member of a different genus of cellular slime mold (*Polysphondylium pallidum*) was also shown by Rosen *et al.* (1975) to possess proteins with similar but distinguishably different properties, and these proteins were called pallidins. Because of the similarity of these proteins to the plant lectins which cause erythrocyte agglutination, these proteins have also been called lectins by their discoverers. To avoid confusion with the lectins of plant origin the term 'slime mold lectin' or 'SM lectin' will be used here.

1.4.2 Evidence for the role of slime mold lectins as recognition and adhesion proteins

Because of their ability to bind to and to agglutinate specific types of erythrocytes and because of the inhibitory effect of specific carbohydrates, the discoidins and pallidins are excellently suited for the role of adhesion proteins acting between specific oligosaccharide components of glycoprotein receptor molecules in the membranes of amoebae during development (Fig. 1.16). Moreover, since different species or genera of slime mold produce slime mold lectins with differing specificity, it is possible that they could act as species recognition proteins ensuring the formation of aggregates of the same species from mixed populations.

However, because the discovery and isolation of SM lectins used extracts of whole cells, it would be dangerous to assume without good evidence that they are located on the external surface of the cell. The evidence is as follows:

(a) The location of at least some of the discoidin and pallidin on the cell surface has been demonstrated by mixing cohesive cells with erythrocytes. Mixed aggregates or rosettes are thereby formed. It is found that the specificity of the sugars that dissociate these rosettes is the same as that inhibiting the agglutination of the erythrocytes by purified discoidins or pallidins (Rosen *et al.*, 1974, 1975).

(b) Chang *et al.* (1975) injected purified discoidin into rabbits and obtained rabbit anti-discoidin antibodies. Using this antibody preparation to label *D. discoideum* amoebae, and then following this label by the subsequent addition of ferritin- (or

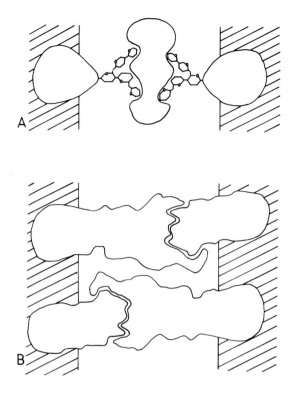

Fig. 1.16 Schematic representation of possible cell docking systems. Type A, which is a possible configuration for the slime mold lectins (discoidins and pallidins), shows a central lectin protein binding to the saccharide moieties of two glycoprotein receptors on adjacent cells. Type B shows a configuration where two membrane-bound proteins or glycoproteins mutually interact. As such type B interactions are probably asymmetric and all the amoebae are thought to possess identical binding capabilities, adjacent cells are shown possessing both forms of the interacting molecule on their membranes. Type B is a possible model for contact sites A.

fluorescein) labelled anti-rabbit immunoglobulin, they showed that the surface of cohesive but not vegetative cells became ferritin- (or fluorescein) labelled. With living cells, patches and then caps of the surface label were seen.

(c) Rosen *et al.* (1974) showed that the sugars that inhibited the erythrocyte agglutination reaction with pallidin (namely D-galactose or lactose) were also the ones that were active in inhibiting the normal endogenous cohesiveness of *P. pallidum* amoebae. Curiously, this important result has not been achieved with discoidin and *D. discoideum* amoebae (Rosen *et al.*, 1973).

(d) It has been demonstrated that cohesive *P. pallidum* amoebae are prevented from re-associating by the presence of purified pallidins (Frazier, 1976) or monovalent antibodies (Fab) prepared against such pallidins (Rosen et al., 1976). Again this phenomenon has not been shown for discoidin. Rosen et al. (1973) could find no conditions under which addition of purified discoidins influenced the aggregation of cohesive amoebae of *D. discoideum* in roller culture. Low concentrations of discoidin did not augment aggregation and high concentrations did not inhibit it. Discoidins did, however, increase the agglutinability of glutaraldehyde-treated *D. discoideum* amoebae, a treatment thought to destroy endogenous discoidin (Reitherman et al., 1975).

(e) To be useful to the cell for discriminating between different slime mold species during aggregation of mixed populations, the lectins must be capable of differential binding to cell surface sugars. In a study with six species of cellular slime molds Rosen et al. (1975) found that various monosaccharides (in soluble form) had different inhibitory abilities in the erythrocyte agglutination assay using lectins from the six different species. This is clearly suggestive of a species-specific role for the slime mold lectins, provided that the effect in the erythrocyte-binding assay is a true measure of the effects of the lectins on slime mold cell cohesiveness *in vivo*.

(f) More recently Siu et al. (1976) labelled the plasma membranes of 12 hour cohesive *D. discoideum* amoebae with ^{125}I by the lactoperoxidase method, then solubilized the cells with NP40 detergent and selectively bound the discoidin with a rabbit anti-discoidin serum. This material in turn, was precipitated by serum containing goat anti-rabbit immunoglobulin. The precipitate was removed from the solution, solubilized with sodium dodecyl sulfate and the components separated by gel electrophoresis. It was found that two predominant species of ^{125}I-labelled proteins were present, one of 56 000 daltons and the other (corresponding in its molecular weight to discoidin) of 26 000. Neither protein was present in controls using normal rabbit serum in place of anti-discoidin serum. Since the lactoperoxidase method only labels surface proteins this is good evidence for a surface location of discoidin. Moreover, using this technique, Siu et al. (1976) showed that discoidin appeared on the surface of bacterially grown amoebae after only 5 hours of development and remained associated with the cell surface through the culmination stage at 20 hours. Using the radioimmune assay for discoidin, they further showed that the discoidin antigen was present in a very low concentration in 4 hour cells and represented only about 0.1% of the soluble protein fraction. Between 4 and 10 hours, however, it accumulated rapidly and then remained roughly constant for at least 6 hours at its maximum level representing about 1% of the soluble cell protein.

It may fairly be concluded from the above evidence, that the case for discoidin and pallidin being associated *at least in part* with the external surface of the amoebae is well supported. The relative amounts of lectin on the external membrane and inside the cell in the soluble fraction have yet to be determined with any certainty. Siu et al. have calculated, however, that at its peak in 10 hours amoebae, discoidin would represent about 1.2×10^6 molecules per cell. As Reitherman et al. (1975) can detect about 5×10^5 receptors per cell for discoidin on *D. discoideum* (see below) the number of discoidin molecules per cell is consistent with their proposed role as adhesive agents linking the receptors on adjacent cells.

1.4.3 Properties of slime mold lectins

Let us now consider the properties of the slime mold lectins in more detail, as revealed by the work of Rosen *et al.* (1973), Simpson *et al.* (1974) and Frazier *et al.* (1975) for discoidin and by Rosen *et al.* (1974), Simpson *et al.* (1975) and Frazier *et al.* (1976) for pallidin.

Discoidins, which by definition are those lectins from *D. discoideum*, were found, after purification on the galactose-containing affinity absorbent Sepharose 4-B, to consist of two distinct proteins: discoidins I and II. Discoidin I is the major component, with discoidin II being present at less than 10% of the activity of discoidin I in axenically grown cells. However, discoidin II can represent as much as 35% of the total SM lectin activity at early stages of differentiation of bacterially grown *D. discoideum* NC4 wild type cells.

Discoidin I is a tetramer with subunits having a molecular weight of 26 000. Discoidin II is also multimeric with subunits of 25 000 daltons and, since it contains substantially more serine and lysine residues per polypeptide chain than discoidin I, it appears unlikely that it is derived from discoidin I by proteolysis. The discoidins also differ in (a) their agglutination reactions; while both discoidin I and II agglutinate fixed rabbit erythrocytes, discoidin I agglutinates fixed sheep erythrocytes but discoidin II does not, (b) their isolectric points, (c) their tryptic peptide maps, (d) their pattern of inhibition by sugars (*N*-acetyl-D-galactosamine being the best inhibitor of discoidin I and lactose of discoidin II), and (e) their time of appearance during differentiation. Neither one is present in vegetative amoebae but discoidin I activity increases linearly from 6–12 hours of development while discoidin II reaches a higher percentage of its peak (12 hour) activity at earlier times than discoidin I. The ratio of the two activities therefore changes substantially during aggregation.

Neither of the discoidins contains any bound sugar. They are not therefore, in this respect, like most of the plant lectins which are usually glycoproteins (Lis and Sharon, 1973). (However, one of the best known of the plant lectins, concanavalin A also lacks carbohydrate.) The SM lectins do resemble some other plant lectins in being rich in aspartic acid and 3-hydroxyamino acids.

The pallidins (by definition the lectins from *P. pallidum*) have also been purified by affinity chromatography using formalin-treated human type 0 and rabbit erythrocytes. The pallidins are multimeric (like the discoidins) and, after purification, form aggregates of about 250 000 daltons and higher. There are probably at least three subunit forms, two of molecular weight 25 000 (separable by their differential binding to either the human 0 or rabbit erythrocytes) and a third of 18 000 daltons. The latter is smaller than all the other SM lectins and it is a weak erythrocyte agglutinin, possibly because it normally exists as a dimer. The amino acid compositions and hence isoelectric points of the pallidins are similar but clearly distinguishable from those of the discoidins, and like the discoidins the pallidins contain no detectable sugars. They also resemble the discoidins by being present in considerable quantities in the cell and comprise more than 1% of the soluble proteins obtained by sonication of *P. pallidum.*

1.4.4 Membrane receptors for slime mold lectins

If the SM lectins function as cohesive agents between cells, then the external surfaces of the cells must possess carbohydrate-containing molecules with which the SM lectin binds. Such receptor molecules (probably glycoproteins) have been tentatively identified by Reitherman *et al.* (1975) in *D. discoideum* and *P. pallidum.* The endogenous SM lectins were destroyed by glutaraldehyde fixation of the cells, apparently leaving the receptors intact. Discoidins and pallidins were found to bind avidly to these fixed cells, the kinetics of binding indicating the presence of about 5×10^5 receptors per cell. The number of receptors varied little from vegetative amoebae to cohesive cells but the affinity increased twenty-fold. While the dissociation constant for 0 hours (vegetative) amoebae was approximately 20 nM, that for 9 hour cohesive cells was about 1 nM. They also found that receptors on the surface of cohesive *D. discoideum* or *P. pallidum* amoebae had a higher affinity for the lectins from the same species than for those of the other species. The nature of the change in the receptors is unknown. The higher affinity receptors seen in cohesive cells could represent the formation of a new glycoprotein and its incorporation into the membrane, or they could result from the modification of the low affinity receptor by, for example, addition or deletion of sugars of the oligosaccharide part of the molecule. The apparent constancy of the total number of receptors from the growth to developmental phase makes the latter scheme seem plausible. I might also speculate that the change in affinity may be due to association of pre-existing molecules in the plane of the membrane as observed for the plasma membrane particles seen in electron microscopy studies by Aldrich and Gregg (1973) (Fig. 1.19 and Section 1.6.6). Such association could produce a high affinity for the SM lectin by a positive co-operative interaction of the low affinity forms.

A possible candidate for the receptor protein in *D. discoideum* has been proposed by Siu *et al.* (1976). This 56 000 dalton protein molecule fits the role of the discoidin receptor in that it is labelled with ^{125}I by external lactoperoxidase and significantly, it is co-precipitated along with discoidin by specific anti-discoidin antiserum. Further work is needed however, to ensure that this protein is not simply a dimer of the 26 000 dalton protein.

1.4.5 Cell-binding lectins in other systems

It is of wider interest to note that lectin-like molecules (capable of agglutinating trypsin-treated rabbit erythrocytes) have recently been detected in a number of tissues of multicellular systems including the electric organ of the eel *Electrophorus electricus*, embryonic and adult chick pectoral muscle and adult rat soleus and diaphragm muscles (Teichberg *et al.*, 1974). Although not all reports are in agreement with the role of the lectins in developing cell lines (Den *et al.*, 1976), some reports have suggested a role for lectins in the cell interaction involved in the fusion of developing muscle cells from both chick and rat tissues. For example, lectin activity was found to increase several-fold in differentiating L_6 myoblasts derived from rat skeletal muscle during the time that the myoblasts were fusing to form multi-nucleated myotubes, and such lectins were shown to be associated with the outside of the cells by their sensitivity to competing sugars and to brief trypsin treatment (Gartner and Podleski, 1975; Nowak *et al.*, 1976).

One may conclude that lectins could generally be of great importance during development. While their surface role seems to be well-supported in the case of slime molds, it is less so, as yet, in other systems. The elucidation of that precise role is an exciting topic for future investigation.

1.5 CONTACT SITES A

One of the methods that can be used to reveal subtle changes in macromolecular shape or composition is to use the surface membrane molecules as antigens in an immunological reaction by injecting the surface components into rabbits. Antisera prepared from whole or fractionated amoebae from different stages of development can then reveal the appearance of any new surface components recognisable by their characteristic conformation. Early work by Gregg (1956) and Gregg and Trygstad (1958) using antisera raised against vegetative and cohesive amoebae from several species of slime mold, as well as against aggregation-defective mutants, indicated differences in reactivity of the antisera between the different species and between the wild types and mutants. Further immunohistochemical work by Takeuchi (1963) clearly showed that the cell surface of *D. mucoroides* gained antigenic groups during aggregation. The antigenic determinants, raised against amoebal homogenates of *D. discoideum* strain NC4 were investigated by Sonneborn, Sussman and Levine (1964) who found that antiserum against cohesive cells (absorbed out with vegetative cells) could prevent aggregation by starving amoebae without causing a loss of cell viability.

Due to the studies of Gerisch and his colleagues, this approach has now yielded a great deal of detailed information about the nature of the antigenic groups involved in making amoebae cohesive. By injecting amoebae of *D. discoideum* strain V12 into rabbits, Beug and Gerisch (1969) and Beug *et al.* (1970) raised antisera to the surface of the cohesive amoebae. The rabbit antibodies active against components of the cell surface that were present at all stages of development were absorbed out by the addition of excess vegetative amoebae. It was then shown that the remaining rabbit antibodies were active against surface components of the amoebae (which were called contact sites A) by their ability to cause agglutination of cohesive but not vegetative amoebae. This agglutination reaction occurred because of the multivalent character of the antibodies in the rabbit antiserum. If the antibodies were split with papain, then agglutination no longer occurred. However, such monovalent (Fab) rabbit antibodies completely inhibited the normal aggregation of cohesive amoebae, presumably by binding to the contact sites A essential for maintaining cell contact. These sites, however, represent only a small fraction (about 2%) of the cell surface, so it is unlikely that the inhibition of aggregation by the Fab was caused by an artifactual covering of the surface of cohesive cells (Beug *et al.*, 1973). This conclusion was strengthened by the finding that Fab prepared against heated cells (the heating apparently denaturing the contact sites A) could not inhibit aggregation of normal cohesive amoebae despite the fact that the surface was in this case much more densely covered by the unabsorbed non-specific Fab molecules. Using contact-site-A-specific Fab, it was estimated that there were roughly 3×10^5 contact sites A per amoeba. Addition of ferritin or fluorescein-labelled anti-contact site A immunoglobulin revealed that these sites were arranged uniformly on the periphery of the amoebal cells (Gerisch *et al.*, 1974a).

If contact sites A represent the two halves of an interacting system (Fig. 1.16), then one could predict that mutants should be found that lacked one or both parts. Using Fab absorption as a test for the presence of either part in aggregation-deficient mutants, Gerisch *et al.* (1974a) indeed found that some mutants isolated could only partially absorb the activity of a contact site A Fab preparation and others could absorb none. However, pairs of mutants that completely absorbed the Fab activity when both were used (i.e. lacking complementary components of the system) have not yet been reported.

More recently, the purification of contact sites A has been reported by Huesgen and Gerisch (1975) from desoxycholate-solubilized membranes of the axenic mutant AX2 of *D. discoideum*. Using DE-cellulose chromatography and Sephadex G200 filtration the contact sites A were purified 160-fold relative to the activities measured at the surface of intact cells. They were able to separate the contact sites A from membrane phosphodiesterase which was also solubilized and found that the contact site fraction eluted in Sephadex G200 as one symmetrical peak corresponding to a molecular weight of 120–130 000.

An interesting finding resulting from this purification was that the purified contact sites were inactive in agglutinating formalin-treated sheep erythrocytes, although the unfractionated desoxycholate-solubilized membranes did so. Moreover, a large excess of the purified contact sites could not inhibit the agglutination of sheep erythrocytes by a small amount of discoidin. These results (Huesgen and Gerisch, 1975) clearly suggest that the discoidin of Rosen *et al.* (1973) is different from the purified contact site A.

One may therefore, at present, conclude that contact sites A and slime mold lectins such as discoidin are probably different species of molecule, which serve ostensibly similar functions. It will be interesting to learn whether further work reveals a special molecular niche for each of them.

1.6 OTHER SURFACE COMPONENTS IMPLICATED IN CELLULAR INTERACTION

1.6.1 Divalent metal ion receptors

Because aggregation in *D. discoideum* is known to require the presence of Ca^{2+} ions at concentrations above 1 μM (Mason, Rasmussen and Dibella, 1971), it is likely that some form of receptor site exists for it inside or on the outside of aggregating amoebae. It is noteworthy, for example, that the spontaneous pulsatile signalling system requires Ca^{2+} ions, because such signalling (but not signal relay) is reversibly blocked if Ca^{2+} ions are removed with EDTA or EGTA (Gerisch *et al.*, 1975c).

Recently, Sussman and Boschwitz found evidence for an interesting receptor that increases in its Ca^{2+} binding activity about 15 to 20-fold during the period of aggregation when amoebae become cohesive. The binding of radioactive $^{45}Ca^{2+}$ to a 'cell ghost' preparation from vegetative or developing amoebae was measured using a membrane filter assay. Two types of Ca^{2+}-binding activity were detected in this way, one type showing a binding activity for Ca^{2+} that was not competed for by Mn^{2+} and which did not change appreciably in activity during the vegetative or developmental state and another type that bound Ca^{2+} that was competable by unlabelled Mn^{2+} ions. The Mn^{2+}-competable binding rose from an activity of about 1 pmol of Ca^{2+} bound μg^{-1} protein in vegetative cells to about 15–20 pmol of Ca^{2+} bound μg^{-1} protein at the plateau reached after roughly twelve hours of development on filters. The location of the Ca^{2+} receptor site is not, however, accurately known. The cell ghost preparations used for the assays were osmotically insensitive broken whole cells that had been washed extensively to remove the soluble cell proteins and small organelles. They did contain mitochondria, although most were apparently damaged and unable to make ATP (Sussman and Boschwitz, 1975). The washed cell ghosts prepared from cohesive cells were also found to agglutinate spontaneously only in the presence of low concentrations of divalent cations such as Ca^{2+}, Mn^{2+}, Zn^{2+} or Cu^{2+}, but this agglutination did not occur with ghosts from vegetative amoebae. A tentative connection can therefore be suggested between the ability to bind divalent cations by the Ca^{2+}/Mn^{2+} receptors and the ability to form cell-cell contacts required for agglutination and presumably for the normal docking process at the end of aggregation.

1.6.2 Glycoproteins binding concanavalin A

With the present availability of plant lectins that can bind specific saccharides on the surface of cells, one approach to the problem of investigating the structure of interacting cells is to determine the extent of binding of particular plant lectins and

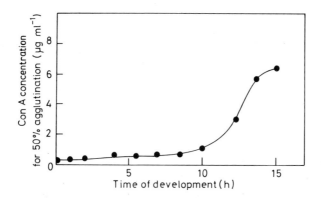

Fig. 1.17 The concentration of Con A required to produce 50% agglutination of developing AX2 amoebae of *D. discoideum*. Aggregation streams were noted at 8 hours and slugs were formed at 16 hours. Cells were harvested into Bonner's salts solution, dissociated mechanically and tested for Con A-dependent agglutination (After Weeks and Weeks, 1975).

the effect of such binding on development of the slime mold. In this way, the roles of particular subsets of glycoproteins possessing particular saccharides may be more readily investigated. Currently, the most popular of the plant lectins is concanavalin A (Con A) which selectively binds to glycoproteins possessing α-D-mannopyrannose, α-D-glucopyranose or D-fructofuranose (Lis and Sharon, 1973).

Weeks (1973), using the axenic strain AX2, found that Con A caused agglutination of vegetative amoebae (in log phase) at considerably lower concentrations than were necessary to agglutinate stationary phase or developing cells (Fig. 1.17). This agglutination was inhibited by the specific Con A inhibitor α-methyl glucoside. This was of interest particularly as malignant cells had been shown to be more easily agglutinated by Con A than normal cells (Sharon and Lis, 1972). Subsequently, Weeks (1975), using binding of ^{125}I-labelled Con A, showed that the decrease in Con A-induced agglutinability during development is due to the appearance of low affinity receptors on the external cell membrane (dissociation constant, $K_d = 5 \times 10^{-6}$ M) in place of the high affinity receptors ($K_d = 5 \times 10^{-7}$ M) seen during the growth phase. A small fraction of the high affinity sites is still apparent, nevertheless, late in the aggregation phase, with the result that Scatchard plots have a biphasic character (Fig. 1.18). The interpretation of such plots is rendered somewhat complex with the recent finding of Geltosky *et al.* (1976) that there are about 15 Con A-binding glycoproteins on the surface of *Dictyostelium* amoebae that contribute to the total Con A binding activity. By separating the glycoproteins on SDS polyacrylamide gels, however, Geltosky *et al.* found that only one of these glycoproteins (of approximate molecular weight 150 000 daltons) unequivocally increased in activity during late aggregation or early development. It was of interest that at the same time as the 150 000 dalton glycoprotein was appearing Geltosky *et al.* found that another

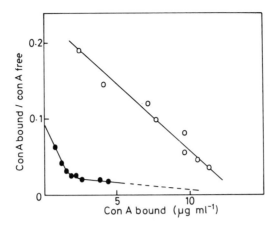

Fig. 1.18 Scatchard plot of the α-methyl glucoside-sensitive binding of ^{125}I-Con A to *D. discoideum* AX2 vegetative amoebae (○) and early aggregation (11 h) amoebae (●). (After Weeks, 1975).

glycoprotein (180 000 daltons was disappearing and there may have been a precursor-product relationship between the two glycoproteins, as could also be the case for the high and low affinity receptors found by Weeks. As only saturating concentrations of Con A were used by Geltosky *et al.*, their glycoprotein bands cannot, unfortunately, be directly correlated with the glycoprotein receptors of Weeks which were distinguished on the basis of their Con A affinity.

What may possibly be the same increase in glycoprotein activity was studied by Wilhelms *et al.* (1974). A strongly Con A-binding glycoprotein (called antigen II) was found to be almost absent during the vegetative phase and appeared as a strong antigenic component of phenol-water extracts of cohesive cells. Its surface location was suggested by its immunoprecipitation from extracts of cell ghosts. The protein component was rich in acidic and hydroxy amino acids and the carbohydrate moiety, which contained *N*-acetylglucosamine, fucose and mannose in the ratio of 12:15:6, was linked glycosidically via the *N*-acetylglucosamine units to the hydroxy group of the hydroxy amino acids. It may yet be found in due course that everyone is looking at the same glycoprotein from different experimental angles.

Another interesting aspect of the Con A story is the effect of Con A on aggregation and development once it has bound to the amoebae Gillette and Filosa reported in 1973 that Con A at concentrations of 50–100 μg ml^{-1} delayed aggregation of bacterially grown *D. discoideum* and induced premature appearance of membrane-bound phosphodiesterase. Weeks and Weeks (1975) confirmed this result, but they could only observe it if the cells were under particular environmental conditions, with the Con A dissolved in the 2% agar on which the amoebae were plated. (Under other conditions, such as incubation on moistened filter paper, even very high concentrations of Con A (800 μg ml^{-1}) had no observable effect.) Moreover, if the axenic mutant AX2 was used for study, they found that only high concentrations

(500 μg ml^{-1}) produced any delay and concentrations of up to 200 μg ml^{-1} were without effect. In a recent study using the AX2 mutant in liquid culture, Darmon and Klein (1976) found that, at low concentrations of Con A (50 μg ml^{-1} or less), there was a slight stimulation of development. The amount of Con A bound (determined by ^{125}I-labelled binding) under such conditions was just at or less than the maximal amount of high affinity binding of about 1.5×10^7 molecules per cell. At higher Con A concentrations, aggregation was progressively delayed such that at 200 μg ml^{-1} the delay was over 20 hours.

A possible, and thought-provoking, explanation for some of the effects of Con A on development was provided by Darmon and Klein (1976) who suggested that this multivalent lectin binds to receptors on the amoebal cell surface and causes their redistribution. At low concentrations of Con A this may be stimulatory. At higher concentrations, however, other membrane components such as cAMP receptors may be redistributed and cause inhibition of development until the Con A is removed from the surface (probably after patching and capping) (Gillette et al., 1974). They also suggest that the observed stimulation of phosphodiesterase formation observed by Gillette and Filosa may in fact be due to a general decrease in the permeability of the membrane to macromolecules in the presence of Con A that leads to an inhibition of the excretion of the phosphodiesterase (ePD) and hence its apparent accumulation in the cell and the cell surface membrane. In support of this notion, they find that the rise in cellular phosphodiesterase seen in the presence of Con A after a short incubation is just offset by a decrease in the amount of ePD excreted into the medium and also that the excretion of other enzymes such as N-acetylglucosaminidase is strongly inhibited.

1.6.3 Glycosphingolipids

At least some of the oligosaccharides on the cell surface are linked to lipids (rather than proteins) and may be highly important for cell interaction. It is possibly unfortunate for our knowledge of cell interaction that the field of sphingolipids inspires awful dread in so many biochemists who have, at the same time, learnt to live happily with the equally complex field of protein chemistry.

One group, however, (Wilhelms et al., 1974), who show no such reaction to these compounds, have isolated a particularly interesting glycosphingolipid after phenol-water extraction of whole cells of D. discoideum (strain V12.M2). They showed that this antigenic molecule resided on the outside of the cells by the finding that the D. discoideum amoebae were agglutinated by antibodies prepared against the purified glycolipid. The lipid part of the molecule, which (presumably) normally sits in the plasma membrane, was shown to contain behenic acid and 4-hydroxysphinganine (C_{18}-phytosphingosine) in equal molar ratio and in addition phosphate and ethanolamine. The oligosaccharide 'tail' (presumably the antigenic region exposed to the outside of the cell) contained N-acetylglucosamine with fucose and mannose in the ratio 12:5:2, the fucose being the terminal sugar.

Although this glycolipid was present in both vegetative and cohesive cells, it was noteworthy that an aggregateless mutant (*aggr*20–2) possessed an altered form that had a markedly reduced content of fucose. (The N-acetylglucosamine/fucose/mannose ratio was 12:1:3). From the finding that treatment of the purified mutant glycolipid with purified N-acetylglucosaminidase liberated N-acetylglucosamine, whereas no sugars were liberated by this treatment from the wild-type glycolipid, it was concluded that N-acetylglucosamine, rather than fucose was the terminal sugar in the aggregateless mutant. Although this is unlikely to be the primary defect in this mutant (which showed loss of other aggregation functions in a pleiotropic manner), it does perhaps indicate a role for such glycolipids in cell interaction in normal aggregation.

1.6.4 Purified plasma membrane components

A direct approach to the study of cell surface components is the isolation of pure plasma membranes from various stages of development followed by the characterization of the proteins or glycoproteins present with a view to looking for additions or deletions during the course of development. The most usual method of tackling the purification problem is to disrupt the cells thoroughly (for example by freeze-thawing with liquid nitrogen) to produce small fragments of membrane which are then separated by density gradients. (Green and Newell, 1974; Rossomando and Cutler, 1975). More recently, other methods have been used, such as attaching colloidal glass beads to the cell surface (McMahon, personal communication) or agglutinating vegetative amoebae with Con A followed by lysis with Triton X-100 detergent (Parish and Müller, 1976). In the latter, rather promising method, the Con A protects the plasma membrane of the vegetative cells from being completely lysed by the detergent, but allows the cell contents to escape. In this respect, it resembles the method of cell ghost preparation of Sussman and Boschwitz, but apparently allows cleaner plasma membranes to be isolated because of the detergent treatment.

The field of membrane purification, unfortunately, can be like quicksands, unless great caution is exercised. Many enzymes used as plasma membrane markers in other cell systems such as the oubain-inhibited Na^+/K^+, ATPase seem to be lacking in the plasma membranes of $D.$ *discoideum* (Green and Newell, 1974) as is true also in *Acanthamoeba* (Klein and Breland, 1966). Other enzymes such as $5'$ nucleotidase or phosphodiesterase are poor plasma membrane markers because they are present on other internal membranes in the cell, besides being on the plasma membrane and there is also a large soluble or easily solubilized fraction to contend with (Green and Newell, 1974; Rossomando and Cutler, 1975; Lee *et al.*, 1975; Green, 1976; Gilkes and Weeks, 1976). Even though such enzymes as alkaline phosphatase may appear to be suitable (and easily assayed) markers for the plasma membrane, it has recently been found by means of biochemical and cytochemical tests (Green, 1976) that during development the majority of alkaline phosphatase activity is not on the external plasma membrane but resides on the inside surface of internal (non-phagocytic) vesicles, together with a considerable fraction of a phosphodiesterase. Such vesicles roughly resemble plasma membranes in their protein composition (but see below) and, unfortunately, generally co-purify with them, but they can be separated with the aid of sequential ficol and sucrose gradients. The only reliable marker for the plasma membrane appears to be ^{125}I-labelling of the cell surface before membrane purification. Using this system to monitor cell surface plasma membrane, together with a number of enzyme assays to detect and eliminate contaminating intracellular membranes, the plasma membranes of vegetative (Green and Newell, 1974) and cohesive amoebae (Green, 1976) can be isolated in pure form.

Using such purified plasma membrane preparations, the protein components have been separated using SDS polyacrylamide electrophoresis. Several striking changes in the Coomasie blue staining proteins and Periodate–Schiff (PAS) staining glycoproteins were noticed between preparations from vegetative and cohesive amoebae (Green, 1976). The most obvious protein that increased in activity was of molecular weight 42 000 and probably corresponded to actin that is known to be very actively synthesized during the first few hours of development (Spudich, 1974; Tuchman *et al.,* 1974). It is interesting to note that, although gels of the alkaline phosphatase-enriched vesicles (mentioned above) strongly resembled those of the plasma membranes in most of the bands, the 42 000 dalton protein was conspicuously absent from them, suggesting that the association of the actin protein with the plasma membrane is specific and occurs only with the plasma membrane actually on the surface of the cell. The most dramatic decrease in the Coomasie blue staining bands was a 31 000 dalton protein, but the explanation of the function of this and the many other minor changes must await further detailed studies.

1.6.5 ^{125}Iodine surface-labelled components

Because of the problems connected with plasma membrane purification, the technique of using ^{125}I surface labelling followed directly by protein separation without prior plasma membrane purification has been adopted by some workers.

Smart and Hynes (1974) treated the surface of vegetative and developing amoebae with the Na–^{125}I-lactoperoxidase system in order to label proteins selectively on the cell surface that possessed exposed tyrosine groups. The components of the whole cells were then separated by SDS polyacrylamide electrophoresis and the position of the surface-labelled components found by autoradiography. They found that the most noticeable change was the appearance of a component of 130 000 daltons molecular weight that was formed (or became available to the lactoperoxidase reagent) at a time in development when firm aggregates were formed. This component was presumably not involved in cohesion however, because if amoebae were allowed to become cohesive in suspension culture (which allows the attainment of cohesiveness but generally blocks further development) then this protein did not appear. They also noticed increases in the components of molecular weight: 160 000; 110 000 and 55 000 daltons and a fairly dramatic decrease in the 135 000 dalton component

As a method for looking at cell surface componets, the ^{125}I surface-labelling technique seems to have much to offer, mainly because it short cuts the extensive procedures needed for the more conventional plasma membrane purifications described in Section 1.6.4. It is also a very convincing and useful technique when only those components on the external face of the plasma membrane are being investigated. It suffers, however, as do all methods that also use the SDS polyacrylamide gel separation in that the final product is an inactive monomer the molecular weight of which can be estimated but activity only guessed at.

Probably the most hopeful extension of the surface-labelling technique is therefore that used by Siu *et al.* (1976). These authors took a partially purified membrane fraction (made by the two-phase interface method of Brunette and Till, 1971) of ^{125}I surface-labelled amoebae and then solubilized the membranes using the non-ionic detergent NP40. This detergent allowed subsequent precipitation of the protein discoidin by a specific antiserum made against purified soluble discoidin. In this way, immunologically active protein, labelled with ^{125}I, was recovered which had clearly been present on the cell surface. In another study, Geltosky *et al.* (1976) passed the radioiodinated NP40-solubilized membranes over a Con A-Sepharose affinity column and thereby specifically selected the Con A-binding glycoproteins. Such use of less denaturing detergents and selection of components with specific antisera or affinity columns is a powerful and very promising combination of techniques.

1.6.6 Plasma membrane particles

A rather different approach to the search for changes in the cell surface during development is actually to look at plasma membranes at various stages of development using the tools of electron microscopy and freeze-etching.

Such an investigation by Aldrich and Gregg (1973) has produced two interesting and possibly highly important findings. Firstly, particles initially observable in the membranes (of *D. discoideum* NC4) during vegetative growth or the initial stages of starvation as being about 6 nm in diameter can be seen to measure about 10 nm in diameter a few hours later (Figs. 1.19 and 1.20). This change is accompanied by a reduction in the total number of particles and probably represents a fusion of mobile proteins (or glycoproteins) 'floating' in the fluid plasma membrane. Secondly, it was also found that the presence in the agar (on which the amoebae develop) of 1 nM cAMP or 100 μM CaCl$_2$ could mimic this effect within a period of 2 hours (Gregg and Nesom, 1973). A spurious and trivial mode of action was made less likely by the finding that in two aggregation-deficient mutants tested (67 and 20-2 from V12.M2) which were chemotactically unresponsive to cAMP, the cAMP had no action on the membrane particles (Gregg and Yu, 1975).

There is undoubtedly a problem in correlating such studies with other work that looks at more functional aspects of the cell surface. However, if such a correlation can be made it should provide a very powerful exploratory tool. One might speculate that a reason for the association of such particles could be that it results in co-operative interactions leading (for example) to increased receptor affinities as

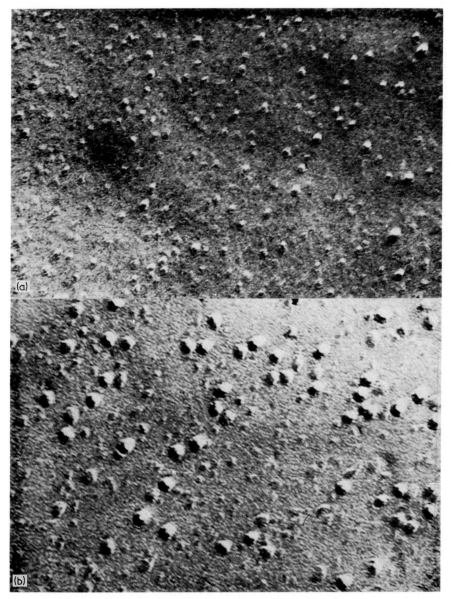

Fig. 1.19 Electron micrographs of freeze-fractured plasma membranes of
D. discoideum amoebae showing membrane particles (× 240 000). (a) Control
vegetative amoebae. (b) Amoebae exposed for 2 h to 1 mM cAMP.
(Photographs by courtesy of Drs James Gregg and Margaret Nesom).

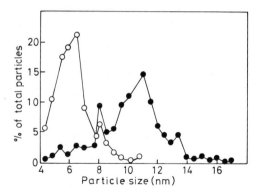

Fig. 1.20 The percentage frequency distribution of particle sizes occurring in *D. discoideum* V12/M2 wild type and the aggregation-deficient mutant Agg67, after 66 hours of cultivation. The particle size has increased in the wild type from a mean value of approximately 6 nm initially (not shown) to a mean value of approximately 10 nm in pre-spore cells of the migrating slug stage. (●). Amoebae of the aggregation-deficient mutant Agg67 show particles that remain small (○). (After Gregg and Yu, 1975).

discussed for the SM lectin receptors in Section 1.4.4 above.

1.7 FURTHER STUDIES

Of several areas of research that seem destined for further study, I will mention here only two that particularly catch my eye. One involves the roles of folates and prostaglandins in the signalling system and the other involves genetic studies of aggregation.

It seems very likely that the signal oscillation system is of far greater significance in the life of the slime mold than we can demonstrate at present. It is for this reason that two preliminary reports indicating effects of folic acid (Wurster, 1976b; Gerisch *et al.*, 1976) and of prostaglandin A_1 (Coe and Kuo, 1976) are of particular fascination. The unexpected finding of Wurster was that folic acid (known to be a chemotactic attractant for *Dictyostelium* during the vegetative phase; Pan *et al.*, 1972, 1975) could, in some ways mimic cAMP in triggering oscillating cAMP emissions in starving suspensions of amoebae. The folic acid was only active if added in pulses and, like cAMP, it could accelerate the onset of autonomous oscillations and the developmental processes. Folic acid differed from cAMP, however, in its coupling to the oscillatory machinery. It could not mimic cAMP in causing phase shifts in amoebae in the pre-oscillating condition. As soon as autonomous oscillation occurred, the cells lost their sensitivity to the folic acid. One could be drawn, on the basis of these findings, to speculate that the folic acid and cAMP receptors could be the same molecules in different guises. Whatever the basis of these effects, however, an exploration of the role of folic acid could be very revealing.

In view of the effects of prostaglandins in other systems, the report of Coe and Kuo (1976) concerning the effects of these compounds in *Dictyostelium* is of particular interest. These authors found that addition of a 1 μM pulse of prostaglandin A_1 to suspensions of (5-hour) starving amoebae could elicit a wave of cAMP emission in a similar manner to an exogenous pulse of 100 nm cAMP. The response from prostaglandin addition, however, differed in that the wave of cAMP emitted was delayed compared with that produced by exogenous cAMP (or was '180°out of phase' with it). This finding is made more significant by the other observation of these authors, that prostaglandins repel amoebae rather than attracting them if drops containing the compounds are placed alongside drops of amoebae on agar plates. Whether the cellular slime molds actually make use of these compounds remains to be established. Coe and Kuo produced evidence, however, that the cellular slime molds did contain prostaglandins (or compounds that resembled them chromatographically) and further studies to probe their role in these organisms seems worthwhile.

With the advent of genetic analysis in the cellular slime molds (see Mosses *et al.*, 1975 for references) it has become feasible to carry out a genetic study of the events of aggregation. Diploids can be made by fusing well-marked haploid strains that carry aggregation-deficiency mutations. From the extent of genetic complementation between pairs of such mutations, the total number of genes involved in the aggregation process and their dominance or recessiveness can also be determined.

Three groups have so far independently initiated such genetic complementation studies, one study being with the strain *P. violaceum* carried out by Warren *et al.* (1975; 1976) and two being with *D. discoideum* (Coukell, 1975; Williams and Newell, 1976). The results so far indicate that about 40–50 genes are probably essential for aggregation (such that their mutation completely halts it), with possibly a hundred more whose mutation affects the process in some way but does not stop it. Only 4 or 5 genes have so far been identified in terms of genetic complementation groups by each group of workers and none so far have been positively correlated with the enzymes or active proteins thought to function during aggregation. It is an understatement to say that considerable genetic work is still required to complete the complementation studies and determine the genes involved in aggregation. Correlation of the aggregational genes with their biochemical mode of action will require both insight and persistence, not only because of the probable complexity of this deceptively simple process, but also because of the pleiotropic nature of most aggregation mutants steming from the sequential nature of the steps of aggregation. The combination of genetics and biochemistry is, however, a powerful one and should greatly strengthen the analysis of the aggregation process that is currently under way.

1.8 SUMMARY

The cellular slime mold is a lower eukaryote that can exist either as a population of independent amoebae or, when nutrients are exhausted, as a multicellular organism capable of movement over a substratum and of differentiation to form spores and stalk cells. The transition between the two forms is achieved by aggregation of the amoebae to a central collecting point in response to chemotatic signals.

This aggregation phase presents an amenable system for studying cellular communication. Aggregation is initiated by some of the starving amoebae emitting rhythmic pulses of the chemical attractant cAMP. The pulsed signals are perceived by the surrounding amoebae by means of cAMP receptors on the amoebal surface, and receipt of signals initiates three responses by the amoebae. Firstly, the amoebae move toward the source of the signal (chemotaxis). Secondly, they synthesize and emit new signals so that the incoming signals are relayed outward from the center to other cells and thirdly, specific macromolecular receptors are produced and located in the plasma membrane so as to enable amoebae to dock together. This signalling system produces visible concentric or spiral waves emanating from the signalling center that later break down to give centrally converging streams of amoebae. The aggregation phase is terminated by the amoebae collecting as heaps of cells that bind tightly together to form a co-ordinated motile organism.

The cAMP receptors used during the process are formed early in the aggregation phase and are highly specific for cAMP. It has been found that they exhibit complex ligand binding properties, suggesting multiple binding sites or negative co-operative interactions between subunits.

The macromolecules associated with the cell docking process seem to be of two types: 'lectin proteins' and 'contact sites A'. The slime mold lectins (such as, for example, the discoidins from *Dictyostelium* species and the pallidins from *Polysphondylium* species) have been purified and studied. Using immunochemical techniques, evidence has been found for their cell surface location, their formation during the aggregation process and for their role as specific cell adhesion agents. The contact sites A have also been partially purified and shown to be formed during aggregation by a process that is probably regulated by the cAMP signalling system.

Other cell surface molecules that appear to be involved in cell interaction in this system include calcium/manganese ion receptors, concanavalin A-binding glycoproteins, and glycosphingolipids. The protein and glycoprotein components involved in amoebal interaction during aggregation have been investigated by gel electrophoresis of purified plasma membranes and by ^{125}I surface-labelling techniques. Changes are also noticeable in the distribution of plasma membrane particles seen by electron microscopy, such that the events of aggregation or exogenously supplied cAMP apparently stimulate the association of particles to form larger complexes in the plasma membrane.

Further biochemical and genetic investigations into the mechanism and components of the amoebal interaction system involved in slime mold aggregation are discussed.

REFERENCES

Alcântara, F. and Bazill, G.W. (1976), Extracellular cyclic AMP-phosphodiesterase accelerates differentiation in *Dictyostelium discoideum*. *J. gen. Microbiol.*, **92**, 351–368.

Alcântara, F. and Monk, M. (1974), Signal propagation during aggregation in the slime mould *Dictyostelium*. *J. gen. Microbiol.*, **85**, 321–334.

Aldrich, H.C. and Gregg, J.H. (1973), Unit membrane structural changes following cell association in *Dictyostelium*. *Exp. Cell Res.*, **81**, 407–412.

Arndt, A. (1937), Untersuchungen über *Dictyostelium mucoroides* Brefeld. *Wilhelm Roux Arch. Entwicklungsmech. Org.*, **136**, 681–747.

Beug, H. and Gerisch, G. (1969), Univalente Fragmente von Antikörpern zür Analyse von Zellmembran-Funktionen. *Naturwissenshaften*, **56**, 374.

Beug, H., Gerisch, G., Kempff, S., Riedel, V. and Cremer, G. (1970), Specific inhibition of cell contact formation in *Dictyostelium* by univalent antibodies. *Exp. Cell Res.*, **63**, 147–158.

Beug, H., Katz, F.E. and Gerisch, G. (1973), Dynamics of antigenic membrane sites relating to cell aggregation in *Dictyostelium discoideum*. *J. Cell Biol.*, **56**, 647–658.

Bonner, J.T. (1944), A descriptive study of the development of the slime mold *Dictyostelium discoideum*. *Am. J. Bot.*, **31**, 175–182.

Bonner, J.T. (1949), The demonstration of acrasin in the later stages of the development of the slime mold *Dictyostelium discoideum*. *J. exp. Zool.*, **110**, 259–271.

Bonner, J.T. (1967), *The Cellular Slime Molds*, Princeton University Press, Princeton, New Jersey.

Bonner, J.T. (1970), Induction of stalk cell differentiation by cyclic AMP in the cellular slime mold *Dictyostelium discoideum*. *Proc. natn. Acad. Sci. U.S.A.*, **65**, 110–115.

Bonner, J.T. (1971), Aggregation and differentiation in the cellular slime molds. *Ann. Rev. Microbiol.*, **25**, 75–92.

Bonner, J.T. (1973), Nonvertebrate Neuroendocrinology and Ageing. In: *Humoral Control of Growth and Differentiation* Vol. II, Academic Press Inc. New York, pp. 81–98.

Bonner, J.T., Hall, E.M., Noller, S., Oleson, F.B. and Roberts, A.B. (1972), Synthesis of cyclic AMP and phosphodiesterase in various species of cellular slime molds and its bearing on chemotaxis and differentiation. *Dev. Biol.*, **29**, 402–409.

Brunette, D.M. and Till, J.E. (1971), A rapid method for the isolation of L-cell surface membranes using an aqueous two-phase polymer system *J. Mem. Biol.*, **5**, 215–225.

Chang, Y.Y. (1968), Cyclic 3′, 5′-adenosine monophosphate phosphodiesterase produced by the slime mold *Dictyostelium discoideum*. *Science*, **160**, 57–59.

Chang, C.M., Reitherman, R.W., Rosen, S.D. and Barondes, S.H. (1975), Cell surface location of discoidin, a developmentally regulated carbohydrate-binding protein from *Dictyostelium discoideum*. *Exp. Cell Res.*, **95**, 136–142.

Chassy, B.M. (1972), Cyclic nucleotide phosphodiesterase in *Dictyostelium*: Interconversion of two enzyme forms. *Science*, **175**, 1016–1018.

Chia, W.K. (1975), Induction of stalk cell differentiation by cyclic-AMP in a susceptible variant of *Dictyostelium discoideum*. *Dev. Biol.*, **44**, 239–252.

Coe, E.L. and Kuo, W.J. (1976), Possible involvement of prostaglandins in the cAMP signalling response of the cellular slime mold *Dictyostelium discoideum*. *Fed. Proc.*, **35**, 1367.

Cohen, M.H., and Robertson, A. (1971), Wave propagation in the early stages of aggregation of cellular slime molds. *J. Theor. Biol.* **31**, 101–118.

Coukell, B. (1975), Parasexual genetic analysis of aggregation-deficient mutants of *Dictyostelium discoideum*. *Mol. gen. Genet.*, **142**, 119–135.

Darmon, M. and Klein, C. (1976), Binding of Concanavalin A and its effects on the differentiation of *Dictyostelium discoideum*. *Biochem. J.*, **154**, 743–750.

Darmon, M., Brachet, P., Pereira da Silva, L.H. (1975), Chemotactic signals induce cell differentiation in *Dictyostelium discoideum*. *Proc. natn. Acad. Sci. U.S.A.*, **72**, 3163–3166.

Den, H., Malinzak, D.A. and Rosenberg, A. (1976), Lack of evidence for the involvement of a β-D-galactosyl-specific lectin in the fusion of chick myoblasts. *Biochem. biophys. Res. Comm.*, **69**, 621–627.

DeMeyts, P., Gavin, J.R., Roth, J. and Neville, D.M. (1974), Concanavalin A: An inhibitor of co-operative interaction among receptors. *Diabetes*, **23**, (Suppl. 1) 355.

DeMeyts, P., Bianco, A.R. and Roth, J. (1976), Site-site interactions among insulin receptors. *J. biol. Chem.*, **251**, 1877–1888.

Durston, A.J. (1973), *Dictyostelium discoideum* aggregation fields as excitable media, *J. Theor. Biol.*, **42**, 483–504.

Durston, A.J. (1974a), Pacemaker activity during aggregation in *Dictyostelium discoideum*. *Dev. Biol.*, **37**, 225–235.

Durston, A.J. (1974b), Pacemaker mutants of *Dictyostelium discoideum*. *Dev. Biol.*, **38**, 308–319.

Edelman, G.M. (1974), Inhibition of ligand-induced cluster formation by Concanavalin A. In: *Control of Proliferation in Animal Cells* (Clarkson, B. and Baserga, R., eds.), Cold Spring Harbor Laboratory pp. 357–377.

Farnham, C.J.M. (1975), Cytochemical localization of adenylate cyclase and 3', 5'-nucleotide phosphodiesterase in *Dictyostelium discoideum*, *Exp. Cell Res.*, **91**, 36–46.

Francis, D. (1975), Cyclic AMP-induced changes in protein synthesis in a cellular slime mould *Polysphondylium pallidum*. *Nature*, **258**, 763–765.

Frazier, W.A. (1976), The role of cell surface components in the morphogenesis of the cellular slime molds. *Trends Biochem. Sci.*, **1**, 130–133.

Frazier, W.A., Rosen, S.D., Reitherman, R.W. and Barondes, S.H. (1975), Purification and comparison of two developmentally regulated lectins from *Dictyostelium discoideum*. *J. biol. Chem.*, **250**, 7714–7721.

Frazier, W.A., Rosen, S.D., Reitherman, R.W. and Barondes, S.H. (1976), In: *Cell surface Receptors* (Bradshaw, R.A. ed.), Plenum Press, New York (In press).

Gartner, T.K. and Podleski, T.R. (1975), Evidence that a membrane-bound lectin mediates fusion of L6 myoblasts. *Biochem. biophys. Res. Comm.*, **67**, 972–978.

Geltosky, J.E., Siu, C.H. and Lerner, R.A. (1976), Glycoproteins of the plasma membrane of *Dictyostelium discoideum* during development. *Cell,* **8**, 391–396.

Gerisch, G. (1965), Stadienspezifische Aggregationsmuster von *Dictyostelium discoideum*. *Roux' Arch. Entwicklungsmech.,* **156**, 127–144.

Gerisch, G. (1968), Cell aggregation and differentiation in *Dictyostelium*. In: *Current Topics in Developmental Biology* Vol. 3, (Moscona, A. and Montroy, A., eds.), Academic Press, New York, pp. 157–197.

Gersich, G. (1976), Extracellular cAMP phosphodiesterase regulation in agar plate cultures of *Dictyostelium discoideum*. *Cell Differentiation,* **5**, 21–25.

Gerisch, G., and Hess, B. (1974), Cyclic-AMP controlled oscillations in suspended *Dictyostelium* cells: their relation to morphogenetic cell interactions. *Proc. natn. Acad. Sci. U.S.A.,* **71**, 2118–2122.

Gerisch, G. and Wick, U. (1975), Intracellular oscillations and release of cyclic AMP from *Dictyostelium* cells. *Biochem. biophys. Res. Comm.,* **65**, 364–370.

Gerisch, G., Malchow, D., Riedel, V., Müller, E. and Every, M. (1972), Cyclic AMP phosphodiesterase and its inhibitor in slime mold development. *Nature New Biology,* **235**, 90–92.

Gerisch, G., Beug, H., Malchow, D., Schwarz, H. and Stein, A.v., (1974a), In: *Biology and Chemistry of Eucaryotic Cell Surfaces* (Lee, E.Y.C. and Smith, E.E., eds.), Vol. 7, Miami Winter Symposia, pp. 49–66.

Gerisch, G., Malchow, D. and Hess, B. (1974b), Cell communication and cyclic-AMP regulation during aggregation of the slime mold *Dictyostelium discoideum*. In: *Mosbacher Colloquium der Gesellschaft für Biologische Chemie,* Vol. 25, pp. 279–298.

Gerisch, G., Fromm, H., Huesgen, A. and Wick, U. (1975a), Control of cell-contact sites by cAMP pulses in differentiating *Dictyostelium* cells. *Nature,* **255**, 547–549.

Gerisch, G., Hülser, D., Malchow, D. and Wick, U. (1975b), Cell communication by periodic cyclic AMP pulses. *Phil. Trans. R. Soc. Lond. B.,* **272**, 181–192.

Gerisch, G., Malchow, D., Huesgen, A., Nanjundiah, V., Roos, W. and Wick, U. (1975c), Cyclic AMP reception and cell recognition in *Dictyostelium discoideum*. In: *ICN-UCLA Symposium of Developmental Biology* (MacMahon, D. and Fox, F. eds.), Benjamin Inc. pp. 76–88.

Gerisch, G., Malchow, D., Roos, W., Wick, U. and Wurster, B. (1976), Periodic cAMP signals and membrane differentiation in *Dictyostelium*. In: *Sigrid Juselius Symposium on Cell Interactions in Cell Differentiation,* Helsinki, Academic Press, (In press).

Gilkes, N.R. and Weeks, G. (1976), The purification and characterization of *Dictyostelium discoideum* plasma membranes. *Biochim. biophys. Acta,* **464**, 142–156.

Gillette, M.U., Denger, R.E. and Filosa, M.F. (1974), Localization and fate of concanavalin A in amebas of the cellular slime mold *Dictyostelium discoideum*. *J. exp. Zool.,* **190**, 234–248.

Gillette, M.U. and Filosa, M.F. (1973), Effect of concanavalin A on cellular slime mold development: premature appearance of membrane-bound cyclic AMP phosphodiesterase. *Biochem. biophys. Res. Comm.,* **53**, 1159–1166.

Gingle, A.R. (1976), Critical density for relaying in *Dictyostelium discoideum* and its relation to phosphodiesterase secretion into the extracellular medium, *J. Cell Sci.*, **20**, 1–20.

Gingle, A.R. and Robertson, A. (1976), The development of the relaying competance in *Dictyostelium discoideum*. *J. Cell Sci.*, **20**, 21–27.

Goldbeter, A. (1975), Mechanism for oscillatory synthesis of cyclic AMP on *Dictyostelium discoideum*. *Nature*, **253**, 540–542.

Green, A.A. (1976), The role of the cell membrane in differentiation of the cellular slime mould *Dictyostelium*. Doctoral Thesis, University of Oxford.

Green, A.A. and Newell, P.C. (1974), Isolation and subfractionation of plasma membranes from the cellular slime mould *Dictyostelium discoideum*. *Biochem. J.*, **140**, 313–322.

Green, A.A. and Newell, P.C. (1975), Evidence for the existence of two types of cAMP binding sites in aggregating cells of *Dictyostelium discoideum*. *Cell*, **6**, 129–136.

Gregg, J.H. (1956), Serological investigations of cell adhesion in *Dictyostelium discoideum*. *J. gen. Physiol.*, **39**, 813–820.

Gregg, J.H. (1971), Developmental potential of isolated *Dictyostelium myxamoebae*. *Dev. Biol.*, **26**, 478–485.

Gregg, J.H. and Nesom, M.G. (1973), Response of *Dictyostelium* plasma membranes to adenosine $3'5'$-cyclic monophosphate. *Proc. natn. Acad. Sci. U.S.A.*, **70**, 1630–1633.

Gregg, J.H. and Trygstad, C.W. (1958), Surface antigen defects in aggregateless variants of *Dictyostelium discoideum*. *Exp. Cell Res.*, **15**, 358–366.

Gregg, J.H. and Yu, N.Y. (1975), *Dictyostelium* aggregateless mutant plasma membranes. *Exp. Cell Res.*, **96**, 283–286.

Gross, J.D., Peacey, M.J. and Trevan, D.J. (1976), Signal emission and signal propagation during early aggregation in *Dictyostelium discoideum J. Cell Sci.*, **22**, 645–656.

Hamilton, I.D. and Chia, W.K. (1975), Enzyme activity changes during cAMP-induced stalk cell differentiation in P4 a variant of *Dictyostelium discoideum*. *J. gen. Microbiol.*, **91**, 295–306.

Henderson E.J. (1975), The cyclic adenosine $3',5'$-monophosphate receptor of *Dictyostelium discoideum*. *J. biol. Chem.*, **250**, 4730–4736.

Huesgen, A. and Gerisch, G. (1975), Solubilized contact sites A from cell membranes of *Dictyostelium discoideum*. *FEBS Letters*, **56**, 46–49.

Klein, C. and Darmon, M. (1975), The relationship of phosphodiesterase to the developmental cycle of *Dictyostelium discoideum*. *Biochem. biophys. Res. Comm.*, **67**, 440–447.

Klein, C. and Darmon, M. (1976), A differentiation-stimulating factor induces cell sensitivity to $3'5'$ cyclic AMP pulses in *Dictyostelium discoideum*. *Proc. natn. Acad. Sci. U.S.A.*, **73**, 1250–1254.

Klein, R.L. and Breland, A.P. (1966), Active cation transport and ATP hydrolysis in *Acanthamoebae sp*. *Comp. Biochem. Physiol.*, **17**, 39–47.

Kobilinsky, L., Weinstein, B.I. and Beattie, D.S. (1976), The induction of filopodia in the cellular slime mold *Dictyostelium discoideum* by cyclic adenosine monophosphate. *Dev. Biol.*, **48**, 477–481.

Konijn, T.M. (1972a), Cyclic AMP as a first messenger. In: *Advances in Cyclic Nucleotide Research* (Greengard, P., Robison, G.A. and Paoletti, R., eds.), Vol 1, Raven Press, New York, pp. 17–31.

Konijn, T.M. (1972b), Cyclic AMP and cell aggregation in the cellular slime molds. *Acta Protozoologica*, **11**, 137–143.

Konijn, T.M. (1973), The chemotactic effect of cyclic nucleotides with substitutions in the base ring. *FEBS Letters*, **34**, 263–266.

Konijn, T.M. and Jastorff, B. (1973), The chemotactic effect of 5'amido analogues of adenosine cyclic 3' 5' monophosphate in the cellular slime molds. *Biochim. biophys. Acta*, **304**, 774–780.

Lee, A., Chance, K., Weeks, C. and Weeks, G. (1975), Studies on the alkaline phosphatase and 5'nucleotidase of *Dictyostelium discoideum*. *Arch. Biochem. Biophys.*, **171**, 407–417.

Limbird, L.E., DeMeyts, P. and Lefkowitz, R.J. (1975), β-adrenergic receptors: Evidence for negative co-operativity. *Biochem. biophys. Res. Comm.*, **64**, 1160–1168.

Lis, H. and Sharon, N. (1973), The biochemistry of plant lectins. *Ann. Rev. Biochem.*, **42**, 541–574.

Loomis, W.F. (1975), *Dictyostelium discoideum: a developmental system*. Academic Press Inc. New York.

Malchow, D. and Gerisch, G. (1973), Cyclic AMP binding to living cells of *Dictyostelium discoideum* in presence of excess cyclic GMP. *Biochem. biophys. Res. Commun.*, **55**, 200–204.

Malchow, D. and Gerisch, G. (1974), Short-term binding and hydrolysis of cAMP by aggregating *Dictyostelium* cells. *Proc. natn. Acad. Sci. U.S.A.*, **71**, 2423–2427.

Malchow, D., Nagele, B., Schwarz, H. and Gerisch, G. (1972), Membrane-bound cyclic AMP phosphodiesterase in chemotactically responding cells of *Dictyostelium discoideum*. *Eur. J. Biochem.*, **28**, 136–142.

Malchow, D., Fuchila, J. and Jastorff, B. (1973), Correlation of substrate specificity of cAMP phosphodiesterase in *Dictyostelium discoideum* with chemotactic activity of cAMP analogues. *FEBS Letters*, **34**, 5–9.

Malchow, D., Fuchila, J. and Nanjundiah, V. (1975), A plausible role for a membrane-bound cyclic AMP phosphodiesterase in cellular slime mold chemotaxis. *Biochim. biophys. Acta*, **385**, 421–428.

Malkinson, A.M. and Ashworth, J.M. (1973), Adenosine 3' 5' cyclic monophosphate concentration and phosphodiesterase activities during axenic growth and differentiation of cells of the cellular slime mould *Dictyostelium discoideum*., *Biochem. J.*, **134**, 311–319.

Mason, J.W., Rasmussen, H. and Dibella, F. (1971), 3' 5' cyclic AMP and Ca^{2+} in slime mold aggregation. *Exp. Cell Res.*, **67**, 156–160.

Mato, J.M. and Konijn, T.M., (1975a), Chemotaxis and binding of cyclic AMP in cellular slime molds. *Biochim. biophys. Acta.*, **385**, 173–179.

Mato, J.M. and Konijn, T.M. (1975b), Enhanced cell aggregation in *Dictyostelium discoideum* by ATP activation of cyclic AMP receptors. *Dev. Biol.*, **47**, 233–235.

Mato, J.M., Losada, A., Nanjundiah, V. and Konijn, T.M. (1975), Signal input for a chemotactic response in the cellular slime mold *Dictyostelium discoideum*. *Proc. natn. Acad. Sci. U.S.A.*, **72**, 4991–4993.

Mockrin, S.C. and Spudich, J.A. (1976), Calcium control of actin-activated myosin adenosine triphosphatase from *Dictyostelium discoideum*. *Proc. natn. Acad. Sci. U.S.A.*, **73**, 2321–2325.

Moens, P.B. and Konijn, T.M. (1974), Cyclic AMP as a cell surface activating agent in *Dictyostelium discoideum*. *FEBS Letters*, **45**, 44–46.

Morris, J.E. and Moscona, A.A. (1970), Induction of glutamine synthetase in embryonic retina: its dependence on cell interactions. *Science*, **167**, 1736–1738.

Morris, J.E. and Moscona, A.A. (1971), The induction of glutamine synthetase in aggregates of embryonic neural retina cells: correlation with differentiation and multicellular organization. *Dev. Biol.*, **25**, 420–444.

Mosses, D., Williams, K.L. and Newell, P.C. (1975), The use of mitotic crossing-over for genetic analysis in *Dictyostelium discoideum*. Mapping of linkage group II. *J. gen. Microbiol.* **90**, 247–259.

Nanjundiah, V. and Malchow, D. (1976), A theoretical study of the effects of cyclic AMP phosphodiesterase during aggregation in *Dictyostelium*. *J. Cell Sci.*, **22**, 49–58.

Nanjundiah, V., Hara, K. and Konijn, T.M. (1976), Effect of temperature on morphogenetic oscillations in *Dictyostelium discoideum*. *Nature*, **260**, 705.

Neet, K.E. and Ainslie, G.R. (1976), Co-operativity and slow transitions in the regulation of oligomeric and monomeric enzymes. *Trends Biochem. Sci.*, **1**, 145–147.

Newell, P.C. (1971), The development of the cellular slime mold *Dictyostelium discoideum*: A model system for the study of cellular differentiation. In: *Essays in Biochemistry,* Vol. 7, Academic Press, London, pp. 87–126.

Newell, P.C. (1975), Cellular communication during aggregation of the slime mold *Dictyostelium*. In: *Microbiology* (Dworkin, M. and Shapiro, L., eds.), American Society for Microbiology pp. 426–433.

Newell, P.C., Longlands, M. and Sussman, M. (1971), Control of enzyme synthesis by cellular interaction during development of the cellular slime mold *Dictyostelium discoideum*. *J. Mol. Biol.*, **58**, 541–554.

Nowak, T.P., Haywood, P.L. and Barondes, S.H. (1976), Developmentally regulated lectin in embryonic chick muscle and a myogenic cell line. *Biochem. biophys. Res. Comm.*, **68**, 650–657.

Pan, P., Hall, E.M. and Bonner, J.T. (1972), Folic acid as a second chemotactic substance in the cellular slime moulds. *Nature New Biology*, **237**, 181–182.

Pan, P., Hall, E.M. and Bonner, J.T. (1975), Determination of the active portion of the folic acid molecule in cellular slime mold chemotaxis. *J. Bact.*, **122**, 185–191.

Pannbacker, R.G. and Bravard, L.J. (1972), Phosphodiesterase in *Dictyostelium discoideum* and the chemotactic response to cyclic adenosine monophosphate. *Science,* **175**, 1014–1015.

Parish, R.W. and Müller, U. (1976), Isolation of plasma membranes from the cellular slime mold *Dictyostelium discoideum* using Con A and Triton X-100. *FEBS Letters,* **63**, 40–44.

Reitherman, R.W., Rosen, S.D., Frazier, W.A. and Barondes, S.H. (1975), Cell surface species specific high affinity receptors for discoidin: Developmental regulation in *Dictyostelium discoideum*. *Proc. natn. Acad. Sci., U.S.A.*, **72**, 3541–3545.

Riedel, V. and Gerisch, G. (1971), Regulation of extracellular cyclic AMP phosphodiesterase activity during development of *Dictyostelium discoideum*. *Biochem. biophys. Res. Comm.*, **42**, 119–124.

Riedel, V., Malchow, D., Gerisch, G. and Nagele, B. (1972), Cyclic AMP phophodiesterase interaction with its inhibitor of the slime mold *Dictyostelium discoideum*. *Biochem. biophys. Res. Comm.*, **46**, 279–287.

Robertson, A. and Grutsch, J. (1974), The role of cAMP in slime mold development. *Life Sci.*, **15**, 1031–1043.

Robertson, A., Drage, D.J. and Cohen, M.H. (1972), Control of aggregation in *Dictyostleium discoideum* by an external periodic pulse of cyclic adenosine monophosphate. *Science*, **175**, 333–335.

Roos, W. and Gerisch, G. (1976), Receptor-mediated adenylate cyclase activation in *Dictyostelium discoideum*. *FEBS Letters*, **68**, 170–172.

Roos, W., Nanjundiah, V., Malchow, D. and Gerisch, G. (1975), Amplification of cyclic AMP signals in aggregation cells of *Dictyostelium discoideum*. *FEBS Letters*, **53**, 139–142.

Rosen, S.D., Kafka, J.A., Simpson, D.L. and Barondes, S.H. (1973), Developmentally regulated carbohydrate-binding protein in *Dictyostelium discoideum*. *Proc. natn. Acad. Sci. U.S.A.*, **70**, 2554–2557.

Rosen, S.D., Simpson, D.L., Rose, J.E. and Barondes, S.H. (1974), Carbohydrate-binding protein from *Polysphondylium pallidum* implicated in intercellular adhesion. *Nature*, **252**, 128–151.

Rosen, S.D., Reitherman, R.W. and Barondes, S.H. (1975), Distinct lectin activities from six species of cellular slime molds. *Exp. Cell Res.*, **95**, 159–166.

Rosen, S.D., Haywood, P.L. and Barondes, S.H. (1976), Inhibition of intercellular adhesion in a cellular slime mould by univalent antibody against a cell-surface lectin. *Nature*, **263**, 425–427.

Rossomando, E.F. and Cutler, L.S. (1975), Localization of adenylate cyclase in *Dictyostelium discoideum*. *Exp. Cell Res.*, **95**, 67–78.

Rossomando, E.F. and Sussman, M. (1972), Adenyl cyclase in *Dictyostelium discoidin*; A possible control element of the chemotactic system. *Biochem. biophys. Res. Comm.*, **47**, 604–610.

Rossomando, E.F., Steffek, A.J., Mujwid, D.K. and Alexander, S. (1974), Scanning electron microscope observations on cell surface changes during aggregation of *Dictyostelium discoideum*. *Exp. Cell Res.*, **85**, 73–78.

Shaffer, B.M. (1957), Aspects of aggregation in cellular slime molds. *Am. Nat.*, **91**, 19–35.

Shaffer, B.M. (1958), Integration in aggregating cellular slime moulds. *Q. J. Microscop. Sci.*, **99**, 103–121.

Shaffer, B.M. (1962), The Acrasina. In: *Advances in Morphogenesis* (Abercrombic, M. and Brachet, J., eds.), Vol. 2, pp. 109–182.

Shaffer, B.M. (1975), Secretion of cyclic AMP induced by cyclic AMP in the cellular slime mould *Dictyostelium discoideum*. *Nature*, **255**, 549–552.

Sharon, N. and Lis, H. (1972), Lectin: Cell-agglutinating and sugar-specific proteins. *Science*, **177**, 949–959.

Simpson, D.L., Rosen, S.D. and Barondes, S.H. (1974), Discoidin, a developmentally regulated carbohydrate-binding protein from *Dictyostelium discoideum*. Purification and characterization. *Biochemistry*, **13**, 3487–3493.

Simpson, D.L., Rosen, S.D. and Barondes, S.H. (1975), Pallidin. Purification and characterization of a carbohydrate binding protein from *Polysphondylium pallidum* implicated in intercellular adhesion. *Biochim. biophys. Acta*, **412**, 109–119.

Singer, S.J. and Nicolson, G.L. (1972), The fluid mosaic model of the structure of cell membranes. *Science*, **175**, 720–731.

Siu, C.H., Lerner, R.A., Ma, G., Firtel, R.A. and Loomis, W.F. (1976), Developmentally regulated proteins of the plasma membrane of *Dictyostelium discoideum*. The carbohydrate-binding protein. *J. mol. Biol.*, **100**, 157–178.

Smart, J.E. and Hynes, R.O. (1974), Developmentally regulated cell surface alterations in *Dictyostelium discoideum*. *Nature*, **251**, 319–321.

Sonneborn, D.R., Sussman, M. and Levine, L. (1964), Serological analysis of cellular slime mold development. *J. Bact.*, **87**, 1321–1329.

Spudich, J.A. (1974), Biochemical and structural studies of actomyosin-like proteins from non-muscle cells. *J. biol. Chem.*, **249**, 6013–6020.

Sussman, M. and Boschwitz, Ch. (1975), An increase of calcium/manganese binding sites in cell ghosts associated with the aquisition of aggregative competence in *Dictyostelium discoideum*. *Exp. Cell Res.*, **95**, 63–66.

Takeuchi, I. (1963), Immunological and immunohistochemical studies on the slime mold *D. mucoroides*. *Dev. Biol.*, **8**, 1–26.

Teichberg, V.I., Silman, I., Beutsch, D.D. and Resheff, G. (1974), A β-D-galactoside binding protein from electric organ tissue of Electrophorus electricus. *Proc. natn. Acad. Sci. U.S.A.*, **72**, 1383–1387.

Town, C., Gross, J. and Kay, R. (1976), *Nature*, **262**, 717–719.

Tuchman, J., Alton, T. and Lodish, H. (1974), Preferential synthesis of actin during early development of the slime mold *Dictyostelium discoideum*. *Dev. Biol.*, **40**, 116–129.

Warren, J.A., Warren, W.D. and Cox, E.C. (1975), Genetic complexity of aggregation in the cellular slime mold *Polysphondylium violaceum*. *Proc. natn. Acad. Sci. U.S.A.*, **72**, 1041–1042.

Warren, J.A., Warren, W.D. and Cox, E.C. (1976), Genetic and morphological study of aggregation in the cellular slime mold *Polysphondylium violaceum*. *Genetics*, **83**, 25–47.

Weeks, C. and Weeks, G. (1975), Cell surface changes during the differentiation of *Dictyostelium discoideum*. *Exp. Cell Res.*, **92**, 372–382.

Weeks, G. (1973), Agglutination of growing and differentiating cells of *Dictyostelium discoideum* by concanavalin A. *Exp. Cell Res.*, **76**, 467–470.

Weeks, G. (1975), Studies of the cell surface of *Dictyostelium discoideum* during differentiation. The binding of ^{125}I-concanavalin A to the cell surface. *J. biol. Chem.*, **250**, 6706–6710.

Wilhelms, O-H., Luderitz, O., Westphal, O. and Gerisch, G. (1974), Glycosphingolipids and glycoproteins in the wild type and in a non-aggregating mutant of *Dictyostelium discoideum*. *Eur. J. Biochem.*, **48**, 89–101.

Williams, K.L. and Newell, P.C. (1976), A genetic study of aggregation in the cellular slime mould *Dictyostelium discoideum* using complementation analysis. *Genetics*, **82**, 287–307.

Wurster, B. (1976a), Temperature dependence of biochemical oscillations in cell suspensions of *Dictyostelium discoideum*. *Nature,* **260**, 703–704.

Wurster, B. (1976b), Stimulation of cell development in *Dictyostelium discoideum* by folic acid pulses. Workshop on *Dictyostelium*, p. 52. Cold Spring Harbor Laboratory.

Wurster, B., Pan, P., Tyan, G.G. and Bonner, J.T. (1976), Preliminary characterization of the acrasin of the cellular slime mold *Polysphondylium violaceum*. *Proc. natn. Acad. Sci. U.S.A.,* **73**, 795–799.

Yu, N.Y. and Gregg, J.H. (1975), Cell contact-mediated differentiation in *Dictyostelium*. *Dev. Biol.,* **47**, 310–318.

2 Bacterial Chemotaxis

GERALD L. HAZELBAUER and
JOHN S. PARKINSON

2.1	Introduction	page	61
2.2	The chemotactic response		61
	2.2.1 Measuring techniques		62
	2.2.2 Behavioral basis of chemotaxis		63
2.3	The swimming mechanism		63
	2.3.1 Flagellar structure		64
	(a) *The filament*, 65, (b) *The hook*, 65, (c) *The basal body*, 65		
	2.3.2 Flagellar rotation		65
	(a) *Properties of the rotary motor*, 66, (b) *Energy source*, 67		
	2.3.3 Swimming and tumbling		67
	2.3.4 The tumble generator		69
2.4	Chemosensors		70
	2.4.1 Physiological analysis		72
	(a) *Temporal sensitivity*, 72, (b) *Drastic temporal stimulation*, 73, (c) *Gradient sensing*, 74		
	2.4.2 Genetic and biochemical analysis		75
	(a) *Reception and transport*, 77, (b) *Receptor properties*, 80, (c) *Biochemistry of sugar receptors*, 80, (d) *Amino acid receptors*, 84, (e) *Repellent receptors*, 85, (f) *Ion receptors*, 85		
2.5	Signalling		86
	2.5.1 Stimulus transduction		86
	2.5.2 Interaction of signals with the tumble generator		88
	2.5.3 The role of methionine		89
2.6	Conclusions		90
	References		91

Acknowledgements
The unpublished studies cited here were supported by Public Health Service Grants GM 19559 from The National Institute of General Medical Studies (to J.S.P.) and AI 12858 from The National Institute of Allergy and Infectious Diseases and a grant from the Swedish Natural Sciences Research Council (to G.L.H.).

Microbial Interactions
(*Receptors and Recognition,* Series B, Volume 3)
Edited by J.L. Reissig
Published in 1977 by Chapman and Hall, 11 New Fetter Lane, London EC4P 4EE
© Chapman and Hall

2.1 INTRODUCTION

Motile bacteria exhibit stimulus-response behavior not unlike that found in more sophisticated creatures. Chemotaxis, the movement of an organism toward or away from chemicals, is an especially well-studied example. The study of bacterial chemotaxis began in the latter part of the nineteenth century with the work of Engelmann (1883a,b), Pfeffer (1883, 1904) and Rothert (1901). Pfeffer (1904) concluded that attraction of bacteria to specific compounds was the result of the particular chemical qualities of the attractants. The view that those qualities are detected through their effect on general cell functions, such as energy production, is no longer tenable. The experimental proof by Julius Adler (1969) that bacterial chemotaxis is mediated by specific receptors marks the beginning of the modern study of bacterial chemotaxis. It is now clear that bacterial chemoreceptors are functionally analogous to chemosensor systems of higher organisms in that *changes* in chemical concentration elicit *transient* behavioral or biochemical responses. The suitability of micro-organisms for genetic and biochemical studies affords the opportunity to examine such behavioral machinery at the molecular level.

Chemotaxis serves bacteria primarily as a food-finding mechanism and as a means of avoiding toxic or otherwise unfavorable environments. Although chemotaxis occurs in many kinds of bacteria, it has been extensively studied in only a few, notably *Escherichia coli* and *Salmonella typhimurium*. These two closely related, Gram-negative organisms will be the main subjects of this review. More recently, the Gram-positive strain *Bacillus subtilis* has come under study as well, and that work will be cited wherever useful comparisons or contrasts can be made to the *E. coli–S. typhimurium* system. Both the nineteenth century work (Berg, 1975a) and the modern studies of chemotaxis (Alder, 1975; Berg, 1975b; Koshland, 1974, 1976; Parkinson, 1975) have been reviewed recently. Here we will consider the progress made toward a molecular description of bacterial chemotaxis with particular emphasis on chemoreception.

2.2 THE CHEMOTACTIC RESPONSE

E. coli and *S. typhimurium* are attracted to many amino acids and sugars (Mesibov and Adler, 1972; Adler, Hazelbauer and Dahl, 1973) and are repelled by a variety of substances, including some amino acids (Tsang, Macnab and Koshland, 1973; Tso and Adler, 1974). Other attractants are oxygen (Adler, 1966; Adler and Dahl, 1967) and some divalent and monovalent cations (Zukin and Koshland, 1976). *B. subtilis* is attracted to an even greater number of amino acids (van der Drift and deJong, 1974; Ordal and Gibson, 1976); however, sugars and other potential attractants have not yet been systematically cataloged in this species. Many compounds such as local anesthetics and other membrane-active agents are strong repellents of *B. subtilis* (Ordal and Goldman, 1975, 1976).

2.2.1 Measuring techniques

The most widely used method for quantitating chemotactic responses is the capillary assay (Adler, 1973), in which a microcapillary containing a test solution is placed in a bacterial suspension. Motile cells respond to the gradient diffusing from the capillary mouth by accumulating around and in the capillary. Response is quantitated by determining the number of cells trapped in the capillary at various concentrations of the test chemical. The effect of one substance on the response to a gradient of another can be determined by placing the first compound at an equal concentration in both the capillary and the suspension (Adler, 1969; Mesibov and Adler, 1972; Mesibov, Ordal and Adler, 1973). This procedure has been called a 'jamming assay' to avoid using words like inhibition or suppression, which carry connotations of specific biological mechanism (Ordal, Villani and Gibson, 1977).

The capillary assay is extremely useful when comparisons are desired e.g., between mutant and parent, two attractants, or two jammers. However, although the diffusion equations for the gradient have been solved (Brokaw 1958; Futrelle and Berg, 1972), it is difficult to extract from capillary assay data a value representing the actual affinity constant of a particular receptor. A modified assay, called the 'sensitivity' assay, allows determination of such values. When attractant is placed in the cell suspension at a concentration that is a fixed fraction of the concentration in the capillary, the resulting gradient has the same relative steepness at different absolute concentrations. Cell accumulation in response to such gradients can be fitted to a simple mass action model, and a value comparable to a K_m determined (Mesibov *et al.*, 1973). The sensitivity assay has been combined with the jamming procedure (with the jammer identical to or different from the attractant) to identify individual, high affinity sites in the presence of many interacting receptors (Ordal *et al.*, 1977).

Difficulties inherent in the capillary assay include dependence on comparable and reproducible motility and the possibility of bacterial metabolism disturbing the gradient. Use of low cell densities reduces the latter problem (Adler, 1973), but since cell densities are necessarily high in regions where bacteria accumulate, the possibility of metabolism-caused changes in gradients cannot be altogether eliminated.

Although useful in studying receptor physiology (see Section 2.4), the capillary assay reveals nothing about the actual behavior of the responding individuals. Other measurements employing spatial gradients include 'tracking' of individual cells with an automatic (Berg, 1971) or manual (Lovely *et al.*, 1974) tracking microscope, and observations of movements of bands of bacteria in a defined (Dahlquist, Lovely and Koshland, 1972) or metabolically generated (Adler, 1966) gradient. These two latter procedures can be used to identify or select taxis mutants (Hazelbauer, Mesibov and

Adler, 1969; Aswad and Koshland, 1975a; Armstrong, Adler and Dahl, 1967; Ordal and Adler, 1974a; Parkinson, 1976) since cells incapable of responding to the gradient will not move with the band.

2.2.2 Behavioral basis of chemotaxis

When presented with a spatial gradient of attractant, a population of *E. coli* or *S. typhimurium* will migrate toward the attractant source at a measurable rate. The average drift velocity up the gradient is, however, much less than the swimming speeds of the individual organisms. For *E. coli* swimming with a velocity of 25 $\mu m\ s^{-1}$, Adler and Dahl (1967) observed a migration rate toward oxygen of about 2 $\mu m\ s^{-1}$. Similar findings were reported for *S. typhimurium* in serine gradients (Dahlquist, Lovely and Koshland, 1972). Observations of individual bacteria in chemical gradients (Berg and Brown, 1972; Macnab and Koshland, 1972; Tsang *et al.*, 1973; Brown and Berg, 1974) clearly showed that the behavioral basis of chemotaxis is modulation of turning frequency, a strategy properly termed klinokinesis (Fraenkel and Gunn, 1940; Gunn, 1975).

In the absence of stimuli, the individual bacteria swim in a three-dimensional random walk (Berg and Brown, 1972, 1974). Directional changes, called 'twiddles' or 'tumbles', occur with constant probability (i.e., randomly) during smooth swimming 'runs'. Each tumbling episode results in a nearly random reorientation of swimming direction so that there is little correlation between the directions of successive runs. In a spatial gradient of attractant, however, tumble probability is modulated as a function of swimming direction (Berg and Brown, 1972). Bacteria headed up-gradient exhibit greater mean run lengths (i.e., decreased tumble probability) than bacteria moving down-gradient. Swimming velocity during a run is essentially constant over a wide range of attractant concentrations: chemotactic migration is achieved by controlling run length rather than run velocity.

How elaborate is the chemotaxis machinery? Evidence from the physiological studies just described shows that perhaps three basic functions are involved: stimulus detection, stimulus transduction and locomotor response. In the sections to follow, we will consider each of these processes individually. Since most assays of stimulus detection and transduction involve the motor response, we will begin with a description of the locomotory apparatus.

2.3 THE SWIMMING MECHANISM

Flagella are the motor organelles of *E. coli* and many other kinds of motile bacteria. Unlike their eukaryotic counterparts, bacterial flagella have a rather simple structure which nevertheless produces an efficient, biologically unique form of locomotion. Recent summaries of biochemical and genetic studies of flagellar structure are available (Hilmen, Silverman and Simon, 1974; Hilmen and Simon, 1976) and these topics will not be extensively discussed in the present review. Instead, we will focus our attention on the way in which flagella cause *E. coli* to swim and tumble, and on the components needed to carry out this behavior during chemotaxis.

2.3.1 Flagellar structure

DePamphilis and Adler (1971a, 1971b, 1971c) purified 'intact' flagella from *E. coli* and investigated their structure by electron microscopy. Their findings and those of other investigators are summarized in Fig. 2.1 which depicts the basal portion of the flagellum. The intact flagellum is composed of three main portions: a long slender filament, a hook connected to the proximal end of the filament and a basal body embedded in the cell wall and membranes to which the hook is joined.

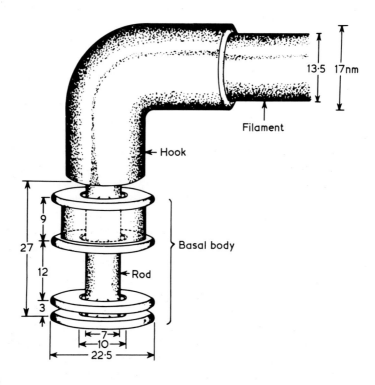

Fig. 2.1 A model of the basal body-hook complex of the *E. coli* flagellum (adapted from DePamphilis and Adler, 1971b). The structure is based on electron microscopic studies with approximate dimensions given in nanometers. The four rings of the basal body are attached to various layers of the cell wall and membranes as discussed in the text.

(a) *The filament*

The filament, which accounts for more than 95% of the flagellar mass (Hilmen *et al.*, 1974), is composed of a single type of protein, flagellin. The filament is typically several μm in length and has a characteristic helical shape which is evidently crucial to proper swimming because flagellin mutants with aberrant filament shapes, such as straight or curly, are non-motile. The filament is built by the addition of flagellin subunits to the free, distal end. The subunits are evidently transported through the hollow core of the growing filament to reach the tip; however, the details of this process are not understood.

(b) *The hook*

The filament is connected to the basal body through a short hook which is slightly larger in diameter than the filament. The hook is composed of a single type of protein (not flagellin) whose structural gene is not known. The assembly of hook subunits is controlled by at least one additional gene, *flaE,* which somehow regulates hook length. Mutants defective in *flaE* function form 'polyhooks' several μm in length (Silverman and Simon, 1972).

(c) *The basal body*

The basal body is the most interesting but least understood of the flagellar components. It serves to anchor the flagellum to the cell and as the motor that drives the hook and filament. It is also likely that the basal body contains some components required for chemotaxis and that these are the ultimate target of sensory information from the chemoreceptors (see Section 2.3.4 below). Genetic studies have shown that some 20 gene products might be involved in the assembly or final structure of the basal body (Silverman and Simon, 1973a, 1973b).

The gross structure of basal bodies revealed by the electron microscope is perhaps a deceptively simple one. At the cell surface, the hook is attached to a rod that spans the periplasmic space between the inner and outer membrane. In *E.coli*, this rod bears four rings or discs that appear to associate with different layers of the cell envelope. The outermost ring is attached to the other membrane; the adjacent ring is probably attached to the peptidoglycan or rigid layer of the cell wall (DePamphilis and Adler, 1971c). Since these two outer rings are not present in *B. subtilis*, a Gram-positive strain, it has been suggested that they may not be directly involved in flagellar movement (DePamphilis and Adler, 1971c; Berg, 1974). The two inner rings are seen in both Gram-positive and Gram-negative organisms and therefore probably play a major role in basal body function. The innermost ring is associated with the cytoplasmic membrane; however, the other ring has no demonstrated attachments.

2.3.2 Flagellar rotation

Bacterial flagella actually rotate very much like the propellers of a ship. Rotary motion was until recently seriously entertained by only a few workers in the field of bacterial movement. However, Berg and Anderson (1973), in examining the available evidence, concluded that flagellar rotation was mechanically and energetically the most plausible mechanism of flagellar motion. Silverman and Simon (1974) provided a direct demonstration of flagellar rotation in a straight flagella mutant of *E.coli* by attaching to the filaments small antibody-linked beads. The movements of the beads, unlike those of flagella, could be readily followed in the light microscope. They observed chains of beads extending from the cells and revolving in unison about a common axis, presumably the invisible flagellar filament. In similar fashion, they showed that polyhook mutants, attached to one another or to beads by means of antihook antibodies, also exhibited rotation. These findings imply that rotary motion is generated by the basal body and transmitted through the hook to the filament.

(a) *Properties of the rotary motor*

Most studies of flagellar rotation, including those discussed below, employ the cell-tethering method for visualizing and characterizing rotational behavior. In this technique, antibody molecules are used to anchor flagellar filaments to a microscope slide or cover slip. Providing that a cell is tethered by only one flagellum, it will rotate around the point of attachment and can be readily observed. Cell rotation can be observed directly (Silverman and Simon, 1974; Larsen *et al.*, 1974a; Parkinson, 1976) or followed with an automatic tracking microscope (Berg, 1974; Berg and Tedesco, 1975). For convenience in comparing the results of different laboratories, the following convention will be employed: the sense of *flagellar* rotation—CW or CCW—is defined as though the observer were looking along the filament axis toward the cell. This will correspond to the direction of *cell* rotation that is seen when the observer views a tethered cell from above.

In the absence of chemotactic stimuli, wild type cells frequently change their direction of rotation (Silverman and Simon, 1974; Larsen *et al.*, 1974a; Berg, 1974). Reversals from CW to CCW and from CCW to CW are random events (Berg and Tedesco, 1975). Because the motion is very smooth, it seems unlikely that the motor moves in a small number of discrete steps (Berg, 1974).

The rotational behavior of tethered bacteria can be temporarily altered by chemotactic stimuli. Addition of attractant and removal of repellent, which suppress tumbling in untethered cells, cause tethered bacteria to rotate only in the CCW direction (Larsen *et al.*, 1974a; Berg and Tedesco, 1975). If no further stimulus is applied, rotational behavior returns to the pre-stimulus state with decay times that are proportional to the magnitude of the initial stimulus (Berg and Tedesco, 1975). Conversely, stimuli that cause tumbling evoke transient periods of CW rotation. In both cases, the frequency of rotation remains constant. It appears, therefore, that swimming corresponds to CCW rotation and tumbling to CW rotation. Tethering studies with chemotaxis mutants that tumble incessantly or not at all (see Section 2.3.4 below) are consistent with this view (Larsen *et al.*, 1974a; Parkinson, 1976).

The cell-tethering technique, as it is generally used, has a potentially serious drawback. In order to ensure that cells are tethered by only one filament, the number of flagella per cell is experimentally reduced either by growth under conditions of catabolite repression (cf. Larsen et al., 1974a) or at elevated temperature (Parkinson, unpublished results). Each procedure has a disadvantage, the first because many chemoreceptors are catabolite-repressed, the second because temperature *per se* affects tumbling behavior (Maeda et al., 1976). One solution to this problem is to grow cells under normal conditions, employing flagellar mutants to prevent possible artefacts caused by free filaments (Silverman and Simon, 1974; Parkinson, 1976). Unfortunately, most of the physiological studies have not been done under these conditions and their results should be accepted with caution.

(b) *Energy source*
Larsen et al., (1974b) investigated the energy requirements for motility in *E.coli* and concluded that ATP is neither necessary nor sufficient. They found, for example, that depletion of internal ATP levels by arsenate treatment did not inhibit motility, provided that the bacteria were supplied an oxidizable substrate that permitted the electron transport chain to function. However, electron transport *per se* is not required so long as ATP is generated by glycolysis, because wild type cells are motile under anaerobic conditions. These observations suggest that an intermediate in oxidative phosphorylation is the energy source for motility. Such an intermediate may be generated either by electron transport or, under anaerobic conditions, from ATP via the membrane-bound Ca^{2+}, Mg^{2+}-dependent ATPase. As further support for this notion, Larsen et al. (1974b) showed that mutants lacking the membrane-bound ATPase activity were non-motile under anaerobic conditions.

The nature of the oxidative phosphorylation intermediate needed for motility is not known; however, there is good evidence (reviewed by Harold, 1972) that many membrane-associated processes are powered by an electrochemical gradient of protons as described in the chemiosmotic hypothesis of Mitchell (1966). According to this model, both electron transport and the membrane-bound ATPase are capable of establishing a pH gradient across the cytoplasmic membrane (interior alkaline) by extruding protons into the extracellular medium. The total proton-motive force available for work is a combination of both the chemical potential (pH difference) and the electrical potential difference across the membrane. Uncouplers of oxidative phosphorylation such as carbonylcyanide *m*-chlorophenylhydrazone (CCCP) are thought to act by conducting protons across the membrane, thereby dissipating the trans-membrane proton gradient. Larsen et al. (1974b) found that CCCP rapidly inhibited motility of *E. coli* even though ATP was still present in the cells. It would appear that the energy source for flagellar rotation is either the proton-motive force itself or some other property associated with the energized membrane state.

2.3.3 Swimming and tumbling

E.coli, S. typhimurium and *B. subtilis* typically have 5–10 flagella per cell arranged in a peritrichous rather than polar fashion. Because the rotational axes of the different filaments originate from separate points on the cell surface, swimming and tumbling are necessarily complex events and are not yet fully understood. Analyses of the geometric and hydrodynamic factors involved (Anderson, 1975; Macnab, 1977) and darkfield observations of motile cells (Macnab and Ornston 1977) have led to the following view of the swimming-tumbling sequence. Flagellar filaments are semi-rigid helices with a left-handed pitch. During CCW rotation, the filaments are able to form a compact bundle and act in a co-ordinated pattern to push the cell body through the medium. Tumbling is initiated by reversal to CW rotation and is accompanied by violent disruption of the bundle. The jerky 'unco-ordinated' movements of tumbling are apparently caused by a gross conformational change in the filaments induced by the reversed torque of CW rotation. The helical sense becomes right-handed, causing the bundle to disperse. The right-handed filaments exhibit a reduced amplitude and wavelength and apparently, under normal conditions, cannot form a bundle. The disruption of the flagellar bundle causes the cell body to move in an unco-ordinated manner, producing a random reorientation. When CCW rotation recommences, the helix pitch relaxes to the normal left-handed condition, the bundle reforms and the next run begins.

2.3.4 The tumble generator

The tumble generator concept was introduced by Berg and Brown (1972) to account for the spontaneous directional changes that punctuate swimming in *E.coli* and *S. typhimurium*. A similar swimming pattern is also found in many other kinds of bacteria, suggesting that machinery for generating spontaneous directional changes is a common feature of bacterial motility (Berg, 1975a). Since tumbling is caused by episodes of CW flagellar rotation (Larsen *et al.*, 1974a) that are random over time (Berg and Tedesco, 1975), the tumble generator must include both a switching mechanism that determines the direction of flagellar rotation and a device for triggering reversals of the switch.

In *E. coli*, the trigger operates in agreement with Poisson statistics (Berg and Brown, 1972, 1974). In one class of models, the trigger is thought to be a molecule or chemical modification the concentration of which determines the direction of flagellar rotation (Macnab and Koshland, 1972; Ordal, 1976b). Fluctuations in the effective level of trigger substance would then account for random behavior in the switch. In *B. subtilis*, for example, Mg^{2+} ions are required for smooth swimming (Ordal, 1976b) and Ordal and Fields (1977) have suggested that the concentration of Mg^{2+} ions in the vicinity of the basal body might control tumbling behavior. A second possibility is that the trigger is a change in the proton potential or overall potential of the cell membrane. Several lines of evidence are at least consistent with this model. In *B. subtilis*, agents that perturb the cells' membrane potential also cause alterations in tumbling behavior (Ordal and Goldman, 1976). Macnab and Kosland (1974) found that short pulses of high intensity blue light caused tumbling in *S. typhimurium*; longer exposures irreversibly destroyed tumbling ability. The action spectrum of these effects resembled that of a flavin, suggesting that perhaps tumbling is controlled by the level of energy available for motility. Recently, Szmelcman and Adler (1976) demonstrated, with the aid of a lipid-soluble cationic probe, that the membrane potential of *Escherichia coli* undergoes a transient hyperpolarization upon addition of attractants or repellents. This effect depends upon having both a functional chemosensor and a continuous supply of methionine, which is known to be required for chemotaxis (see Section 2.5.3 below). Since generally non-chemotactic mutants (see following Section) also showed membrane potential changes when stimulated with attractants or repellents, the role of this phenomenon in chemotaxis is not yet clear. On the one hand, it could represent the signal from the chemoreceptors to the flagella, but on the other, it might simply be a consequence of stimulus-induced changes in the membrane that accompany chemotaxis. One class of mutants that failed to show this effect are the *mot* mutants which have morphologically normal, but paralyzed, flagella. The *mot* product may comprise some sort of ion gate in the cytoplasmic membrane, and this phenomenon could provide much needed insight about the role of this protein in flagellar rotation.

It is possible that a single trigger event can affect all of a cell's flagella simultaneously. The tumble probability in swimming cells is the same as the probability of a CCW to CW reversal in tethered cells (Berg and Tedesco, 1975). Assuming that reversal of a single filament would have an observable effect on swimming behavior, this result implies that all the flagella reverse at the same time. In *Spirillum volutans*, a large polarly flagellated bacterium, the flagellar bundles at each end reverse in unison, and Berg (1975b) has argued that the synchrony is too exact to be caused by a diffusable substance. If there is also flagellar synchrony in *E. coli*, the most likely trigger would be a change in the electrical properties of the cell membrane. Localized chemical triggers could be ruled out unless they were somehow co-ordinated by a generalized cellular process. The use of antibody-coated beads to monitor several flagella on the same cell should provide a means of testing for flagellar synchrony.

Genetic studies in *E. coli* (Armstong, Adler and Dahl, 1967; Parkinson, 1974) and in *S. typhimurium* (Vary and Stocker, 1973; Aswad and Koshland, 1975a; Collins and Stocker, 1976; Warrick, Taylor and Koshland, 1977) have turned up chemotaxis mutants with aberrant tumbling patterns. Most such mutants fail to tumble and exhibit exclusively CCW rotation in cell-tethering experiments; the rest tumble incessantly and rotate predominantly in the CW direction (Larsen *et al.*, 1974a; Parkinson, 1976). Almost all the mutants map in areas of the genome that contain flagellar structural genes (Fig. 2.2) (Armstrong and Adler, 1969a; Parkinson, 1976; Collins and Stocker, 1976; Warrick *et al.*, 1977). Many of the mutants define genes (e.g., *cheA* and *cheB* in *E. coli*) whose products are essential for tumbling rather than flagellar assembly (Armstrong and Adler, 1969b; Parkinson, 1976). A few mutants represent specific alterations of known flagellar genes (Silverman and Simon, 1973b; Parkinson, 1976; Collins and Stocker, 1976; Warrick *et al.*, 1976). The latter class is especially interesting because they probably represent flagellar components that are part of the switch mechanism. The role of tumble-specific genes like *cheA* and *cheB* is less certain; however, their proximity to flagellar genes suggests they may make components of the switch or trigger that are part of the basal body.

2.4 CHEMOSENSORS

In this section, we will consider 'chemosensors' following the nomenclature suggestion of Adler (1975) that chemosensors detect specific compounds and include at least a 'chemoreceptor' that directly binds the chemical and a 'signaller', linking the receptor to the rest of the chemotaxis machinery. Excellent discussions of the proof of the receptor hypothesis and definition of the number and specificity of chemosensors are available in a number of reviews (Adler, 1974a, 1974b, 1975, 1976; Hazelbauer,

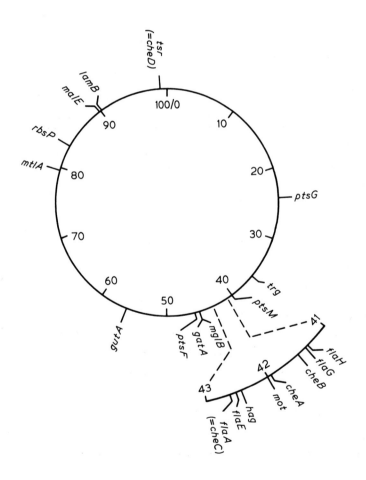

1974), as well as in the original papers (Adler, 1969; Mesibov and Adler, 1972; Adler, Hazelbauer and Dahl, 1973) and these questions will not be considered here in any detail. We will first discuss the functions of chemosensors, followed by the characteristics of identified chemoreceptors and finally will offer suggestions of how chemosensors might be organized and thus perform their functions. The bulk of our information about chemosensors comes from studies of *E. coli* and *S. typhimurium* which appear to have almost identical complements of chemosensors. Recent descriptions of chemosensor systems in Gram-positive bacteria (Ordal *et al.*, 1977; Ordal, 1976a; Van der Drift *et al.*, 1975; De Jong, Van der Drift and Vogels, 1975) indicate that the number and specificity of chemosensors can vary drastically in different species. Consideration of the differences may assist in defining general and species-specific features of chemosensors.

Fig. 2.2 Genetic map of *E.coli* showing the location of chemotaxis genes discussed in the text (map positions and gene symbols are described by Bachmann, Low and Taylor, 1976).

Chemotaxis function	Gene symbol	Approximate map position (min)
Receptor proteins		
glucose	*ptsG*	24
mannose	*ptsM*	40
galactose	*mglB*	45*
galactitol	*gatA*	45*
fructose	*ptsF*	46
glucitol	*gutA*	56
mannitol	*mtlA*	80
ribose	*rbsP*	83
maltose	*malE*	90
Signalling components		
galactose-ribose	*trg*	(37)
serine-repellents	*tsr* (*cheD*)	99
Flagellar components		
tumble generator	*cheA*; *cheB*:	42
motility	*mot*	42
flagellin	*hag*	42
basal body	*flaA* (*cheC*)	43
	flaG; *flaH* (?)	41
polyhooks	*flaE*	42

* The relative order of these genes has not been established.

2.4.1 Physiological analysis

(a) *Temporal sensitivity*

In some of the earliest published observations of bacterial taxis, Engelmann (1883a,b) described the response of *Bacterium photometricum* to temporal changes in light intensity. When illumination was dimmed uniformly, all cells reversed direction, thus mimicking the reversal observed when a cell encountered a spatial boundary between light and dark. *E. coli* and *S. typhimurium* also respond to temporal stimuli: both sudden and gradual changes in attractant or repellent concentration elicit swimming responses that parallel those observed in spatial gradients. Behavior in spatial gradients appears to be controlled solely by temporal changes sensed by the bacteria as they swim. Temporal sensing allows bacteria to average out local fluctuations by making comparisons over a relatively long time, and thus, for a swimming cell, over a greater concentration diffference. The upper limit of comparison time should not be greater than the mean run duration. The mean run time, in turn, must be no longer than a few seconds, otherwise rotational diffusion, rather than tumbling, will dominate changes in direction (Berg and Brown, 1974). The intrinsic tumble frequencies of *E. coli* and *S. typhimurium* fall within the theoretically optimum range dictated by the above factors (Berg and Brown, 1972, 1974; Macnab and Koshland, 1972, 1973).

Temporal sensitivity has been demonstrated in two ways. Brown and Berg (1974) used an enzyme–substrate system to generate gradual (i.e., physiological) changes in attractant concentration. Increasing attractant concentration caused longer mean run lengths due to a reduction in tumble rate. Decreasing attractant levels produced no significant change. These responses are comparable to those observed in spatial gradients. Macnab and Koshland (1972) used a rapid mixing device to subject bacteria to sudden, large concentration changes ('drastic temporal stimulation'– see Section 2.4.1(d). In this case, tumbling can be totally suppressed for several minutes by attractant increases or repellent decreases. Stimuli of opposite sign elicit short periods of increased tumbling activity. Thus the chemotaxis machinery is capable of raising as well as lowering tumble frequency, although under 'normal' conditions (i.e., spatial gradients) enhancement of tumble probability has, at best, a minor role in chemotactic behavior (Berg and Brown, 1974). The asymmetry of responses to tumble-suppressing and tumble-evoking stimuli remains unexplained but could be caused by differential adaptation rates to the two sorts of stimuli. Repellent responses in *B. subtilis* also exhibit asymmetry, but of opposite sign (Ordal and Goldman, 1976). In effect, asymmetry results in positive feedback for individuals headed one way in a spatial gradient, and in negative feedback for those headed in the opposite direction (Macnab and Koshland, 1973).

(b) *Drastic temporal stimulation*

Mixing experiments in which the concentration of an active compound changes greatly, often by many orders of magnitude, subject bacteria to drastic and unphysiological stimuli, and the cells behave with a fittingly dramatic response (Macnab and Koshland, 1972; Larsen *et al.*, 1974a; Parkinson, 1974; Berg and Tedesco, 1975; Spudich and Koshland, 1975). Yet the response parallels normal chemotaxis behavior, differing only in degree. For this reason, drastic temporal stimulation has become an important technique in the study of chemotaxis. For many purposes, the stimulus can be applied by simply mixing a small volume of a test solution with a bacterial suspension without the need of a special mixing chamber or delivery system and determining the response by immediate observation of the cells with a microscope (Larsen *et al.*, 1974a; Parkinson, 1974; Ordal and Goldman, 1976). In addition to their general simplicity and convenience, temporal gradient procedures allow precise definition of the gradient and eliminate concerns about metabolic perturbation. Temporal stimulation has proved quite useful in rapid characterization of potential taxis mutants (Aswad and Koshland, 1975a; Parkinson, 1976) as well as a basis for quantitative assays of taxis response (Spudich and Koshland, 1975; Berg and Tedesco, 1975).

Sensitivity to changes in concentration, not absolute concentration, means that bacteria respond to new concentrations as compared to old, but then adapt to new concentrations by returning to a normal behavior pattern. The pattern of behavior in response to drastic temporal stimulation can be interpreted to mean that the taxis system is functionally saturated by the magnitude of the signals, and the relaxation time required to return to normal behavior reflects the time necessary to 'process' the accumulated signal, or to 'adapt' to the altered environment (Spudich and Koshland, 1975; Berg and Tedesco, 1975). The physiological significance of adaptation times following temporal stimutlation is based on the fitting of concentration dependence of adaptation times to a model in which adaptation time (termed transition time (Berg and Tedesco, 1975) or recovery time (Spudich and Koshland, 1975)) is proportional to the change in amount of receptor bound (Mesibov *et al.*, 1973). Assuming that active compounds are in an association-dissociation equilibrium with their receptors then the curve fit to adaptation times can be used to determine an apparent dissociation constant. A study of temporal stimulation of a tumbly mutant of *S. typhimurium* by ribose and its analog allose (Spudich and Koshland, 1975) showed that concentration dependence of recovery times was quite similar to the binding curve of ribose to the pure receptor protein and to the binding curve calculated from the inhibition constant determined for allose (Aksamit and Koshland, 1972, 1974). The relative response of normal cells to the two compounds is also consistent with those curves (Aksamit and Koshland, 1974).

Drastic temporal stimulation of tethered cells results in unidirectional rotation of the cells (Larsen *et al.*, 1974a) for as long as hundreds of seconds (Berg and Tedesco, 1975). The concentration dependence of transition times after stimulation by α-methylaspartate is in good agreement with an apparent receptor-ligand dissociation constant determined from capillary sensitivity assays (Mesibov *et al.*, 1973). Data from capillary assays, equilibrium dialysis binding to purified receptor, and transition time determinations on tethered cells, all generate the same apparent dissociation constant for the complex of maltose and its receptor (Hazelbauer, unpublished observations).

(c) *Gradient sensing*

Bacterial chemotaxis can be envisioned as a two-step process: detection of changes in concentration of stimulus compounds, and suppression or activation of the tumble generator. Chemosensors play a central role in the first step, that of gradient sensing, but are not directly involved in tumble regulation, since chemosensor mutants have normal tumbling behavior. Gradient sensing requires the ability to detect the presence of a compound and to determine its relative concentration. In addition, there must be a means of comparing past and present environments in order to determine whether concentration is increasing or decreasing. Chemosensors, by definition, are responsible for compound recognition and concentration measurements, however, they probably don't possess the sort of 'memory' needed for temporal discrimination. As we will discuss later in this section, 'memory' appears to be a property of the shared adaptation mechanism and not of individual chemosensors.

The receptor portion of a chemosensor is a protein that binds a closely related family of compounds at a stereospecific site. Correlations of *in vitro* receptor–ligand affinities with chemotactic responses to the same compounds indicate that the specificity and the sensitivity range of the chemosensory system reside in the receptor binding site. Good correlations have been observed with the water-soluble receptor proteins for galactose, maltose and ribose (Hazelbauer and Adler, 1971; Hazelbauer, 1975b; Aksamit and Koshland, 1974). Other sugar sensors employ hydrophobic membrane protein receptors, the *in vitro* ligand affinity of which might well be modified upon removal from their native environment as is the case for sugar phosphorylating activity of purified enzymes II (Epstein and Curtis, 1972).

Studies employing temporal stimulation have shown that changes in swimming behavior are directly related to changes in occupancy at the receptor site (Spudich and Koshland, 1975; Berg and Tedesco, 1975). However, signals from different chemosensors are functionally equivalent and additive (Tsang *et al.*, 1973; Adler and Tso, 1974; Berg and Tedesco, 1975). For this reason, the adaptation machinery cannot operate at the level of the individual chemosensors. If that were the case, each chemosensor would adapt at its own intrinsic rate independent of other stimuli and transition times for responding to mixed stimuli would not be additive since they would reflect only the slowest adapting chemosensor. Thus there may be no time-component in chemosensor signals; instead, 'memory' resides in the adaptation system that functions at a general level. The receptor and signaller proteins that constitute a chemosensor probably transmit information about site occupancy only, and not about changes in site occupancy. As site occupancy changes, the chemosensor signal also changes causing suppression or activation of tumbling that persists until adaptation is complete.

The view that chemosensor signals are generated by binding at receptor sites is also favored by evidence that signals pass through the taxis machinery from receptors whose percentage site occupancy by ligand does not change i.e., no gradient is present. For example, methionine is continually consumed by the chemotaxis machinery (Armstrong, 1972a; Springer et al., 1975), and low levels of methionine are more rapidly depleted by cells in the presence of constant, saturating levels of attractant than by cells in buffer (Springer et al., 1975). The rate of methionine depletion appears to be a function of percentage occupancy of the attractant receptor. This indicates that, while no taxis behavior is induced by the isotropic attractant concentration, occupancy at receptor sites results in some part of the taxis machinery 'running faster'. Further support for this notion derives from the fact that the presence of a constant concentration of ribose or galactose can inhibit (Adler et al., 1973), and, in some situations, eliminate (Strange and Koshland, 1976) response to a gradient of the other sugar, although the inhibiting sugar does not bind to the functioning receptor. Thus a constant occupancy at one receptor inhibits response to a gradient detected at a separate receptor. A similar effect is exerted on the sensitivity of many receptors by high concentrations of serine (Mesibov and Adler, 1972) and, in fact, 10 mM serine, present in an isotropic solution, suppresses the frequency of tumbles (Berg and Brown, 1972). The effects of serine will be considered in greater detail in a later section.

2.4.2 Genetic and biochemical analysis

There are approximately 25 known chemosensors in *E. coli* (Tables 2.1, 2.2 and 2.3). Taxis sensitivities that have been described for *S. typhimurium* (Macnab and Koshland, 1973; Tsang et al., 1973; Aksamit and Koshland, 1974; Aswad and Koshland, 1975b; Strange and Koshland, 1976) are consistent with a similar collection of sensors in that organism. However, the particular chemosensor set is not common to all bacteria, since recent characterizations of responses by Gram-positive organisms show a vastly different pattern (Van der Drift et al., 1975; De Jong et al., 1975; Ordal, 1976a; Ordal et al., 1976). Specific mutants have been found for over half the sensors of *E. coli* (Tables 2.1 and 2.3), and in many cases a specific location for the mutation can be assigned on the *E. coli* genetic map (shown in Fig. 2.2). There is no indication of clustering of chemosensor loci nor of any common control. Most chemosensors for a sugar are induced by growth on that particular compound (Adler et al., 1973). Methods of mutant selection have been based on separating cells that do not respond to a gradient of a particular compound from those that do. However, in many instances, receptor mutants have been identified by direct testing of responses of clones isolated from mutagenized cultures or by characterization of taxis patterns of mutants selected for other defects, e.g., transport.

Table 2.1 Sugar chemosensors in enteric bacteria

Sensor	Ligands	Type of receptor protein*	Receptor mutants	Signaller mutants	Isolated purified receptor	In vivo/in vitro affinity correlation
Galactose	Glucose Galactose Glycerol-β-galactoside Fucose	Shockable BP	Yes	Yes	Yes	Yes
Maltose	Maltose	Shockable BP	Yes	Yes	Yes	Yes
Ribose	Ribose	Shockable BP	Yes	Yes	Yes	Yes
Glucose	Glucose Methyl-α-glucoside Methyl-β-glucoside	Enz II	Yes		Yes	
Mannose	Mannose Glucose 2-deoxy-glucose	Enz II	Yes		Yes	
Fructose	Fructose	Enz II	Yes		Yes	
N-acetylglucosamine	N-acetylglucosamine	Enz II	Yes			
Mannitol	Mannitol	Enz II	Yes			
Glucitol	Glucitol	Enz II	Yes			
Galactitol	Galactitol	Enz II	Yes			
Aryl-β-glucosides	Arbutin Salicin	Enz II				
Trehalose	Trehalose	Unknown				

* **BP** = binding protein; Enz II = enzyme II of phosphotransferase transport system.

Table 2.2 Amino acid attractants of enteric bacteria

Group	Amino acid detected	Approximate chemotaxis threshold † μM
Serine* signal group	ser	0.3
	gly	30
	ala	70
	thr	1
Aspartate* signal group	asp	0.06
	glu	5
	met	300
Others	asn	2
	cys	4

* There are probably at least two separate receptors in the group, each having different affinities for the various amino acids.
† From Mesibov and Adler (1972), determined by capillary assays at high cell density (7×10^7 ml^{-1}).

(a) *Reception and transport*

Each of the identified receptor molecules is a component of a transport system for its ligand(s) as well as a component of a chemosensor, but the process of transport and taxis are separable, at least for some compounds. In the case of galactose (Hazelbauer and Adler, 1971; Ordal and Adler, 1974b) and maltose (Hazelbauer, 1975a), there are mutations that affect transport but not taxis, as well as mutations that affect taxis but not transport. Mutations in the β-methylgalactoside transport system (the galactose system of interest here) (Ordal and Adler, 1974a; Robbins, 1975; Robbins, Guzman and Rotman, 1976) and in the maltose transport system (Hofnung, 1974; Hofnung, Hatfield and Schwartz, 1974) have been mapped and analyzed by complementation. Mutations in *mglA* or *mglC* alter galactose transport, but seldom galactose taxis (Ordal and Adler, 1974a). Galactose-binding protein is the product of *mglB*, and most mutations of that gene affect both transport and taxis (Ordal and Adler, 1974a; Boos, 1972). One exceptional *mglB* mutant is normal in transport but defective in taxis (Hazelbauer and Adler, 1971). Mutations in *malK* or *malF* eliminate maltose transport but have little effect on maltose taxis (Hazelbauer, 1975a). *malE* codes for the maltose-binding protein (Kellerman and Szmelcman, 1974) and most mutations in that gene completely eliminate detectable maltose binding activity, maltose transport and maltose taxis. Three *malE* missense mutations eliminate transport but not taxis, and two of those three do not alter maltose-binding activity of the protein (Hazelbauer, 1975a). Mutations eliminating ribose-binding protein and ribose taxis also reduce ribose transport. When a *S. typhimurium* strain carrying such a mutation reverts to production of binding protein, normal taxis and transport are also regained (Aksamit and Koshland, 1974). It may be that some

Table 2.3 Repellent sensors in E. coli *

Sensor	Ligands	Sensor mutants	Affected by *tsr* mutation
Fatty acids	Acetate Propionate *n*-Butyrate *iso*-Butyrate	No	Yes
Aliphatic alcohols	Ethanol *iso*-Butanol	No	No
Hydrophobic Amino acids	*l*-Leucine *l*-Isoleucine *l*-Valine *l*-Tryptophan	Yes	Yes
Indole and analogues	Indole	Yes	Yes
Benzoate	Benzoate	No	Yes
Salicylate	Salicylate	Yes	Slightly
H^+	H^+	No	Yes
OH^-	OH^-	No	No
Metallic cations	$CoSO_4$ $NiSO_4$	No	No

* Adapted from Tso and Adler (1974).

strains characterized as defective in growth on ribose, but normal for binding protein and taxis, represent transport-defective, taxis-normal mutants. Mutations mapping at loci separated from the binding protein genes can result in defective response to galactose and ribose without affecting transport (Ordal and Adler, 1974b; Strange and Koshland, 1976).

These observations indicate that galactose-binding, maltose-binding, and ribose-binding proteins, are receptors for both a sensor of a chemotaxis system and a transport system (Hazelbauer and Adler, 1971; Kalckar, 1971; Adler *et al.*, 1973; Hazelbauer, 1975a). No other component participates in a similar manner in both processes. Additional taxis and transport components, independent of each other, must also exist, since as stated above, it is possible to inactivate taxis by mutation without affecting transport, and vice versa. The exceptions to this model are a few apparent missense mutations, located by complementation in *mglA* or *mglC*, that greatly reduce galactose taxis (Ordal and Adler, 1974a,b), and one deletion in *malK* and *lamB* that completely eliminates maltose taxis although it complements transport defects of *malE* mutants (Hazelbauer, 1975a). Their effects might be explained by indirect inhibitory effects of those particular mutant transport components on normal functioning of the binding protein. Genetic polarity effects are not likely since none of the mutations are operator proximal to the respective binding protein genes.

The known membrane-associated receptor proteins are identical with enzymes II, substrate-specific components of the phosphotransferase transport system (cf. Roseman, 1972). Mutations in enzymes II for glucose, mannose, fructose, mannitol, glucitol, or galacitol, all exhibit defective taxis to the respective compounds (Adler and Epstein, 1974; Lengeler, 1975a). There are also enzymes II for N-acetylglucosamine (White, 1970) and aryl-β-glucosides that may well serve as receptors for those compounds (Adler and Epstein, 1974). The phosphotransferase system transports sugars by vectorial phosphorylation, in which the movement of a molecule across the cytoplasmic membrane occurs in conjunction with addition of a phosphate group donated by phosphoenolpyruvate. Phosphorylation and the transport process require two soluble proteins, enzyme I and Hpr, as well as an enzyme II complex containing enzyme IIA (sugar-specific) and enzyme IIB (probably non-specific) (Roseman, 1972). The whole of this system is also required for normal taxis toward a compound recognized by an enzyme II chemoreceptor. Enzyme I or Hpr mutants exhibit only weak responses to glucose. Thus the transport function and the chemoreceptor function of the phosphotransferase system cannot be separated. However, it has been convincingly argued that the role of this system in taxis is to recognize the various sugars, not simply to transport or phosphorylate them (Adler and Epstein, 1974). The requirement of an intact phosphorylation system for normal taxis may reflect a requirement of enzyme II for phosphorylated Hpr or direct phosphorylation to maintain high affinity for substrate (Roseman, 1972; Adler and Epstein, 1974). The respective enzymes II may also mediate catabolite repression exerted by their substrates via inhibition of adenylate cyclase (Roseman, 1972; Harwood *et al.*, 1976). It is not known whether the mechanism of that inhibition is in any way related to regulation of the tumble generator by those receptors but it is unlikely to be directly related since other receptors, i.e., periplasmic binding proteins, have not been implicated in catabolite repression.

Sharing of a binding protein by an attractant chemosensor and a transport system makes evolutionary sense. Ability to move towards higher concentrations of a compound is an advantage only if the compound can be efficiently taken into the cell and metabolized. Thus, it is reasonable to suppose that transport systems predated chemotaxis systems. The binding protein of a transport system would fit the receptor requirement of an evolving chemosensor. This reasoning would predict that all attractant chemoreceptors would be ligand-binding components of transport systems, but not all transport-binding proteins would serve as chemoreceptors. With about two-thirds of the attractant receptors identified the first prediction is holding. The second is clearly true. For example, binding proteins involved in transport of the branched chain amino acids (Oxender, 1972), arginine (Wilson and Holden, 1969a,b) or lysine (Wilson and Holden, 1969) are not chemoreceptors (Mesibov and

Adler, 1972; Tso and Adler, 1974). Neither is the lactose permease (Adler *et al.*, 1973). The collection of chemosensors of a given bacterial species should be a function of the environment in which that organism evolved. Characterization of taxis sensitivities and chemosensors in Gram-positive species demonstrates the expected interspecies variation.

(b) *Receptor properties*

Very little is known about the steps intervening between ligand binding by chemoreceptor and the flagellar response. We might obtain some insight into chemosensors by considering features common to all identified receptors. If it is assumed that the chemoreceptor, i.e., the actual binding molecule, of an enzyme II complex is the sugar-specific enzyme IIA, then all sugar receptors are water-soluble proteins. Enzymes IIA can be released as active, buffer-soluble proteins by a relatively mild treatment of membrane fragments that neither disrupts the integrity of the membrane nor solubilizes the hydrophobic enzyme IIB (Kundig and Roseman, 1971; Roseman, 1972). In some cases, enzyme IIA is replaced by Factor III, a soluble protein that appears to be very similar to an enzyme IIA (Roseman, 1972). These proteins, like periplasmic sugar-binding proteins, are thus peripheral membrane proteins, although their association with an integral membrane protein (enzyme IIB) is clearly much more stable than the postulated association of a periplasmic protein (e.g., galactose-binding protein) with a membrane component of its transport system (the *mglA* or *mglC* product). It would be interesting to know whether some small proportion of an enzyme IIA is found *in vivo* in the cytoplasm or periplasm.

Both the phosphotransferase system and transport systems mediated by periplasmic binding proteins contain multiple protein components, one of which provides a specific sugar-binding site, the others of which are presumably involved in actual translocation of ligand across the membrane (Robbins and Rotman, 1975) and in supplying energy necessary for transport. The role of energy-supplying proteins is well defined for the phosphotransferase system (Roseman, 1972) but still hypothetical for the other systems. In contrast, the lactate dehydrogenase-linked transport systems, such as the lactose permease, appear to have only a single protein for all functions (Kaback, 1974). If a chemoreceptor must interact with a chemotaxis component that is independent of its transport complex it may be required to detach from one place and attach somewhere else. Restriction on such movements may help to explain why the binding site of the lactose permease of *E. coli* is not employed as a chemoreceptor.

(c) *Biochemistry of sugar receptors*

The receptors for galactose (Anraku, 1968), maltose (Kellerman and Szmelcman, 1974) and ribose (Willis and Furlong, 1974) have been purified from *E. coli* as have the galactose (Strange and Koshland, 1976) and ribose (Aksamit and Koshland, 1972) receptors from *S. typhimurium*. The proteins from the two species are closely related (Table 2.4). The molecules are relatively heat-stable and ligand binding is not sensitive to sulfhydryl reagents (Boos, 1974). In normal buffers they all exist as

Table 2.4 Properties of receptor proteins

Receptor	Organism	Molecular weight	High affinity ligands	K_d (μM) at 4°C	K_d (μM) at 23–25°C	Binding site/ monomer	Fluorescence quenched by ligand binding	Reference
Galactose	E. coli	33–36 000	Glucose Galactose	0.2 0.5	1 1	1 1	Yes Yes	Anraku (1968); Boos et al. (1972); Zukin et al. (1976)
	S. typhimurium	33 000	Glucose Galactose	0.2 0.4	1 2	1 1	Yes	Strange and Koshland (1976); Zukin et al. (1976)
Maltose	E. coli	38 000	Maltose	3.5	1.2	1	Yes	Kellerman and Szmelcman (1974); Szmelcman et al. (1976); Schwartz et al. (1976)
Ribose	E. coli	29 500	Ribose	0.1	—	1	*	Willis and Furlong (1974)
	S. typhimurium	29 500	Ribose	—	0.3	1	*	Aksamit and Koshland (1972)

* Protein does not contain tryptophan

monomers of the molecular weights listed in Table 2.4.

Early equilibrium dialysis studies indicated that binding of galactose to *E. coli* galactose-binding protein is 'biphasic' and that two molecules of sugar can bind to each monomer (Boos and Gordon, 1971; Boos *et al.*, 1972). It was suggested that the two apparent dissociation constants might be a reflection of two binding sites which would be important in the mechanisms of chemotaxis (Hazelbauer, 1974) and transport (Boos and Gordon, 1971; Boos, 1974). The binding experiments were controversial (Kepes and Richarme, 1972; Richarme and Kepes, 1974; Silhavy and Boos, 1975). Studies on the binding of maltose to maltose-binding protein were interpreted to suggest two binding sites, although the protein bound only one sugar molecule at saturation (Kellerman and Szmelcman, 1974; Hazelbauer, 1975a,b). However, other sugar-binding proteins, including galactose-binding protein from *S. typhimurium,* exhibit binding patterns consistent with one dissociation constant as well as one site per monomer (Table 2.4). Binding to both the maltose-binding and the galactose-binding proteins appears linear when assayed by fluorescence quenching (see below). Recent investigations of galactose binding to the *E. coli* galactose-binding protein strongly suggest single-site, single-affinity association (Zukin *et al.*, 1976). Detailed equilibrium dialysis studies of maltose-binding protein (Schwartz *et al.*, 1976) indicate that the previously observed deviations from linear binding are not statistically significant. Thus the protein most probably has a single affinity constant for maltose.

Galactose-binding (McGowan, Silhavy and Boos, 1974), and maltose-binding (Szmelcman *et al.*, 1976) proteins both exhibit an intrinsic fluorescence characteristic of tryptophan residues. The ribose-binding protein does not contain tryptophan (Aksamt and Koshland, 1972; Willis and Furlong, 1974). Fluorescence is quenched by ligand binding, and apparent affinities of protein for ligand, determined by quenching, agree with values derived from equilibrium dialysis experiments (Boos *et al.*, 1972; McGowan *et al.*, 1974; Schwartz *et al.*,1976; Szmelcman *et al.*, 1976). Different spectral shifts are produced by binding of the various ligands (maltotriose, etc.) to the maltose-binding protein (Szmelcman *et al.*, 1976), and by binding galactose or glucose to the galactose protein (McGowan *et al.* 1974). For both proteins, there appears to be a common site for all ligands. Ultraviolet difference and solvent perturbation difference spectroscopy of galactose-binding protein in the presence and absence of ligand, clearly demonstrate that the tryptophan residue detected in the fluorescence quenching studies is in the active binding site (McGowan *et al.*, 1974). The data give no indication of significant conformational alterations removed from the active site. However, substrate does protect from fluorescence quenching by KI, probably indicating a small change in surface charge (McGowan *et al.*, 1974). These results are consistent with those obtained by a range of other physical and spectroscopic techniques (Boos *et al.*, 1972) that indicate no gross conformational change upon galactose binding to the protein.

However, it is attractive to think of taxis signals being generated by a conformational change in the receptor upon binding, and there are some indications of a slight alteration in the binding protein upon ligand binding: it appears that substrate binding to galactose-binding protein results in increased migration in an electrophoretic field (Boos and Gordon, 1971; Boos et al., 1972). The implication of an alteration in surface charge is consistent with the fluorescence-quenching protection cited above (McGowan et al., 1974). Altered migration is not observed for maltose binding (Hazelbauer, unpublished results) or ribose-binding (Willis and Furlong, 1974) proteins in similar experiments. Interaction of galactose-binding protein with specific antibody increases the retention of labeled galactose (Rotman and Ellis, 1972). This change could mimic an *in vivo* process. During storage in saturated amonium sulfate, the galactose-binding protein forms stable dimers, but binding properties remain unaltered (Rasched, Shuman and Boos, 1976).

Many studies of periplasmic binding proteins have included the observation that the rate of exit of radioactive ligand from a dialysis bag into buffer is greatly reduced by the presence of binding protein (Hogg and Engelsberg, 1969; Rosen and Heppel, 1973; Kellerman and Szmelcman, 1974; Willis and Furlong, 1974; Silhavy and Boos, 1975). This retention of ligand is not due to any unique or unusual property of binding proteins, but simply reflects a high binding-site concentration relative to the dissociation constant (Silhavy et al., 1975). The kinetics of exit can be correctly predicted from equations based on the law of mass action (Silhavy et al., 1975). There is no implication of an unusually slow dissociation rate constant, and fluorescence studies demonstrate that dissociation of ligand from galactose-binding (Silhavy and Boos, 1975) or maltose-binding (Szmelcman et al., 1976) protein is rapid. The retention effect can also account for the retention of ligand when binding protein is chromatographed (Anraku, 1968; Richarme and Kepes, 1974), precipitated (Kepes and Richarme, 1972; Rotman and Ellis, 1972), electrophoresed (Kellerman and Szmelcman, 1974) or retained on filters (Schleif, 1969; Boos, 1974).

However, there may be an important analogy between a dialysis bag and the periplasm. If an upper limit for the volume of the periplasm is assumed to be a few percent of total cell volume (thus about 10^{-14} ml) then 10^4 to 10^5 copies of binding protein present in an induced cell (Silhavy and Boos, 1975) are at a concentration in excess of 0.5 mM. Thus, binding site concentration is high relative to ligand concentration and dissociation constant, and the retention effect could occur. The magnitude of the effect would depend upon rates of ligand exit due to transport and loss back through the outer membrane. The derivations of Silhavy et al. (1975) predict that ligand concentration in the periplasm (where it would all be essentially bound) would be relatively insensitive to decreases in exterior ligand concentration. However, a derivation based on somewhat different assumptions predicts insensitivity to both increases and decreases (H. Berg, personal communication). The effect may contribute to a damping of local statistical fluctuations in concentration that are large for an individual bacterium in the sensitivity range for most attractants (Macnab and Koshland, 1972; Koshland, 1974). It is also interesting that bacteria are much more sensitive to increases in attractant concentration than to decreases (Section 2.4.1(a)).

For enteric bacteria, repellent receptors are active only at relatively high concentrations of ligand and so retention effects would not occur. The enteric species exhibit a greater response to decrease than to increase in repellent concentration, but the differential sensitivity is not as great as for attractants (Tsang *et al.*, 1973). *B. subtilis* is sensitive to low concentrations of a special class of repellents and is much more sensitive to increase in concentration of those compounds than to decrease (Ordal and Goldman, 1975, 1976; Ordal, 1976a).

For maltose, it can be directly shown that its chemoreceptor senses the concentration of attractant in the periplasm, not simply the concentration in the surrounding medium. The outer membrane protein originally identified as the receptor for phage λ (Randall-Hazelbauer and Schwartz, 1973) functions in passage of maltose and higher maltodextrins across the outer membrane (Szmelcman and Hofnung, 1975; Hazelbauer, 1975c; Szmelcman *et al.*, 1976; Braun and Hantke, Chapter 3). Impaired transport of those sugars caused by defective λ receptor protein results in lowered taxis response to maltose, although the maltose-binding protein is normal (Hazelbauer, 1975a,c). The λ receptor does not have a direct role in maltose chemoreception, but normal movement of maltose and maltodextrins across the outer membrane and into the periplasm, mediated by the λ receptor, is a prerequisite for interaction of those ligands with binding protein.

If defective entry into the periplasm can affect apparent sensitivity of a chemoreceptor, defective exit, in the form of reduced movement of ligand into the cell because of a transport mutation, might also affect apparent sensitivity. Apparent increased sensitivity of chemoreceptors has often been observed for strains defective in the relevant transport system, and has been explained as a result of reduced metabolic destruction of the spatial gradient (Adler, 1969; Adler *et al.*, 1973; Aksamit and Koshland, 1974; Ordal and Adler, 1974b). However, if that were the complete explanation, accumulations in capillary assays at low cell concentrations would be almost the same for a transport mutant and its parent. That is not always the case (Adler *et al.*, 1973; Aksamit and Koshland, 1974; Ordal and Adler, 1974b; Hazelbauer, 1975c).

(d) *Amino acid receptors*

There is no biochemical information about amino acid receptors even though significant effort has been expended searching for amino acid receptor proteins (Mesibov and Adler, 1972; Adler, 1975; Aksamit, Howlett and Koshland, 1975). Much of the difficulty is due to lack of detailed physiological and genetic information about these receptors. Neither of the two classes of mutants originally thought to represent amino acid receptor mutants appears to be defective in a specific receptor protein, but instead in components involved in signal transduction (see Section 2.5). It is probable that grouping of amino acid attractants under an 'aspartate receptor' and a 'serine receptor' reflects organization at the level of signal transduction, not reception. It seems likely that there are more than two amino acid receptors, and that those receptors may well resemble sugar receptors in their narrow specificities

and relation to transport systems. The sensitivity over many orders of magnitude to
the best amino acid attractants (Mesibov et al., 1973) may reflect recognition, over
different concentration ranges, by more than one receptor. It would be extremely
useful to re-examine the pattern of jamming of response to amino acids. Sensitivity
and jamming sensitivity assays (described in Section 2.2.1) have been used recently in
the study of *B. subtilis* amino acid taxis to identify highest affinity receptors in the
presence of multiple receptors (Ordal et al., 1977). In that organism, it appears that
there are many independent amino acid receptors and many secondary interactions.
Similar work could be profitably done with the enteric species.

(e) *Repellent receptors*

There are at least nine receptor classes for repellents of *E. coli* (Tso and Adler, 1974).
S. typhimurium appears to have at least some of the same receptors (Tsang et al., 1973).
The apparent ligand affinity of repellent receptors in enteric bacteria is three to four
orders of magnitude weaker than attractant receptors (Tso and Alder, 1974) and thus
it would be extremely difficult to identify a receptor molecule by direct binding
assays. As mentioned earlier, the Gram-positive *B. subtilis* is repelled by many
membrane-active compounds including ones known to uncouple oxidative phosphorylation as well as some that act as local anesthetics (Ordal and Goldman, 1975, 1976;
Ordal, 1976a). The compounds are active at concentrations comparable to those at
which good attractants are active for the enteric bacteria. The active concentrations
of uncouplers as repellents correspond to concentrations that uncouple electron
transport from ATP production; however, the former effect is only observed in a
temporal or spatial gradient of the compounds while the latter occurs in constant
concentrations. There is evidence that different uncouplers act as repellents at sites
functionally separable from each other (Ordal, 1976a), but direct demonstration
that those sites are saturable in the usual Michaelis—Menton sense is lacking. It has
been suggested that these compounds are repellents because they are uncouplers,
implying that tumbling rates are altered directly by perturbations of membrane
potential, not by passage of signals from receptors to tumble generator. The
observation that a temporal stimulus of an uncoupler does not sum algebraically
with an attractant stimulus, as repellent and attractant stimuli do in enteric bacteria,
supports the idea of an unorthodox pathway for uncoupler repellent action (Ordal
and Goldman, 1976). However, that same observation makes it difficult to decide how
the response induced by membrane-active compounds is related to the conventional
mechanism of chemotaxis. It could be that the putative molecule possessing a site for
binding an uncoupler could interact with two separate entities, one affecting oxidative
phosphorylation, the other tumble generation. As in the case of transport and taxis
mediated by binding proteins considered above, one process is affected by absolute
levels of ligand, and the other by relative changes.

(f) *Ion receptors*

Recently, the membrane-bound Mg^{2+}, Ca^{2+}-dependent adenosine triphosphatase
(Mg^{2+}, Ca^{2+}-ATPase) has been implicated in the chemoreception of divalent cations

(Ca^{2+}, Mg^{2+}, Zn^{2+}, Mn^{2+}, and Co^{2+}) in *E. coli* and *S. typhimurium* (Zukin and Koshland, 1976). The ATPase complex phosphorylates ADP to ATP using the energized membrane state produced by the respiratory chain, and can also supply energy for transport of certain compounds by the reverse of that reaction (Dahl and Hokin, 1974; Simoni and Postma, 1975). There is a strong phylogenetic conservation of ATPase structure and thus it is interesting that, in some species, although not yet in bacteria, ATPases have been identified as cation transporters (Dahl and Hokin, 1974). The ATPase activity can be removed from the membrane complex as a water-soluble oligomer termed F1, leaving hydrophobic protein components in the membrane fraction. Taxis phenotypes of Mg^{2+}, Ca^{2+}-ATPase mutants indicate that the cation receptor is carried on F1 (Zukin and Koshland, 1976). Thus, although the Mg^{2+}, Ca^{2+}-ATPase is not simply a conventional transport system, its properties as a membrane-bound complex consisting of separable hydrophobic and hydrophilic components parallel the generalizations about transport systems and chemoreceptors outlined above. A chemoreceptor role for F1 is yet another example of the fact that chemoreceptors are binding proteins that function in at least two separate processes.

2.5 SIGNALLING

2.5.1 Stimulus transduction

Chemosensor information appears to be passed to the flagella through a hierarchical organization of signalling components. Each such component seems to receive several different inputs, effectively condensing stimulus information from different chemosensors. Only a few of these components have been detected in mutant studies; however, there is physiological evidence to suggest that the array of shared signallers is an extensive one.

Ordal and Adler (1974a,b) have isolated mutants that are defective in both ribose and galactose taxis, and which do not map near any of the known genes for the ribose or galactose chemosensors (see *trg* locus in Fig. 2.2). The *trg* product could either be a component common to each of these chemosensors, or a separate signalling element that is used by both receptors. Studies of attractant jamming support the latter possibility. In *E. coli*, for example, the response to gradients of galactose or ribose is reduced by high, isotropic concentrations of the other sugar, even though the two are detected by different chemoreceptors (Adler *et al.*, 1973). This effect is especially dramatic in an *S. typhimurium* strain which has a ten-fold excess of ribose-binding over galactose-binding protein: ribose entirely eliminates the galactose response (Strange and Koshland, 1976). This effect does not occur in a ribose-binding protein mutant, suggesting that inhibition of galactose taxis is mediated through the ribose chemosensor, since neither sugar binds detectably to the heterologous binding protein (Strange and Koshland, 1976). Thus the ribose and galactose receptors appear to compete for a common component, presumably the *trg* product. The limited amount or capacity of this component can result in reduction of sensitivity to one receptor when the other is saturated.

Reduction of response to a sugar due to the presence of saturating amounts of a second sugar recognized by a heterologous receptor is quite common (Adler et al., 1973). For example, 10 mM glucose totally jams taxis towards glucitol (Adler et al., 1973; Lengeler, 1975a), although glucose is not detectably transported by enzyme II^{gut}, the glucitol chemoreceptor (Lengeler, 1975a). Mutations in glucose receptor enzyme II^{glu} eliminate this jamming by glucose, suggesting that it is mediated by interaction of glucose with its own receptor (Lengeler, 1975a). Mannitol is not transported via enzyme II^{glu} nor glucose via enzyme II^{mtl} (Lengeler, 1975b), yet the presence of 10 mM of one sugar inhibits response to a gradient of the other (Adler et al., 1973). Under some induction conditions, maltose jams response to galactose or ribose (Hazelbauer, unpublished results) although maltose does not bind to the receptor for either of those sugars (Hazelbauer and Adler, 1971; Adler et al., 1973).

It appears that there is extensive interaction of signals from different amino acid receptors in *B. subtilis* (Ordal and Gibson, 1977; Ordal et al., 1977). Probably much of the complex jamming pattern of amino acid attractants observed in *E. coli* (Mesibov and Adler, 1972) is due to interaction of signals from separate receptors, but a study of such relationships must await characterization of specific receptors.

A footnote should be added to the above consideration of signal transduction components. If the *trg* product interacts directly with both galactose-binding and ribose-binding protein then the site of that interaction should have a broad enough specificity to bind two antigenically unrelated proteins (Willis and Furlong, 1974). The arabinose-binding protein of *E. coli* shares antigenic determinants with the galactose protein, although the two are clearly different (Parsons and Hogg, 1974). The arabinose protein is not involved in the modest arabinose response mediated by weak binding of arabinose to galactose-binding protein. If the *trg* product is a gateway to the taxis system, then perhaps selection for a strong response to arabinose would yield a strain with an arabinose-binding protein able to interact with the *trg* product. An *E. coli* strain selected for arabinose taxis appears to depend on the arabinose-binding protein for the improved response, and thus may be an example of the postulated mutation (Hazelbauer, unpublished studies).

Components, like the *trg* product, which collect signals from a small group of chemosensors, could interact directly with the tumble generator, although there is no evidence that this does actually happen. More likely, there are additional transduction components through which stimulus information converges further before it reaches the tumbling machinery. An example of such a component is the *tsr* product, which is required for responses to a variety of compounds including serine (an attractant) and fatty acids, hydrophobic amino acids, indole, benzoate and hydrogen ions (all repellents) (Hazelbauer et al., 1969; Tso and Adler, 1974). Most *tsr* mutations eliminate these responses, but have no effect on tumbling behavior or on responses to other compounds. The key to understanding the role of the *tsr* product is provided by a second class of *tsr* mutants with a very different phenotype (*cheD* mutants) (Parkinson, 1974, 1975). *cheD* mutants are smooth swimmers and therefore generally nonchemotactic. A variety of genetic evidence shows that they are alleles of the *tsr* locus (Parkinson, in preparation). The tumbling defect of *cheD* strains is dominant to wild type, but can be reversed by additional mutations such as deletions that completely inactivate the *tsr* gene. Thus the inability of *cheD* mutants to tumble is caused by an altered *tsr* product that inhibits the tumbling apparatus. The target of inhibition appears to be the *cheB* product, since *cheD* mutants can also revert by acquiring suppressor mutations in the *cheB* gene. These findings suggest that the *tsr* product normally interacts with the *cheB* component of the tumble generator to modulate tumbling activity in response to serine and certain repellents. It is not yet known whether additional transduction components are interposed between the serine and repellent receptors and the *trs* product; however, it is clear that the *tsr* product represents the final link between a large number of chemosensors and the tumble generator.

The postulated interaction of the *tsr* product with the tumble generator might explain a puzzling property of *tsr* mutants observed by Hazelbauer *et al.* (1969) and by Mesibov and Adler (1972). They found that, with the exception of responses mediated directly by the *tsr* product, chemotactic responses are significantly better in a *tsr* strain than in wild type. It is possible that the tumble generator can assume a more effective conformation in the absence of *tsr* product, thereby allowing it to detect and respond better to chemosensor signals.

2.5.2 Interaction of signals with the tumble generator

Experiments employing various stimulus combinations indicate that signals from different chemosensors are algebraically summed by some element in the chemotaxis system (Tsang *et al.*, 1973; Adler and Tso, 1974; Berg and Tedesco, 1975; Spudich and Koshland, 1975). This element may be the *cheB* product (Parkinson, 1974). Genetic studies indicate that the *cheB* component of the tumble generator is the eventual target of sensory information from the different receptors. One class of *cheB* mutants, for example, has very high tumble frequencies and increased response thresholds to drastic temporal stimuli (Parkinson, 1974, 1976). It appears as though these mutants are variously defective in receiving or responding to many different signal inputs. If *cheB* tumbly mutants are, in fact, uncoupled from the normal control mechanism, their high tumbling rate implies that chemoreceptor signals are inhibitory rather than excitatory and that the tumble generator is under negative control.

The number of different inputs to the *cheB* element is at least three, since some *cheB* mutants are specifically defective in aspartate taxis (Mesibov and Adler, 1972; Kort, Reader and Adler, in preparation). The three inputs are thus aspartate signals, *tsr*-mediated signals, and all others. It is likely that additional genetic studies will subdivide the last group as more mutants become available. For this reason, it seems most unlikely that these signals will prove to be electrical in nature. If membrane potential is used as a signalling device, it probably operates after the *cheB* step. For example, the *cheB* product could be a component of the trigger the firing rate of which is modulated by chemosensor inputs. Alternatively, the trigger itself could be independent of chemosensor control, and changes in tumbling rate might be effected by regulating the response of the flagellar switch mechanism to the trigger. The latter possibility is an attractive one because reversion studies indicate that the *cheB* component may be in physical contact with the *cheC* product, a likely component of the flagellar rotational machinery (Parkinson and Parker, in preparation).

2.5.3 The role of methionine

Methionine auxotrophs that are otherwise normal for the chemotaxis machinery cease tumbling and rapidly lose chemotactic ability when deprived of methionine. Upon restoration of exogenous methionine, both tumbling and chemotaxis are quickly regained. This effect has been observed in *E. coli* (Adler and Dahl, 1967; Armstrong, 1972a), *S. typhimurium* (Aswad and Koshland, 1974) and *B. subtilis* (Ordal, 1976c). Since no other amino acid produces comparable effects, the methionine requirement for chemotaxis is specific as well as continuous.

Springer *et al.* (1975) have examined the effect of methionine starvation on various tumbly mutants of *E. coli*. Whereas wild-type strains stop tumbling soon after removal of methionine, tumbly mutants required much longer periods of methionine starvation before losing their ability to tumble. In general, the time needed to achieve complete loss of tumbling was inversely proportional to the chemotactic ability of the mutant strains. Those with poor chemotactic responses continued tumbling for several hours; one mutant, in fact, never ceased tumbling. The exhaustion of the organisms' internal reserve of methionine available for tumbling was accelerated both by temporal stimuli and by incubating the bacteria in unchanging levels of attractant compounds. Thus, static attractant concentrations, which do not ordinarily affect tumbling behavior, do elicit increased consumption of methionine by the chemotaxis machinery as discussed previously (Section 2.4.1).

Kort (1975) has isolated mutants that remain chemotactic when starved for methionine, which suggests that a normal reserve of internal methionine is not essential for chemotaxis. Perhaps methionine is not directly required for tumbling but rather for adaptation to tumble-suppressing signals (Aswad and Koshland, 1974). This could account for tumbly strains that continue to tumble in the absence of methionine since mutants with severe defects in responding to chemotactic signals should tumble even when the adaptation machinery is not functioning.

Studies by Armstrong (1972b) and by Aswad and Koshland (1975b) showed that *S*-adenosylmethionine (SAM), a compound derived from methionine and ATP, was somehow involved in the methionine effect. Moreover, arsenate, which depletes intracellular ATP levels, causes a progressive loss of tumbling (Larsen *et al.*, 1974b). These results suggest that SAM, not methionine, is the crucial compound. Since SAM is a methyl donor in many reactions, Kort *et al.* (1975) looked for membrane proteins whose methylation pattern was correlated with chemotactic activity. One such protein (MCP) exhibited aberrant methylation levels in various chemotaxis mutants.

Chemotactic stimuli also enhanced the extent of methylation of this protein. Since the MCP methyl groups seemed to turn over with a time course approximating the rate of adaptation to chemotactic stimuli, it is possible that MCP is a component of the tumble generator or adaptation system.

2.6 CONCLUSIONS

Chemotactic behavior in *Escherichia coli* is achieved through modulation, by chemosensors, of the direction of rotation of flagella, the motor organelles. The chemosensors, which detect and transmit information about the chemical environment, are perhaps the best understood components of the chemotactic machinery, and may prove to closely resemble chemosensor devices of higher organisms, in molecular design as well as function.

The primary recognition element of the chemosensor is a protein that is capable of binding to a family of structurally related molecules. The chemoreceptor proteins for most of the attractant sugars have been studied in detail; however, amino acid receptors have not yet been identified. The galactose, maltose and ribose receptors are water-soluble binding proteins that can be removed from the cells by osmotic shock treatments. The receptors for glucose, mannose and most other sugars are binding proteins that are more intimately associated with the cytoplasmic membrane, although they, too appear to be peripheral, rather than integral membrane proteins. In both cases, the binding proteins also function as part of the active transport systems for their various ligands. Genetic studies show that other components of the transport and chemo-reception machinery are not shared. Whether individual binding protein molecules are compartmentalized, or whether they function equally well in both processes, is not known.

There is essentially no biochemical information about the other components of a chemoreceptor; however, they are thought to be involved in signal transduction. Both genetic and physiological evidence indicates that some of these components may be shared by several different receptors. Since different binding proteins, when complexed with ligand, appear to compete for limiting amounts of these shared elements, it is reasonable to suppose that the binding proteins interact with common components only upon binding of ligand. Thus the receptor part of a chemosensor could be physically separate from other chemosensor elements in the absence of ligand.

The remainder of the chemotaxis machinery is even less well understood. A rudimentary 'map' of the information flow from chemosensors to flagella has been constructed from genetic and physiological studies. The present picture is one of converging pathways that eventually reach a common component which could serve as a device for integrating different stimuli. The ultimate target must, of course, be the rotational machinery of the flagella. The overall complexity of the communication network is not known nor have all the elements in any chemosensor-flagella link been convincingly demonstrated. Much of our present ignorance could be eliminated if we only knew the form — electrical, mechanical or chemical — in which this information is transmitted. One approach would be to determine whether flagella act independently or in synchrony; another would be to show whether chemosensors are physically associated with the flagellar basal body.

Another component of interest is the adaptation system that enables bacteria to 'remember' past environments. The adaptation machinery probably operates at a general level where different chemosensor signals are functionally equivalent and additive. Methionine, or more specifically, methylation of a membrane protein, appears to play a central role in the adaptation process.

In summary, the current state of knowledge in the field of bacterial chemotaxis is tantalizingly incomplete, but nevertheless it is now possible to begin to ask meaningful questions about the system. The prospect of some day being able to describe even this simple behavior in molecular terms is an exciting one, but the task is proving to be a difficult challenge. If the machinery of so 'simple' a behavior proves to be as complex as it now appears, what must that of more 'sophisticated' creatures be like?

REFERENCES

Adler, J. (1966), Chemotaxis in bacteria. *Science,* **153**, 708—716.
Adler, J. (1969), Chemoreceptors in bacteria. *Science,* **166**, 1588—1597.
Adler, J. (1973), A method for measuring chemotaxis and use of the method to determine optimum conditions for chemotaxis by *Escherichia coli. J. gen. Microbiol.,* **74**, 77—91.
Adler, J. (1974a), Chemoreception in bacteria. *Antibiotics and Chemotherapy,* **19**, 12—20.
Adler, J. (1974b), Chemotaxis in bacteria. *Hoppe Seylers Z. Physiol. Chem.,* **355**, 105.
Adler, J. (1975), Chemotaxis in bacteria. *Ann Rev. Biochem.,* **44**, 341—356.
Adler, J. (1976), Chemotaxis in bacteria. In: *The Taxes and Tropisms of Microorganisms and Cells.* (M.S. Carlile, ed.), Academic Press, London.
Adler, J. and Dahl, M. (1967), A method for measuring the motility of bacteria and for comparing random and non-random motility. *J. gen. Microbiol.,* **46**, 161—173.
Adler, J. and Epstein, W. (1974), Phosphotransferase-system enzymes as chemoreceptors for certain sugars in *Escherichia coli* chemotaxis. *Proc. natn. Acad. Sci. U.S.A.,* **71**, 2895—2899.
Adler, J., Hazelbauer, G.L. and Dahl, M.M. (1973), Chemotaxis toward sugars in *Escherichia coli. J. Bact.,* **115**, 824—847.
Adler, J. and Tso, W-W. (1974), 'Decision'-making in bacteria: Chemotactic response of *Escherichia coli* to conflicting stimuli. *Science,* **184**, 1292—1294.
Aksamit, R. and Koshland, D.E., Jr. (1972), A ribose-binding protein of *Salmonella typhimurium. Biochem. biophys. Res. Commun.,* **48**, 1348—1353.
Aksamit, R.R. and Koshland, D.E., Jr. (1974), Identification of the ribose-binding protein as the receptor for ribose chemotaxis in *Salmonella typhimurium. Biochemistry,* **13**, 4473—4478.

Aksamit, R.R., Howlett, B.J. and Koshland, D.E., Jr. (1975), Soluble and membrane-bound aspartate-binding activities in *Salmonella typhimurium. J. Bact.*, **123**, 1000–1005.

Anderson, R.A. (1975), Formation of the bacterial flagellar bundle. In: *Swimming and Flying in Nature*, (T.Y.T. Wu, C.J. Brokaw, and C. Brennan, eds.), Vol. 1, Plenum, New York, pp. 45–56.

Anraku, Y. (1968), Transport of sugars and amino acids in Bacteria I: Purification and specificity of the galactose- and leucine-binding proteins. *J. biol. Chem.*, **243**, 3116–3122.

Armstrong, J.B. (1972a), Chemotaxis and methionine metabolism in *Escherichia coli. Can. J. Microbiol.*, **18**, 591–596.

Armstrong, J.B. (1972b), An S-adenosylmethionine requirement for chemotaxis in *Escherichia coli. Can. J. Microbiol.*, **18**, 1695–1701.

Armstrong, J.B. and Adler, J. (1969a), Location of genes for motility and chemotaxis on the *Escherichia coli* genetic map. *J. Bact.*, **97**, 156–161.

Armstrong, J.B. and Adler, J. (1969b), Complementation of nonchemotactic mutants of *Escherichia coli. Genetics*, **61**, 61–66.

Armstrong, J.B., Adler, J. and Dahl, M.M. (1967), Nonchemotactic mutants of *Escherichia coli. J. Bact.*, **93**, 390–398.

Aswad, D. and Koshland, D.E., Jr. (1974), Role of methionine in bacterial chemotaxis. *J. Bact.*, **118**, 640–645.

Aswad, D. and Koshland, D.E., Jr. (1975a), Isolation, characterization and complementation of *Salmonella typhimurium* chemotaxis mutants. *J. mol. Biol.*, **97**, 225–235.

Aswad, D.W. and Koshland, D.E., Jr. (1975b), Evidence for an S-adenosylmethionine requirement in the chemotactic behavior of *Salmonella typhimurium. J. mol. Biol.*, **97**, 207–223.

Bachmann, B.L., Low, K.B. and Taylor, A.L. (1976), Recalibrated linkage map of *Escherichia coli* K12. *Bact., Rev.*, **40**, 116–167.

Berg, H.C. (1971), How to track bacteria. *Rev. Sci. Instrum.*, **42**, 868–871.

Berg, H.C. (1974). Dynamic properties of bacterial flagellar motors. *Nature*, **249**, 77–79.

Berg, H.C. (1975a), Chemotaxis in bacteria. *Ann. Rev. Biophysics Bioengineering*, **4**, 119–136.

Berg, H.C. (1975b), Bacterial behavior. *Nature*, **254**, 389–392.

Berg, H.C. and Anderson, R.A. (1973), Bacteria swim by rotating their flagellar filaments. *Nature*, **245**, 380–382.

Berg, H.C. and Brown, D.A. (1972), Chemotaxis in *Escherichia coli* analyzed by three-dimensional tracking. *Nature*, **239**, 500–504.

Berg, H.C. and Brown, D.A. (1974), Chemotaxis in *Escherichia coli* analyzed by three-dimensional tracking: Addendum. *Antibiotics and Chemotherapy*, **19**, 55–78.

Berg, H.C. and Tedesco, P.M. (1975), Transient response to chemotactic stimuli in *Escherichia coli. Proc. natn. Acad. Sci. U.S.A.*, **72**, 3235–3239.

Boos, W. (1972), Structurally defective galactose-binding protein isolated from a mutant negative in the β-methyl-galactoside transport system of *Escherichia coli. J. biol. Chem.*, **247**, 5414–5424.

Boos, W. (1974), The properties of the galactose-binding protein, the possible chemoreceptor for galactose chemotaxis in *Escherichia coli*. *Antibiotics and Chemotherapy*, **19**, 21–54.

Boos, W. (1974), Bacterial transport. *Ann. Rev. Biochem.*, **43**, 123–146.

Boos, W. and Gordon, A.S. (1971), Transport properties of the galactose-binding protein of *Escherichia coli*: occurrence of two conformational states. *J. biol. Chem.*, **246**, 621.

Boos, W., Gordon, A.S., Hall, R.E. and Price, H.D. (1972), Transport properties of the galactose-binding protein of *Escherichia coli*. Substrate induced conformational change. *J. biol. Chem.*, **247**, 917–924.

Brokaw, C.J. (1958), Chemotaxis of bracken spermatozoids. *J. exp. Biol.*, **35**, 197–212.

Brown, D.A. and Berg, H.C. (1974), Temporal stimulation of chemotaxis in *Escherichia coli. Proc. natn. Acad. Sci. U.S.A.*, **71**, 1388–1392.

Collins, A.L. and Stocker, B.A.D. (1976), *Salmonella typhimurium* mutants generally defective in chemotaxis. *J. Bact.*, **128**, 754–765.

Dahl, J.L. and Hokin, L.E. (1974), The sodium-potassium adenosine-triphosphatase. *Ann. Rev. Biochem.*, **43**, 327–356.

Dahlquist, F.W., Lovely, P. and Koshland, D.E., Jr. (1972), Quantitative analysis of bacterial migration in chemotaxis. *Nature*, **236**, 120–123.

DeJong, M.H., Van der Drift, C. and Vogels, G.D. (1975), Receptors for chemotaxis in *Bacillus subtilis. J. Bact.*, **123**, 824–827.

DePamphilis, M.L. and Adler, J. (1971a), Purification of intact flagella from *Escherichia coli* and *Bacillus subtilis. J. Bact.*, **105**, 376–383.

DePamphilis, M.L. and Adler, J. (1971b), Fine structure and isolation of the hook-basal body complex of flagella from *Escherichia coli* and *Bacillus subtilis*. *J. Bact.*, **105**, 384–395.

DePamphilis, M.L. and Adler, J. (1971c), Attachment of flagellar basal bodies to the cell envelope: specific attachment to the outer, lipopolysaccharide membrane and the cytoplasmic membrane. *J. Bact.*, **105**, 396–407.

Engelmann, T.W. (1883a), *Bacterium photometricum*. Ein Beitrag zur vergleichenden Physiologie des Licht-und Farbensinnes. *Pflugers Archiv für gesammte Physiologie*, 95–124.

Engelmann, T.W. (1883b), Prufung des Diathermanitat einiger Medien mittelst *Bacterium photometricum. Pfugler's Archiv für gesammte Physiologie*, 125–128.

Epstein, W. and Curtis, S.J. (1972), Genetics of the phosphotransferase system. *Role of Membranes in Secretory Processes.* (L. Bolis, R.D. Keynes, W. Wilbrandt, eds.), North Holland, Amsterdam, p. 98–112.

Fraenkel, G.S. and Gunn, D.L. (1940), The orientation of animals: kineses, taxes, and compass reactions. Clarendon Press, Oxford.

Futrelle, R.P. and Berg, H.C. (1972), Specification of gradients used for studies of chemotaxis, *Nature*, **239**, 517–518.

Gunn, D.L. (1975), The meaning of the term 'klinokinesis' *Anim. Behav.*, **23**, 409–412.

Harold, F.M. (1972), Conservation and transformation of energy by bacterial membranes. *Bact. Rev.*, **36**, 172–230.

Harwood, J.P., Cazdar, C., Prasad, C., Peterkofsky, A., Curtis, S.J. and Epstein, W. (1976), Involvement of the glucose enzymes II of the sugar phosphotransferase system in the regulation of adenylate cyclase by glucose in *Escherichia coli*. *J. biol. Chem.*, **251**, 2462–2468.

Hazelbauer, G.L. (1974), Chemoreception in *Escherichia coli*. In: *Transduction Mechanisms in Chemoreception*. (T.M. Poynder, ed.), Information Retrieval Ltd. London, p. 149–157.

Hazelbauer, G.L. (1975a), The maltose chemoreceptor of *Escherichia coli*. *J. Bact.*, **122**, 206–214.

Hazelbauer, G.L. (1975b), The binding of maltose to 'virgin' maltose-binding protein is biphasic. *Eur. J. Biochem.*, **60**, 445–449.

Hazelbauer, G.L. (1975c), Role of the receptor for bacteriphage λ in the functioning of the maltose chemoreceptor of *Escherichia coli*. *J. Bact.*, **124**, 119–126.

Hazelbauer, G.L. and Adler, J. (1971), Role of the galactose-binding protein in chemotaxis of *Escherichia coli* toward galactose. *Nature New Biol.*, **230**, 101–104.

Hazelbauer, G.L., Mesibov, R.E. and Adler, J. (1969), *Escherichia coli* mutants defective in chemotaxis toward specific chemicals. *Proc. natn. Acad. Sci. U.S.A.*, **64**, 1300–1307.

Hilmen, M., Silverman, M. and Simon, M. (1974), The regulation of flagellar formation and function, *J. supramol. Struct.*, **2**, 360–371.

Hilmen, M. and Simon, M. (1976), Motility and the structure of bacterial flagella. In: *Cell Motility* (Goldman, R., Pollard, T. and Rosebaum, J., eds.), book A, pp. 35–45, Cold Spring Harbor Laboratory, Cold Spring Harbor.

Hofnung, M. (1974), Divergent operons and the genetic structure of the maltose B region in *Escherichia coli* K 12. *Genetics*, **76**, 169–184.

Hofnung, M., Hatfield, D. and Schwartz, M. (1974), *malB* region in *Escherichia coli* K12: characterization of new mutations. *J. Bact.*, **117**, 40–47.

Hogg, R.W. and Engelsberg, E. (1969), L-arabinose binding protein from *Escherichia coli* B/r *J. Bact.*, **100**, 423–432.

Kaback, H.R. (1974), Transport studies in bacterial membrane vesicles. *Science*, **186**, 882–892.

Kalckar, H.M. (1971), The periplasmic galactose-binding protein of *Escherichia coli*. *Science*, **174**, 557–565.

Kellerman, O. and Szmelcman, S. (1974), Active transport of maltose in *Escherichia coli* K12: involvement of a 'periplasmic' maltose-binding protein. *Eur. J. Biochem.*, **47**, 139–149.

Kepes, A. and Richarme, G., (1972), Interactions between galactose- and galactose-binding protein of *Escherichia coli*. *Fed. Eur. Biochem. Soc. Proc. 8th Meeting*, Amsterdam (Van den Bergh *et al.*, eds.) North Holland, Amsterdam, **28**, 327.

Kort, E.N. (1975), Information processing: the role of methionine in bacterial chemotaxis, Ph.D. thesis University of Wisconsin, Madison, Wisconsin.

Kort, E.N., Goy, M.F., Larsen, S.H. and Adler, J. (1975), Methylation of a membrane protein involved in bacterial chemotaxis. *Proc. natn. Acad. Sci. U.S.A.*, **72**, 3939–3943.

Koshland, D.E., Jr. (1974), Chemotaxis as a model for sensory systems. *FEBS Letters*, **40**, (suppl), S3–S9.

Koshland, D.E., Jr. (1976), Sensory response in bacteria. *J. Adv. Neurochem.* (in press).

Kundig, W. and Roseman, S. (1971), Sugar transport II. Characterization of constitutive membrane-bound enzymes II of the *Escherichia coli* phosphotransferase system. *J. biol. Chem.*, **246**, 1407–1418.

Larsen, S.H., Reader, R.W., Kort, E.M., Tso, W.W. and Adler, J. (1974a), Change in direction of flagellar rotation is the basis of the chemotactic response in *Escherichia coli. Nature*, **249**, 74–77.

Larsen, S.H., Adler, J., Gargus, J.J. and Hogg, R.W. (1974b), Chemomechanical coupling without ATP: the source of energy for motility and chemotaxis in bacteria. *Proc. natn. Acad. Sci. U.S.A.*, **71**, 1239–1243.

Lengeler, J. (1975a), Mutations affecting transport of the hexitols *D*-mannitol, *D*-glucitol and galacitol in *Escherichia coli* K12: Isolation and mapping. *J. Bact.*, **124**, 26–38.

Lengeler, J. (1975b), Nature and properties of hexitol transport systems in *Escherichia coli. J. Bact.*, **124**, 39–47.

Lovely, P., Dahlquist, F.W., Macnab, R. and Koshland, D.E., Jr. (1974), An instrument for recording the motions of micro-organisms in chemical gradients. *Rev. Sci. Instrum.*, **45**, 683–686.

Macnab, R.M. (1977), Bacterial flagella rotating in bundles: a study in helical geometry. *Proc. natn. Acad. Sci. U.S.A.*, **74**, 221–225.

Macnab, R.M. and Koshland, D.E., Jr. (1972), The gradient-sensing mechanism in bacterial chemotaxis. *Proc. natn. Acad. Sci. U.S.A.*, **69**, 2509–2512.

Macnab, R. and Koshland, D.E., Jr. (1973), Persistence as a concept in the motility of chemotactic bacteria. *J. Mechanochem. Cell Motility*, **2**, 141–148.

Macnab, R. and Koshland, D.E., Jr. (1974), Bacterial motility and chemotaxis: Light-induced tumbling response and visualization of individual flagella. *J. mol. Biol.*, **84**, 399–406.

Macnab, R. and Oruston, M.K. (1977), Polymorphic flagellar transitions in bacterial motility: the induction of discrete changes in polymeric structure by mechanical force. *J. Mol. Biol.*, (In press).

Maeda, K., Imae, Y., Shioi, J-I. and Oosawa, F. (1976), Effect of temperature on motility and chemotaxis of *Escherichia coli. J. Bact.*, **127**, 1039–1046.

McGowan, E.B., Silhavy, T.J. and Boos, W. (1974), Involvement of a tryptophan residue in the binding site of *Escherichia coli* galactose-binding protein. *Biochemistry*, **13**, 993–999.

Mesibov, R. and Adler, J. (1972), Chemotaxis toward amino acids in *Escherichia coli. J. Bact.*, **112**, 315–326.

Mesibov, R., Ordal, G.W. and Adler, J. (1973), The range of attractant concentrations for bacterial chemotaxis and the threshold and size of response over this range. *J. gen. Physiol.*, **62**, 203–223.

Mitchell, P. (1966), Chemiosmotic coupling in oxidative and photosynthetic phosphorylation. *Biol. Rev.*, **41**, 445–502.

Ordal, G.W. (1976a), Recognition sites for chemotactic repellents of *Bacillus subtilis. J. Bact.*, **126**, 72–79.

Ordal, G.W. (1976b), Control of tumbling in bacterial chemotaxis by divalent cation. *J. Bact.*, **126**, 706–111.

Ordal, G.W. (1976c), Effect of methionine on chemotaxis by *Bacillus subtilis*, *J. Bact.*, **125**, 1005–1012.

Ordal, G.W. and Adler, J. (1974a), Isolation and complementation of mutants in galactose taxis and transport, *J. Bact.*, **117**, 509–516.

Ordal, G.W. and Adler, J. (1974b), Properties of mutants in galactose taxis and transport. *J. Bact.*, **117**, 517–526.

Ordal, G.W. and Fields, R.B. (1977), A biochemical mechanism for bacterial chemotaxis. *J. Theor. Biol.*, (In press).

Ordal, G.W. and Gibson, K.J. (1977), Chemotaxis toward amino acids by *Bacillus subtilis*. *J. Bact.*, **129**, 151–155.

Ordal, G.W. and Goldman, D.J. (1975), Chemotaxis away from uncouplers of Oxidative phosphorylation in *Bacillus subtilis*. *Science*, **189**, 802–804.

Ordal, G.W. and Goldman, D.J. (1976), Chemotactic repellents of *Bacillus subtilis*. *J. mol. Biol.*, **100**, 103–108.

Ordal, G.W., Vilani, D.P. and Gibson, K.J. (1977), Amino acid chemoreceptors of *Bacillus subtilis*. *J. Bact.*, **129**, 156–165.

Oxender, D.L. (1972), Membrane transport. *Ann. Rev. Biochem.*, **41**, 777–814.

Parkinson, J.S. (1974), Data processing by the chemotaxis machinery of *Escherichia coli*. *Nature*, **252**, 317–319.

Parkinson, J.S. (1975), Genetics of chemotactic behavior in bacteria. *Cell*, **4**, 183–188.

Parkinson, J.S. (1976), *cheA*, *cheB* and *cheC* genes of *Escherichia coli* and their role in chemotaxis. *J. Bact.*, **126**, 758–770.

Parsons, R.G. and Hogg, R.W. (1974), A comparison of the L-arabinose- and D-galactose-binding proteins of *Escherichia coli* B/r. *J. biol. Chem.*, **249**, 3608–3614.

Pfeffer, W. (1883), Locomotorische Richtungsbewegungen durch chemische Reize. *Berichte der Deutschen Botanischen Gesellschaft* **1**, 524–533.

Pfeffer, W. (1904), *Pflanzenphysiologie*, Vol. 2. Wilhelm Engelman, p. 798–814.

Randall-Hazelbauer, L.L. and Schwartz, M. (1973), Isolation of the bacteriophage lambda receptor from *Escherichia coli*. *J. Bact.*, **116**, 1436–1446.

Rasched, I., Shuman, H. and Boos, W. (1976), The dimer of the *E. coli* galactose-binding protein. *Eur. J. Biochem.* **69**, 545–550.

Richarme, G. and Kepes, A. (1974), Release of glucose from purified galactose-binding protein of *Escherichia coli* upon addition of galactose. *Eur. J. Biochem.* **45**, 129–133.

Robbins, A.R. (1975), Regulation of the *Escherichia coli* methylgalactoside transport system by gene *mglD*. *J. Bact.*, **123**, 69–74.

Robbins, A.R., Guzman, R. and Rotman, B. (1976), Roles of individual *mgl* gene products in the β-methylgalactoside transport system of *Escherichia coli*. K12. *J. biol. Chem.*, **251**, 3112–3116.

Robbins, A.R. and Rotman, B. (1975), Evidence for binding protein-independent substrate translocation by the methyl-galactoside transport system of *Escherichia coli* K12. *Proc. natn. Acad. Sci. U.S.A.*, **72**, 423–427.

Roseman, S. (1972), Carbohydrate transport in bacterial cells, *Metabolic Pathways*, 3rd edn., *Vol Vi, Metabolic Transport*. (L.E. Hokin, ed.), Academic Press, New York.

Rosen, B.P. and Heppel, L.A. (1973), Present status of binding proteins that are released from Gram-negative bacteria by osmotic shock. *Bacterial Membranes and Walls.* (L. Leive, ed.), Marcel Dekker, New York.

Rothert, W. (1901), Beobachtungen und Betrachtungen uber Tactische Reizerscheinungen, *Flora,* **88**, 391–421.

Rotman, B. and Ellis, J.H., Jr. (1972), Antibody-mediated modification of the binding properties of a protein related to galactose transport. *J. Bact.,* **111**, 791–796.

Schleif, R. (1969), An L-arabinose binding protein and arabinose permeation in *Escherichia coli. J. mol. Biol.,* **46**, 185–196.

Schwartz, M., Kellermann, O., Szmelcman, S. and Hazelbauer, G.L. (1976), Further studies of the binding of maltose to the maltose-binding protein of *Escherichia coli. Eur. J. Biochem.,* **71**, 167–170.

Silhavy, T.J. and Boos, W. (1975), The 'hidden ligand' of the galactose-binding protein. *Eur. J. Biochem.,* **54**, 163–167.

Silhavy, T.J., Szmelcman, S., Boos, W. and Schwartz, W. (1975), On the significance of the retention of ligand by protein. *Proc. natn. Acad. Sci. U.S.A.,* **72**, 2120–2124.

Silverman, M. and Simon, M. (1972), Flagellar assembly mutants in *Escherichia coli. J. Bact.,* **112**, 986–993.

Silverman, M. and Simon, M. (1973a), Genetic analysis of flagellar mutants in *Escherichia coli. J. Bact.,* **113**, 105–113.

Silverman, M. and Simon, M. (1973b), Genetic analysis of bacteriophage Mu-induced flagellar mutants in *Escherichia coli. J. Bact.,* **116**, 114–122.

Silverman, M. and Simon, M. (1974), Flagellar rotation and the mechanism of Bacterial motility. *Nature,* **249**, 73–74.

Simoni, R.D. and Postma, P.W. (1975), The energetics of bacterial active transport. *Ann. Rev. Biochem.,* **44**, 523–554.

Springer, M.S., Kort, E.N., Larsen, S.H., Ordal, G.W., Reader, R.W. and Adler, J. (1975), Role of methionine in bacterial chemotaxis; requirement for tumbling and involvement in information processing. *Proc. natn. Acad. Sci. U.S.A.,* **72**, 4640–4644.

Spudich, J.L. and Koshland, D.F., Jr. (1975), Quantitation of the sensory response in bacterial chemotaxis. *Proc. natn. Acad. Sci. U.S.A.,* **72**, 710–713.

Strange, P.G. and Koshland, D.E., Jr. (1976), Receptor interactions in a signalling system: competition between ribose receptor and galactose receptor in the chemotaxis response. *Proc. natn. Acad. Sci. U.S.A.,* **73**, 762–766.

Szmelcman, S. and Adler, J. (1976), Change in membrane potential during bacterial chemotaxis. *Proc. natn. Acad. Sci. U.S.A.,* **73**, 4387–4391.

Szmelcman, S. and Hofnung, M. (1975), Maltose transport in *Escherichia coli* K12: involvement of the bacteriophage λ-receptor. *J. Bact.,* **124**, 112–118.

Szmelcman, S., Schwartz, M., Silhavy, T.J. and Boos, W. (1976), Maltose transport in *Escherichia coli* K12. A comparison of transport kinetics in wild-type and λ-resistant mutants with the dissociation constants of the maltose-binding protein as measured by fluorescence quenching. *Eur. J. Biochem.,* **65**, 13–19.

Tsang, N., Macnab, R. and Koshland, D.E., Jr. (1973), Common mechanism for repellents and attractants in bacterial chemotaxis. *Science,* **181**, 60–63.

Tso, W.W. and Adler, J. (1974), Negative chemotaxis in *Escherichia coli*, *J. Bact.*, **118**, 560–576.

Van der Drift, C. and De Jong, M.H. (1974), Chemotaxis toward amino acids in *Bacillus subtilis*. *Arch. Microbiol.*, **96**, 83–92.

Van der Drift, C., Duiverman, J., Bexkens, H. and Krijnen, A. (1975), Chemotaxis of a motile streptococcus toward sugars and amino acids. *J. Bact.*, **124**, 1142–1147.

Vary, P.S. and Stocker, B.A.D. (1973), Nonsense motility mutants in *Salmonella typhimurium*. *Genetics*, **73**, 229–245.

Warrick, H.M., Taylor, B.L. and Koshland, D.E. Jr. (1977), The chemotactic mechanism of *Salmonella typhimurium*: mapping and characterization of mutants. *J. Bact.*, (in press).

White, R.J. (1970), The role of the phosphoenolpyruvate phosphotransferase system in the transport of N-acetyl-D-glucosamine by *Escherichia coli*. *Biochem. J.*, **118**, 89–92.

Willis, R.C. and Furlong, C.W. (1974), Purification and properties of a ribose-binding protein from *Escherichia coli*. *J. biol. Chem.*, **249**, 6926–6929.

Wilson, O.H. and Holden, J.T. (1969a), Stimulation of arginine transport in osmotically shocked *Escherichia coli* W cells by purified arginine-binding protein fractions. *J. biol. Chem.*, **244**, 2743–2749.

Wilson, O.H. and Holden, J.T. (1969b), Arginine transport and metabolism in osmotically shocked and unshocked cells of *Escherichia coli* W. *J. biol. Chem.*, **244**, 2737–2742.

Zukin, R.S. and Koshland, D.F., Jr. (1976), Mg^{2+}, Ca^{2+}-dependent adenosine triphosphatase as receptor for divalent cations in bacterial sensing. *Science*, **193**, 405–408.

Zukin, R.S., Strange, P.G., Heavy, L.R. and Koshland, D.E., Jr. (1976), Properties of the galactose binding protein of *Salmonella typhimurium* and *Escherichia coli*. *Biochemistry*, **16**, 381–386.

3 Bacterial Receptors for Phages and Colicins as Constituents of Specific Transport Systems

V. BRAUN and K. HANTKE

3.1	Bacterial receptors: an attempt at definition *page*	101
3.2	The outer membrane of *E. coli* and its role as permeability barrier	103
3.3	Receptor systems connected with the *ton*B function	105
	3.3.1 Receptors for bacteriophages T1, T5 and ϕ80: the *ton*A protein	105
	3.3.2 Relationships between bacteriophage receptors, colicin receptors and outer membrane uptake systems. *The ton A protein; ferrichrome uptake and colicin M, 109 Ferric-enterochelin and colicins B, I and V. III Summary of inter-relationships*	109
3.4	Receptor system for bacteriophage lambda and maltose	116
3.5	Receptor system for bacteriophage BF23, vitamin B_{12} and the E colicins	119
3.6	Receptor system for bacteriophage T6, nucleosides and colicin K	119
3.7	Comparative biochemistry of receptor-dependent uptake systems	120
	3.7.1 Specificity	120
	3.7.2 Dependence on molecular size	122
	3.7.3 Regulation	125
3.8	Concluding remarks and generalizations	128
	References	130

Acknowledgments

The authors work was supported by the Deutsche Forschungsgemeinschaft (SFB 76). We thank Dr R.E.W. Hancock for reading the manuscript.

Microbial Interactions
(*Receptors and Recognition*, Series B, Volume 3)
Edited by J.L. Reissig
Published in 1977 by Chapman and Hall, 11 New Fetter Lane, London EC4P 4EE
© Chapman and Hall

3.1 BACTERIAL RECEPTORS: AN ATTEMPT AT DEFINITION

The term receptor is often applied rather loosely to cellular components, usually proteins, which bind biologically active molecules. A stricter definition of cell surface receptors takes into account their role as intermediaries between environmental factors and cellular activity. The receptors act as a cognitive device capable of discrimination between various effectors. Thus one of their properties is selectivity, whilst a second function is transduction of the signal or the effector itself to the target. Membrane receptors have to bridge the membrane barrier in order to transduce information from the cell's environment to the interior of the cell. To do this, they should be asymmetrically constructed and disposed in the membrane with a specific combining site for the effector at the cell surface and a transducing device across the membrane. Do bacteria, in fact, have such receptors? Indeed, elaborate receptor systems facilitate the exchange of genetic material between bacterial cells during mating, transformation or transduction (Achtman, 1973 and Achtman and Skurray, this volume). Another example is provided by chemotaxis. Certain bacteria can detect changes in the concentration of certain chemicals, and respond to them by swimming preferentially either along or against the concentration gradient (Adler, 1975; Koshland, 1974; Berg, 1975; Hazelbauer and Parkinson, this volume). A sensory system (receptors and other molecules) apparently detects temporal gradients and transmits a stimulus to an analysing device which generates a behavioral response.

Bacteria can be killed by phage envelopes, called ghosts, which contain no nucleic acid (Duckworth, 1970). Phage-resistant strains — i.e. strains lacking specific phage receptors — are also resistant to the ghosts of the same phages, indicating that there exists a common binding site at the cell surface for both ghosts and the corresponding intact phage particles. The ghost proteins probably do not enter the cell and so must act via the specific membrane receptors. Colicins constitute another category of lethal agents. These proteins, synthesized by certain bacterial strains, act only on closely related species containing the corresponding colicin-specific receptor protein in their membrane. Different colicins with different modes of action, or certain phages and colicins, can share the same receptor (Hardy, 1975; Davies and Reeves, 1975; Lindberg, 1973). Binding to a common receptor does not imply that the same pathway of lethal activity will be followed: the opposite seems to be true. For example, colicins E2 and E3 bind to the same receptor protein (Sabet and Schnaitman, 1973), yet colicin E2 causes the degradation of cellular DNA, whereas colicin E3 inhibits ribosomal protein synthesis in a way described in detail recently (reviewed by Nomura *et al.*, 1974). Phage binding leads to the release of the nucleic acid from the phage head and its transfer into the cytoplasm of the cell. It is unlikely that the colicin is transferred to its target within the cell envelope or cytoplasm by the same mechanism used for the translocation of the phage nucleic acid. In male *E. coli* cells, appendages called pili serve as binding sites for some small DNA and RNA phages, which in turn transfer their nucleic acids into the cell by an unknown mechanism (Brinton, 1971). The same protein pili are important for mating pair formation between donor and recipient cells, which leads to the transfer of DNA. In *E. coli*, a major outer membrane protein of molecular weight 33 000 (the *con* protein) apparently serves in the recipient cell (F^-) for the attachment of the donor cell. The same protein appears to be the receptor for a phage (Achtman and Skurray, this volume, Skurray *et al.*, 1974; Henning *et al.*, 1976).

One conclusion can be already drawn: in contrast to the common situation in eucaryotic cells, where a single type of receptor serves as binding site for only one kind of effector molecule, in *E. coli* a receptor can apparently bind very different agents. After binding, the effector signals are channelled to distinct cellular targets. This is unlike the situation in eucaryotic cells. Thus, in the case of the extensively studied peptide hormones, binding of the various hormones to their highly selective receptors leads to the stimulation or inhibition of the same target, namely adenylate cyclase (Cuatrecasas 1974). In view of this, is it appropriate to also use the designation of receptor for these *E. coli* outer membrane proteins? Could they not, instead, be simply components of membrane pores which passively allow the passage of nucleic acids or colicins into the cell? Cells which lack one protein are deficient in this specific type of pore and are therefore resistant to phages or colicins, or are unable to form mating pairs. Does, for example, the fixation of colicins occur at a target further inside the cells, rather than at a receptor at the cell's surface? Before we go into the more detailed description of selected 'receptor systems' it is fair to say that in most, although perhaps not in all, cases in which the term receptor has been used, this use appears justified even according to the more strict definition offered at the beginning of this chapter. Binding of phages and colicins to isolated outer membranes or to membrane components has been demonstrated (Lindberg, 1973; Braun and Wolff, 1973; Braun *et al.*, 1973; Bradbeer *et al.*, 1976; Konisky and Lin, 1974; Sabet and Schnaitman, 1971; Schwartz, 1975). Initial binding can thus be dissected from subsequent physiological events.

In cases like those of phages T1, ϕ80 and T5 (Lanni, 1968; Garen, 1954; Hancock and Braun, 1976a) and of colicins E2, E3 and K (Nomura *et al.*, 1974; Jetten and Jetten, 1975), at least two stages of interaction have been demonstrated to occur at the membrane level. For a short interval after initial binding of T1, ϕ80, E2, E3 and K, no detectable physiological damage occurs and the cells can be rescued. In this first stage, the phages are reversibly bound and killing of cells by the colicins can still be prevented by treatment with trypsin. During the second stage, the phages become irreversibly bound and transfer their DNA into the cells. In this same stage, irreversible damage (killing) of the cells by colicins has been initiated. Despite the different nature of the events triggered by the phages and the colicins, the irreversible step requires, as a common element, an energized state of the membrane (Jetten and Jetten, 1975; Hancock and Braun, 1976a). In the cases studied in more detail, the receptor protein has been localized in the outer membrane (Braun *et al.*, 1973; Bradbeer *et al.*, 1976; Konisky and Liu, 1974; Sabet and Schnaitman, 1971), although energizing occurs at the level of the cytoplasmic membrane. One could propose the existence of a physical connection between the receptor and the cytoplasmic membrane, such that the receptor function is influenced by the energy state of that membrane, possibly by the induction of a conformational change. It is equally likely that either a second component on the pathway to the target, or the actual target itself, is regulated by the energy state of the cytoplasmic membrane. The existence of intermediate steps in the transmission of signals to the targets is suggested by studies with colicin-tolerant* mutants, which bind colicin as effectively as colicin-sensitive strains (Zwaig and Luria, 1967; Nomura and Witten, 1967), and by studies with an *E. coli* mutant (*pel*) which adsorbs phage-λ normally but fails to inject its DNA (Scandella and Arber, 1974). Earlier work showed that at $0°C$ the binding of several phages is independent of DNA penetration into the cell. Binding to a specific cell surface component can be differentiated from transfer through the membrane and action on the target. From the above descriptions, there can be no doubt about the existence of cell surface receptors. The subsequent steps in the transfer of macromolecular nucleic acids and colicins into the cell are unknown. It is, however, evident that entry of naked DNA into intact cells is an extremely unlikely event as compared with phage-mediated DNA entry. Binding of the phages to the cell surface not only triggers the release of the phage nucleic acid from the phage capsid, but also causes membrane alterations which facilitate the entry of the nucleic acid (Hantke and Braun, 1974).

3.2 THE OUTER MEMBRANE OF *E. coli* AND ITS ROLE AS PERMEABILITY BARRIER

In order to understand the problems faced by macromolecules, like nucleic acids or colicins, or by small molecules in entering the *E.coli* cell, a short excursion into the present day view of the cell envelope is in order. The term cell envelope is taken to include the inner cytoplasmic membrane, the outer membrane and the rigid layer called the murein or peptidoglycan (Weidel and Pelzer, 1964), localized between the outer and inner membrane. The outer membrane of wild-type cells contains phospholipids in an amount sufficient to cover one half of the membrane bilayer (Smit *et al.*, 1975). A glycolipid, called lipopolysaccharide, is localized exclusively in the outer leaflet of

---

* The following terms are used: resistant strains, for those unable to bind colicins or phages; tolerant strains, for those which bind colicins or phages with no physiological effect; insensitive strains, when a classification into the previous categories has not been undertaken.

the outer membrane (Mühlradt *et al.*, 1974). The outer membrane, in principle, is constructed like other biological membranes. The lipid components potentially form a permeability barrier for hydrophilic molecules. Proteins should serve as valves for the exchange of substances between the cell and its environment. About four so-called major proteins are present in the outer membrane; the number depends to some extent on the strain, the culture conditions and the resolution of the sodium dodecyl sulfate-containing polyacrylamide gel electrophoresis systems used to separate the proteins (Schnaitman, 1974; Garten *et al.*, 1975; Bragg and Hou, 1972; Lugtenberg *et al.*, 1975). The protein composition can be extensively manipulated by mutation, without any harmful effect on the viability of cells under laboratory conditions (Henning and Haller, 1975; Alphen *et al.*, 1976). Limited information exists about the position of some proteins within the outer membrane; e.g. the so-called matrix protein is probably non-covalently bound to the murein and it appears to span the outer membrane because it governs the permeability of this layer in relation to small hydrophilic molecules (Rosenbusch, 1974; Nakae, 1976a,b). The matrix protein also seems to bind phages (Schnaitman *et al.*, 1975; Schmitges and Henning, 1976). The lipoprotein is covalently bound to the murein (Braun, 1975) and also exists detached from the murein layer (Inouye, 1975). No phages or colicins are known which use the lipoprotein as receptor and it is only an immunogen and antigen in mutants lacking sugar residues in the cell surface lipopolysaccharide (Braun *et al.*, 1976a). Thus it is clear that the matrix protein spans the whole outer membrane from the innermost murein to the cell surface, whereas the evidence available for the lipoprotein argues against surface exposure in wild-type cells. Another protein, originally called *con*, now designated *omp*A protein (also p*Omp*A) must also be exposed at the cell surface since it binds phages (K3, Tu II$^+$) and is involved in mating pair formation (Achtman and Skurray, this volume). About thirty different polypeptides, the so-called minor proteins, are also present in the outer membrane (Haller *et al.*, 1975). The term 'minor' is misleading since, under certain growth conditions, the functions of some of these proteins become essential and they are made in quantities similar to the 'major' proteins (see Fig. 3.9). Besides pilin and flagellin, 6 'minor' outer membrane proteins, which serve as phage and/or colicin receptors, have been identified to date (Table 1: *feu*B, *ton*A, *cir, bfe, lam*B, *tsx*). All of them, as well as additional proteins with no known receptor function, are concerned with the specific passage of some small molecules through the outer membrane. Whenever energy is required, it is generated in the cytoplasmic membrane where the electron transport chain, the adenosine triphosphatase and the transport proteins are localized.

With respect to low molecular weight substances, the outer membrane constitutes a permeability barrier for most hydrophilic compounds having a molecular weight of about 600 or upwards (Nakae and Nikaido, 1975). Substances below this critical molecular weight diffuse freely across the outer membrane (but see discussion concerning maltose). For some larger substrates specific pores are present, as will be discussed later.

Small hydrophobic molecules usually diffuse through the outer membrane without being restricted by size. This was shown for a set of antibiotics and dyes, whose permeation properties could be ordered, with few exceptions, according to their degree of hydrophobicity (Nikaido, 1976). The permeability of the outer membrane for hydrophobic molecules increases dramatically in deep rough mutants which lack most of the sugar residues usually encountered in the lipopolysaccharide (*Rd, Re* strains). These mutants often contain 60% less protein and, as compensation, 70% more phospholipids (Koplow and Goldfine, 1974; Ames *et al.*, 1974; Smit *et al.*, 1975). The increase in the amount of phospholipids explains the enhanced solubility and transfer of the hydrophobic substances across the outer membrane. The lack of the hydrophilic sugar residues at the cell surface may also play a role, since this deficiency would facilitate access to the hydrophobic membrane (Nikaido, 1976).

3.3 RECEPTOR SYSTEMS CONNECTED WITH THE *ton*B FUNCTION

3.3.1 Receptors for bacteriophages T1, T5 and ϕ80: the *ton*A protein

In a review of the biochemistry of the bacterial cell envelope (Braun and Hantke, 1974), we have emphasized the usefulness of studying phage–host interactions as models for membrane–ligand interactions in general. In this review, we shall focus on some of the most recent developments which have brought to light functions for phage and colicin receptors not previously envisaged. Infection of *E. coli* cells by phage T5 is especially suitable as an introduction to phage–host interactions because it has been dissected into various steps at the levels of the whole cell, the supramolecular structure of the isolated receptor particle and the receptor protein.

The process of phage T5 infection is interesting because the transfer of the viral deoxyribonucleic acid occurs in a unique way: A piece of DNA, comprising 8% of the total DNA molecule, is transferred first into the cell upon adsorption of the virus. Transfer of the rest of the DNA requires protein synthesis and one can take advantage of this requirement in order to study each of the steps separately (Lanni, 1968). When the infection process is arrested at the first stage of DNA transfer, the virus capsid can be removed from the rest of the DNA simply by centrifugation at 6000 g (Labedan *et al.*, 1973). Electron micrographs (Fig. 3.1) show the naked DNA fixed to the bacterial cell at one end, probably by the DNA piece first injected into the cell. Sometimes the capsid is still attached at the other end of the DNA. The naked 92% piece of DNA is taken up by the cell when protein synthesis is allowed to proceed (Labedan and Legault-Demare, 1973). About 60% of the uncoiled intact DNA molecules can be transferred into the cell and give rise to infectious centers. The phage capsid plays no role in the transfer of the 92% piece.

The first step in the infection process, adsorption of the phage and release of the DNA, has already been studied (1953) in an *in vitro* system. A fragment of the *E. coli* envelope, which exhibited phage-inactivating activity, was extracted with

Fig. 3.1 *E.coli*–phage T5 complex arrested after the first step of DNA-transfer has taken place. The phage head (marked by an arrow) remains attached to the distal end of the phage chromosome. Another chromosome with its attached phage head passes over the former. Magnification × 18 000 (From Labedan *et al.*, 1973).

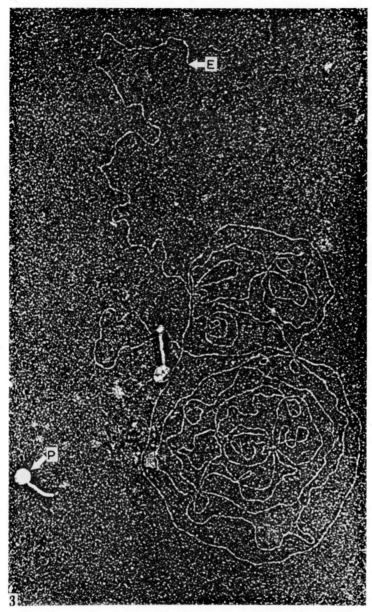

Fig. 3.2 Intact phage T5 (marked by P) and empty phage particle bound to the receptor particle by the distal end of the phage tail (center of the picture), having released its DNA, as visualized by the Kleinschmidt spreading technique. E marks the free end of the DNA (From Frank *et al.*, 1963).

weak alkali (Weidel, 1958). Weidel and Kellenberger (1955) showed by electron microscopy that the phage tail was bound to the receptor particle and that the DNA was completely released from the phage head. A more recent electron micrograph is shown in Fig. 3.2 (Frank *et al.*, 1963). When methods became available to study the composition of the membrane receptor particle, it was found that it contained lipopolysaccharide, phospholipids, and some, but not all, of the proteins of the outer membrane (Braun *et al.*, 1973; Braun, 1974). After separation of the outer and cytoplasmic membrane of *E. coli*, receptor activity was found exclusively in the outer membrane (Braun and Wolff, 1973). The components of the receptor particle stick together very strongly. After considerable effort they were separated under conditions which preserved the phage-inactivating function. Binding ability could be ascribed to a protein with a molar mass of 80 000 daltons (Braun and Wolff, 1973) called *ton*A. From genetic studies of phage-resistant mutants, it has long been known that the structural gene for this protein maps at 3 min on the new genetic map of *E. coli* (Bachmann *et al.*, 1976). After identification of the T5 receptor protein, it was found that nearly all T5-resistant mutants of *E. coli* are deficient for it (Braun *et al.*, 1976b).

Although the virus—cell interaction has been reduced to the binding of the phage to the receptor protein, no experimental data exist to show how phage binding triggers the release of the phage nucleic acid, or how it opens the way for the transfer of the DNA in both the outer membrane and the cytoplasmic membrane. We assume that structural transitions occur in both reaction partners. The situation becomes even more complicated when we consider the dependency of the *ton*A function on the *ton*B function in the infection of the phages T1 and ϕ80. The gene was designated *ton*A, *ton* being a mnemonic for T-one resistance. *Ton*B is required for the irreversible binding of the phages T1 and ϕ80. Both these phages bind only reversibly to *ton*B mutant cells (Garen, 1954) and to the isolated *ton*A receptor protein (Hancock and Braun, 1976a). The irreversible binding and subsequent infection requires energy generated either by ATP hydrolysis or by the electron transport chain, as shown by the use of mutants which lack either a functional adenosine triphosphatase or cytochromes (Hancock and Braun, 1976a). Likewise, T5 adsorption and the transfer of the first 8% DNA piece requires only the *ton*A protein, but then energy is required for the expression of functions, coded by the first transfer DNA piece with the help of the cell, for the transfer of the remaining T5 DNA. With both types of phage, DNA transfer into the cell seems to be an energy-consuming process and the victim provides the energy and other functions required to kill it. Phage DNA is 10 times longer than the bacterial cell, and the mechanism by which the nucleic acid is transferred into the cell remains unknown, not only for the T5, T1 and ϕ80 phages described, but for all phages.

3.3.2 Relationships between bacteriophage receptors, colicin receptors and outer membrane uptake systems

In the last three years, it has become apparent that the phage and colicin receptors located in the outer membrane of *E. coli* are also involved in the uptake of certain low molecular weight substrates into the cell. In 1973, while using colicin E1 as a tool to inhibit the energy-dependent transport of vitamin B_{12}, it was found that this colicin also prevents the initial binding of vitamin B_{12} to its specific receptor in the outer membrane of the *E. coli* envelope (Di Masi *et al.*, 1973). Although this was, at first, probably considered by many people as a curiosity, other phage and colicin receptors were soon shown to be constituents of uptake systems (Hantke and Braun, 1975a,b; Szmelcman and Hofnung, 1975; Hantke, 1976). To date, five surface receptor-dependent uptake systems have been identified. These systems will be described in the following sections.

(a) *The tonA protein; ferrichrome uptake and colicin M*

Two observations led to the discovery of the requirement of the *tonA/tonB* functions for the uptake of ferric iron as ferrichrome complex. Both were first made in *Salmonella typhimurium* for which no outer membrane receptor proteins were known. Therefore the involvement of proteins in the translocation of certain substrates across the outer membrane could not be detected until studies were undertaken with *E. coli*. First, in *S. typhimurium* the *car* mutants (for carbohydrate utilisation, now called *chr*), which map near the tryptophan operon (*trp*) at 52 min (Sanderson, 1972), were found to be unable to grow on agar plates with different carbohydrates as the carbon source unless citrate was also supplied. These mutants grew perfectly well on the same carbohydrates in liquid media. Chromium (Cr^{3+}) was found to be the inhibiting agent in the agar (Corwin *et al.*, 1966). The same phenomenon was observed in *E. coli* mutants with deletions extending from *tonB* into the *trp* region (at 27 min on the recalibrated *E. coli* map, Bachman *et al.*, 1976). Further studies of Wang and Newton (1971) revealed that Cr^{3+} forms a co-polymer with Fe^{3+} which significantly lowers the concentration of free iron in the medium. In addition, these authors showed that *tonB* mutants have a ten-fold higher apparent Michaelis–Menten constant (K_m) for iron uptake than the wild-type (Wang and Newton, 1971). This makes the mutants very sensitive to low iron concentrations. The chromium-dependent polymerization of Fe^{3+} is reverted by complexing agents like citrate which explains the citrate dependence of *tonB* mutants on chromium-containing agar plates. The second observation was that *S. typhimurium* effectively takes up iron supplied as ferrichrome complex (Luckey *et al.*, 1972). *Salmonella* as well as other Enterobacteriaceae are unable to produce the iron-complexing agent desferri-ferrichrome (Fig. 3.3) which belongs to the hydroxamate-type siderochromes (sideramines) commonly synthesized by fungi. Mutants in ferrichrome uptake could easily be isolated using the antibiotic albomycin δ_2, which has a structure similar to ferrichrome (Fig. 3.3). Albomycin-resistant mutants were called *sid* (from *sid*erochrome) and

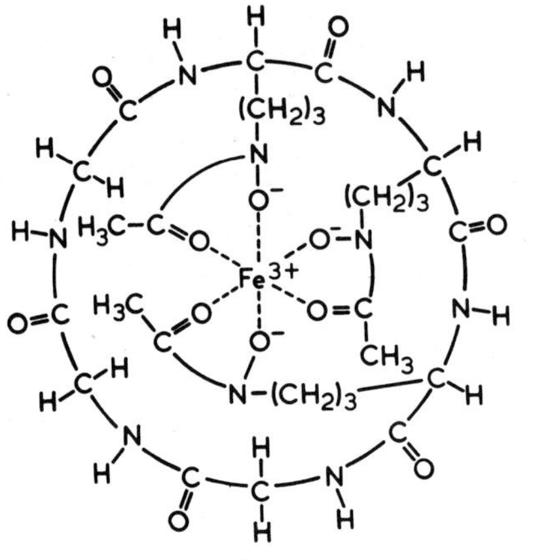

Fig. 3.3 Structure of ferrichrome and albomycin. Detailed information concerning these structures and other siderchromes (sideramines) can be drawn from papers by Luckey et al. (1972) and Diekmann (1973).

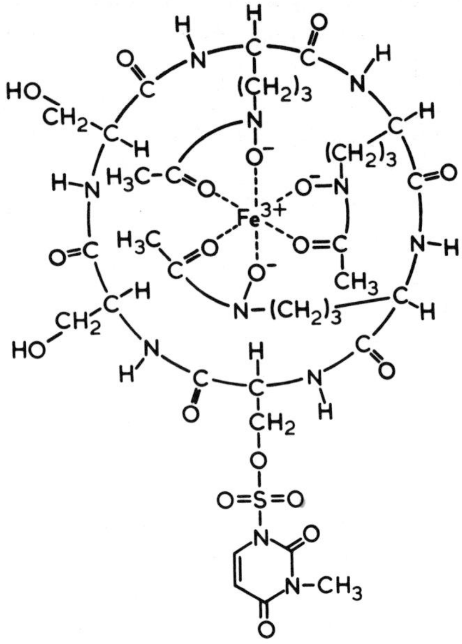

most of them mapped near 4 min on the genetic map of *S. typhimurium*. This locus is close to the *ton*A position on the genetic map of *E. coli*.

In the course of our studies on the T5 receptor protein (*ton*A gene) (Braun et al., 1973; Braun and Wolff, 1973) in *E. coli*, we looked for a function for the T5 receptor beneficial for the cell. We were guided by the new discovery of the correlation between the receptor for the E-colicins and vitamin B_{12} transport and by the equivalent map position of most of the *sid* mutations in *Salmonella* relative to the *ton*A gene in *E. coli*. It was of major relevance that the *ton*B function had been related to iron transport (Wang and Newton, 1973) and to T1 infection, the latter in co-operation with the *ton*A receptor protein. We therefore compared T5, T1 and albomycin-resistant *E. coli* mutants and studied the uptake of iron complexed with ferrichrome (Hantke and Braun, 1975a). All T5- and T1-resistant *E. coli* mutants obtained were also found to be albomycin resistant and unable to take up iron in form of the ferrichrome complex. T5-resistant mutants (*ton*A gene) were impaired in ferrichrome-iron uptake, but not in citrate or enterochelin iron transport. Competition experiments point to the *ton*A protein as a common receptor: ferrichrome prevents inhibition of cell growth by albomycin (Zähner et al., 1960; Alexanian et al., 1972), plaque formation of phage ϕ80 on sensitive cells (Luckey et al., 1975; Wayne and

Neilands, 1975) and colicin M action (Hantke and Braun, 1975a). Competition could also be shown to occur between colicin M and T5 for the common receptor (Braun et al., 1973). Equilibrium dialysis experiments with isolated outer membranes of T5-sensitive cells revealed binding of ferrichrome with an apparent K_m of 50 nM. No binding was observed with outer membranes of tonA mutants lacking the tonA protein (K. Hantke, unpublished). Since we had previously identified the tonA protein (Braun and Wolff, 1973), we were able to compare outer membrane protein patterns of wild-type and tonA mutants by SDS-gel electrophoresis. Out of 25 tonA mutants studied, 24 had lost the tonA protein. The evidence that the tonA protein serves as receptor for the phages T5, T1 and ϕ80, the colicin M, ferrichrome and albomycin is therefore convincing. When ferrichrome occupies the binding site, the cells cannot be killed by the phages, the colicin or the antibiotic. The relationship between the tonA and tonB functions will be discussed later. They are illustrated in Fig. 3.7.

The following questions remain unanswered. How did the highly specialized ferrichrome uptake system evolve in cells unable to synthesize desferri-ferrichrome? How and where is the iron mobilized from the very stable ferrichrome complex? What is the molecular mechanism of ferrichrome transport?

Desferri-ferrichrome can be used repeatedly to transfer iron into the cell (Braun et al., 1976b), therefore the ligand is not structurally altered during the transport. Leong and Neilands (1976) reached the same conclusion for *E. coli*. Either the ligand is not taken up, or it is expelled rapidly after release of the iron. It is quite possible that the redox potential of the respiratory chain in the cytoplasmic membrane is used for the release of the Fe^{3+} as Fe^{2+} from the complex.

S. typhimurium is resistant to phage T5. However, plaque formation of a general transducing phage of *S. typhimurium*, ES18, could be inhibited by about 3 μM ferrichrome. It was also demonstrated that a *sid* gene specifies a receptor for ferrichrome utilization and adsorption of phage ES18 (Luckey and Neilands, 1976). We have also isolated albomycin-resistant mutants, some of which are ES18 resistant. The protein profiles of these mutants are analogous to those of the tonA or tonB mutants of *E. coli*. In Fig. 3.4, rows 4 and 5 show the outer membrane proteins of two independently isolated *S. typhimurium* mutants lacking a protein band which corresponds with the tonA protein of *E. coli* (row 6). The protein profile shown in row 1 is similar to the protein pattern seen in tonB, mutants of *E. coli* (Braun et al., 1976b). In tonB mutants, iron shortage in the cells induces three outer membrane proteins which we designated, according to their molecular weight, as 74 K, 81 K, 83 K (K = Kilo). These observations suggest that *S. typhimurium* also contains a gene (region) equivalent to tonB of *E. coli* and which may be identical with the *chr* locus mentioned before.

(b) *Ferric-enterochelin and colicins B, I and V*

Fredericq and Gratia (1960) found that T1-resistant tonB mutants are also insensitive to colicins B, I and V. Wang and Newton (1969) showed that tonB mutants

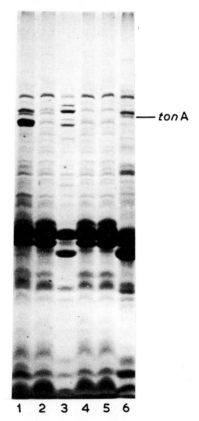

Fig. 3.4 Gel electropherogram of the outer membrane proteins of
S. typhimurium and *E. coli*. Row 1, *S. typhimurium* SL 1027/24 is an
albomycin- and phage ES 18-resistant derivative of the wild-type SL 1027
shown in row 2. Although grown under the same conditions as the wild-type,
the proteins in the 70–80 K molecular weight region are markedly induced in
the mutant (*ton*B-like pattern). Row 3 shows the protein pattern of *E. coli* F 470,
a T5-resistant strain which lacks the *ton*A protein. The equivalent mutant of
S. typhimurium SL 1027/1, resistant against albomycin and phage ES 18,
also lacks one protein band in the molecular weight region of the *E. coli*
*ton*A protein (row 4). Another independently isolated mutant, SL 1027/23,
of the same phenotype, is shown in row 5. For comparison, the protein
profile of an *E. coli* wild-type K12 strain is shown in row 6. The outer
membranes were prepared according to the method of Osborn *et al.* (1972),
and run on sodium dodecyl sulfate polyacrylamide gels as described by
Lugtenberg *et al.* (1975).

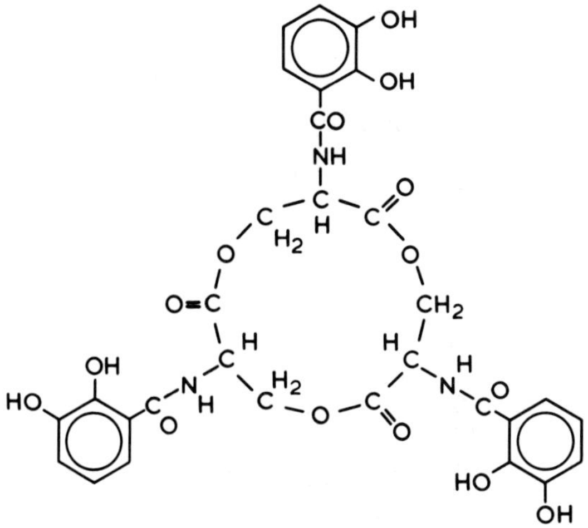

Fig. 3.5 Structure of enterochelin. Enterochelin is a cyclic trimer of 2, 3-dihydroxy-*N*-benzoyl-*l*-serine (O' Brien and Gibson (1970), Pollack *et al.* (1970)).

hyperexcreted dihydroxybenzoyl serine (DBS), which later proved to be the degradation product of the iron chelator enterochelin (Fig. 3.5) which is synthesized . by *E. coli* under low iron conditions (O'Brien and Gibson, 1970). Enterochelin is identical with enterobactin produced by *S. typhimurium* (Pollack and Neilands, 1970; see review of Rosenberg and Young, 1974).Guterman (1971, 1973) identified enterochelin as the colicin antagonist, excreted by colicin B-insensitive *E. coli* strains, which protects sensitive cells from colicin B action. One subgroup of these colicin B-insensitive mutants, *exb*A, (probably originally meant *ex*cretion of a colicin *B* inhibiting substance) is identical with, or maps near to, the *ton*B mutants (Guterman and Dann, 1973; compare Gratia, 1964). The other subgroup, *exb*B, requires methionine and maps close to *ser*A. The *exb* mutants are largely insensitive to colicin B even in the absence of enterochelin. For example, the wild-type strain was inhibited by a 1 : 4096 dilution of the original colicin B solution, whereas inhibition of the *exb* derivative required a 32-fold higher colicin B concentration. Enterochelin protected *exb* mutants against colicin B. Since tolerance and enterochelin protection are superimposed in the *exb* mutants, it was difficult to trace the protection of enterochelin. As will be shown below, our studies on colicin B resistant strains show clearly that enterochelin competes for the same binding site in the outer membrane of *E. coli* at which colicin B adsorbs. Enterochelin protects cells against colicin B by preventing colicin B adsorption. The colicin B-resistant strains we studied lack an outer membrane protein. They are unable to bind enterochelin and thus cannot transport iron in the form of the ferric-enterochelin complex (Hancock *et al.*, 1976).

In addition to the colicin B tolerant *exb*A (*ton*B), *exb*B mutants, *exb*C and *cbt* (colicin B-tolerant) mutants (Pugsley and Reeves, 1976a) have been studied. Interestingly they are all impaired in ferric-enterochelin uptake. Translocation of colicin B and ferric-enterochelin across the outer membrane involved common elements in addition to the adsorption process. The defects of the tolerant mutants have not been identified biochemically. This may become a difficult task since tolerance phenomena are usually of pleiotrophic nature, albeit of differing degrees (Zwaig and Luria, 1967; Nomura and Witten, 1967). The complex relationship of a single gene product with various membrane functions can be appreciated with *ton*B mutants. They are insensitive to the phages T1 and φ80, the colicins B, I and V and are unable to take up iron via the known active transport systems in complex with ferrichrome, enterochelin and citrate. The *exb* and *cbt* functions have been shown to exhibit tolerance against B-type colicins and impaired uptake of ferric-enterochelin. The scope of the *ton*B gene seems to be much broader, but a note of caution is appropriate here since the mapping of most mutants studied has not been precise enough to assure that they involve single-gene differences.

Under conditions of iron limitation, the colicin-tolerant mutants overproduce the same proteins as the wild-type (Pugsley and Reeves, 1976a, b). The increased amount of proteins correlates with enhanced binding activity for colicin B and enterochelin.

Following the lead provided by the observation that enterochelin protects sensitive cells against colicin B, we isolated true resistant mutants as opposed to tolerant or target site mutants, which were unaffected by colicin B even in the absence of enterochelin (Hantke and Braun, 1975b). We called these mutants *feu* since they were impaired in *f*erric-*e*nterochelin *u*ptake. Later we classified them according to their different resistance patterns into two subsets (*feu*A and *feu*B); *feu*A mutants being resistant to colicin I and *feu*B mutants being resistant to colicin B (Hancock and Braun, 1976b; Hancock *et al.*, 1976). The outer membrane of *feu*A mutants lacks a protein of mol. wt. 74 000, while that of *feu*B mutants lacks a protein of mol. wt. 81 000 (Fig. 3.6, rows 2–4). *Feu*B mutants lack the colicin B receptor protein and show a markedly reduced capacity to transport ferric-enterochelin. Pugsley and Reeves (1976c) reported the same observation with their *cbr* mutants devoid of a functional *c*olicin *B r*eceptor. However, they claimed that their mutants still contained the receptor protein. Two other proteins have electrophoretic mobilities similar to the colicin B receptor, the citrate-induced protein and the 83 K protein (Hancock *et al.*, 1976, Braun *et al.*, 1976b). The Pugsley and Reeves electrophoretic system does not separate proteins with closely similar mobilities, as ours does. Conceivably, their strains lacking colicin B receptor activity do lack the protein (as do our strains), but Pugsley and Reeves may not have detected the protein's absence because of interference from other proteins in the same region of the gel.

Mutants in the *feu*B gene are defective in ferric-enterochelin uptake. We originally surmised the same for *feu*A mutants, but after rechecking we have discovered that this does not hold true. Although the synthesis of the colicin I receptor protein

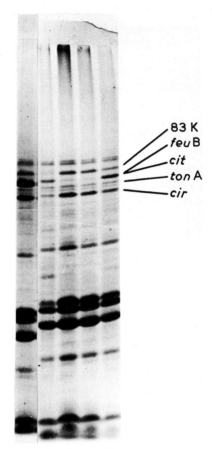

Fig. 3.6 Gel electropherogram of outer membrane proteins of *E. coli*. In row 1 is shown a *ton*B mutant of *E. coli* K12 AB 2847 in which the proteins 83 K, *feu*B, *cit* and *cir* are induced (for a key to protein designations, see Fig. 3.7). In the *feu*B mutant one can observe the induction of the *cit* protein by growth in the presence of citrate (row 2), the other proteins are only present in low amounts due to sufficient iron being supplied by the citrate complex. When the *feu*B mutant is grown in the absence of citrate, neither *cit* nor *feu*B proteins are synthesized, and the 83 K protein and the *cir* protein are induced (row 3). The same phenomenon is shown in row 4 with another *feu*B mutant grown without added citrate. Induction of the proteins due to iron shortage is also shown by wild-type *E. coli* K12 grown under conditions of iron limitation (row 5).

(*feu*A gene) is induced by iron shortage, its relationship to the known iron chelate transport systems is not clear (Hancock and Braun, 1976b, Konisky *et al.*, 1976). Ferric-enterochelin protects sensitive cells against colicin I (Wayne *et al.*, 1976), but protection against colicin I requires a 3000-fold higher concentration than protection against colicin B (Pugsley and Reeves, 1976b). The *exb*B mutants described above are completely insensitive to colicin I in the presence of ferric-enterochelin, but partially sensitive in its absence. Since this protective effect is confined to ferric-enterochelin, and neither enterochelin nor any other iron complex tested has such an effect, a relationship must exist between colicin I action and ferric-enterochelin. This relationship has yet to be elucidated. Since *feu*A mutants are not defective in ferric-enterochelin transport but are colicin I resistant, the term *feu*A should be dropped in favor of *cir* (colicin *I r*esistant) (Konisky and Lin, 1974).

(c) *Summary of inter-relationships*
The present state of knowledge on the receptor-dependent iron uptake in *E. coli* is schematically diagrammed in Fig. 3.7. The *exb*B, *exb*C and *cbt* functions have been left out since little is known about them at the biochemical level. The localization of the *ton*B function is also rather arbitrary and only reflects that it is required for all the translocation processes except T5 DNA entry. In *aro*B mutants unable to synthesize dihydroxybenzoate (DHB), the *ton*B character can be circumvented by the addition of low amounts (5 μM) of DBH but not by equivalent amounts of enterochelin. Higher concentrations of DHB inhibit growth (Frost and Rosenberg, 1975) and iron transport (Braun *et al.*, 1976b). Ferric-enterochelin uptake is rather low in *ton*B mutants, but the enterochelin produced inside the cell seems to bypass the *ton*B function by transferring iron adsorbed at the membrane to the membrane permease (*fep* function), which translocates the ferric-enterochelin complex across the cytoplasmic membrane. Higher concentrations of DHB lead to synthesis of enterochelin in large amounts and to its excretion into the medium, where it chelates the iron. The ferric-enterochelin complex is then unable to overcome the outer membrane barrier without the *ton*B function (Frost and Rosenberg, 1975). The *cit* protein is induced by growth of *E. coli* K12 in the presence of citrate and has been localized in the outer membrane (Hancock *et al.*, 1976). Since a citrate-inducible iron uptake system exists in *E. coli*, the *cit* protein may be required for the translocation of the high molecular iron–citrate complex across the outer membrane.

3.4 RECEPTOR SYSTEM FOR BACTERIOPHAGE LAMBDA AND MALTOSE

The location of the structural gene of the lambda phage receptor in the *mal*B operon, coding for maltose transport functions, was puzzling until it was found that the lambda receptor protein is also involved in maltose uptake (Szmelcman and Hofnung, 1975). *Lam*B mutants affected at the receptor gene are unable to utilize maltose as a

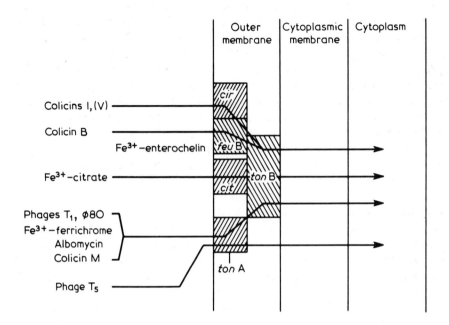

Fig. 3.7 Present model of the *E. coli* outer membrane proteins involved in the uptake of ferric iron. The lack of the *cir* protein confers colicin *I* resistance. This protein is induced by iron shortage, but no uptake of an iron complex has been related so far to the *cir* protein. Mutants missing the *feu*B protein are deficient in *ferric-enterochelin uptake*, and resistant to colicin *B*. The *cit* protein is induced by growth of *E. coli* in the presence of citrate. The *ton*A (T-one resistance) protein is necessary for ferrichrome uptake. The *ton*B function is not identified biochemically and is involved in all the processes as indicated. The picture is only meant to suggest inter-relations, rather than actual localization in a physical sense. The *cir*, *feu*B, *cit* and *ton*A proteins have been localized in the outer membrane (see e.g. Fig. 3.6).

carbon source when it is present at low concentrations ($< 10 \mu M$). When cells are treated with antibodies against the receptor protein, the transport of maltose at low concentrations is reduced. With increasing amounts of maltose in the medium, the effect of mutations in the *lam*B gene becomes progressively smaller until it is completely abolished at a maltose concentration of 1 mM (Szmelcman and Hofnung, 1975). Three additional gene products, those of genes *mal*E, *mal*F and *mal*K, among them the maltose binding protein, are essential for maltose transport (Kellermann and Szmelcman, 1974). In cells with a functional lambda receptor, the K_m for maltose transport was found to be 1 μM, which is identical to the K_d of binding to the binding protein (Szmelcman *et al.*, 1976). Mutations in the receptor for phage lambda increase the K_m for maltose transport by a factor of 100–500. The maximal rate of transport at saturating maltose concentrations is the same as in the wild-type. Thus, in the presence of the lambda receptor the access of maltose to the maltose-binding protein is not limited, but in receptor-deficient strains maltose diffusion through the outer membrane becomes the rate-limiting step for transport at low maltose concentrations. Mutations in *lam*B that increase the K_m of maltose transport also result in an apparent decrease in the sensitivity of the maltose chemoreceptor (Hazelbauer, 1975). This effect can be explained as the result of decreased access of low concentrations of maltose to maltose-binding protein, its chemoreceptor, because of defective passage of the sugar through the outer membrane. It is not necessary to postulate any direct function of the lambda receptor in maltose chemoreception.

When testing with maltotriose, the effect of the lambda receptor is more pronounced than with maltose. Strains without receptors do not transport maltotriose (Szmelcman *et al.*, 1976). In the presence of lambda receptor, the K_m of maltotriose transport (2 μM) is 13-fold higher than the K_d of binding to the binding protein, suggesting that diffusion through the outer membrane is the rate-limiting step. Since reciprocal plots of the initial rate of uptake versus substrate concentration are linear over a large concentration range, and no binding of maltose or maltotriose to the receptor could be demonstrated, it is apparent that the lambda receptor is part of a passive pore for maltose and higher homologs. If it should be found that this receptor only functions for sugars structurally similar to maltose, then explanations involving interaction of the substrates with the receptor protein would be required. It is possible that special measurements will be necessary to demonstrate substrate binding. It is interesting that irreversible phage binding to the isolated receptor requires peculiar experimental conditions (Randall-Hazelbauer and Schwartz, 1973). Ethanol or chloroform have to be added to the assay mixture in order to obtain inactivation of the phage by the receptor. However, host range mutants such as λho inactivate wild-type receptor (R). Interestingly enough, receptor (R⁺) isolated from certain bacterial strains, e.g. *E. coli* 6371, is able to inactivate wild-type lambda without addition of the organic solvents. R receptor protects lambda phage against R⁺ receptor inactivation, suggesting that the R receptor adsorbs phage by a freely reversible reaction (Schwartz, 1975). The lambda receptor—maltose transport system offers several advantages for the investigator, when compared with the other known receptor/transport systems. The amount of receptor protein can be regulated by maltose. When cells are grown without maltose, most cells contain less than 30 copies of receptor protein. Upon induction by maltose, 6000 active receptor molecules are manufactured per cell (Schwartz, 1976; but see discussion in Section 3.7.3). The protein can be readily isolated in pure form (Randall-Hazelbauer and Schwartz, 1973). Apparently, the other components of the outer membrane, especially proteins and lipopolysaccharide, are not so tightly attached to the lambda receptor as, for instance, in the case of phage T5 and E-colicin receptors. In *lam*B missense mutants, the initial rate of maltose transport is higher than in nonsense mutants (Szmelcman and Hofnung, 1975), which allows a detailed analysis of the structural requirements for phage binding and maltose passage through the outer membrane.

3.5 RECEPTOR SYSTEM FOR BACTERIOPHAGE BF23, VITAMIN B_{12} AND THE E COLICINS

The uptake of vitamin B_{12} (cyanocobalamin) by whole cells of *Escherichia coli* is a biphasic process consisting of an initial rapid phase which is independent of the energy metabolism of the cell, followed by a slower energy-dependent step (Di Girolamo and Bradbeer, 1971). The sites for the initial binding of vitamin B_{12} have been localized in the outer membrane (White et al., 1973), numbering about 200 per cell. Prompted by the observation that colicin E1 inhibits not only the energy-coupled second phase of vitamin B_{12} transport, but also intial binding, Di Masi et al., (1973) found that the vitamin B_{12} binding site is identical with the receptor for the E colicins and phage BF23. Vitamin B_{12}, colicins E1 and E3 and phage BF23 were shown to compete for the same binding sites on the cell surface (Di Masi et al., 1973, Bradbeer et al., 1976). Vitamin B_{12} protects sensitive cells against colicins E1, E3 and phage BF23. A direct interaction between the E colicins, phage BF23 and vitamin B_{12} has been excluded. Mutual exclusion between the E colicins, phage BF23 and vitamin B_{12} for binding to isolated outer membranes has been demonstrated. Colicin E- and phage BF23-resistant mutants are deficient in the initial step in vitamin B_{12} binding. In such mutants uptake of vitamin B_{12}, measured at low substrate concentration (4 nM), was less than 5% of the wild-type level for either the initial or the secondary phase of uptake. Genetic studies suggest the identity of gene *btu*B, the locus controlling the initial phase of B_{12} uptake (Kadner and Liggins, 1973), and *bfe*, the locus specifying the receptor for the three E colicins and for the bacteriophage BF23 (Buxton, 1971). The utilization of the same receptor by vitamin B_{12}, bacteriophage BF23 and the E colicins does not necessarily mean a complete identity of the binding site. Sabet and Schnaitman (1973) have obtained some evidence that colicin E1 requires additional constituents for binding, which are not required by colicin E3. Using rough mutants of *S. typhimurium*, which are able to absorb and propagate BF23, the gene *bfe* could be studied. It maps in the same position as in *E. coli* (Guterman et al. 1975; Mojica-a and Garcia, 1976). However, the *S. typhimurium* studies are handicapped by the fact that this organism is tolerant to the E colicins.

As in the case of the other receptor-dependent outer membrane translocation systems discussed in this paper, the need for a vitamin B_{12} specific channel through the outer membrane is understandable on the basis of the size of this vitamin, which has a molecular weight of 1355.

3.6 RECEPTOR SYSTEM FOR BACTERIOPHAGE T6, NUCLEOSIDES AND COLICIN K

In 1949, it was shown that strains sensitive to colicin K were also sensitive to phage T6 (Fredericq, 1949). Mutants selected as resistant to one of these agents were very often cross-resistant to both, suggesting a common receptor. By immunological (Michael, 1968) and biochemical methods (Weltzien and Jesaitis, 1971) it was shown that at least part of the receptor is proteinaceous. After degradation with pronase the receptor activity for colicin K was lost, while the receptors for phage T6 were still active. With amino acid reagents specific for tryptophan, the receptor activity for both phage T6 and colicin K was destroyed. These results may provide a basis for the further characterization of this receptor protein.

In T6 resistant mutants (*tsx*) an outer membrane protein of about 25 000 daltons is missing (Mannings and Reeves, 1976; Hantke, 1976). It was shown that this protein is an important component for nucleoside transport in whole cells (Hantke, 1976). Several specialized transport systems for nucleosides have been described which seem to be regulated by the *cyt*R and *deo*R gene loci, and which are under the control of catabolite repression. The thymidine transport system is the only one which is not repressed by glucose (McKeown *et al.*, 1976). The uptake of the following nucleosides was lowered in *tsx* mutants as compared to the wild-type: thymidine, uridine, adenosine and deoxyadenosine. This low specificity is unusual as compared with that of the nucleoside transport system localized in the cytoplasmic membrane (von Dippe *et al.*, 1973).

3.7 COMPARATIVE BIOCHEMISTRY OF RECEPTOR-DEPENDENT UPTAKE SYSTEMS

Certain proteins, located at the cell surface of *Escherichia coli* and which were originally characterized as receptors for killing agents like bacteriophages and colicins, have a beneficial function for the cell in that they serve to take up substrates which are present in low amounts in the medium. For the description of their functions we would like to coin the term *receptor-facilitated transport*. A thorough comparison of the various aspects of all receptor-dependent uptake systems is at present hampered by the fact that they have been detected only quite recently and therefore have not generally been studied in detail. However the following common features become apparent:

(1) the receptors seem to be very specific;
(2) the receptors seem to be needed mainly for substrates present in low concentrations or having molecular weights greater than 700;
(3) the receptor requirement can be overcome by supplying high concentrations of the substrate to the medium;
(4) the number of receptor molecules in the outer membrane is regulated by the substrate.

3.7.1 Specificity

Substrate specificity has been most extensively studied for the iron uptake systems. For the ferrichrome, enterochelin and citrate iron complexes, different receptors are present which do not cross-react with the various chelates (Braun et al., 1976b). Among the iron(III)-trihydroxamate type chelates (Fig. 3.3), only ferrichrome analogs like ferricrocin and the iron-substituted aluminium- and chromium-desferrichromes compete with phage T5 for binding to the partially purified receptor (Luckey et al., 1975). Rhodotorulic acid, Desferal and ferrichrome A are ineffective. A serious problem of specificity arises when one considers the very different agents which apparently bind to one receptor protein. That binding really takes place was shown for ferrichrome, phages T5, T1 and ϕ80 and colicin M using isolated outer membranes or partially purified receptor preparations (Weidel, 1958; Braun et al., 1973; Hancock and Braun, 1976a; Luckey et al., 1975). Outer membranes from T5-resistant mutants are unable to bind any of the five agents. Nearly all of these mutants have lost the T5 receptor protein (Braun et al., 1976b). However one mutant which retains the protein, has lost ferrichrome uptake and is, in addition, resistant against all the killing agents, phages T5, T1 and ϕ80, colicin M and the antibiotic albomycin (Braun et al., 1976b). Since the analysis of more than 50 mutants isolated as T5-resistant revealed in every case a total deficiency of receptor-binding activity towards all five agents, we undertook a study of revertants to see whether any among them had regained the wild-type phenotype for only some of the agents. To date, most revertants have wild-type properties in all respects, although a few function partially towards all 5 agents. This behavior is different from that of the lambda receptor system, where most lambda-resistant missense mutants transport maltose much better and grow faster on maltotriose than nonsense mutants (Szmelcman and Hofnung, 1975; Szmelcman et al., 1976; Braun and Krieger, unpublished). In addition we found that maltose or maltotriose do not prevent binding of lambda to cells, isolated outer membranes or purified receptor protein. Maltose and lambda apparently have different binding sites on the receptor protein. We were also unable to show inhibition of T6 binding by high concentrations of nucleosides (K. Hantke, unpublished).

The concept of narrow specificity that one has come to expect for enzyme—substrate binding is hardly applicable to the interaction between ferrichrome/albomycin, the phages T5, T1 and ϕ80 and colicin M with the receptor. It seems more appropriate to think along the lines of antigen—antibody reactions which feature overlapping binding regions with rather different binding constants. The affinities of the five agents to the receptor are unknown but could well be very different. In addition, the antibody molecule is flexible and responds by structural changes to the binding of antigens (Poljak, 1975; Richards et al., 1975). In the case of phage receptor we are far from understanding in molecular terms the apparently *multifunctional* combining regions.

For the initial phase of vitamin B_{12} (cyanocobalamin) uptake (binding to the receptor), competitive inhibition of cyanocobalamin was found with aquocobalamin, cyanocobinamide, methylcobalamin and deoxyadenosyl cobalamin with K_i values ranging from 0.3 to 4.2 nM. There was no detectable inhibition by cyanocobalamin 5'-phosphate (White et al., 1973).

We are not aware of any study on the sterospecificity of the lambda-dependent maltose transport. Maltose (4-α-glucosyl-glucose) and higher homologues of maltose (maltotriose, maltotetrose etc.) are taken up by this system. Mutations in the structural gene of the lambda receptor protein (*lam*B), or antibodies bound to the lambda receptor, reduce maltose transport but not transport of thiomethyl β-D-galactoside (Szmelcman and Hofnung, 1975). In a preliminary survey, we found that when cells are grown in a medium with trehalose (1-α-glucosyl-1-α-glucoside) or melibiose (6-α-galactosyl-glucose) as sole carbon source, the maltose transport system and the lambda receptor protein are induced. After growth on glucose or lactose (4-β-galactosyl-glucose), no induction of either lambda receptor protein or maltose transport was detected. The initial rate of maltose uptake was 20% in trehalose-grown cells and 7% in melibiose-grown cells of the rate obtained for maltose-grown cells. However, a 1000-fold excess of trehalose or melibiose over maltose in the maltose transport assay did not inhibit maltose transport (Hantke and Krieger, unpublished). Either induction is less specific than transport or, more likely, some maltose is synthesized in the cell after cleavage of trehalose and melibiose. Apart from the α-glycosidic bond, trehalose and melibiose are structurally quite different to maltose. Closer analogs of maltose will have to be tested to ascertain the stereospecificity of maltose transport. After completion of our experiments, we became aware of a study (Schwartz, 1967) in which it was shown that trehalose induces the synthesis of the lambda receptor. However, our results failed to confirm the inducing effect of lactose reported in that paper.

3.7.2 Dependence on molecular size

Scrutiny of the receptor-dependent uptake systems known to date reveals that they transport a class of hydrophilic substrates ranging in size from 250 (thymidine) to 1357 daltons (vitamin B_{12}). The smaller the molecular weight of the substrate, the less essential the receptor seems to be. For example at the maltose concentration of 3.5 μM maltose transport is severely impaired in *lam*B missense mutants and barely detectable in *lam*B nonsense mutants. At a maltose concentration of 35 μM the missense mutants transport as effectively as the wild-type, and only the nonsense mutants show markedly reduced transport. At a maltose concentration of 1 mM no difference in transport can be recognized between all three types of cells (Szmelcman and Hofnung, 1975). In contrast to the maltose transport, no transport of maltotriose could be measured at a substrate concentration of 1 mM in *lam*B mutants (Szmelcman *et al.*, 1976). In missense mutants, the altered protein functions partially in the transfer of maltose (344 daltons) but not in the transfer of maltotriose (504 daltons) through the outer membrane. We compared the growth of wild-type and of *lam*B nonsense mutants at substrate concentrations of 1 mM, and found that the mutants grew on maltose as well as the wild-type, while their growth on maltotriose was much slower (Fig. 3.8). Among 18 lambda resistant mutants in our strain collection, there is one exception: *E. coli* W 3110 λv^r 21, a λvir-resistant derivate of *E. coli* W 3110. This mutant failed to adsorb lambda; and we found no lambda-inactivating activity

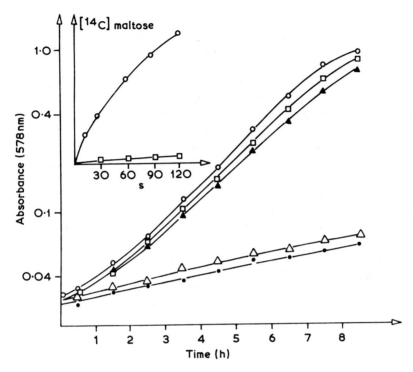

Fig. 3.8 Growth of *E. coli* on maltose and maltotriose, and maltose transport. Wild-type *E. coli* W 3110 (○ – ○) and lambda-resistant derivatives W 3110 λ v^r 21 (□ – □) were grown on 3 mM maltose; or on 3 mM maltotriose (W 3110 λv^r 4, △ – △; W 3110 λv^r 10, ● – ●; W 3110 λv^r 21, ▲ – ▲). Note that the exceptional lambda-resistant strain W 2110 λv^{r}21 grows on maltotriose as fast as on maltose, like the wild-type strain, despite the lack of the lambda receptor protein. The inset shows transport of [^{14}C] maltose at a 3.5 μM concentration in the wild-type (○ – ○) and in the lambda-resistant derivative W 3110 λv^r 21 (□ – □).

in the deoxycholate–EDTA extract of the cell envelope, no protein band at the position of the receptor after SDS-polyacrylamide gel electrophoresis of outer membranes of maltose-induced cells and no receptor protein after separation of outer membrane proteins on a column of QAE-Sephadex. However, despite the fact that apparently no lambda receptor protein is present in this mutant, it grows on maltotriose at the same rate as the wild-type strain grows on maltose or maltotriose (Fig. 3.8). We have, at present, no experimentally supported explanation for the unusual behavior of this mutant.

The correlation between substrate size and the presence of receptor-facilitated transport systems is further exemplified by the ferrichrome–iron (III) complex. Ferrichrome is relatively large (740 daltons) and in the absence of a functional phage T5 receptor protein in the outer membrane there is no transport and no growth with ferrichrome as sole iron source at a concentration of 1 μM (Braun et al., 1976b).

Some T5-sensitive revertants of T5-resistant mutants showed a reduced growth rate, as compared to the wild-type, when ferrichrome served as sole iron source. The ferrichrome transport capacity was as low as in the original tonA mutants. No receptor protein could be detected after the outer membrane proteins had been separated by electrophoresis on SDS-polyacrylamide gels. If numbers of input phage and concentrations of colicin M are higher than normal, plaques are obtained on these partial revertants. This most probably indicates that in these cells a few receptor molecules are still present, but cannot be detected electrophoretically. Another revertant, designated 4a, even displays normal growth rate on ferrichrome as sole iron source, as well as a plating efficiency for phage T5 and a sensitivity to colicin M which are indistinguishable from those of the wild-type; yet it contains no detectable receptor protein. Its rate of T5 adsorption, however, is only 5% of the wild-type rate (Braun et al., 1976b). A small number of receptor molecules can apparently insure that the uptake of the ferrichrome–iron complex is not limiting for growth. Bradbeer et al. (1976) also came to the conclusion that very few receptors per cell suffice to give an adequate supply of vitamin B_{12}. A quantitative analysis of a colicin-resistant mutant (btuB69) with a greatly reduced adsorption rate for phage BF23, revealed only one to two B_{12} receptors per cell. This mutant (metH) can transport B_{12} adequately for growth in the absence of methionine.

Exact figures concerning the concentration of iron in the medium required to support full growth of receptor-deficient mutants (cir, feuB) of E. coli cannot be given. The percentage of iron solubilized by chelating agents in the medium (organic acid, sugars etc.) which can be taken up by aerobically grown cells at pH 7, is unknown and hard to estimate. TonA, tonB, cir and feuB mutants all grow without the addition of chelating agents at wild-type rates in media containing 50 μM iron(III). The requirement for an outer membrane receptor protein for the passage of substrates is not an all or none phenomenon. When supplied in sufficiently high concentrations, all these substrates flow into the cell fast enough to support full growth in receptor-lacking mutants. The receptor requirement become more strict with increasing molecular weight of the substrates. Uptake of maltotriose (504 daltons), ferrichrome (740 daltons), enterochelin (746 daltons) and vitamin B_{12} (1357 daltons) become totally dependent on the presence of receptors at low substrate concentrations. Payne and Gilvarg have proposed an outer membrane permeability barrier dependent on the size of the substrate to be taken up by the cell. They showed, for example, that lysine homologs up to tetralysine readily support growth of E. coli in contrast with pentalysine, which cannot act as a nutrient (Payne and Gilvarg, 1968). The molecular sieve effect was ascribed to the outer membrane by the studies of Nakae and Nikaido. With intact plasmolyzed cells, peptidoglycan-defective cells and isolated outer membranes, they showed a 50% penetration of stachyose (666 daltons). Saccharides with molecular weights higher than 900 to 1000 were essentially excluded (Nakae and Nikaido, 1975; Nakae, 1976a; b). The hydrophilic channels through the outer membrane apparently have the properties of a molecular sieve.

In addition to the general channels for all of the low molecular weight substrates (amino acids, sugars, salts) specific channels have been developed by the cells for some substrates, and certainly for more than are now known, with molecular weights where the exclusion of the sieve starts to operate. Their effectiveness can be recognized even with substrates of small size. Most probably developed for the uptake of starch and glycogen hydrolysis products, mostly oligosaccharides, the lambda receptor is also operative in the uptake of the small molecule maltose, in that it lowers the apparent K_m of transport by a factor of 100–500 (Szmelcman et al., 1976).

3.7.3 Regulation

Maltose induces synthesis of the lambda receptor protein as well as the products of the other three genes involved in maltose uptake which are located in the *mal*B operon (Hofnung, 1974). As shown in Fig. 3.9 the lambda receptor protein becomes a major protein of the outer membrane after induction. The number of outer membrane proteins have been estimated to be in the range of 100 000 copies per cell (Braun and Rehn, 1969; Braun et al., 1970; Henning et al., 1973). Schwartz (1976) came to the conclusion that only 6000 lambda receptor copies are present in induced cells. The discrepancy may be explained by the assay method used, since the number of phage particles inactivated by cell extracts was determined under conditions such that receptor molecules were most likely aggregated. In iron-deficient medium, *E. coli* K12 synthesizes three outer membrane proteins with molecular weights of 74 000, 81 000 and 83 000. As we have seen (Table 3.1, Figs. 3.4, 3.6, 3.7) the 74 K protein is involved in binding of colicin I whilst the 81 K protein is the receptor for colicin B and takes part in enterochelin-iron transport (Braun et al., 1976b; Hancock and Braun, 1976b; Hancock et al., 1976). The receptor for phage T5 (*ton*A gene product) only weakly responds to the variation of iron supply. An additional protein is made when citrate is in the medium (80 000–81 000 daltons). It is at present unknown, and also difficult to establish since the presence of iron can not be strictly ruled out, whether citrate alone or in complex with iron induces synthesis of this protein. At present, we know of no iron chelate which requires the *cir* protein and the 83 K protein.

In addition to the outer membrane proteins, enzymes concerned with the synthesis and degradation of enterochelin are also regulated by the available iron (Rosenberg and Young, 1974). It would be of interest to know how all the various iron uptake systems are regulated by the level of intracellular iron. Some of the corresponding genes are known to map in different regions on the *E. coli* chromosome, e.g. the genes of enterochelin synthesis (*ent*) transport (*fep*) and degradation (*fes*) map together at 13 min, whilst the *feu*B gene (outer membrane protein 81 K) maps at 73 min and the *cir* gene at 45 min.

Other workers have noticed differences in the amount of two proteins of the outer membrane of *E. coli* due to changes in the concentration of iron in the medium (McIntosh and Earhart, 1976; Mizushima et al., personal communication) or in

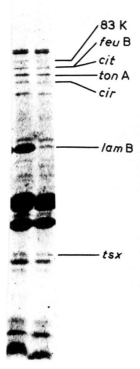

Fig. 3.9 Induction of the lambda receptor protein by growth of *E. coli* on maltose as sole carbon source. Sodium dodecyl sulfate polyacrylamide gel electropherogram of outer membrane proteins of *E. coli* HfrH grown on maltose (1) or glucose (2) as sole carbon source. The lambda receptor protein (*lam*B) appears as a new 'major' protein. For comparison, the position of other phage and colicin receptor proteins is given (see Figs. 3.6 and 3.7 for protein designations; cf. Table 3.1).

mutants in the *ton*B and *cbt* genes impaired in iron enterochelin uptake (Davies and Reeves, 1975; Pugsley and Reeves, 1976a). Under iron-limiting conditions, the two proteins are overproduced by a factor of 10 to 15 (Pugsley and Reeves, 1976a,b). Our finding that three, rather than two, proteins are induced by iron starvation may be understandable on the basis of the better resolution of the electrophoresis system we use or, possibly, on the basis of strain differences. The differences in the molecular weights tentatively assigned by McIntosh and Earhart and by ourselves, may arise from differences in the degree of gel cross-linking or of buffer systems used. We presume that identical proteins have been identified by all three groups. In *S. typhimurium*,

Table 3.1 Proteins of the outer membrane of *E. coli* K12 involved in substrate uptake

MW*	Designation	Function	Receptor for phage and colicin	References †
83	83K	Inducible by low iron	–	[1]
81	*feu*B	Fe^{3+}-enterochelin uptake	Colicin B	[2]
80.5	*cit*	Fe^{3+}-citrate uptake	–	[2]
78	*ton*A	Ferrichrome uptake	T5, T1, ϕ80, colicin M	[1]
74	*cir*	Inducible by low iron	Colicins I and V	[3]
60‡	*bfe*	Vitamin B_{12} transport	BF23, E colicins	[4]
55	*lam*B	Maltodextrin transport	Lambda	[5]
37 36	Matrix protein	Hydrophilic pore	–	[6]
27	*tsx*	Nucleoside transport	T6, colicin K	[7]
10	*lpm*	Structural component ?	–	[8]

* Molecular weights ($\times 10^{-3}$) are given as they appear in the gel system of Lugtenberg *et al.*, (1975) – the values determined in other gel electrophoretic systems are often higher.

† [1] Braun *et al.* (1976b), [2] Hancock *et al.* (1976), [3] Hancock and Braun (1976b), [4] Sabet and Schnaitman (1971), [5] Hazelbauer and Schwartz (1973), Brauer-Krieger and Braun, unpublished, [6] Schmittges and Henning (1976), Nakae (1976), Schnaitman *et al.* (1975), [7] Mannings and Reeves (1976), Hantke (1976), [8] Braun (1975). For additional information, see text.

‡ This value was determined by Sabet and Schnaitman (1973) for the colicin E3 receptor which was missing from a mutant resistant to colicins E1, E2 and E3.

three outer membrane proteins have also been found to be induced by low iron concentrations in the growth medium (Bennett and Rothfield, 1976). The isolation of a constitutive mutant which expresses all three proteins independently of the iron concentration, suggests that these three proteins — designated OM-1 to OM-3 — are regulated in a co-ordinate fashion.

3.8 CONCLUDING REMARKS AND GENERALIZATIONS

As has been discussed, receptor-deficient strains are resistant to the corresponding colicins and phages, and require higher concentrations of the corresponding substrates. Ostensibly, macromolecular agents are excluded from normal cells whilst low molecular weight substrates can pass across the outer membrane. In the latter case, a gradation of exclusion is observed with increasing size of the substrate, as pointed out previously for maltose/maltotriose, ferrichrome, enterochelin and vitamin B_{12}.

Binding of phage particles provokes more than just the fixation of the phage to the cell. It triggers an alteration of the supramolecular arrangement of phage proteins and the consequent release of the nucleic acid. It also causes membrane alterations which allow the penetration of the nucleic acid across the outer and the cytoplasmic membranes. Membrane changes can be monitored by fluorescent dyes and they reflect some features of the infection process (Hantke and Braun, 1974; Braun, 1974). For example, the only two phages with a known two-step transfer of the DNA, T5 and BF23, cause a two-step curve of fluorescence increase. The second step of fluorescence increase can be prevented by measures which prevent the second DNA transfer. The fluorescence response is correlated in time with the infection process and with the number of phage particles per cell, reaching a maximum level at a multiplicity of approximately 10 phages per cell. The phenomenon of super-infection exclusion can also be monitored by the fluorescence response curve. Super-infecting T4 phages cause only a slight increase of fluorescence. The fluorescence increase caused by the binding of colicin E1 to sensitive cells is clearly different from the response caused by colicins E2 and E3. This seems to reflect their different mode of action: E1 impairs energy-dependent processes in the cytoplasmic membrane, while E2 leads to DNA degradation and E3 inhibits protein synthesis (Hardy, 1975). The three colicins retain their acceptor-binding activities even when bound to Sephadex G-25 beads. Sephadex-bound colicin E3 also retains its ability to inhibit protein synthesis *in vitro*, but analogous complexes of either E2 or E3 exhibit negligible activity on whole cells. This is in contrast with Sephadex-immobilized colicin E1, which kills 90% of the cells (Lan and Richards, 1976). These results are interpreted to mean that only E1, which acts on the cytoplasmic membrane, can reach its target in spite of being immobilized on beads; whereas E2 and E3, which have to penetrate the cytoplasmic membrane at least partially in order to kill the cells, can not approach the target under those conditions.

The mechanism by which nucleic acids and colicins penetrate the outer membrane is still unknown, but it is clear that 'channels' are specific and strictly related to the receptor proteins. Mechanistic considerations suggest two alternatives. Either each receptor protein is an essential component of a specific 'pore' (or it alone forms the pore), or the 'pores' are unspecific and the different receptor proteins are constituents of the general pore. In the latter case, each receptor protein could function like a specific *gate* thus providing the required selectivity.

Adhesion zones between the cytoplasmic and the outer membrane at which adsorbed phages were preferentially localized (Bayer, 1975) have been visualized by electron microscopy. The adhesion zones were also shown to be the export sites of newly synthesized lipopolysaccharide from the cytoplasmic membrane to the cell surface (Bayer, 1975; Mühlradt *et al.*, 1973). The adhesion zones could thus be the communicating element between the cytoplasmic and the outer membrane. The alteration of the outer membrane expected to occur when phages and colicins bind to the receptors, should almost certainly be followed by structural transitions in the cytoplasmic membrane, such that the latter becomes permeable for the nucleic acids and the colicins respectively. But just as events at the outer membrane appear to influence the functional state of the cytoplasmic membrane, there is also evidence for effects exerted in the opposite direction.

The mutual inter-relationship between the outer and the cytoplasmic membrane is revealed by the energy requirement for the irreversible adsorption of the phages T1 and ϕ80 (Hancock and Braun, 1976a). The state of energization at the cytoplasmic membrane apparently governs the events which follow the initial contact between the phage and the cell. All receptor-dependent iron chelate transport systems require an intact *ton*B-function. In addition *ton*B mutants are also deficient in the energy-dependent phase of vitamin B_{12} transport (Bassford *et al.*, 1976). These observations point to a more general role for the *ton*B function in the energy-requiring step of transport systems utilizing outer membrane components. It is possible that the *ton*B function is localized at the adhesion zones discussed above.

For receptor-facilitated transport of low molecular weight substrates, it is difficult to imagine that they simply travel along a water-filled channel which spans the outer membrane. To explain the narrow specificity of the 'channel', the receptor protein must at least have the property of a gate. However, this cannot be the whole story since binding of low molecular weight substrates to cells or purified receptor preparations has been demonstrated, except in the case of maltose, maltotriose and nucleosides. So one is left with the question: channel or carrier? The hydrodynamic properties of only one receptor have been determined in any detail (Konisky and Liu, 1974). This receptor, having a molecular weight of 387 000, a Stokes radius of 73.5 Å, a frictional ratio of 1.5 and an axial ratio of 8 to 9, could easily span the outer membrane. Taking these dimensions at face value (and some doubt remains since the receptor preparation still contained Triton X-100) it is easier to imagine that the receptor forms a pore for transport than to suppose that it can flip across or rotate within the membrane.

The involvement of one receptor protein in several functions, for example the *ton*A receptor in the binding of phages T5, T1 and ϕ80 and colicin M and in the uptake of albomycin and ferrichrome, does not imply that these different agents use this protein in the same way. Additional components may govern the consequent events which follow the initial binding reaction. In addition to the above-mentioned cases, the *ton*B function is also required for vitamin B_{12} transport and ferric enterochelin uptake. The *exb* and *cbt* functions are involved in the killing action of colicin B and ferric-enterochelin transport; the maltose-binding protein channels maltose via either the transport or the chemotactic system.

Outer membrane proteins may generally be components of pores through which molecules of various sizes traverse the outer membrane. The receptors may just be the first components of these pores to be detected. A pore function has already been ascribed to the matrix protein (Nakae, 1976). The questions then arise whether the other outer membrane proteins specify selective pores for certain substrates, or groups of structurally similar molecules, or whether they form general transmembrane channels for all kinds of molecules up to a certain size.

The concept of pores through the outer membrane satisfactorily takes into account the data collected to date with regard to receptor-facilitated translocation across the outer membrane. The specificity problem discussed above requires more detailed studies on the way in which receptors function. In this respect, we are in the good company of others who study receptor systems and transport phenomena in eukaryotes (Shamoo, 1975; Helmreich, 1976). However, the field has developed to the point at which questions which lead to sensible experiments can be asked. This has already given rise to considerable advances.

REFERENCES

Achtman, M. (1973), Genetics of the F sex factor in Enterobacteriaceae. *Curr. Top. Microbiol. Immunol.,* **60**, 79–123.

Adler, J. (1975), Chemotaxis in bacteria. *Ann. Rev. Biochemistry,* **44**, 341–356.

Alexanian, A., Diekmann, H., Zähner, H. (1972), Stoffwechselprodukte von Mikroorganismen. 94. Mitt. Vergleich der Wirkung von Sideraminen als Wuchsstoffe und als Antagonisten der Sideromycine. *Arch. Mikrobiol.,* **82**, 55–65.

Alphen, W. van, Lugtenberg, B. and Berendsen, W. (1976), Heptose-deficient mutants of *Escherichia coli* K12 deficient in up to three major outer membrane proteins. *Molec. gen. Genet.,* **147**, 263–269.

Ames, G.F., Spudich, E.N. and Nikaido, H. (1974), Protein composition of the outer membrane of *Salmonella typhimurium*: effect of lipopolysaccharide mutations. *J. Bact.,* **117**, 406–416.

Bachmann, B.J., Low, K.B. and Taylor, A.L. (1976), Recalibrated linkage map of *Escherichia coli* K12. *Bact. Rev.,* **40**, 116–167.

Bassford, P.J., Bradbeer, C., Kadner, R.J. and Schnaitman, C.A. (1976), Transport of vitamin B_{12} in *ton*B mutants of *Escherichia coli*. *J. Bact.,* **128**, 242–247.

Bayer, M. (1975), Role of adhesion zones in bacterial cell-surface function and biogenesis, In: *Membrane Biogenesis* (Tzagoloff, A., ed.), Plenum Press, New York, pp. 393–427.

Bennett, R.L. and Rothfield, L.J. (1976), Genetic and physiological regulation of intrinsic proteins of the outer membrane of *Salmonella typhimurium*. *J. Bact.*, **127**, 498–504.

Berg, H.C. (1975), Chemotaxis in bacteria. *Ann. Rev. Biophys. Bioeng.*, **4**, 119–136.

Bradbeer, C., Woodrow, M.L. and Khalifah, L.I. (1976), Transport of vitamin B_{12} in *Escherichia coli*: common receptor system for vitamin B_{12} and bacteriophage BF23 on the outer membrane of the cell envelope. *J. Bact.*, **125**, 1032–1039.

Bragg, P.D. and Hou, C. (1972), Organization of proteins in the native and reformed outer membrane of *Escherichia coli*. *Biochim. biophys. Acta*, **274**, 478–488.

Braun, V. (1974), Murein-Lipoprotein und Rezeptor für den Phagen T5 und das Colicin M: Definierte Struktur-Funktionsbereiche der äussere Membran von *Escherichia coli*. *Zbl. Bakt. Hyg. Abt. Orig. A*, **228**, 233–240.

Braun, V. (1975), Covalent lipoprotein from the outer membrane of *Escherichia coli*. *Biochem. biophys. Acta*, **415**, 335–377.

Braun, V., Bosch, V., Klumpp, E.R., Neff, I., Mayer, A. and Schlecht, S. (1976a), Antigenic determinants of murein lipoprotein and its exposure at the surface of Enterobacteriaceae. *Eur. J. Biochem.*, **62**, 555–566.

Braun, V., Hancock, R.E.W., Hantke, K. and Hartmann, A. (1976b), Functional organization of the outer membrane of *Escherichia coli*. Phage and colicin receptors as components of iron uptake systems. *J. supramolec. Struct.*, **5**, 37–58.

Braun, V. and Hantke, K. (1974), Biochemistry of bacterial cell envelopes. *Ann. Rev. Biochem.*, **43**, 89–121.

Braun, V. and Rehn, K. (1969), Chemical characterization, spatial distribution and function of a lipoprotein (murein-lipoprotein) of the *E. coli* cell wall. *Eur. J. Biochem.*, **10**, 426–438.

Braun, V., Rehn, K. and Wolff, H. (1970), Supramolecular structure of the rigid layer of *Salmonella, Serratia, Proteus,* and *Pseudomonas fluorescens*. Number of lipoprotein molecules in a membrane layer. *Biochemistry*, **9**, 5041–5049.

Braun, V., Schaller, K. and Wolff, H. (1973), A common receptor protein for phage T5 and colicin M in the outer membrane of *Escherichia coli* B. *Biochem. biophys. Acta*, **323**, 87–97.

Braun, V. and Wolff, H. (1973), Characterization of the receptor protein for phage T5 and colicin M in the outer membrane of *E. coli* B. *FEBS Letters*, **34**, 77–80.

Brinton, C.C. (1971), The properties of sex pili, the viral nature of 'conjugal' genetic transfer systems, and some possible approaches to the control of bacterial drug resistance. *Crit. Rev. Microbiol.*, **1**, 105–160.

Buxton, R.S. (1971), Genetic analysis of *Escherichia coli* K12 mutants resistant to bacteriophage BF23 and the E-group colicins. *Mol. Gen. Genet.*, **113**, 154–156.

Corwin, L.M., Fanning, G.R., Feldman J. and Margolin, P. (1966) Mutation leading to increased sensitivity to chromium in *Salmonella typhimurium*. *J. Bact.*, **91**, 1509–1515.

Cuatrecasas, P. (1974), Membrane receptors. *Ann. Rev. Biochem.*, **43**, 169–214.

Davies, J.K. and Reeves, P. (1975), Genetics of resistance to colicins in *Escherichia coli* K12: cross-resistance among colicins of group A. *J. Bact.*, **123**, 102–117; Cross-resistance among colicins of group B. *J. Bact.*, **123**, 96–101.

Diekmann, H. (1973), Siderochromes Iron (III)-Trihydroxamates. *Handbook of Microbiology*, Vol. 3, Microbial products (Laskin, A.I. and Lechevalier, H.A., eds.), CRC Press, Cleveland, pp. 449–457.

Di Girolamo, P.M. and Bradbeer, C. (1971), Transport of vitamin B_{12} in *Escherichia coli. J. Bact.*, **106**, 745–750.

Di Masi, D.R., White, J.C., Schnaitman, C.A. and Bradbeer, C. (1973), Transport of vitamin B_{12} in *Escherichia coli*: common receptor sites for vitamin B_{12} and the E colicins on the outer membrane of the cell envelope. *J. Bact.*, **115**, 506–513.

Dippe, P.J. von, Roy-Burman, S. and Wisser, D.W. (1973), Transport of uridine in *Escherichia coli* B and a Showdomycin-resistant mutant. *Biochem. biophys. Acta*, **318**, 105–112.

Duckworth, D.H. (1970), Biological activity of bacteriophage ghosts and 'take over' of host functions by bacteriophages. *Bact. Rev.*, **34**, 344–363.

Frank, H., Zarnitz, M.-L. and Weidel, W. (1963), Über die Rezeptorsubstanz für den Phagen T5. Elektronenoptische Darstellung und Längenbestimmung der aus T5/R5-Komplexen freigesetzten DNA. *Z. Naturforsch.*, **18b**, 281–284.

Fredericq, P. (1949) Sur la résistance croisée entre colicine K et bacteriophage III. *C. r. Soc. biol.*, **143**, 1014.

Fredericq, P. and Gratia, J.P. (1960), Recherches génétiques sur la résistance croisée à la colicine B et au bacteriophage T1 chez *Escherichia coil. C. R. Soc. Biol.*, **154**, 2150–2154.

Frost, G.E. and Rosenberg, H. (1975), Relationship between the *ton*B locus and iron transport in *Escherichia coli. J. Bact.*, **124**, 704–712.

Garen, A. (1954) Thermodynamic and kinetic studies on the attachment of T1 bacteriophage to bacteria. *Biochem. biophys. Acta*, **14**, 163–172.

Garten, W., Hindenach, I. and Henning, U. (1975), The major proteins of the *Escherichia coli* outer cell envelope membrane. Characterization of proteins II[+] and III, comparison of all proteins. *Eur. J. Biochem.*, **59**, 215–221.

Gratia, J.P. (1964), Résistance à la Colicine B chez *E. coli. Ann. Inst. Past.*, **107**, 132–151.

Guterman, S.K. (1971), Inhibition of Colicin B by enterochelin. *Biochem. biophys. Res. Comm.*, **44**, 1149–1155.

Guterman, S.K. (1973), Colicin B: mode of action and inhibition by enterochelin. *J. Bact.*, **114**, 1217–1224.

Guterman, S.K. and Dann, L. (1973) Excetion of Enterochelin by *exb*A and *exb*B mutants of *Escherichia coli. J. Bact.*, **114**, 1225–1230.

Guterman, S.K., Wright, A. and Boyd, D.H. (1975), Genes affecting coliphage BF23 and E colicin sensitivity in *Salmonella typhimurium. J. Bact.*, **123**, 1351–1358.

Haller, I., Hoehn, B. and Henning, U. (1975), Apparent high degree of asymmetry of protein arrangement in the outer *Escherichia coli* cell envelope membrane. *Biochemistry*, **14**, 478–484.

Hancock, R.E.W. and Braun, V. (1976a), Nature of the energy requirement for the irreversible adsorption of bacteriophages T1 and ϕ80 to *Escherichia coli. J. Bact.*, **125**, 409–415.

Hancock, R.E.W. and Braun, V. (1976b), The colicin I receptor of *Escherichia coli* K12 has a role in enterochelin-mediated iron transport. *FEBS Letters*, 208–210.

Hancock, R.E.W., Hantke, K. and Braun, V. (1976), Iron transport in *Escherichia coli* K12: The involvement of the colicin B receptor and of a citrate-inducible protein. *J. Bact.*, **127**, 1370–1375.

Hantke, K. (1976), Phage T6 – colicin K receptor and nucleoside transport in *Escherichia coli. FEBS Letters*, **70**, 109–112.

Hantke, K. and Braun, V. (1974), Fluorescence studies of first steps of phage-host interactions. *Virology*, **58**, 310–312.

Hantke, K. and Braun, V. (1975a), Membrane receptor-dependent iron transport in *Escherichia coli. FEBS Letters*, **49**, 301–305.

Hantke, K. and Braun, V. (1975b), A function common to iron-enterochelin transport and action of colicins B, I, V, in *Escherichia coli. FEBS Letters.*, **59**, 277–281.

Hardy, K.G. (1975), Colicinogeny and related phenomena. *Bact. Rev.*, **39**, 464–515.

Hazelbauer, G.L. (1975), Role of the receptor for bacteriophage lambda in the functioning of the maltose-chemoreceptor of *Escherichia coli. J. Bact.*, **124**, 119–126.

Helmreich, E.J.M. (1976) Hormone–receptor interactions. *FEBS Letters*, **61**, 1–5.

Henning, U. and Haller, I. (1975), Mutants of *Escherichia coli* K12 lacking all 'major' proteins of the outer cell envelope membrane. *FEBS Letters*, **55**, 161–164.

Henning, U., Hindenach, I. and Haller, I. (1976), The major proteins of the *Escherichia coli* outer cell envelope membranes: evidence for the structural gene of protein II$^+$. *FEBS Letters*, **61**, 46–48.

Henning, U., Höhn, B. and Sonntag, I. (1973), Cell envelope and shape of *Escherichia coli* K12. The ghost membrane. *Eur. J. Biochem.*, **39**, 27–36.

Hofnung, M. (1974), Divergent operons and the genetic structure of the maltose B region in *Escherichia coli. Genetics*, **76**, 169–184.

Inouye, M. (1975), In: *Membrane Biogenesis, Mitochondria, Chloroplasts, Bacteria*, (Tzagoloff, A., ed.), Plenum Press, New York, pp. 351–391.

Jetten, A.M. and Jetten, M.E.R. (1975), Energy requirement for the initiation of colicin action in *Escherichia coli. Biochem. biophys. Acta*, **387**, 12–22.

Kadner, R.J. and Liggins, G.L. (1973), Transport of vitamin B_{12} in *Escherichia coli*: Genetic studies. *J. Bact.*, **115**, 514–528.

Kellermann, O. and Szmelcman, S. (1974), Active transport of maltose in *Escherichia coli* K12: involvement of a 'periplasmic' maltose binding protein. *Eur. J. Biochem.*, **47**, 139–149.

Konisky, J. and Liu, C.-T. (1974), Solubilization and partial characterization of the colicin I receptor of *Escherichia coli. J. biol. Chem.*, **249**, 835–840.

Konisky, J., Soucek, S., Frick, K., Davies, J.K. and Hammond, C. (1976), A relationship between the transport of iron and the amount of specific colicin Ia membrane receptors in *Escherichia coli* K-12. *J. Bact.*, **127**, 249–257.

Koplow, J. and Goldfine, H. (1974), Alterations in the outer membrane of the cell envelope of heptose-deficient mutants of *Escherichia coli. J. Bact.*, **117**, 527–543.

Koshland, D.E. (1974), Chemotaxis as model of sensory systems. *FEBS Letters*, **40**, S. 3–S. 9.

Labedan, B. and Legault-Demare, J. (1973), Penetration into host cells of naked, partially injected (post-FST) DNA of bacteriophage T5. *J. Virol.*, **12**, 226–229.

Labedan, B., Crochet, M., Legault-Demare, J. and Stevens, B.J. (1973), Location of the first step transfer fragment and single-strand interruptions in T5 st0 bacteriophage DNA. *J. mol. Biol.*, **75**, 213–234.

Lan, C. and Richards, F.M. (1976), Behavior of colicins E1, E2, and E3 attached to Sephadex beads. *Biochemistry*, **15**, 666–671.

Lanni, Y.T. (1968), First-step transfer deoxyribonucleic acid of bacteriophage T5. *Bact. Rev.*, **32**, 227–242.

Leong, J. and Neilands, J.B. (1976), Mechanism of siderophore iron transport in enteric bacteria. *J. Bact.*, **126**, 823–830.

Lindberg, A. (1973), Bacteriophage receptors. *Ann. Rev. Microbiol.*, **27**, 205–241.

Luckey, M. and Neilands, J.B. (1976), Iron transport in *Salmonella typhimurium* LT-2: Prevention, by ferrichrome, of adsorption of bacteriophages ES18 and ES18.hl to a common cell envelope receptor. *J. Bact.*, **127**, 1036–1037.

Luckey, M., Pollack, J.R., Wayne, R., Ames, B.N. and Neilands, J.B. (1972), Iron uptake in *Salmonella typhimurium*: utilization of exogenous siderochromes as iron carriers. *J. Bact.*, **111**, 731–738.

Luckey, M., Wayne, R. and Neilands, J.B. (1975), *In vitro* competition between ferrichrome and phage for the outer membrane T5 receptor complex of *Escherichia coli*. *Biochem. biophys. Res. Comm.*, **64**, 687–693.

Lugtenberg, B., Meijers, J., Peters, R., van der Hoek, P. and van Alphen, L. (1975), Electrophoretic resolution of the 'major outer membrane protein' of *Escherichia coli* K12 into four bands. *FEBS Letters.*, **58**, 254–258.

Manning, P.A. and Reeves, P. (1976), Outer membrane of *E. coli* K-12: *Tsx* mutants (resistant to bacteriophage T6 and colicin K) lack an outer membrane protein. *Biochem. biophys. Res. Comm.*, **71**, 466–471.

McIntosh, U.A. and Earhart, C.F. (1976), Effect of iron on the relative abundance of two large polypeptides of *Escherichia coli* outer membrane. *Biochem. biophys. Res. Comm.*, **70**, 315–322.

McKeown, M., Kahn, M. and Hanawalt, P. (1976), Thymidine uptake and utilization in *Escherichia coli*: a new gene controlling nucleoside transport. *J. Bact.*, **126**, 814–822.

Michael, J.G. (1968), The surface antigens and phage receptors in *Escherichia coli* B. *Proc. Soc. exp. Biol. Med.*, **128**, 434–438.

Mojica-a, T. and Garcia, E. (1976), Growth of coliphage BF23 on rough strains of *Salmonella typhimurium*: the *bfe* locus. *Molec. gen. Genet.*, **147**, 195–202.

Mühlradt, P., Menzel, J., Golecki, J.R. and Speth, V. (1974), Lateral mobility and surface density of lipopolysaccharide in the outer membrane of *Salmonella typhimurium*. *Eur. J. Biochem.*, **43**, 533–539.

Nakae, T. (1976a), Identification of the outer membrane protein of *E. coli* that produces transmembrane channels in reconstituted vesicle membranes. *Biochem. biophys. Res. Comm.*, **71**, 877–884.

Nakae, T. (1976b) Outer membrane of *Salmonella*. Isolation of protein complex that produces transmembrane channels. *J. biol. Chem.*, **251**, 2176–2178.

Nakae, T. and Nikaido, H. (1975), Outer membrane as diffusion barrier in *Salmonella typhimurium*. Penetration of oligo- and polysaccharides into isolated outer membrane vesicles and cells with degraded peptidoglycan layer. *J. biol. Chem.*, **250**, 7359–7365.

Nikaido, H. (1976), Outer membrane of *Salmonella typhimurium*. Transmembrane diffusion of some hydrophobic substances. *Biochim. biophys. Acta*, **433**, 118–132.

Nomura, M., Sidikaro, J., Jakes, K. and Linder, N. (1974), Effects of colicin E3 on bacterial ribosomes. In: *Ribosomes* (Nomura, M., Tissières and Lengyel, P., eds.), Cold Spring Harbor Laboratory, pp. 805–814.

Nomura, M. and Witten, C. (1967), Interaction of colicins with bacterial cells. Colicin-tolerant mutations in *Escherichia coli. J. Bact.*, **94**, 1093–1111.

O'Brien, J.G. and Gibson, F. (1970), The structure of enterochelin and related 2,3-dihydroxy-N-benzoylserine conjugates from *Escherichia coli. Biochim. biophys. Acta*, **215**, 393–402.

Osborn, M.J., Gander, J.E., Parisi, E. and Carson, J. (1972), Mechanism of assembly of the outer membrane of *Salmonella typhimurium*. Isolation and characterization of cytoplasmic and outer membrane. *J. biol. Chem.*, **247**, 3962–3972.

Payne, J.W. and Gilvarg, C. (1968), Size restriction of peptide utilization in *Escherichia coli. J. biol. Chem.*, **243**, 2691–2699.

Poljak, R.J. (1975), Three-dimensional structure, function and genetic control of immunoglobulins. *Nature*, **256**, 373–376.

Pollack, J.R., Ames, B.N. and Neilands, J.B. (1970), Iron transport in *Salmonella typhimurium*: mutants blocked in the biosynthesis of enterobactin. *J. Bact.*, **104**, 635–639.

Pugsley, A.P. and Reeves, P. (1976a), Iron uptake in colicin B-resistant mutants of *Escherichia coli. J. Bact.*, **126**, 1052–1062.

Pugsley, A.P. and Reeves, P. (1976b), Characterization of group B colicin-resistant mutants of *Escherichia coli* K12: Colicin resistance and the role of enterochelin. *J. Bact.*, **127**, 218–228.

Pugsley, A.P. and Reeves, P. (1976c), Increased production of the outer membrane receptors for colicins B, D and M by *Escherichia coli* under iron starvation. *Biochem. biophys. Res. Comm.*, **70**, 846–853.

Randall-Hazelbauer, L. and Schwartz, M. (1973), Isolation of the bacteriophage lambda receptor from *Escherichia coli. J. Bact.*, **166**, 1436–1446.

Richards, F.F., Konigsberg, W.H., Rosenstein, R.W. and Varga, J.M. (1975), On the specificity of antibodies. Biochemical and biophysical evidence indicates the existence of polyfunctional antibody-combining regions. *Science*, **187**, 130–137.

Rosenberg, H. and Young, I.G. (1974), Iron transport in the enteric bacteria. In: *Microbial iron metabolism* (Neilands, J.B., ed.), Academic Press, New York, pp. 67–82.

Rosenbusch, J.P. (1974), Characterization of the major envelope protein of *Escherichia coli*. Regular arrangement on the peptidoglycan and unusual dodecyl sulfate binding. *J. biol. Chem.*, **249**, 8019–8029.

Sabet, S.F. and Schnaitman, C.A. (1971), Localization and solubilization of colicin receptors. *J. Bact.*, **108**, 422–430.

Sabet, S.F. and Schnaitman, C.A. (1973), Purification and properties of the colicin E3 receptor of *Escherichia coli. J. biol. Chem.*, **248**, 1797–1806.

Sanderson, K.E. (1972), Linkage Map of *Salmonella typhimurium*, Edition IV *Bact. Rev.*, **36**, 558–586.

Scandella, D. and Arber, W. (1974), An *Escherichia coli* mutant which inhibits the injection of phage λ DNA. *Virology,* **58**, 504–513.

Schmitges, J. and Henning, U. (1976), The major proteins of the *Escherichia coli* outer cell-envelope membrane; Heterogeneity of Protein I. *Eur. J. Biochem.,* **63**, 47–52.

Schnaitman, C. (1974), Outer membrane proteins of *Escherichia coli*. Differences in outer membrane proteins due to strain and cultural differences. *J. Bact.,* **118**, 454–464.

Schnaitman, C., Smith, D. and Forn de Salsas, M. (1975), Temperate bacteriophage which causes the production of a new major outer membrane protein by *Escherichia coli. J. Virol.,* **15**, 1121–1130.

Schwartz, M. (1967), Sur l'existence chez *Escherichia coli* K12 d'une régulation commune à la biosynthèse des recepteurs du bacteriophage λ et au métabolisme de maltose. *Ann. Inst. Pasteur. Paris,* **113**, 685–704.

Schwartz, M. (1975), Reversible interaction between coliphage lambda and its receptor protein. *J. mol. Biol.,* **99**, 185–201.

Schwartz, M. (1976), The adsorption of coliphage lambda to its host: effect of variations in the surface density of receptor and in phage-receptor affinity. *J. mol. Biol.,* **103**, 521–536.

Shamoo, A.E. ed. (1975), Carriers and channels in biological systems. *Ann. N.Y. Acad. Sci.,* **264**.

Skurray, R.A., Hancock, R.E.W. and Reeves, P. (1974), Con$^-$ mutants: class of mutants in *Escherichia coli* K12 lacking a major cell wall protein and defective in conjugation and adsorption of a phage. *J. Bact.,* **119**, 726–735.

Smit, J., Kamio, Y. and Nikaido, H. (1975), Outer membrane of *Salmonella typhimurium*: Chemical analysis and freeze-fracture studies with lipopolysaccharide mutants. *J. Bact.,* **124**, 942–958.

Szmelcman, S. and Hofnung, M. (1975), Maltose transport in *Escherichia coli* K12: Involvement of the bacteriophage lambda receptor. *J. Bact.,* **124**, 112–118.

Szmelcman, S., Schwartz, M., Silhavy, T.J. and Boos, W. (1976), Maltose transport in *Escherichia coli* K12. A comparison of transport kinetics in wild type and λ-resistant mutants with the dissociation constants of the maltose-binding protein as measured by fluorescence quenching. *Eur. J. Biochem.,* **65**, 13–19.

Wang, C.C. and Newton, A. (1969), Iron transport in *Escherichia coli*: relationship between chromium sensitivity and high iron requirement in mutants of *Escherichia coli. J. Bact.,* **98**, 1135–1141.

Wang, C.C. and Newton, A. (1971), An additional step in the transport of iron defined by the *ton*B locus of *Escherichia coli. J. biol. Chem.,* **246**, 2147–2151.

Wayne, R., Frick, K. and Neilands, J.B. (1976), Siderophore protection against colicins M, B, V and Ia in *Escherichia coli. J. Bact.,* **126**, 7–12.

Wayne, R. and Neilands, J.B. (1975), Evidence for a common binding site for ferrichrome compounds and bacteriophage ϕ80 in the cell envelope of *Escherichia coli. J. Bact.,* **121**, 497–503.

Weidel, W. (1958), Bacterial viruses (with particular reference to adsorption/penetration). *Ann. Rev. Microbiol.,* **12**, 28–48.

Weidel, W. and Kellenberger, E. (1955), The *E. coli* B-receptor for the phage T5, II. Electron microscopic studies. *Biochim. biophys. Acta,* **17**, 1–9.

Weidel, W. and Pelzer, H. (1964), Bag-shaped macromolecules — a new outlook on bacterial cell walls. *Adv. Enzymol.*, **26**, 193–232.

Weltzien, H.U. and Jesaitis, M.A. (1971), The nature of the colicin K receptor of *Escherichia coli* Cullen. *J. exp. Med.*, **133**, 534–553.

White, J.C., Di Girolamo, P.M., Fu, M.L., Preston, Y.A. and Bradbeer, C. (1973), Transport of vitamin B_{12} in *Escherichia coli*. Location and properties of the initial B_{12}-binding site. *J. biol. Chem.*, **248**, 3978–3986.

Zähner, J., Hütter, R., Bachmann, E. (1960), Zur Kenntnis der Sideromycinwirkung. *Arch. Mikrobiol.*, **36**, 325–349.

Zwaig, R.S. and Luria, S. (1967), Genetics and physiology of colicin-tolerant mutants of *Escherichia coli*. *J. Bact.*, **94**, 1112–1123.

4 The Attachment of Bacteria to the Surfaces of Animal Cells

GARTH W. JONES

4.1	Introduction	page	141
	4.1.1 Definitions of terms		141
4.2	Adhesive activities of bacteria		142
	4.2.1 Detection of adhesive properties		142
	4.2.2 Activities, functions and morphologies of selected bacterial adhesins		142
	(a) *Enterobacteriaceae*, 143, (b) *Vibrios*, 145, (c) *Neisseriaceae*, 146, (d) *Bordetella*, 147, (e) *Streptococci*, 147, (f) *Lactobacilli*, 148, (g) *Corynebacteria*, 148, (h) *Mycoplasma*, 149, (i) *Norma flora*, 149		
4.3	Chemical and physical properties of adhesins		150
	4.3.1 Proteinaceous adhesins		150
	(a) *Chemical compositions and molecular weights*, 150, (b) *Structural characteristics*, 152, (c) *Physical properties*, 152		
	4.3.2 Non-proteinaceous adhesins		153
	(a) *Chemical composition*, 153, (b) *Mode of action*, 154		
4.4	The nature of the surface of the eukaryotic cell		155
	4.4.1 Lipid bilayer		155
	4.4.2 Membrane proteins		155
	4.4.3 Membrane carbohydrates		155
4.5	Cell surface receptors of bacterial adhesins		156
	4.5.1 Detection of animal cell receptors		156
	(a) *Adhesion inhibition tests*, 156, (b) *Modification of the animal cell surface*, 157		
	4.5.2 Chemical composition of receptors		157
	(a) *Carbohydrate receptors*, 157, (b) *Receptors on the lipid bilayer*, 160		
4.6	The interaction of bacteria with animal cell surfaces		161
	4.6.1 Cell surface potential		161
	4.6.2 The DLVO theory		162
	4.6.3 Factors that may may influence bacterial adhesion		163
	(a) *Diffuse double layers*, 163, (b) *Surface potential of bacteria*, 163, (c) *The influence of the substratum*, 164,		

Contents

    4.6.3 Factors that may influence bacterial adhesion (*continued*)
           (d) *Radius of curvature, 164,* (e) *Reversible and irreversible adhesion, 165,* (f) *Chemical bonds involved in cell–cell interactions, 165*

	4.6.4 Conclusions	166
4.7	Concluding remarks	166
	4.7.1 Advantages of adhesion	166
	4.7.2 Possible origins of adhesins	167
	References	168

Acknowledgements
It should be noted that there is much reference to personal communications and to unpublished data in the text. My thanks to all those persons who entrusted me with their hard-earned results and ideas and my apologies to others who may feel slighted because I have inadvertently omitted their work from reference in this chapter.

Microbial Interactions
(*Receptors and Recognition,* Series B, Volume 3)
Edited by J.L. Reissig
Published in 1977 by Chapman and Hall, 11 New Fetter Lane, London EC4P 4EE
© Chapman and Hall

4.1 INTRODUCTION

It has been known for some time that bacteria possess adhesive properties, but it is only within the last few years that this ubiquitous phenomenon has become widely known and accepted as one of some consequence. Bacteria are found attached to the surfaces of clays (Stotzky, 1974) and glass (Marshall *et al.*, 1971a,b), the root hair surfaces of plants (Menzel *et al.*, 1972), the hyphae of fungi (Lockwood, 1968), the exoskeleton of copepods (Kaneko and Colwell, 1975), the gut of nematodes (Tannock and Savage, 1974), the surfaces of protozoans (Cleveland and Grimestone, 1964) and even the surfaces of other bacterial species (Davis and Baird-Parker, 1959; Jones, 1972). From these examples and from the following limited account of the attachment of bacteria to animal cells, it should be noted that adhesive bacteria are as diverse in character as are the surfaces to which they attach.

Adhesion is probably an initial event in the colonization of a habitat by a bacterial species and the means and ways by which a bacterium attaches to a surface, in consequence, are of considerable ecological significance. The object of this chapter is to examine the properties of the adhesive substances of bacteria, the nature of the animal cell receptors with which they react and to consider in general terms the phenomenon of prokaryotic—eukaryotic cell interactions.

4.1.1 Definitions of terms

The term *fimbriae* (Duguid *et al.*, 1955, 1966; Duguid and Anderson, 1967) is used to describe non-flagellar filaments on the surface of bacteria that are known, or thought to be, the organelles responsible for the adhesive properties of bacteria. The term pili (Brinton, 1965) is reserved exclusively for those filamentous non-flagellar appendages that are involved directly in the transmission of genetic information (Ottow, 1975). *Fibrillae* is used to describe surface structures that are known or believed to have adhesive properties; although in some cases this term is applied to structures which bear little morphological, nor in some cases chemical, resemblance to the original descriptions (Houwink and van Iterson, 1950 and above). Like fimbriae (for examples see Duguid *et al.*, 1955; Duguid and Gillies, 1957; Brinton, 1965) they may appear to be filaments but, unlike fimbriae, they are irregular in form and dimension with indiscrete points of origin on the bacterial surface (for examples of fibrillae see Stirm *et al.*, 1967b; Swanson *et al.*, 1969; Yanagawa and Otsuki, 1970; Liljemark and Gibbons, 1972; Fuller and Brooker, 1974). It is very likely that the filaments are artefacts created by the preparative procedures. The term *adhesin* (Duguid, 1959) is used to describe any substance on the bacterial surface that mediates the attachment of the bacterium to a surface. *Haemagglutinin* is used for any component of the bacterial surface that causes the agglutination of erythrocytes. Haemagglutinin and adhesin are often synonyms. *Receptors* are those components of the animal cell surface that react specifically with bacterial adhesins and haemagglutinins, and are responsible for the adhesion of bacteria to animal cell surfaces.

4.2 ADHESIVE ACTIVITIES OF BACTERIA

4.2.1 Detection of adhesive properties

Two basic techniques are used to detect the adhesive properties of bacteria *in vitro*: first, semi-quantitative haemagglutination tests (Duguid *et al.*, 1955; Jones and Rutter, 1974) which measure the ability of adhesive bacteria or cell-free adhesins to bind erythrocytes together; second, adhesion tests, in which isolated epithelial cells (Gibbons and van Houte, 1971; Jones, 1972; Wilson and Hohman, 1974), tissue culture cells (Swanson, 1973), erythrocytes, isolated cell membranes (Jones *et al.*, 1976), slices of epithelial tissue (Freter, 1969; Jones and Rutter, 1972) or organ cultures (Collier and Baseman, 1973; Ward *et al.*, 1974) are incubated with bacteria and quantitative measurements of adhesion obtained either microscopically or from counts of viable bacteria or of bacteria labeled with radioisotope.

Usually, the same adhesive activity of a culture is measured with both the haemagglutination and the adhesion test (Duguid, 1964; Jones and Rutter, 1974; Jones *et al.*, 1976; Jones and Freter, 1976) but there are exceptions (Freter and Jones, 1976). Moreover, some bacterial cultures produce more than one adhesin and the adhesion of other bacteria to different surfaces may be mediated by distinct mechanisms. Furthermore, there are differences in the reactivity of cells from different animal species with the same bacterial adhesin and of the same type of animal cell with different adhesive bacteria (e.g., Duguid, 1964). Indeed, some species yield cells that vary in reactivity depending upon the individual animal (Jones, 1972; Jones and Rutter, 1974; Sellwood *et al.*, 1975), whereas other cells require chemical modification before they interact with bacterial adhesins.

The conditions of culture markedly influence production of adhesins (e.g. Duguid and Wilkinson, 1961; Jones and Rutter, 1972 and 1974) or bacteria may become less adhesive because the adhesins and haemagglutinins are lost from the bacterial surface (e.g. Lankford and Legsomburana, 1965; Jones *et al.*, 1976). Finally, the adhesive activity of bacteria in tests can be influenced by both physical and chemical conditions of the test system (Duguid, 1964; Stirm *et al.*, 1967b; Jones and Rutter, 1974; Jones *et al.*, 1976).

4.2.2 Activities, functions and morphologies of selected bacterial adhesins

This section is concerned with some of the better documented examples of bacteria that adhere to animal cells.

(a) *Enterobacteriaceae*

Many different adhesive properties are exhibited by bacteria that belong to the majority of the genera of this family. For convenience, the classification of the fimbriae produced by these bacteria is that devised by Duguid (Duguid *et al.*, 1966; Duguid and Anderson, 1967). The existance and use of this classification in no way implies the identity of fimbriae of any particular type nor indeed that such identity exists.

Type 1 fimbriae

Type 1 fimbriae and the associated adhesive activities are produced by most cultures of *E. coli* (Duguid *et al.*, 1955; Duguid, 1964; Duguid, 1968) and by *Enterobacteria cloacea* (Constable, 1956; Duguid, 1959), *Shigella flexneri* (Duguid and Gillies, 1957), *Klebiella* spp. (Duguid, 1959; Cowan *et al.*, 1960; Thornley and Horne, 1962; Duguid, 1968), *Serratia marcescens* (Duguid, 1959) and by many serotypes of *Salmonella* (Duguid *et al.*, 1966). Up to 400 fimbriae are found per cell in a peritrichous arrangement (Duguid, 1964). They average 70 Å in diameter and 2 μm in length (Thornley and Horne, 1962; Brinton *et al.*, 1964; Duguid, 1964; Brinton, 1965; Duguid, 1968).

The adhesive activities of type 1 fimbriae are similar, although antigenic differences (Gillies and Duguid, 1958; Duguid and Campbell, 1969) imply chemical and perhaps structural differences. Type 1 fimbriae confer on the bacterial cells the ability to grow as a pellicle on the surface of static broth cultures (Duguid and Gillies, 1957; Duguid *et al.*, 1966; Old *et al.*, 1968) and to attach to a variety of animal and plant cells. Perhaps the most unifying characteristic of type 1 fimbriae is the inhibition of their adhesive activities (see references above and Old, 1972) and pellicle-forming ability (Old *et al.*, 1968) by D-mannose.

The function of type 1 fimbriae remains an enigma. They are produced by saprophytes (*K. aerogenes*), commensals (*E. coli*) and pathogens (salmonellae) but whether or not their function is the same in each of these ecologically different groups is unclear. There is no substantial evidence that the adhesive properties of fimbriae promote pathogenicity (Duguid, 1964; Duguid, 1968) except perhaps for renal infections caused by *Pr. mirabalis* (Silverblatt, 1974). Inter-bacillary adhesion caused by type 1 fimbriae, however, may be of importance. Thus, type 1 fimbriae production increases the rate of transfer of ColI (Mulczyk and Duguid, 1966; Meynell and Lawn, 1967) possibly by increasing the stability of mating pairs (Meynell and Lawn, 1967). Inter-bacillary adhesion is at least in part responsible for the formation of pellicles by fimbriate bacteria grown in static broth culture. Under such conditions when available oxygen is limited, fimbriate bacteria rapidly outgrow non-fimbriate ones (Brinton, 1965; Old and Duguid, 1970) and the transition from non-fimbriate to fimbriate phenotype results in a secondary growth phase of fimbriate bacteria (Old *et al.*, 1968). In contradiction of earlier results (Maccacaro, 1957; Dettori and Maccacaro, 1959; Maccacaro and Turrin, 1959; Maccacaro and Dettori, 1959), Wohlhieter *et al.* (1962) found that respiratory activity and fimbriation are unrelated characteristics and it is probable that the increased oxidative activity of fimbriate bacteria is the result of pellicle formation and location at the oxygen-rich medium/air interface (Old et al., 1968; Old and Duguid, 1970).

Type 2 fimbriae

Type 2 fimbriae, produced by some salmonellae (Duguid and Gillies, 1958; Duguid et al., 1966; Duguid, 1968; Old et al., 1968), are devoid of known adhesive and pellicle-forming activities. Morphologically indistinguishable from type 1 fimbriae, they are sufficiently related antigenically for Old and Payne (1971) to suggest that type 2 fimbriae are non-adhesive forms of type 1 fimbriae.

Type 3 fimbriae

Type 3 fimbriae, produced by *K. aerogenes* and *Serratia marcescens* (Duguid, 1959; Cowan et al., 1960; Duguid et al., 1966; Duguid, 1968) are 48 Å in diameter (Thornley and Horne, 1962). They mediate bacterial adhesion to fungal cells, plant cells, cellulose fibers and glass, but not to animal cells unless these are first modified by chemical or physical means (Duguid, 1959). The unique substrate affinity of type 3 fimbriae led Duguid (1959, 1968) to suggest they may confer some advantage on bacteria that grow as saprophytes in the soil.

Type 4 fimbriae

The type 4 fimbriae of *Proteus* spp. (Duguid and Gillies, 1958; Coetzee et al., 1962; Sheddon, 1962; Hoeniger, 1965; Silverblatt, 1974) are about 40 Å in diameter and cause mannose-resistant haemagglutination of selected erythrocytes (Duguid and Collee, 1960; Coetzee et al., 1962; Sheddon, 1962; Duguid et al., 1966; Duguid, 1968). The *Pr. mirabilis* fimbriae designated as type 3 pili by Silverblatt (1974) are probably the type 4 fimbriae of the Duguid classification (Duguid et al., 1966; Duguid, 1968). The adhesive properties of the type 4 fimbriae produced by *Pr. mirabilis* may contribute to the organisms' ability to cause pyelonephritis (Silverblatt, 1974).

Other fimbriae

A colonization factor (CF) discovered in a strain of *E. coli* pathogenic for man is associated with fimbriae measuring 80–90 Å in diameter (Evans et al., 1975). CF promotes successful bacterial colonization of the intestine of laboratory animals and isolated fimbriae cause mannose-resistant haemagglutination of guinea-pig erythrocytes (Evans, personal communication). Unpublished data (Evans, personal communication) indicates that CF fimbriae may occur in other strains of enteropathogenic *E. coli* and it appears likely, therefore, that CF is an important ecological determinant similar in function to K88 and other *E. coli* adhesins (see below).

Pr. mirabilis produce a second type of fimbria 70 Å in diameter. The dimensions of these fimbriae are similar to those of type 1 fimbriae, but in the absence of information on the mannose sensitivity of their adhesive properties it is not possible to identify them as such. *In vivo*, this form replaces the type 4 fimbriae and is thought to play a significant role in infectivity by allowing the bacteria to adhere to the renal pelvic epithelium (Silverblatt, 1974).

Fibrillar adhesins and haemagglutinins

Non-fimbrial adhesins and haemagglutinins are produced by a minority of *E. coli* and *Salmonella* serotypes and may be produced by cultures that also produce fimbrial adhesins (Duguid, 1964; Duguid *et al.*, 1966; Jones and Rutter, 1974). Fibrillar adhesins are of irregular morphology (Stirm *et al.*, 1967b; Wilson and Hohman, 1974; Jones, 1975; Burrows *et al.*, 1976) and cause mannose-resistant haemagglutination.

The most thoroughly investigated of these fibrillar haemagglutinins and adhesins is the K88 antigen produced by a limited number of serotypes of *E. coli* of porcine origin. K88 haemagglutinins have many characteristics which may be considered typical of this type of activity, including mannose resistance and temperature sensitivity (Stirm *et al.*, 1967b; Jones and Rutter, 1974). However, only haemagglutination is temperature-sensitive. Adhesion to the normal substrate such as the intestinal mucosal surfaces (Arbuckle, 1970; Jones and Rutter, 1972; Hohman and Wilson, 1975), isolated epithelial cells (Jones, 1972; Wilson and Hohman, 1974) and brush-border membranes of such cells (Sellwood *et al.*, 1975) occurs at temperatures up to 37°C. K88 is known to be produced *in vivo* and to function as an adhesin which allows the bacteria to colonize the intestinal mucosa of the host (Jones and Rutter, 1972). Bacteria that do not produce K88 fail to adhere to the mucosa (Jones and Rutter, 1972) and hence fail to colonize and cause disease (Smith and Linggood, 1971; Jones and Rutter, 1972).

Non-fimbrial adhesive properties, designated K99 antigen by serological means (Ørskov *et al.*, 1975), are present in cultures of bovine strains of *E. coli* (Burrows *et al.*, 1976). From these studies and those of Smith and Linggood (1972), it appears that K99 may function *in vivo* in a manner similar to that of K88. It is tempting to suggest that the adhesive properties of other *E. coli* cultures detected *in vitro* (Duguid *et al.*, 1955; Duguid, 1964; see Jones, 1975; McNeish *et al.*, 1975) and *in vitro* (Bertschinger *et al.*, 1972; Jones, 1975; Hohman and Wilson, 1975; Nagy *et al.*, 1976) may play a role in colonization similar to that of K88. Although morphologically similar (Jones, 1975; Hohman and Wilson, 1975; Burrows *et al.*, 1976) it is known that some of the fibrillar adhesins, for example, K88 and K99, differ both antigenically and in the range and specificity of their adhesive activities. Indeed, it is very possible that a particular fibrillar adhesin is restricted to a discrete pathogenic type which can be characterized by its propensity to adhere to or to infect a particular type of animal cell.

(b) *Vibrios*

V. cholerae interacts intimately with mucosal surfaces of the intestine and this phenomenon appears to be of importance in the pathogenesis of cholera (Freter, 1969). However, the ways and means by which *V. cholerae* associates with mucosae appears to involve more than one mechanism (Freter and Jones, 1976); namely, interaction with the mucus (Schrank and Verwey, 1976; Jones *et al.*, 1976) and adhesion to the brush-borders of the epithelial cells (Jones *et al.*, 1976; Nelson *et al.*, 1976). Motility is a prerequisite of the former (Guentzel and Berry, 1975; Freter and Jones, 1976) whereas the presence of flagella is necessary for the latter (Jones and Freter, 1976), although the polar flagellum does not appear to act as the adhesive appendage (Nelson *et al.,* 1976). The interaction with mucus may be caused by entrapment within the mucus (Schrank and Verwey, 1976) or may involve adhesion of vibrios to mucus receptors (Freter and Jones, 1976). However, as recent observations indicate, a more probable explanation of the latter is that mucosal extracts disorientate a chemotactic response of the vibrios to substances released by the mucosa (Allweiss *et al.,* in press).

Recently, the adhesiveness of *V. parahaemolyticus* (Kaneko and Colwell, 1975) has been related to the production of unsheathed lateral flagella by vibrios in response to environmental factors (De Boer *et al.,* 1975; De Boer, 1975; Scheffers *et al.,* 1976). Similar phase variation in *V. cholerae* is unknown, but lateral strands have been observed on the surfaces of vibrios attached to mucosal surfaces (Nelson *et al.,* 1976). These strands are not too unlike the lateral flagella of *V. parahaemolyticus* (Nelson, personal communication) or the lateral fimbriae of some vibrios (Tweedy *et al.,* 1968).

The haemagglutinating activity of *V. cholerae* has been reviewed by Finkelstein (1973). Haemagglutinating activity is associated with the ability to adhere to the surfaces of epithelial cells (Jones and Freter, 1976) but, unlike the haemagglutinins of *E. coli*, the haemagglutinins of some vibrios appear to be only partially inhibited by D-mannose (Barua and Mukherjee, 1965; Tweedy *et al.,* 1968). Other possibly distinct haemagglutinins are associated with slime production (Lankford and Legsomburana, 1965; Neogy *et al.,* 1966).

(c) *Neisseriaceae*

N. gonorrhoeae

N. gonorrhoeae colony types designated 1 and 2 produce lateral fimbriae (Swanson, 1972) approximately 85 Å in diameter (Swanson *et al.,* 1971), whereas those gonococci that grow as colony types 3 and 4 do not produce fimbriae (Jephcott *et al.,* 1971; Swanson *et al.,* 1971). Subsequent studies with a variety of animal cells established that fimbriate gonococci are significantly more adhesive than non-fimbriate gonococci (Swanson *et al.,* 1971; Punsalang and Sawyer, 1973; James-Holmquest *et al.,* 1974; Ward *et al.,* 1974; Swanson *et al.,* 1975a,b) and that fimbriate gonococci adhere to the mucosal surface of fallopian tubes (Ward *et al.,* 1974) and cause haemagglutination (Punsalang and Sawyer, 1973; Waitkins, 1974; Chan and Wiseman, 1975). Haemagglutinating activity is not inhibited by D-mannose and may be restricted to erythrocytes of human origin (Koransky *et al.,* 1975); isolated fimbriae also cause haemagglutination (Punsalang and Sawyer, 1973; Buchanan and Pearce, 1976).

Antigenic heterogeneity of *N. gonorrhoeae* fimbriae (Novotny and Turner, 1975; Buchanan and Pearce, 1976) suggests probable differences in composition and possibly in structure, whereas dissimilar haemagglutinating activity (Punsalang and Sawyer, 1973; Koransky *et al.,* 1975) indicates differences in the nature of the active site of the adhesins or the presence of more than one adhesin in cultures of gonococci. Indeed, Swanson (Swanson *et al.,* 1974; Swanson *et al.,* 1975a,b) has found that the association of gonococci with human leukocytes is independent of the presence of fimbriae and depends upon the presence of a distinct factor (leukocyte association factor).

Other Neisseria
N. meningitidis produces fimbriae somewhat similar in dimensions to the fimbriae of *N. gonorrhoeae* (Devoe and Gilchrist, 1974). Fimbriae and adhesive activities, however, are not confined to pathogenic neisseriaceae. Species of *Neisseria*, considered to be commensals of mucosal surfaces, produce adhesive surface appendages that exhibit both intra- and inter-species differences in form and adhesive properties (Wistreich and Baker, 1971).

(d) *Bordetella*
Only recently has the significance and nature of the haemagglutinin of *Bordetella pertussis* (Keogh and North, 1948; Fisher, 1950; Sutherland and Wilkinson, 1961) been realized and its separation from the pharmacologically active substance, termed leucocytosis-promoting factor (LPF), been achieved (Arai and Sato, 1976). The principal haemagglutinating activity of *B. pertussis* culture supernates is associated with fine filaments 20 Å in diameter and about 400 Å in length, which are often associated with spherical LPF units 60 Å in diameter (Sato *et al.*, 1973, 1974; Arai and Sato, 1976). Filaments of the same width were observed earlier by Morse and Morse (1970) on the cell surface of *B. pertussis* and other *Bordetella* spp. These and haemagglutinin (Sutherland and Wilkinson, 1961) were lost from the cell surface upon prolonged incubation. Recently, Sato (personal communication) has demonstrated that the attachment of *B. pertussis* to animal cells and epithelial surfaces is mediated by these fimbriae.

(e) *Streptococci*
The oral streptococci have been the subjects of a long series of investigations by Gibbons and colleagues, and form the substance of recent review articles (Gibbons and van Houte, 1975; Gibbons, 1975). The study of these organisms has provided some of the most convincing information on the ecological importance of adhesion and the selectivity of this process. Briefly, four streptococcal species (*S. salivarius, S. sanguis, S. mitis* and *S. mutants*) demonstrate preferences for colonizing particular surfaces of the buccal cavity. *S. salivarius* preferentially colonizes the dorsum of the tongue, *S. sanguis* and *S. mutants* predominate on the surface of teeth and *S. mitis* inhabits the non-keratinized oral mucosa (Liljemark and Gibbons, 1972; Gibbons and van Houte, 1975). The relative distribution of these species *in vivo* corresponds with the distribution they assume after their introduction into the mouth (Liljemark and Gibbons, 1972; van Houte *et al.*, 1971) and their relative affinities for surfaces of the buccal cavity (Hillman *et al.*, 1970; Gibbons and van Houte, 1971; Ørstavik *et al.*, 1974). From the work Gibbons and others (Gibbons and van Houte, 1975), the conclusion that different species and perhaps strains of oral streptococci possess different adhesive mechanisms, and that the adhesive properties dictate the habitat of these bacteria, is convincing.

Fibrillar layers on the surfaces of streptococci have been demonstrated (Gibbons et al., 1972; Liljemark and Gibbons, 1972; Lai et al., 1973; Nalbandian et al., 1974). The close association between the fibrillar material and epithelial cell surface suggests that adhesion is mediated by this surface layer (Gibbons et al., 1972; Liljemark and Gibbons, 1972). Indeed, trypsinized bacteria that are devoid of fibrillar material lack adhesive properties (Gibbons et al., 1972). *S. pyogenes* also produces surface fibrillar material (Swanson et al., 1969; Ellen and Gibbons, 1972) that is associated with the selective adhesive properties of the streptococcal cell (Ellen and Gibbons, 1972, 1974; Beachey and Ofek, 1976; Ofek et al., 1976). Removal of the fibrillar material results in the inability to attach to epithelial cells (Ellen and Gibbons, 1972; Ofek et al., 1975) and loss of the ability to colonize the rodent oral cavity (Ellen and Gibbons, 1972). There is some evidence, however, that the adhesins of different *S. pyogenes* strains are not identical (Ellen and Gibbons, 1974).

Other gram-positive cocci such as *Staphylococcus aureus* and *S. agalactiae* adhere to the epithelial cells of the bovine udder (Frost, 1975) and *S. faecalis* attaches to the surface of the rat tongue (Gibbons et al., 1976).

(f) *Lactobacilli*

The lactobacilli of the crop of the domestic chicken are adhesive and attach to crop epithelial cells *in vitro* (Fuller, 1973) and *in vivo* (Fuller and Turvey, 1971). Only lactobacilli isolated from birds were found to adhere to crop epithelial cells; lactobacilli of mammalian origin did not do so (Fuller, 1973). The adhesive lactobacilli of the mammalian stomach (Savage, 1972) have not been examined with *in vitro* tests.

Adhesive lactobacilli appear to attach to the epithelial surface by means of fibrillar material, the staining characteristics of which suggest that it is predominantly carbohydrate in composition (Takeuchi and Savage, 1973; Fuller and Brooker, 1974; Brooker and Fuller, 1975). The fibrillar layer of adhesive avian lactobacilli is well developed whereas little material can be detected on non-adhesive lactobacilli (Brooker and Fuller, 1975). Whether or not the difference between adhesive and non-adhesive lactobacilli is due to quantitative or qualitative differences in this layer remains to be determined.

(g) *Corynebacteria*

Many species of corynebacteria produce surface appendages that vary in size (20—60 Å in width) and morphology depending upon the species (Yanagawa and Honda, 1976). The adhesive properties of *C. renale* and *C. diphtheriae* are associated with such structures (Honda and Yanagawa, 1974; Yanagawa and Honda, 1976). The haemagglutinin of *C. diphtheriae* agglutinates untreated sheep erythrocytes (Yanagawa and Honda, 1976) whereas the haemagglutinin of *C. renale* acts on trypsinized erythrocytes only (Honda and Yanagawa, 1974). The haemagglutinins produced by the three types of *C. renale* differ in antigenic composition, physico-chemical properties and distribution on the cell surfaces (Yanagawa et al., 1968;

Yanagawa and Otsuki, 1970; Kumazawa and Yanagawa, 1973).

C. renale adheres to tissue culture cells (Honda and Yanagawa, 1975) and to the bladder wall of the mouse (Yanagawa, personal communication) and it is very possible that *C. renale* colonizes the urinary tract of cattle and causes disease in much the same way because of these adhesins.

(h) *Mycoplasma*

Species of mycoplasma can adhere to tissue culture cells, leukocytes, and the surfaces of tracheae in organ culture (Zucker-Franklin et al., 1966a,b; Sobeslavsky et al., 1968; Manchee and Taylor-Robinson, 1969a; Collier and Clyde, 1971; Stanbridge, 1971; Collier and Baseman, 1973; Muse et al., 1976). Mycoplasmas also exhibit haemagglutinating activity (Manchee and Taylor-Robinson, 1968), haemadsorption (Manchee and Taylor-Robinson, 1968) and the ability to attach to and subsequently grow on glass surfaces (Somerson et al., 1967; Taylor-Robinson and Manchee, 1967b; Bredt, 1968). Although haemagglutination and haemadsorption are superficially similar, the results of comparative studies suggest that they are separate processes and distinct from adhesion to glass (Manchee and Taylor-Robinson, 1968; Taylor-Robinson and Machee, 1967b).

Some mycoplasmas have distinct organelles located at one pole of the cell (Maniloff et al., 1965; Collier and Clyde, 1971). The organelle of *M. pneumoniae* is 80–100 x 250–300 nm in size and is composed of an electron-dense core surrounded by a translucent space (Collier and Clyde, 1971; Wilson and Collier, 1976). The organelle of *M. gallisepticum* measures 80 x 130 nm and consists of an electron translucent zone delineated by electron-dense plates at the exterior and interior zones (Maniloff et al., 1965).

Electron microscopy demonstrated that when mycoplasmas attach to tissue cell surfaces, the organelle is located immediately adjacent to the tissue cell (Zucker-Franklin et al., 1966a; Collier and Clyde, 1971; Muse et al., 1976) and probably acts as the adhesin. *M. pulmonis* and *M. neurolyticum* do not appear to attach in this way (Zucker-Franklin et al., 1966a; Jones and Hirsch, 1971). This adhesins of mycoplasmas are not identical; the presence or absence of polar organelles of differing forms, and intra- and interspecies differences in haemagglutinating and other adhesive activities (Taylor-Robinson and Manchee, 1967a,b; Manchee and Taylor Robinson, 1968; Sobeslavsky et al., 1968), all indicate the diversity of this phenomenon in these bacteria.

Adhesive properties probably facilitate colonization of the surfaces of the animal body. A few pathogenic types of mycoplasma have been examined and those that are virulent are invariably adhesive (Lipman and Clyde, 1969; Lipman et al., 1969; Clyde, 1975).

(i) *Normal flora*

It must be mentioned that the normal flora of the oral cavity and the flora of the gastro-intestinal tract may attach in other ways to those mentioned above

(Savage, 1975). Firstly, attachment to other bacteria already present on the surface (Gibbons and van Houte, 1975); secondly, adhesion to or entrapment within mucus (Savage *et al.*, 1971; Davis *et al.*, 1973; Leach *et al.*, 1973) and thirdly, partial penetration of the animal cell to form a hold-fast (Hampton and Rosario, 1965; Takeuchi and Zeller, 1972; Davis and Savage, 1974 and 1976).

4.3 CHEMICAL AND PHYSICAL PROPERTIES OF ADHESINS

There exists little similarity in the special properties of adhesive appendages. For the purpose of this discussion only, adhesins are classified as proteinaceous or non-proteinaceous adhesins.

4.3.1 Proteinaceous adhesins

Included in this category are the type 1 fimbriae and K88 antigens of *E. coli*, the fimbriae of *N. gonorrhoeae* and *B. pertussis* and the fibrillar material of *C. renale*. Much of the data has been derived from studies on single strains. The probable heterogeneity of some of these adhesins has been mentioned and, consequently, the assumption that the findings presented here apply equally well to the adhesins of even closely related organisms is not warranted.

(a) *Chemical composition and molecular weights*

Amino acid compositions of type 1 fimbriae, K88 and *C. renale* adhesin are compared in Table 4.1. The type 1 fimbriae of *E. coli* are composed of almost 100% protein (Brinton *et al.*, 1961; Brinton, 1965) with less than 0.6% nucleic acid, phosphorus and carbohydrates (Brinton and Stone, 1961). Amino acids accounted for over 90% of the dry weight of the K88 of a strain of *E. coli* (Stirm *et al.*, 1967a). The remainder, which is probably composed of cell wall materials, consists of lipids, neutral sugars and inorganic ash. The adhesin of a strain of type 2 *C. renale* is composed of protein, but carbohydrates were not detected (Kumazawa and Yanagawa, 1972). Whether or not the adhesins of the other two types of *C. renale* are identical is questionable because important differences have been noted (Kumazawa and Yanagawa, 1973).

The adhesins of mycoplasmas have been identified variously as lipid or lipoprotein in *M. pneumoniae* (Sobeslavsky *et al.*, 1968) and as 'part of the cell membrane' of *M. hominis* (Hollingdale and Manchee, 1972). More recent electron microscopic studies indicate that the polar organelles of *M. gallisepticum* and *M. pneumoniae*, contain basic proteins (Maniloff, 1972; Wilson and Collier, 1976), and recent work (Baseman, personal communication) has resulted in the isolation of a large molecular weight protein (160 000 daltons) from virulent adhesive *M. pneumoniae* which might be the adhesin.

The molecular weights of the subunits of type 1 fimbriae and *C. renale* adhesins,

Table 4.1 Amino acid composition of three bacterial adhesins and the hydrophobicity of their constituent amino acid side chain

Amino acid	Mol/10^5 g protein			Hydrophobicity §
	K88,*	Type 1 fimbriae†	C. renale adhesin ‡	($\mu°_{organic} - \mu°_{water}$) cals mol^{-1}
Tryptophan	15	0	5	-3400
Phenylalanine	34	48	20	-2500
Tyrosine	33	12	30	-2300
Leucine	62	60	50	-1800
Isoleucine	39	24	29	NV ¶
Valine	66	78	74	-1500
Lysine	35	18	59	ca. -1500
Methionine	10	0	6	-1300
1/2 Cystine	0	12	4	NV
Alanine	80	205	90	-500
Histidine	1	12	15	-500
Proline	19	12	46	NV
Threonine	84	120	85	-400
Glycine	97	102	79	0
Serine	52	60	40	$+300$
Glutamic acid	57	78	117	NV
Aspartic acid	94	120	113	NV
Arginine	28	18	31	NV

* Data from Stirm et al., (1967a).
† Calculated from Brinton (1965).
‡ Taken from Kumazawa and Yanagawa (1972).
§ Data from Nozaki and Tanford (1971) and Tanford (1973) where $\mu°_{organic} - \mu°_{water}$ represents the additional change in free energy (compared with the hydrogen atom side chain of glycine) upon the transfer of the side chain from water to an organic solvent.
¶ NV, no values available.

calculated from amino acid analysis data are 16 600 (Brinton, 1965) and 19 400 (Kumazawa and Yanagawa, 1973) respectively. There is an obvious discrepancy between the minimum molecular weight of K88, which can be calculated from the amino acid composition, and the molecular weight of 2.3 x 10^4 (Jones, 1975) found by polyacrylamide gel electrophoresis. B. pertussis fimbrial fragments (20 x 400 Å) isolated from culture supernates are composed of over 80% proteins, and have a molecular weight of 126 000 even after treatment with sodium dodecyl sulphate and mercaptoethanol (Arai and Sato, 1976).

(b) *Structural characteristics*

Brinton deduced that the subunits of type 1 fimbriae are organized into a right-handed helix (Brinton, 1965; Brinton and Beer, 1967) to form a rigid rod 70 Å in diameter. Each rod is thought to have an axial hole estimated to be 20–25 Å in diameter (Brinton and Beer, 1967). The subunits spontaneously organize into rods under suitable conditions, and rods may be dispersed with a variety of hydrogen-bond splitting agents (Brinton, 1965). The apparent rigidity of *B. pertussis* fimbrial fragments and of the fimbriae of other enterobacteria may indicate a subunit organization in these fimbriae similar to that proposed for type 1 fimbriae of *E. coli*. Fuerst and Hayward (1969) proposed a similar helical structure for the filamentous appendages of pseudomonads. Structurally, the *C. renale* adhesin and K88 appear to be quite different from type 1 fimbriae. Both tend to form bundles or aggregates of varying diameter in the cell-bound (Stirm *et al.*, 1967b; Yanagawa *et al.*, 1968; Yanagawa and Otsuki, 1970) and in the cell-free states (Stirm *et al.*, 1967b; Kumazawa and Yanagawa, 1972). By high resolution electron microscopy, K88 appears to exist as a linear array of subunits that form intertwining filaments on the bacterial surface (cited by Jones, 1975). The high molecular weight polymer of K88 [Svedberg constants 33.5 (Jones, 1972) and 36.7 (Stirm *et al.*, 1967a) and greater than 1×10^6 daltons by gel filtration (Jones, 1975)] and the *C. renale* adhesin (Kumazawa and Yanagawa, 1972) are readily depolymerized by detergents or urea.

(c) *Physical properties*

From Table 4.1 it can be seen that there is a tendency for the amino acid residues that make up type 1 fimbriae and K88 to have apolar side chains. Recent analysis of gonococcus fimbriae also demonstrates a preponderance of amino acids with apolar side chains (Ward, personal communication). In contrast, hydrophilic amino acids are more abundant in the *C. renale* adhesin. However, whether or not these adhesin proteins exhibit hydrophobic patches will depend on sequences in addition to primary structure.

The hydrophobic properties of type 1 fimbriae, demonstrated by their tendency to aggregate in aqueous solution (Brinton, 1965) may account for the attachment of isolated fimbriae to polystyrene beads, for the adhesion of fimbriate bacteria to animal cell surfaces (Brinton, 1965) and for the localization of *E. coli* pellicles at air/water interfaces. However, aggregation results from parallel stacking or angled layering of the fimbrial rods whereas attachment to a surface occurs at one end (Brinton, 1965). Indeed, attachment is very similar to the end-on interaction of *B. pertussis* fimbrial fragments with erythrocyte membranes (Sato *et al.*, 1973). The adhesive properties displayed by the ends of both these appendages might be due to their hydrophobic nature; but whether the same properties operate in holding the subunits together, or whether in the cell-bound state these sticky ends are proximal (and therefore occluded) or distal (and potentially adhesive) to the surface of the bacterial cell, is a matter of conjecture. It should be noted that the *B. pertussis* fimbrial fragments are asymmetrical; only one end attaches to the erythrocyte membrane and haemagglutination results from the interaction of the free ends of fimbriae attached to adjacent surfaces (Sato, personal communication).

A possibly significant characteristic of the fimbriae of *E. coli* and of *N. gonorrhoeae* is the reduction in surface charge conferred on the bacterial cell by these appendages (Brinton *et al.*, 1954; Brinton, 1959 and 1965; Heckels *et al.*, 1976). Consequently, the mutual repulsion between a fimbriate bacterium and an animal cell will be less than that between a non-fimbriate bacterium and an animal cell, and this must facilitate the close approach necessary for adhesion (Section 4.6.3(a) and (b)).

4.3.2 Non-proteinaceous adhesins

Some streptococci and lactobacilli attach to cell surfaces by means that do not appear to include adhesins of a protein nature.

(a) *Chemical composition*

Ellen and Gibbons (1972) noted that streptococci which lacked the surface fibrillar layer and associated M antigen were not adhesive (Ellen and Gibbons, 1972 and 1974). Trypsinization destroyed the fibrillar layer and adhesive properties of *S. pyogenes* (Ellen and Gibbons, 1974) and had similar consequences for other streptococcal species (Gibbons *et al.*, 1972; Liljemark and Gibbons, 1972; Lai *et al.*, 1973). Beachey and Ofek (1976) concluded that *S. pyogenes* still attached to mucosal cells of the buccal cavity after the M protein had been removed, but did not attach unless the fibrillar layer remained intact. Subsequent studies (Beachey, 1975; Ofek *et al.*, 1977; Beachey and Ofek, 1976) demonstrated that the fibrillar layer was destroyed by treatments that also extract lipoteichoic acids (LTA) and that streptococcal adhesion was inhibited if epithelial cells were pre-treated with LTA or with the lipid component of LTA, or if the bacteria were treated with LTA antibody. Their observation that the LTA was bound to the surface of animal cells via the lipid moiety (mainly palmitate) is consistent with the results of others. Consequently, the conclusion that the LTA of streptococcal cell walls, and in particular the lipid moiety, appears to play a major role in the adhesive activity is a reasonable conclusion.

Lipoteichoic acids have been reviewed by Knox and Wickens (1973) and Wickens and Knox (1975). In the cell-bound state, it is thought that the hydrophobic lipid component interacts with the cell membrane (Joseph and Shockman, 1975), whereas the hydrophilic moiety may extend through the cell wall to the outer surface (Knox and Wickens, 1973). Indeed, teichoic acids are found exposed on the outer walls of several gram-positive species (Knox and Wickens, 1973; Birdsell *et al.*, 1975; Wickens and Knox, 1975) and may be excreted into the supernatant culture fluid (Markham *et al.*, 1975).

The adhesive properties of lactobacilli are associated with an extensive polysaccharide-rich fibrillar layer on the cell surface (Brooker and Fuller, 1975). In comparison with the adhesins of streptococci (Gibbons *et al.*, 1972; Ellen and Gibbons, 1974) the adhesive activities of lactobacilli are only slightly impaired by treatment with trypsin, but markedly impaired by periodate oxidation and by pepsinization (Fuller, 1975). Concanavalin A inhibits adhesion and also precipitates carbohydrates released by pepsin, a treatment that renders cells non-adhesive (Fuller, 1975). A carbohydrate that binds Con A, therefore, appears to be involved in the adhesive mechanism, but whether or not some other material which is associated with the carbohydrate layer acts as the adhesin remains to be determined.

(b) *Mode of action*
Obvious questions need to be answered before a mechanism of streptococcal adhesion that involves an LTA adhesin can be accepted. It must be established that lipoteichoic acids constitute part of the surface fibrillar layer and that the fatty acids of LTA are presented on the cell surface in a way that allows their interaction with animal cell surfaces. Consequent upon these, the way in which the fatty acids may be shielded from the aqueous phase assumes importance. Of no less importance are the mechanisms by which hydrophobic interactions may result in the selective adhesion of streptococci to particular types of eukaryotic cells (Ellen and Gibbons, 1974; Gibbons *et al.*, 1976; Gibbons and van Houte, 1975; Ofek *et al.*, 1976). Finally, the possibility that inhibition of adhesion by LTA is an artefact due to interference similar to that proposed for the inhibition by antibody-mediate haemagglutination by lipids (Murray *et al.*, 1950) must be investigated.

The answers to some of these problems may be found in the studies of Springer and Adye (1975) on the cell wall lipopolysaccharide (LPS) of gram-negative bacteria. LPS binds to erythrocytes through the lipid A moiety. Binding is promoted by some fatty acid components (amide linked 3-D-hydroxymyristic acid) and hindered by others. Furthermore, these workers have isolated from erythrocytes, leukocytes and platelets, membrane receptors for LPS (Springer *et al.*, 1973; Springer *et al.*, 1974; Springer and Adye, 1975). Binding of LPS occurs at a different site from that of LTA (Beachey, 1975) and the *O*-stearoyl derivative (Slade and Hammerling, 1968) of streptococcal cell wall carbohydrate antigen (Springer *et al.*, 1970). Thus, it is possible that selective adhesion may be mediated by particular fatty acids of LTA, in particular, palmitate, re-esterification with which partially restores the binding properties of hydrolysed LTA (Oftek *et al.*, 1975). The necessary shielding of the fatty acids from the aqueous phase may be achieved, for example, by the complexing of fatty acids with the fibrillar-associated protein antigens (Swanson *et al.*, 1969; Lai *et al.*, 1973; Fox, 1974). Kimbelberg and Papahadjopoulos (1971) have proposed a model that is relevant in this respect.

It is too early to assume that the LTA adhesin exists and that the adhesive mechanism of all streptococci are similar, although the inhibition of the attachment of *S. pyogenes* to epithelial cells by cell wall fragments of *S. salivarius* (Ellen and Gibbons, 1974) indicates a similarity between these adhesive mechanisms. Explanations for the involvement of glycoproteins (Williams and Gibbons, 1975; Gibbons and van Houte, 1975) and of the attachment of streptococci to hydroxapatite surfaces whether or not they are coated by glycoproteins (Hillman *et al.*, 1970; Ørstavik *et al.*,

1974) need to be forthcoming. The latter, however, may be potentiated by LTA (Markham et al., 1975).

4.4 THE NATURE OF THE SURFACE OF THE EUKARYOTIC CELL

4.4.1 Lipid bilayer

It has been calculated that there is sufficient lipid in the human erythrocyte membrane to account for about 80% of the membrane surface (Guidotti, 1972). Almost 50% of this lipid is in the form of phosphoglycerides (Rouser et al., 1968). The polar head groups (glycerol, inositol, serine, ethanolamine or choline residues) of the phospholipids interact with the aqueous phase, whereas the hydrocarbon chains form the apolar internal domain of the membrane. Phospholipids, however, may be distributed asymmetrically in the lipid bilayer, with phosphatidylcholine (and the hydrophilic hydroxyl groups of the cholesterols) located on the external surface of membranes, and phosphotidylserine and phosphotidylethanolamine on the internal surface (Tanford, 1973).

The external hydrophilic regions of the lipid bilayer, therefore, are obvious sites for the attachment of bacterial adhesins, and there is evidence that this is so. However, the availability of the receptor sites must be influenced by the space occupied by more external components, in particularly the heterosaccharides. Presumably, hindrance of this type will depend on the particular type of cell used as substrate.

4.4.2 Membrane proteins

Although most of the integrated membrane protein molecule is probably sequestered within the hydrophobic domain of the membrane (Singer and Nicolson, 1972; Nicolson, 1974a), parts are exposed on the outer surface of the cell and interact with the aqueous phase (Cook and Stoddart, 1973; Marchesi et al., 1974). The major membrane proteins of erythrocytes are glycoproteins (Guidotti, 1972; Nicolson, 1974a) and, although adhesins may interact with the exposed parts of these proteins, it may be expected that the availability of such sites is influenced greatly by the presence of large branched oligosaccharides.

4.4.3 Membrane carbohydrates

The carbohydrate layer or glycocalyx (Bennett, 1963) appears to be present on the surfaces of most animal cells (Rambourg, 1971) including the exposed brush-border surfaces of intestinal epithelial cells (Ito, 1969; Rambourg, 1971), the surface of the avian crop (Brooker and Fuller, 1975) and the surface of erythrocytes (Rambourg, 1971), to mention but three cell surfaces to which bacteria attach.

The glycocalyx is an integral part of the cell membrane. The heterosaccharides are linked covalently to membrane proteins or lipids. Some immunologically active heterosaccharides may be linked to lipid or protein (Cook and Stoddart, 1973; Hughes, 1973) whereas others are invariably present as glycoproteins (Hughes, 1973). The most common constitutive monosaccharides are L-fucose and D-isomers of mannose, galactose, 2-deoxy-2-acetamido-glucose, 2-deoxy-2-acetamido-galactose and sialic acids; fucose and sialic acid are found only at the terminal non-reducing ends of complete chains. The partial molecular structures of several heterosaccharides are known (for examples see Kiss, 1970; Gottschalk, 1972; Marshall, 1972; Hughes, 1973; Cook and Stoddart, 1973; Lee and Smith, 1974). Upon inspection it rapidly becomes apparent that there is considerable variation in the lengths, degrees of branching, linkages and composition of the various heterosaccharide units. Heterogeneity (Gibbons, 1963; Montgomery, 1972) and gross structural diversity are due to genetic differences (Hughes, 1973), or possible changes due to cell phase (Glick, 1974; Nicolson, 1974a) and maturation (Etzler and Branstrator, 1974).

Clearly, the cell surface heterosaccharides are optimally placed to act as receptors for bacterial adhesins and there is some evidence to support this view. If bacterial adhesins behave like antibody and react with a few sugar residues only, then a single glycoprotein may carry determinants to several bacterial adhesins as they may do for two or more blood group-specific antibodies (Nicolson, 1974a).

4.5 CELL SURFACE RECEPTORS OF BACTERIAL ADHESINS

4.5.1 Detection of animal cell receptors

(a) *Adhesion inhibition tests*

The degree of inhibition that results upon the addition of a particular a substance to the adhesion or haemagglutination test system is measured quantitatively or qualitatively. It is assumed that inhibitory substances are in part structurally similar to the receptors of the animal cell surface and are bound by the adhesin and hence effectively prevent the adhesin reacting with the cell surface receptor. However, non-specific inhibition can result from changes in the animal cell surface (Murray *et al.*, 1950; Ceppallini and Landy, 1963), and possibly from changes in the surface of the bacterial cell, at sites other than the active site of the adhesin. The problem of non-specific inhibition can be partially overcome by demonstrating that the inhibitor is bound by the purified bacteria-free adhesin and by adhesive bacteria and that it is a component of the animal cell membrane. Inhibitors should not bind to any significant extent to non-adhesive mutant bacteria nor to the animal cell. Bacterial aggregation is another source of error (Freter and Jones, 1976) because the extent of adhesion often depends upon the initial concentration of bacteria in the test mixture (e.g. Jones *et al.*, 1976).

The substances most often examined for inhibitory activities are epithelial and other glycoproteins from convenient sources (Gottschalk, 1966; 1972) and their constituent monosaccharides or oligosaccharides (e.g., those from milk, see Kornfeld *et al.*, 1974; Ginsberg, 1972), various synthetic glycosides, sterols and phospholipids. The inhibitory activities of such substances, compared with the inhibitory activities of preparations from animal cells, provide some idea of the nature of the active moiety of the cell surface. Immobilized substances provide the added advantage that their interaction with adhesive bacteria can be monitored by microscopy. This has been achieved with sterols by incorporating them into liposomes (Sato, personal communication) and by covalently attaching a monosaccharide inhibitor to an insoluble carrier (Jones and Freter, 1976).

Finally, it should be noted that inhibition is seldom total. Besides the obvious possibility that this is due to the presence of a second adhesive system in the culture, there is also the possibility that the overall molecular size of the inhibitor is of importance (Springer, 1970) in addition to its precise molecular configuration. Haemagglutination by *V. cholerae*, for example, is inhibited more effectively by fucosides than by the fucose, and the larger the substitutive group the less fucose is required to inhibit (Jones and Freter, 1976).

(b) *Modification of the animal cell surface*
It cannot be assumed that reagents that destroy the receptivities of cells do so because they act directly on the adhesin receptor. Loss of receptivity can be the result of changes in the molecular configuration or distribution of the receptor, or may result from the release of the receptor from the surface. Treatment with trypsin (Honda and Yanagawa, 1974) or sialidase (Sobeslavsky *et al.*, 1968), has been found to increase adhesion. The former is probably due to the aggregation of surface components (and hence concentration of receptors) caused by trypsinization (Tillack *et al.*, 1972; Nicolson, 1974b) and the latter is probably the result of the reduction in surface charge. Blocking of receptors with antibody or with lectins is equally difficult to interpret because such reagents may cause steric hindrance. For example, it was found that lectins with specificities for different carbohydrates (Lis and Sharon, 1973; Nicolson, 1974a) inhibited the adhesion of *E. coli* to brush-border membranes, but inhibition was independent of the sugar specificity of the lectin and only depended upon the ability of the lectin to bind to the brush-border (unpublished observation).

4.5.2 Chemical composition of receptors

For convenience, receptors are classified as carbohydrate (including glycoproteins) and as lipid.

(a) *Carbohydrate receptors*
Adhesins that appear to interact with carbohydrate receptors are the type 1 fimbriae of *E. coli*, a haemagglutin of *Pseudomonas aeruginosa* and possibly K88 antigen.

Also included are mycoplasms, *V. cholerae,* avian lactobacilli and streptococci.

D-*Mannose*

It has been recognized for some years that D-mannose inhibits the adhesive, haemagglutinating and pellicle-forming activities of type 1 fimbriae. This phenomenon was examined in detail by Old (1972). From this study it can be concluded the inhibition of haemagglutination is a highly specific event that depends upon the stereochemistry of D-mannopyranose. The importance of the configuration of hydroxyls at carbons 2, 3, and 4 can be deduced from the observation that D-glucose, D-altrose, and D-talose were not inhibitory. 2-deoxy-D-mannose (i.e. 2-deoxy-D-glucose) and 6-deoxy-D-mannose (i.e. D-rhamnose) were weakly inhibitory, but the substitution of large chemical groups for the hydroxyls of carbons 2 and 6 appeared to prevent binding and hence no inhibition was noted. In contrast, mannosides were inhibitory although α-D-mannosides were more inhibitory than β-D-mannosides.

A high degree of specificity in the attachment of fimbria to mannose residues of the animal cell is suggested by these results. However, other evidence appears contradictory. If D-mannose causes inhibition by effectively competing with receptors that bind fimbriate bacteria firmly, it is to be expected that mannose would also bind firmly to fimbriate bacteria. However, this appears not to be the case (Duguid and Gillies, 1957) although a more precise measure of binding is required before this point can be accepted. Moreover, fimbriate bacteria agglutinate human erythrocytes poorly, or not at all, although mannose is a component of the heterosaccharides of the erythrocytes surface (Cook and Stoddart, 1973). Furthermore, D-fructose also inhibited haemagglutination, yet D-fructose is an unlikely component of animal cell surfaces. Whether or not D-mannose and D-fructose bind to fimbriae or, if they do, whether they bind to the same site is unknown.

L-*Fucose*

The adhesion of *V. cholerae* to brush-border membranes, and the haemagglutination caused by *V. cholerae,* are inhibited by L-fucose, slightly inhibited by L-galactose and not inhibited by D-fucose (Jones and Freter, 1976). Fucosides are more inhibitory than fucose and produce degrees of inhibition that depend upon the molecular size of the fucoside. Fucose appears to be bound to the vibrio cell (R. Freter, personal communication). Sepharose beads to which L-fucose had been covalently linked also bind vibrios and this reaction is inhibited by L-fucose, but not by D-fucose. The conclusion that adhesion to brush-border surfaces is mediated by some unknown adhesin reacting with an L-fucosyl receptor is consistent with the observation that fucose is a component of the brush-borders (unpublished observations) and is present on the surfaces of human erythrocytes (ABO antigens). Adhesion is also inhibited by D-mannose but this sugar does not augment inhibition by L-fucose and it may act therefore on some other adhesive mechanism. The latter adhesive mechanism may be the one described by Tweedy *et al.* (1968).

Neuraminic acid
Gesner and Thomas (1965) demonstrated that the haemagglutination of turkey erythrocytes by *M. gallisepticum* does not occur if the erythrocytes are pre-treated with neuraminidase, or if sialic acid-rich glycoprotein or neuraminic acid are added to the reaction mixture. These observations were confirmed by Sobeslavsky, *et al.* (1968) who employed a cytoadsorption test. However, Manchee and Taylor-Robinson (1969b) found that only three of six *M. gallisepticum* strains completely failed to adsorb erythrocytes and HeLa cells after these cells had been treated with swine influenza virus. Differences in the cytoadsorption and haemagglutination tests exist and it is perhaps unwise to compare results obtained with these test systems. Notwithstanding this difference and the possibility that sialic acid residues are not all equally susceptible to sialidases, there is some reason to believe that sialic acid residues constitute the receptor for the adhesins of some strains of *M. gallisepticum.*

M. pneumoniae is similar to *M. gallisepticum* in that haemadsorption (Sobeslavsky *et al.*, 1968; Lipman and Clyde, 1969) and attachment to tracheal organ cultures (Collier and Baseman, 1973) appear to depend on sialic acid receptors. Again, however, the haemagglutinating activity of only some strains is affected by neuraminidase treatment of the erythrocytes (Manchee and Taylor-Robinson, 1969b) which indicates that haemagglutination by *M. pneumoniae* is not due to a single mechanism.

The adhesion of other serotypes of mycoplasma to animal cells is unaffected by the removal of neuraminic acid (Sobeslavsky *et al.*, 1968; Manchee and Taylor-Robinson, 1969b). Indeed, the haemadsorption of *M. orale* is increased after neuraminidase treatment (Sobeslavsky *et al.*, 1968).

D-Galactose
Gilboa-Garbor *et al.* (1972) described the purification of a *Pseudomonas aeruginosa* haemagglutinin by affinity chromatography on Sepharose columns. The haemagglutinin was inhibited by D-galactose and to a lesser extent by *N*-acetyl galactosamine and lactose (Gilboa-Garber, 1972). It is not known if this adhesive reaction plays any role in the attachment of this organism to eukaryotic cell surfaces.

The binding of K88 to galactose residues is far less certain. K88 haemagglutination is not inhibited by D-galactose or by other monosaccharides and glycosides (Jones, 1972; Gibbons *et al.*, 1975), and K88 does not bind the free carbohydrates (unpublished observations). Haemagglutination is inhibited by some glycoproteins (Jones, 1972; Gibbons *et al.*, 1975), however, and it was tentatively suggested that the β-D-galactosyl-1, 4-*N*-acetyl hexosamine moiety that was common to inhibitory glycoproteins may account for this activity (Jones, 1972). Subsequent studies after sequential chemical degradation of the heterosaccharides of inhibitory and non-inhibitory glycoproteins were inconclusive (Gibbons *et al.*, 1975).

Glycoproteins
Recent evidence (R.A. Gibbons, personal communication) suggests that the receptor for K88, on the porcine intestinal brush-borders, is a glycoprotein. A carbohydrate-rich

fraction has been isolated from brush-borders which reacts with K88-producing bacteria and causes them to aggregate; mutant bacteria that do not produce K88 are not agglutinated by this fraction. Of some significance is the observation that pig brush-borders that lack receptor capacity (Sellwood *et al.*, 1975) do not yield this aggregation factor.

The adhesion of oral streptococci to enamel surfaces is fostered by the presence on the enamel of a glycoprotein pellicle (Ørstavik *et al.*, 1974), but this is not characteristic of all strains of streptococci (Hillman *et al.*, 1970). The adhesion of streptococci to glycoprotein coated enamel (Ørstavik *et al.*, 1974) and to epithelial cells (Williams and Gibbons, 1975) is inhibited in the presence of salivary glycoproteins. One interpretation of such observations is that the epithelial cell receptors of streptococcal adhesins are glycoproteins (Gibbons and van Houte, 1975). However, salivary glycoproteins also agglutinate streptococci (Hay *et al.*, 1971; Ørstavik *et al.*, 1974; Williams and Gibbons, 1975) and this raises the possibility that non-specific inhibition caused by agglutination was a factor that contributed to inhibition (Section 4.5.1a). Some blood group antisera neutralize the inhibitory activity of glycoproteins and block the adhesion of streptococci to epithelial cells (Williams and Gibbons, 1975), whereas monosaccharides inhibit the haemagglutinating activity of oral streptococci (McCabe and Smith, 1976).

The facts that streptococci are agglutinated by glycoproteins and that glycoproteins influence the adhesion of streptococci differently (Ørstavik *et al.*, 1974; Gibbons and van Houte, 1975) are, in themselves, highly significant. This may reflect the existence of two mechanisms (Ørstavik *et al.*, 1974), which is consistent with the postulate that glycoproteins act as receptors on enamel surfaces and that other substances, possibly phospholipids, function as the main receptors of animal cells.

(b) *Receptors on the lipid bilayer*

The most clear demonstration of the interaction of bacterial adhesins with components of the lipid bilayer has been that of Sato and associates (Suzuki *et al.*, 1974; Y. Sato, personal communication). These workers showed that haemagglutination by *B. pertussis* filaments was inhibited by liposomes that contained sterols with a *cis*-hydroxyl at the 3-position of the sterol nucleus, but not by liposomes that contained sterols that lacked the *cis*-hydroxyl. However, both stigmasterol and dehydro-epiandrosterone have 3-*cis*-hydroxyls, but only the former was inhibitory. They differ, however, in terms of their side chains and consequently, it is possible that the side chains of the sterol nucleus influence the inhibitory activity of liposomes by the way this moiety determines insertion into the lipid bilayer.

S. salivarius (Gibbons *et al.*, 1972) and *S. pyogenes* (Ellen and Gibbons, 1974) adhesins may also interact with lipids of the animal cell membranes. The adhesion of *S. salivarius* to epithelial cells was reduced in the presence of sphingomyelin, lecithin, phosphatidyl-L-serine and phosphatidylethanolamine. Although the adhesiveness of *S. pyogenes* was affected in a similar manner, strain variation in sensitivity to phospholipids is apparent. The possible interaction of the lipids of LTA with the phospholipids of the cell membrane is perhaps similar to the mechanism proposed by Springer and Adye (1975) to account for the interaction the LPS with the membranes of erythrocytes. Extracts of human leucocytes and platelets that inhibit the binding of LPS to erythrocytes contained mainly glycerophosphatides with some cholesterol. Of the limited number of commercial compounds tested, phosphatidylethanolamine and, to a lesser extent, cholesterol effected some inhibition. Accordingly, LPS may react with particular cell membrane lipid materials, although these studies suggest that the inhibitor is probably not identical to accepted phospholipids. The important aspect of this work is that interactions of measurable, although not absolute (Springer and Adye, 1975), specificity can occur between lipoid materials. It should be noted that the LPS receptor of erythrocyte membranes is chemically different from the receptor of leukocytes and is probably a lipoglycoprotein (Springer et al., 1973; Springer et al., 1974).

4.6 THE INTERACTION OF BACTERIA WITH ANIMAL CELL SURFACES

The biophysics of eukaryotic cell interactions have been the subjects of study for a number of years. With the exception of a few groups of investigators, there has been little work of a similar nature on prokaryotic—eukaryotic cell interactions. The following is a general account of cell interactions as they may apply to the interaction of bacteria with animal cells.

4.6.1 Cell surface potential

The surfaces of both prokaryotic (Abramson et al., 1942; Davis et al., 1956; Plummer et al., 1962; James and List, 1966; Marshall, 1967; James and Brewer, 1968a,b; Schott and Young, 1972; Stotzky, 1974) and eukaryotic (Ambrose, 1966; Weiss, 1970; Weiss, 1972) cells have a negative potential that results from the ionization of various chemical groups of the surface zone. The surface potential of the human erythrocyte, for example, results mainly from the ionization of the sialic acid carboxyl groups (Eylar et al., 1962). Other surface groups that may contribute to the surface potential of animal cells are discussed by Weiss (1972). The chemical entities responsible for the surface potentials of bacterial cells vary with species, strain and growth conditions (Plummer et al., 1962; Hill and James, 1972a,b). As examples, the teichoic acids of the cell walls of *Staphylococcus aureus* (James and Brewer, 1968a; Marshall and James, 1971), the glucuronic acid of the capsules of *Klebsiella aerogenes* (James and List, 1966), the fimbriae of *E. coli* (Brinton et al., 1954; James and List, 1966) and the hyaluronic acid of *S. pyogenes* capsules (Plummer et al., 1962) markedly influence surface potential. Loss or modification of these surface components may result in a change in surface potential which reflects the influence of other groups (Brinton et al., 1954; Hill and James, 1972b). The possible nature of the ionized groups of bacterial surfaces may be found in a chapter by Richmond and Fisher (1973).

The negative charge of the fixed groups on a surface attract positively charged ions from the surrounding medium, which form a diffuse electrical double layer. The double layer, for all practical purposes, is part of the cell surface. When a cell is caused to move by subjecting it to an electrical field, a thin layer of fluid and the ions of the double layer move with the cell. It is apparent, therefore, that free electrophoresis does not measure the true potential of a surface, but rather the lower zeta (ζ) potential at the hydrodynamic slip plane of the diffuse double layer and the bulk liquid (Weiss, 1972). Under physiological conditions, the surface potential of animal cells probably reflects ionized groups within 10 Å or less of the hydrodynamic slip plane (Heard and Seaman, 1960).

4.6.2 The DLVO theory

The apparent paradox that two bodies, both with negative surface potentials, can attract and then adhere to one another can be explained, in part, by the application of the lyophobic colloid theories of Derjaguin and Landau (1941) and of Verwey and Overbeek (1948). The so-called DLVO theory, which has been treated extensively in several papers (Pethica, 1961; Weiss, 1970; Bangham, 1972; Good, 1972; Weiss, 1972), considers that the energy of interaction of two charged particles of like sign and magnitude is the sum of the electrostatic energy of repulsion V_R and the energy of attraction V_A provided by London–van der Waals' forces. The forms of equations that describe the interaction of two bodies of different geometry, charge and size have been established (Weiss, 1972). The simplest form describes the interaction of two similar spherical particles and is given by:

$$V_R = \tfrac{1}{2} \Psi_0^2 \, r\epsilon \, \ln[1 + e^{-Kd}] \tag{4.1}$$

$$\text{and } V_A = \frac{-Ar}{12d} \tag{4.2}$$

where Ψ_0 is the surface potential of the spheres, r is their radius of curvature and d is their distance of separation; ϵ is the dielectric constant, K is the inverse Debye–Hückel parameter and A is the Hamaker constant.

The forces of repulsion (V_R) and of attraction (V_A) between approaching surfaces vary with the distance of separation d in such a manner that the surfaces at some values of d attract one another and at other values of d repel one another. One energy level that favors surface-surface interactions occurs at the close approach of the particles (primary minimum) and another, the secondary minimum, occurs at greater distances of separation (approximately 10 to more than 100 Å). At distances between those that define the minima, the interacting forces cause repulsion. The interposing energy barrier that causes repulsion between particles may be equated with the mutual repulsion between the diffuse electrical double layers of the surfaces and it is only after this barrier has been overcome that the surfaces can react at the primary minimum. At the primary minimum the attractive forces are considerably stronger than at the secondary minimum (Good, 1972) and the energy levels and the distances of separation correspond to those of molecular interactions.

The DLVO theory describes inadequately cell/cell interactions (Good, 1972; Weiss, 1972; Maroudas, 1975) when these occur at close range, i.e. between adhesin and receptor, and parameters such as r and Ψ lack meaning. However, interactions of a long-range nature are adequately described.

4.6.3 Factors that may influence bacterial adhesion

(a) *Diffuse double layers*

The Debye–Hückel parameter $1/K$, with dimensions of length, may be said to represent the diffuse electrical double-layer (Weiss, 1970). It is related to the ionic strength I at 25°C (Heard and Seaman, 1960) by the equation

$$1/K = 3.05\, I^{-\frac{1}{2}}. \tag{4.3}$$

Accordingly, in an environment of high ionic strength the double layer is effectively contracted and consequently (Equation 4.1) the repulsion between surfaces is reduced (Weiss, 1970). Conversely, at low ionic strength the extent of the double layer may be such that the secondary minimum is reduced as an effective force of attraction (Marshall, 1975). The influence of ionic strength on the interaction of bacteria with animal cells, glass and clay surfaces is illustrated by the studies of James *et al.* (1976), Marshall *et al.* (1971b) and Stotzky (Santaro and Stotzky, 1968; Stotzky, 1974) respectively. Marshall (Marshall *et al.*, 1971b) found that a close relationship existed between ionic strength and the adsorption of bacteria at the secondary minimum to a glass surface. A double layer thickness of 200 Å was sufficient to repel bacteria from the glass surface and prevent adsorption.

(b) *Surface potential of bacteria*

Obviously surface charge and potential have an important effect on the adhesive properties of bacteria. Heckels *et al.* (1976) investigated this aspect of the adhesion of non-fimbriate (non-adhesive) and fimbriate (adhesive) gonococci and found that neutralization of the negative charge of the gonococcal surface resulted in an increase in the adhesion of fimbriate gonococci and a comparable rate of adhesion of non-fimbriate bacteria. Presumably, the nature of this interaction differs from that mediated by gonococcal fimbriae. Neutralization of the positive charge of the gonococcal surface, in contrast, resulted in almost total elimination of adhesiveness.

The lower potential of the fimbriae of *E. coli* and of *N. gonorrhoeae* compared to the bacterial cell potential and its influence on adhesion, has been noted previously (Section 4.3.1). Similarly, in the presence of trivalent cations, the surface potential of bacteria may be reversed (Stotzky, 1974) and adhesion fostered (Stotzky, 1974; James *et al.*, 1976).

There is no recorded evidence of charge mosaics on bacterial surfaces (Marshall and Cruickshank, 1973; Heckels *et al.*, 1976) and therefore no evidence that areas of unlike charges on opposing surfaces cause adhesion. The interaction of the poles of bacteria with glass (Meadows, 1971; Marshall and Cruickshank, 1973), plant roots (Sahlman and Fåhraeus, 1963; Menzel *et al.*, 1972) or fungal surfaces (Bohlool and Schmidt, 1976) may suggest non-random charge distribution, or may indicate equally well the location of specific adhesin sites (Bohlool and Schmidt, 1974; 1976) or of hydrophobic regions (Marshall and Cruickshank, 1973).

(e) *The influence of the substratum*
In contrast to the reported decreased attachment of eukaryotic cells (Maroudas, 1975), bacteria adsorb more readily to non-water wettable surfaces with low negative charge than to water wettable surfaces with higher negative charge (Fletcher, personal communication). Both bacteria (Fletcher, personal communication) and tissue culture cells (Maroudas, 1975) attach readily to positively charged surfaces. These observations may not be true of all bacteria as is indicated by studies on the influence of various proteins on adsorption (Meadows, 1971; Fletcher, 1976). However, a reduction in the negative charge of the animal cell surface may be also expected to promote the attachment of bacteria, and does appear to foster the adhesion of some mycoplasmas (e.g., Sobeslavsky *et al.*, 1968).

(d) *Radius of curvature*
On an electrostatic basis, Bangham and Pethica (1960) suggested that the close approach of two surfaces would be facilitated if it occurred between surface protrusions with low radii of curvature. From Equations 4.1 and 4.2 it can be seen that a reduction in the value of r would decrease the values of both V_A and V_R but decreases need not nullify one another because V_A and V_R are also dependent on the distance of separation d. Thus, when $V_R > V_A$ a greater reduction in the value of V_R is to be expected whereas when $V_A > V_R$ the converse is true (Weiss, 1972). Although small values of r lower the attractive forces at the secondary minimum, more importantly the forces of repulsion are reduced. Consequently, the adhesive filaments may approach the surface more effectively than the bacterial cell proper. Indeed, it is possible that when these filaments are long (for example, type 1 fimbriae of *E. coli* may be up to 2 μm long) the filaments may interact with the animal cell receptor while the bacterial cell is held at a distance that generates little or no interaction with the animal cell. Pseudopodia and microvilli on animal cells may facilitate adhesion in a similar manner (Sellwood *et al.*, 1975; Ward *et al.*, 1975; Ward and Watt, 1975), and it is known that phagocytes that cannot form pseudopodia fail to engulf bacteria and attach to glass surfaces (van Oss, 1971; van Oss *et al.*, 1972).

(e) *Reversible and irreversible adsorption*

Marshall *et al.* (1971a,b) demonstrated that the attachment of bacteria to a surface such as glass consisted of two phases. The first or reversible phase of adhesion occurs at the secondary minimum; the binding forces are weak and the bacteria exhibit Brownian movement. This phase may be equated with adsorption, and is essentially an instantaneous process which depends on general rather than specific surface properties. The second or irreversible phase may be equated with adhesion and occurs when the bacteria produce polymers that bridge the gap between the bacterium and the surface (Corpe, 1970; Hirsh and Pankratz, 1970; Marshall *et al.*, 1971a,b; Fletcher and Floodgate, 1973; Marshall and Cruickshank, 1973; Marshall, 1975). Irreversible adhesion occurs at the primary minimum and is characterized by the firm attachment of the bacteria. It is this phase of interaction that occurs when bacterial adhesins attach to animal cell receptors.

Marshall (1975) computed that the energy of repulsion on a glass surface was greater by several orders of magnitude than the kinetic energy of a motile pseudomonad that nevertheless did adhere to the glass. Subsequent to adsorption at the secondary minimum, the firm adhesion of this organism at the primary minimum occurred after filaments had been produced to bridge the space between glass and bacterium. The surface potential of glass (-17 mV, Ghosh and Bull, 1963) is comparable to that of animal cells (greater than -10 mV, Curtis, 1967) and it may be assumed that barriers of similar magnitude probably exist on both glass and animal cell surfaces. However, it is uncertain if a distinct phase of reversible adsorption need precede adhesion of bacteria to animal cells. The majority of bacteria that attach to animal cells which have been considered worthy of investigation have preformed adhesins, and these may intervene directly in assisting the bacterial cell to adhere rapidly and firmly to the animal cell. The possibility exists, nevertheless, that bacteria in a more natural environment may produce adhesins only after they become reversibly adsorbed to the cell surface.

(f) *Chemical bonds involved in cell–cell interactions*

The nature of chemical bonds and of molecular groups that may mediate cell–cell interactions have been discussed by Pethica (1961) and by Weiss (1970, 1972), and ligand binding involving cations such as calcium has been proposed by Bangham (1972).

Of particular interest are hydrophobic bonds. Marshall and Cruickshank (1973) thought that localized hydrophobic areas at the poles of some bacteria may be responsible for their perpendicular orientation at water/oil interfaces. Changes on the bacterial surface that increase its hydrophobic properties result in increased phagocytosis (van Oss and Gillman, 1972a,b; Stendahl *et al.*, 1973; Cunningham *et al.*, 1975). An interfacial reaction of this type will occur if a favorable change in free energy results from the displacement of water from both the interacting surfaces (Neuman *et al.*, 1974), and such must occur at the primary minimum (Good, 1972).

4.6.4 Conclusions

Perhaps the most interesting conclusion that can be made concerns the characteristics of adhesive appendages. In addition to specific receptor binding sites, it may be predicted that an adhesin will have the form of a long filament with radius of curvature considerably smaller than that of the bacterial cell. Such a form reduces repulsion between it and the animal cell, and accomodates the bacterial cell at a distance from the animal cell that minimizes the unfavorable forces of repulsion. For reasons similar to the former, adhesins may have a reduced surface charge compared with the bacterial cell and/or hydrophobic characteristics. The majority of bacteria examined have adhesive appendages that conform in one way or another with these characteristics. Finally, note should be taken of the influence on adhesion of the surface potential, charge and topography of the animal cell and the of interfacial tension generated between it and the milieu.

4.7 CONCLUDING REMARKS

4.7.1 Advantages of adhesion

A bacterial cell attached to a surface may benefit from the association in several ways (ZoBell, 1972; Gibbons and van Houte, 1975). In media of low nutrient content, organic compounds adsorb to surfaces and create nutritionally enriched environments in which the attached bacteria can grow (ZoBell, 1972). Bacteria that attach specifically to the surface of an animal cell may benefit in additional ways from the secretions of the host cell, from the utilization of the products of the extracellular activity of the host cell enzymes and from host cell surface components as additional sources of substrate. Intestinal bacteria appear to utilize the host intestinal epithelial mucus (Hoskins and Zamcheck, 1968) and produce many enzymes capable of such functions (Hawksworth *et al.*, 1971).

The most frequently stressed advantage gained by a bacterium that attaches to a surface is that this secures the bacterial cell against removal from the environment by the movements of fluids. Gibbons and van Houte (1971) emphasized this beneficial aspect of adhesion after their studies on oral streptococci; its importance may be deduced from other investigations such as those of Jones and Rutter (1972). Irreversible adhesion of an entire microbial population, however, may not be wholly beneficial. Surface layers are often shed continually, and over a period of time this would substantially reduce the microbial population if all cells remained firmly attached to the shed surface (Gibbons and van Houte, 1975). Microbial species may overcome such a disadvantage in several ways. The population may associate and dissociate at intervals (Meadows, 1971), adhesion may be temporary as appears to be the case in the adhesion of *V. cholerae* (Jones *et al.*, 1976), bacteria may be continually released because of individual variation in adhesive properties, or bacteria may alternate between adhesive and non-adhesive phenotypes (Scheffers et al., 1976).

The adhesion of a bacterium to an animal cell is a specific event brought about by the molecular interaction of the adhesin and the receptor. Such specificity is of ecological advantage because it increases the probability that the bacteria becomes located on a surface suitable for colonization. Adhesion may be a random process or it may have a directional component such as chemotaxis (Allweiss et al., 1977). Specific adhesive properties however may not always confer ecological advantages on the recipient cell. For example, K88 enhances the ability of some E. coli to colonize the intestine, but the synthesis of K88 by salmonellae and shigellae severely impairs their virulence (Smith, 1976). In contrast, the production of type 1 fimbriae by some bacteria appears to neither benefit nor disadvantage the host cell.

4.7.2 Possible origins of adhesins

Knowing as we do relatively little about the adhesion of bacteria to surfaces, it may be stated without fear of contradiction that considerably less is known about the evolutionary aspects of this phenomenon. One very obvious obstacle to any simple consideration of the origins of these appendages is that bacterial adhesins, their mode of action, genetic regulation and the nature of the surfaces with which they react are so very different.

Ottow (1975) discussed the similarities of flagella and fimbriae. Briefly these are similarities in morphology, in subunit periodicity, in regeneration from intracellular pools of preformed subunits and in attachment to basal bodies within the cell (Hoeniger, 1965; Morse and Morse, 1970). Some non-flagellar appendages are contractile (Bradley, 1972) and others may cause twitching of the bacterial cell (Henrichsen, 1975; Henrichsen and Blom, 1975a,b). The validity of such comparisons is perhaps enhanced by suggestions that the flagella of some aquatic bacteria may function as adhesins as well as organs of cell movement (Meadows, 1971; de Boer et al., 1975a,b).

Other adhesins may have evolved from cell-bound enzymes that have become modified in such a way that they retain their ability to bind substrates, but lack their original catalytic activity. Consequently, binding of the enzyme to the substrate on the animal cell surface results in the adhesion of the bacterium. It is possible that a haemagglutinin of *Cl. welchii* is of this type. *Cl. welchii* adheres to intestinal mucosae (Arbuckle, 1972) and produces a haemagglutinin (Dafaalla and Soltys, 1953; Wickham, 1956; Collee, 1965) which exhibits sialidase activity under appropriate conditions (Rood and Wilkinson, 1976). There is no evidence that the adhesin and the haemagglutinin/sialidase are the same microbial component, and no cell-bound state of the haemagglutinin is known. However, it is worth noting that the haemagglutinin of *B. pertussis* is cell-free during particular stages of the growth cycle yet it is known that this haemagglutinin is an adhesin (Section 4.2.2d).

The attachment of *S. mutans* and some aquatic bacteria is facilitated by the extracellular deposition of insoluble materials (Corpe, 1970; Marshall *et al.*, 1971a; Gibbons and van Houte, 1975; Slade, 1976). In the event that such deposits adsorb to a surface, they may trap bacteria either mechanically (Slade, 1976) or by reacting with specific surface components of the bacterial cell (Gibbons and van Houte, 1975; Slade, 1976) and bring about adhesion of the bacterium.

The unsubstantiated binding of *S. pyogenes* to cell surfaces via fatty acid moieties of the LTA provides yet another interesting development. It is generally considered that the lipids of LTA anchor the teichoic acids to the plasma membrane of the bacteria and thus form a link between membrane and cell wall. The presence of LTA on the outer cell wall surface may allow it to react with the animal cell plasma membrane in a similar manner and thus provide a second linkage between bacteria and animal cell surfaces.

REFERENCES

Abramson, H.A., Moyers, L.S. and Gorin, M.H. (1942), *Electrophoresis of Proteins and the Chemistry of Cell Surfaces*, Reinhold, New York.
Allweiss, B., Dostal, J., Carey, K.E., Edwards, T.F. and Freter, R. (1977), *Nature*, **266**, 448.
Ambrose, E.J. (1966), *Prog. Biophys. mol. Biol.*, **16**, 243–265.
Arai, H. and Sato, Y. (1976), *Biochim. biophys. Acta*, **444**, 765–782.
Arbuckle, J.B.R. (1970), *J. med. Microbiol.*, **3**, 333–340.
Arbuckle, J.B.R. (1972), *J. Path.*, **106**, 65–72.
Bangham, A.D. (1972), *Ann. Rev. Biochem.*, **41**, 753–776.
Bangham, A.D. and Pethica, B.A. (1960), *Proc. R. phys. Soc. (Edinb.)*, **28**, 43–50.
Barua, D. and Mukherjee, A.C. (1965), *Ind. J. med. Res.*, **53**, 399–400.
Beachey, E.H. (1975), *Trans. Ass. Am. Phys.*, **88**, 285–292.
Beachey, E.H. and Ofek, I. (1976), *J. exp. Med.*, **143**, 759–771.
Bennett, H.S. (1963), *J. Histochem. Cytochem.*, **11**, 14–23.
Bertschinger, H.U., Moon, H.W. and Whipp, S.C. (1972), *Infect. Immun.*, **5**, 595–605.
Birdsell, D.C., Doyle, R.J. and Morgenstern, M. (1975), *J. Bact.*, **121**, 726–734.
Bohlool, B.B. and Schmidt, E.L. (1974), *Science*, **185**, 269–271.
Bohlool, B.B. and Schmidt, E.L. (1976), *J. Bact.*, **125**, 1188–1194.
Bradley, D.E. (1972), *J. gen. Microbiol.*, **72**, 303–319.
Bredt, W. (1968), *Proc. Soc. exp. Biol. Med.*, **128**, 338–340.
Brinton, C.C. (1959), *Nature*, **183**, 782–786.
Brinton, C.C. (1965), *Trans. N.Y. Acad. Sci.*, **27**, 1003–1054.
Brinton, C.C. and Beer, H. (1967), In: *The Molecular Biology of Viruses*, (Colter, J.S. and Paranchych, W., eds.), pp. 251–259, Academic Press, New York.
Brinton, C.C., Buzzell, A. and Lauffer, M.A. (1954), *Biochim. biophys. Acta*, **15**, 533–542.
Brinton, C.C., Gemski, P. and Carnahan, J. (1964), *Proc. natn. Acad. Sci. U.S.A.*, **52**, 776–783.
Brinton, C.C., Gemski, P., Falkow, S. and Baron, L.S. (1961), *Biochem. biophys. Res. Comm.*, **5**, 293–299.

Brinton, C.C. and Stone, M.J. (1961), *Bact. Proc.*, **96**.
Brooker, B.E. and Fuller, R. (1975), *J. ultrastruct. Res.*, **52**, 21–31.
Buchanan, T.M. and Pearce, W.A. (1976), *Infect. Immun.*, **13**, 1483–1489.
Burrows, M.R., Sellwood, R. and Gibbons, R.A. (1976), *J. gen. Microbiol.*, **96**, 269–275.
Ceppallini, R. and Landy, M. (1963), *J. exp. Med.*, **117**, 321–338.
Chan, K. and Wiseman, G.M. (1975), *Br. J. Vener. Dis.*, **51**, 251–256.
Cleveland, L.R. and Grimestone, A.V. (1964), *Proc. R. Soc. Ser. B.*, **159**, 668–686.
Clyde, W.A. (1975), In: *Microbiology 1975*, (Schlessinger, D., ed.), pp. 143–146, American Society for Microbiology, Washington, D.C.
Coetzee, J.N., Pernet, G. and Theron, J.J. (1962), *Nature*, **196**, 497–498.
Collee, J.G. (1965), *J. Path. Bact.*, **90**, 13–29.
● Collier, A.M. and Baseman, J.B. (1973), *Ann. N.Y. Acad. Sci.*, **225**, 277–289.
Collier, A.M. and Clyde, W.A. (1971), *Infect. Immun.*, **3**, 694–701.
Constable, F.L. (1956), *J. path. Bact.*, **72**, 133–136.
Cook, G.M.W. and Stoddart, R.W. (1973), *Surface Carbohydrates of the Eukaryotic Cell*, Academic Press, London and New York.
Corpe, W.A. (1970), In: *Adhesion in Biological Systems.*, (Manly, R.S., ed.), pp. 73–87, Academic Press, New York and London.
Cowan, S.T., Steel, K.J., Shaw, C. and Duguid, J.P. (1960), *J. gen. Microbiol.*, **23**, 601–612.
Cunningham, R.K., Söderström, T.O., Gillman, C.F. and van Oss, C.J. (1975), *Immun. Commun.*, **4**, 429–442.
Curtis, A.S.G. (1967), *The Cell Surface*, Academic Press, New York.
Dafaalla, E.N. and Soltys, M.A. (1953), *Nature*, **172**, 38–39.
Davis, C.P., McAllister, J.S. and Savage, D.C. (1973), *Infect. Immun.*, **7**, 666–672.
Davis, C.P. and Savage, D.C. (1974), *Infect. Immun.*, **10**, 948–956.
Davis, C.P. and Savage, D.C. (1976), *Infect. Immun.*, **13**, 180–188.
Davis, G.H.G. and Baird-Parker, A.C. (1959), *Br. Dent. J.*, **106**, 142–146.
Davis, J.T., Haydon, D.A. and Rideal, E. (1956), *Proc. R. Soc. Ser. B.*, **145**, 375–383.
de Boer, W.E. (1975), Ph. D. Thesis. Delft University of Technology, Delft, The Netherlands.
de Boer, W.E., Golten, C. and Scheffers, W.A. (1975a), *Netherlands J. Sea Res.*, **9**, 197–213.
de Boer, W.E., Golten, C. and Scheffers, W.A. (1975b), Antonie van Leeuwenhoek, *J. Microbiol. Serol.*, **41**, 385–403.
Dettori, R. and Maccacaro, G.A. (1959), *Giorn. Microbiol.*, **7**, 37–51.
Derjaguin, B.V. and Landau, L. (1941), *Acta Physiochem. U.S.S.R.*, **14**, 633–656.
Devoe, I.W. and Gilchrist, J.E. (1974), *Infect. Immun.*, **10**, 872–876.
Duguid, J.P. (1959), *J. gen. Microbiol.*, **21**, 271–286.
Duguid, J.P. (1964), *Rev. lat-amer. Microbiol.*, **7**, Suppl. 13–14, pp. 1–16.
Duguid, J.P. (1968), *Arch. Immun. Ther. Exp.*, **16**, 173–188.
Duguid, J.P. and Anderson, E.S. (1967), *Nature*, **215**, 89–90.
Duguid, J.P., Anderson, E.S. and Campbell, I. (1966), *J. Path. Bact.*, **92**, 107–138.
Duguid, J.P. and Campbell, I. (1969), *J. med. Microbiol.*, **2**, 535–553.
Duguid, J.P. and Collee, J.A. (1960), *Proc. phys. Soc. (Edinb.)*, **28**, 65–69.

Duguid, J.P. and Gillies, R.R. (1957), *J. Path. Bact.*, **74**, 397–411.
Duguid, J.P. and Gillies, R.R. (1958), *J. Path. Bact.*, **75**, 519–520.
Duguid, J.P., Smith, I.W., Dempster, G. and Edmunds, P.N. (1955), *J. Path. Bact.*, **70**, 335–348.
Duguid, J.P. and Wilkinson, J.F. (1961), In: *Microbial Reaction to Environment*, pp. 69–99, Society for General Microbiology, Great Britain.
Ellen, R.P. and Gibbons, R.J. (1972), *Infect. Immun.*, **5**, 826–830.
Ellen, R.P. and Gibbons, R.J. (1974), *Infect. Immun.*, **9**, 85–91.
Elzter, M.E. and Branstrator, M.L. (1974), *J. Cell Biol.*, **62**, 329–343.
Evans, D.G., Silver, R.P., Evans, D.J., Chase, D.G. and Gorbach, S.L. (1975), *Infect. Immun.*, **12**, 656–667.
Eylar, E.H., Madoff, M.A., Brady, O.V. and Oncley, J.L. (1962), *J. biol. Chem.*, **237**, 1992–2000.
Finkelstein, R.A. (1973), In: *Critical Reviews, Microbiol.*, **2**, 553–623.
Fisher, S. (1950), *Aust. J. exp. Biol. Med. Sci.*, **28**, 509–516.
Fletcher, M. (1976), *J. gen. Microbiol.*, **94**, 400–404.
Fletcher, M. and Floodgate, G.D. (1973), *J. gen. Microbiol.*, **74**, 325–334.
Fox, E.N. (1974), *Bact. Rev.*, **38**, 57–86.
Freter, R. (1969), *Texas Rep. Biol. Med.*, **27**, 299–316.
Freter, R. and Jones, G.W. (1976), *Infect. Immun.*, **14**, 246–256.
Frost, A.J. (1975), *Infect. Immun.*, **12**, 1154–1156.
Fuerst, J.A. and Hayward, A.C. (1969), *J. gen. Microbiol.*, **58**, 227–237.
Fuller, R. (1973), *J. appl. Bact.*, **36**, 131–139.
Fuller, R. (1975), *J. gen. Microbiol.*, **87**, 245–250.
Fuller, R. and Brooker, B.E. (1974), *Am. J. clin. Nutr.*, **27**, 1305–1312.
Fuller, R. and Turvey, A. (1971), *J. appl. Bact.*, **34**, 617–622.
Gesner, B. and Thomas, L. (1965), *Science*, **151**, 590–591.
Ghosh, S. and Bull, H.B. (1963), *J. Colloid Interf. Sci.*, **18**, 157–164.
Gibbons, R.A. (1963), *Nature*, **200**, 665–666.
Gibbons, R.A., Jones, G.W. and Sellwood, R. (1975), *J. gen. Microbiol.*, **86**, 228–240.
Gibbons, R.J. (1975), In: *Microbiology 1975*, (Schlessinger, D., ed.), pp. 127–131, American Society for Microbiology, Washington, D.C.
Gibbons, R.J., Spinell, D.M. and Skobe, Z. (1976), *Infect. Immun.*, **13**, 238–246.
Gibbons, R.J. and van Houte, J. (1971), *Infect. Immun.*, **3**, 567–573.
Gibbons, R.J. and van Houte, J. (1975), *Ann. Rev. Microbiol.*, **29**, 19–44.
Gibbons, R.J., van Houte, J. and Liljemark, W.F. (1972), *J. Dent. Res.*, **51**, 424–435.
Gilboa-Garber, N. (1972), *FEBS Letters*, **20**, 242–244.
Gilboa-Garber, N., Mizrahi, L. and Garber, N. (1972), *FEBS Letters*, **28**, 93–95.
Gillies, R.R. and Duguid, J.P. (1958), *J. Hyg.* (Cantab) **56**, 303–318.
Ginsberg, V., ed. (1972), *Meth. Enzymol.*, **28**, Academic Press, New York and London.
Glick, M.C. (1974), In: *Biology and Chemistry of Eukaryotic Cell Surfaces*, (Lee, E.Y.C. and Smith, E.E., eds), pp. 213–240, Academic Press, New York and London.
Good, R.S. (1972), *J. theor. Biol.*, **37**, 413–434.

Gottschalk, A., ed. (1966), *Glycoproteins. Their Composition, Structure and Function,* Elsevier, Amsterdam, London and New York.
Gottschalk, A., ed. (1972), *Glycoprotein. Their Composition, Structure and Function,* 2nd edn., Elsevier, Amsterdam, London and New York.
Guentzel, M.N. and Berry, L.J. (1975), *Infect. Immun.,* **11,** 890–897.
Guidotti, G. (1972), *Ann. Rev. Biochem.,* **41,** 731–752.
Hampton, J.C. and Rosario, B. (1965), *Lab. Invest.,* **14,** 1464–1481.
Hawksworth, G., Drasar, B.S. and Hill, M.J. (1971), *J. med. Microbiol.,* **4,** 451–459.
Hay, D.I., Gibbons, R.J. and Spinell, D.M. (1971), *Caries Res.,* **5,** 111–123.
Heard, D.H. and Seaman, G.V.F. (1960), *J. gen. Physiol.,* **43,** 635–654.
Heckels, J.E., Blackett, B., Everson, J.S. and Ward, M.E. (1976), *J. gen. Microbiol.,* **96,** 359–364.
Henrichsen, J. (1975), *Acta path. microbiol. scand., Sect. B,* **83,** 187–190.
Henrichsen, J. and Blom, J. (1975a), *Acta path. microbiol. scand., Sect. B,* **83,** 103–115.
Henrichsen, J. and Blom, J. (1975b), *Acta path. microbiol. scand., Sect. B,* **83,** 161–170.
Hill, A.W. and James, A.M. (1972a), *Microbios,* **6,** 157–167.
Hill, A.W. and James, A.M. (1972b), *Microbios,* **6,** 169–178.
Hillman, J.D., van Houte, J. and Gibbons, R.J. (1970), *Arch. Oral Biol.,* **15,** 899–903.
Hirsch, P. and Pankratz, S.H. (1970), *Zschr. Allg. Mikrobiol.,* **10,** 589–605.
Hoeniger, J.F.M. (1965), *J. gen. Microbiol.,* **40,** 29–42.
Hohman, A. and Wilson, M.R. (1975), *Infect. Immun.,* **12,** 866–880.
Hollingdale, M.R. and Manchee, R.J. (1972), *J. gen. Microbiol.,* **70,** 391–393.
Honda, E. and Yanagawa, R. (1974), *Infect. Immunity,* **10,** 1426–1432.
Honda, E. and Yanagawa, R. (1975), *Am. J. Vet. Res.,* **36,** 1663–1666.
Hoskins, L.C. and Zamcheck, N. (1968), *Gastroenterology,* **54,** 201–217.
Houwink, A.L. and van Iterson, W. (1950), *Biochim. biophys. Acta,* **5,** 10–44.
Hughes, R.C. (1973), In: *Progress in Biophysics and Molecular Biology,* (Buller, J.A.V. and Noble, D., eds.), Vol. 26, pp. 189–268, Pergamon Press, Oxford.
Ito, S. (1969), *Fedn. Proc. Fedn. Am. Socs. exp. Biol.,* **28,** 12–25.
James, A.M. and Brewer, J.E. (1968a), *Biochem. J.,* **107,** 817–821.
James, A.M. and Brewer, J.E. (1968b), *Biochem. J.,* **108,** 257–262.
James, A.M., Knox, J.M. and Williams, R.P. (1976), *Br. J. Vener. Dis.,* **52,** 128–135.
James, A.M. and List, C.F. (1966), *Biochim. biophys. Acta,* **112,** 307–317.
James-Holmquest, A.N., Swanson, J., Buchanan, T.M., Wende, R.D. and Williams, R.P. (1974), *Infect. Immun.,* **9,** 897–902.
Jephcott, A.E., Reyn, A. and Birch-Anderson, A. (1971), *Acta path. microbiol. scand., Sect. B,* **79,** 437–439.
Jones, G.W. (1972), Ph.D. Thesis, University of Reading, England.
Jones, G.W. (1975), In: *Microbiology 1975,* (Schlessinger, D., ed.), pp. 137–142, American Society for Microbiology, Washington, D.C.
Jones, G.W., Abrams, G.D. and Freter, R. (1976), *Infect. Immun.,* **14,** 232–239.
Jones, G.W. and Freter, R. (1976), *Infect. Immun.,* **14,** 240–245.
Jones, G.W. and Rutter, J.M. (1972), *Infect. Immun.,* **6,** 918–927.
Jones, G.W. and Rutter, J.M. (1974), *J. gen. Microbiol.,* **84,** 135–144.

Jones, S.J. (1972), *Arch. Oral Biol.*, **17**, 613–616.
Jones, T.C. and Hirsch, J.G. (1971), *J. exp. Med.*, **133**, 231–259.
Joseph, R. and Shockman, G.D. (1975), *J. Bact.*, **122**, 1375–1385.
Kaneko, T. and Colwell, R.R. (1975), *Appl. Microbiol.*, **29**, 269–274.
Keogh, E.V. and North, E.A. (1948), *Aust. J. exp. Biol. Med. Sci.*, **26**, 315–322.
Kimbelberg, H.K. and Papahadjopoulos, D. (1971), *Biochim. biophys. Acta*, **233**, 805–809.
Kiss, J. (1970), *Advances in Carbohydrate Chemistry and Biochemistry*, (Wolfrom, M.L., Tipson, R.S. and Hawton, D., eds.), Vol. 24, pp. 382–433, Academic Press, New York and London.
Knox, K.W. and Wickens, H.J. (1973), *Bact. Rev.*, **37**, 215–257.
Koransky, J.R., Scales, R.W. and Kraus, S.J. (1975), *Infect. Immun.*, **12**, 495–498.
Kornfeld, S., Adair, W.L., Gottlieb, C. and Kornfeld, R. (1974), In: *Biology and Chemistry of Eukaryotic Cell Surfaces*, (Lee, E.Y.C. and Smith, E.E., eds.), pp. 291–316, Academic Press, New York and London.
Kumazawa, N. and Yamagawa, R. (1972), *Infect. Immun.*, **5**, 27–30.
Kumazawa, N. and Yanagawa, R. (1973), *Jap. J. Microbiol.*, **17**, 13–19.
Lai, C., Listgarten, M. and Rosan, B. (1973), *Infect. Immun.*, **8**, 475–481.
Lankford, C.E. and Legsomburana, U. (1965), In: *Proc. Cholera Res. Symp. USPHS Publ. 1328*, U.S. Gov. Print. Office, Washington, D.C.
Leach, W.D., Lee, A. and Stubbs, R.P. (1973), *Infect. Immun.*, **7**, 961–972.
Lee, E.Y.C. and Smith, E.E., eds., (1974), *Biology and Chemistry of Eukaryotic Cell Surfaces*, Academic Press, New York and London.
Liljemark, W.F. and Gibbons, R.J. (1972), *Infect. Immun.*, **6**, 852–859.
Lipman, R.P. and Clyde, W.A. (1969), *Proc. Soc. exp. Biol. Med.*, **131**, 1163–1167.
Lipman, R.P., Clyde, W.A. and Denny, F.W. (1969), *J. Bact.*, **100**, 1037–1043.
Lis, H. and Sharon, N. (1973), *Ann. Rev. Biochem.*, **42**, 541–574.
Lockwood, J.L. (1968), In: *The Ecology of Soil Bacteria*, (Gray, T.R.G. and Parkinson, D., eds.), pp. 44–65, Liverpool University Press, Liverpool, England.
Maccacaro, G.A. (1957), *Ann. Rev. Microbiol.*, **7**, 156–164.
Maccacaro, G.A. and Dettori, R. (1959), *Giorn. Microbiol.*, **7**, 52–68.
Maccacaro, G.A. and Turrin, M. (1959), *Giorn. Microbiol.*, **7**, 21–36.
Manchee, R.J. and Taylor-Robinson, D. (1968), *J. gen. Microbiol.*, **50**, 465–478.
Manchee, R.J. and Taylor-Robinson, D. (1969a), *Br. J. exp. Path.*, **50**, 66–75.
Manchee, R.J. and Taylor-Robinson, D. (1969b), *J. Bact.*, **98**, 914–919.
Maniloff, J. (1972), In: *Ciba Found. Symp. Pathogenic Mycoplasmas*, (Birch, J. ed.), pp. 67–91, Elsevier, Amsterdam.
Maniloff, J., Morowitz, H.J. and Barrnett, R.J. (1965), *J. Bact.*, **90**, 193–204.
Marchesi, V.T., Furthmayr, H. and Tomita, M. (1974), In: *Biology and Chemistry of Eukaryotic Cell Surfaces*, (Lee, E.Y.C. and Smith, E.E., eds.), pp. 273–289, Academic Press, New York and London.
Markham, J.L., Knox, K.W., Wicken, A.J. and Hewett, M.J. (1975), *Infect. Immun.*, **12**, 378–386.
Maroudas, N.G. (1975), *J. theor. Biol.*, **49**, 417–424.
Marshall, K.C. (1967), *Aust. J. biol. Sci.*, **20**, 429–438.
Marshall, K.C. (1975), *Ann. Rev. Phytopathol.*, **13**, 357–373.

Marshall, K.C. and Cruickshank, R.H. (1973), *Arch. Microbiol.*, **91**, 29–40.
Marshall, K.C., Stout, R. and Mitchell, R. (1971a), *Can. J. Microbiol.*, **17**, 1413–1416.
Marshall, K.C., Stout, R. and Mitchell, R. (1971b), *J. gen. Microbiol.*, **68**, 337–348.
Marshall, N.J. and James, A.M. (1971), *Microbios*, **4**, 217–225.
Marshall, R.D. (1972), *Ann. Rev. Biochem.*, **41**, 673–702.
McCabe, M.M. and Smith, E.E. (1976), In: *Immunological Aspects of Dental Caries*, (Bowen, W.H., Genco, R.J. and O'Brien, T.C., eds.), pp. 111–119, Information Retrieval Inc., Washington, D.C. and London.
McNeish, A.S., Fleming, J., Turner, P. and Evans, N. (1975), *The Lancet*, **ii**, 946–948.
Meadows, P.S. (1971), *Arch. Microbiol.*, **75**, 374–381.
Menzel, G., Uhlig, H. and Weichel, G. (1972), *Zbt. Bakt. Abr. II*, **127**, 348–358.
Meynell, G.G. and Lawn, A.M. (1967), *Genet. Res.* (Camb.), **9**, 359–367.
Montgomery, R. (1972), In: *Glycoproteins. Their Composition, Structure and Function.*, (Gottschalk, A., ed.), 2nd edn. Vol. A, pp. 518–528, Elsevier, Amsterdam, London and New York.
Morse, J.H. and Morse, S.I. (1970), *J. exp. Med.*, **131**, 1342–1357.
Mulczyk, M. and Duguid, J.P. (1966), *J. gen. Microbiol.*, **45**, 459–477.
Murray, J., Lathe, G.H., Clark, E.C., Ruthven, C.R.J. and Levine, M. (1950), *Br. J. exp. Path.*, **31**, 566–575.
Muse, K.E., Powell, D.A. and Collier, A.M. (1976), *Infect. Immun.*, **13**, 229–237.
Nagy, B., Moon, H.W. and Issaason, R.E. (1976), *Infect. Immun.*, **13**, 1214–1220.
Nalbandian, J., Freedman, M.L., Tanzer, J.M. and Lovelace, S.M. (1974), *Infect. Immun.*, **10**, 1170–1179.
Nelson, E.T., Clements, J.D. and Finkelstein, R.A. (1976), *Infect. Immun.*, **14**, 527–547.
Neogy, K.N., Sanyal, S.N., Mukherjee, M.K. and Nandy, P.K. (1966), *Bull. Cal. Sch. Trop. Med.*, **14**, 1–3.
Neumann, A.W., Gilman, C.F. and van Oss, C.J. (1974), *Electroanalyt. Chem. Interf. Electrochem.*, **49**, 393–400.
Nicolson, G.L. (1974a), *Int. Rev. Cytol.*, **39**, 89–190.
Nicolson, G.L. (1974b), In: *Biology and Chemistry of Eukaryotic Cell Surfaces*, (Lee, E.Y.C. and Smith, E.E., eds.), pp. 103–124, Academic Press, New York and London.
Novotny, P. and Turner, W.H. (1975), *J. gen. Microbiol.*, **89**, 87–92.
Nozaki, Y. and Tanford, C. (1971), *J. biol. Chem.*, **246**, 2211–2217.
Ofek, I., Beachey, E.H., Eyal, F. and Morrison, J.C. (1977), *J. infect. Dis.*, **135**, 267–274.
Ofek, I., Beachey, E.H., Jefferson, W. and Campbell, G.L. (1975), *J. exp. Med.*, **141**, 990–1003.
Old, D.C. (1972), *J. gen. Microbiol.*, **71**, 149–157.
Old, D.C., Corneil, I., Gibson, L.F., Thomson, A.D. and Duguid, J.P. (1968), *J. gen. Microbiol.*, **51**, 1–16.
Old, D.C. and Duguid, J.P. (1970), *J. Bact.*, **103**, 447–456.
Old, D.C. and Payne, S.B. (1971), *J. med. Microbiol.*, **4**, 215–225.
Ørstavik, D., Kraus, F.W. and Henshaw, C. (1974), *Infect. Immun.*, **9**, 794–800.
Ørskov, I., Ørskov, F., Smith, H.W. and Sojka, W.J. (1975), *Acta path. microbiol. scand., Sect. B.*, **83**, 31–36.
Ottow, J.C.G. (1975), *Ann. Rev. Microbiol.*, **29**, 79–108.

Pethica, B.A. (1961), *Exp. Cell Res. Suppl.*, **8**, 123–140.
Plummer, D.T., James, A.M., Gooder, H. and Maxted, W.R. (1962), *Biochim. biophys. Acta*, **60**, 595–603.
Punsalang, A.P. and Sawyer, W.D. (1973), *Infect. Immun.*, **8**, 255–263.
Rambourg, A. (1971), *Int. Rev. Cytol.*, **31**, 57–114.
Richmond, D.V. and Fisher, D.J. (1973), In: *Adv. Microb. Physiol.*, (Rose, A.H. and Tempest, D.W., eds.), Vol. 9, pp. 1–27, Academic Press, London and New York.
Rood, J.I. and Wilkinson, R.G. (1976), *J. Bact.*, **126**, 845–851.
Rouser, G., Nelson, G.J., Fleischer, S. and Simon, G. (1968), In: *Biological Membranes*, (Chapman, D., ed.), Academic Press, New York.
Sahlman, K. and Fåhraeus, G. (1963), *J. gen. Microbiol.*, **33**, 425–427.
Santaro, R. and Stotzky, G. (1968), *Can. J. Microbiol.*, **14**, 299–307.
Sato, Y., Arai, H. and Suzuki, K. (1973), *Infect. Immun.*, **7**, 992–999.
Sato, Y., Arai, H. and Suzuki, K. (1974), *Infect. Immun.*, **9**, 801–810.
Savage, D.C. (1972), *Am. J. clin. Nutr.*, **25**, 1372–1379.
Savage, D.C. (1975), In: *Microbiology 1975*, American Society for Microbiology, (Schlessinger, D., ed.), pp. 120–123, Washington, D.C.
Savage, D.C., McAllister, J.S. and Davis, C.P. (1971), *Infect. Immun.*, **4**, 492–502.
Scheffers, W.A., de Boer, W.E. and Looyaard, A.M. (1976), In: *Symposium on Aquatic Microbiology Soc. Appl. Bacteriol., Soc. Applied Bacteriology.*
Schott, H. and Young, C.Y. (1972), *J. pharm. Sci.*, **61**, 182–187.
Schrank, G.D. and Verwey, W.F. (1976), *Infect. Immun.*, **13**, 195–203.
Sellwood, R., Gibbons, R.A., Jones, G.W. and Rutter, J.M. (1975), *J. med. Microbiol.*, **8**, 405–411.
Sheddon, W.I.H. (1962), *J. gen. Microbiol.*, **28**, 1–7.
Silverblatt, F.J. (1974), *J. exp. Med.*, **140**, 1696–1711.
Simpson, C.F., White, F.H. and Sandhu, T.S. (1976), *Can. J. comp. Med.*, **40**, 1–4.
Singer, S.J. and Nicolson, G.L. (1972), *Science*, **175**, 720–731.
Slade, H.D. (1976), In: *Immunological Aspects of Dental Caries*, (Bowen, W.H., Genco, R.J. and O'Brien, T.C., eds.), pp. 21–38, Informational Retrieval, Inc., Washington, D.C. and London.
Slade, H.D. and Hammerling, V. (1968), *J. Bact.*, **95**, 1572–1579.
Smith, H.W. (1976), In: *Microbiology in Agriculture, Fisheries and Food*, Soc. Applied Bacteriology Symposia No. 4, (Skinner, F.A. and Carr, J.G., eds.), pp. 227–242, Academic Press, London.
Smith, H.W. and Linggood, M.A. (1971), *J. med. Microbiol.*, **4**, 467–485.
Smith, H.W. and Linggood, M.A. (1972), *J. med. Microbiol.*, **5**, 243–250.
Sobeslavsky, O., Prescott, B. and Chanock, R.M. (1968), *J. Bact.*, **96**, 695–705.
Somerson, N.L., James, W.D., Walls, B.E. and Channock, R.M. (1967), *Ann. N.Y. Acad. Sci.*, **143**, 384–389.
Springer, G.F. (1970), In: *Blood and Tissue Antigens*, (Aminoff, D., ed.), pp. 265–287, Academic Press, New York and London.
Springer, G.F. and Adye, J.C. (1975), *Infect. Immun.*, **12**, 979–986.

Springer, G.F., Adye, J.C., Bezhorovainy, A. and Jirgensons, B. (1974), *Biochemistry*, **13**, 1379–1389.
Springer, G.F., Adye, J.C., Bezhorovainy, A. and Murthy, J.R. (1973), *J. Infect. Dis.*, **128**, 5202–5212.
Springer, G.F., Huprikar, S.V. and Neter, E. (1970), *Infect. Immun.*, **1**, 98–108.
Stendahl, O., Tagesson, C. and Edebo, M. (1973), *Infect. Immun.*, **8**, 36–41.
Stirm, S., Ørskov, F., Ørskov, I. and Mansa, B. (1967a), *J. Bact.*, **93**, 731–739.
Stirm, S., Ørskov, F., Ørskov, I. and Birch-Anderson, A. (1967b), *J. Bact.*, **93**, 740–748.
Stotzky, G. (1974), In: *Microbial Ecology*, (Laskin, A. and Lechevalier, H., eds.), pp. 57–135, CRC Press, Cleveland, Ohio, USA.
Suzuki, K., Sato, Y. and Arai, H. (1974), *Jap. J. Bact.*, **29**, 145 (in Japanese).
Sutherland, I.W. and Wilkinson, J.F. (1961), *J. Path. Bact.*, **82**, 431–438.
Swanson, J. (1972), *J. exp. Med.*, **136**, 1258–1271.
Swanson, J. (1973), *J. exp. Med.*, **137**, 571–589.
Swanson, J., Hsu, K.C. and Gotschlich, E.C. (1969), *J. exp. Med.*, **130**, 1063–1091.
Swanson, J., King, G. and Zeligs, B. (1975a), *Infect. Immun.*, **11**, 453–459.
Swanson, J., Kraus, S.J. and Gotschlich, E.C. (1971), *J. exp. Med.*, **134**, 886–906.
Swanson, J., Sparks, E., Young, D. and King, G. (1975b), *Infect. Immun.*, **11**, 1352–1361.
Swanson, J., Sparks, E., Zeligs, B., Siam, M.N., and Parrott, C. (1974), *Infect. Immun.*, **10**, 633–644.
Takeuchi, A. and Savage, D.C. (1973), *Ann. Meet. Am. Soc. Microbiol., Abst.*, p. 116.
Takeuchi, A. and Zeller, J.A. (1972), *Infect. Immun.*, **6**, 1008–1018.
Tanford, C. (1973), *The Hydrophobic Effect: Formation of Micelles and Biological Membranes*, J. Wiley and Sons, New York, London, Sidney, Toronto.
Tannock, G.W. and Savage, D.C. (1974), *Infect. Immun.*, **9**, 475–476.
Taylor-Robinson, D. and Manchee, R.J. (1967b), *J. Bact.*, **94**, 1781–1782.
Taylor-Robinson, D. and Manchee, R.J. (1967a), *Nature*, **215**, 484–487.
Thornley, M.J. and Horne, R.W. (1962), *J. gen. Microbiol.*, **28**, 51–56.
Tillack, T.W., Scott, R.E. and Marchesi, V.T. (1972), *J. exp. Med.*, **135**, 1209–1227.
Tweedy, J.M., Park, R.W.A. and Hodgkiss, W. (1968), *J. gen. Microbiol.*, **51**, 235–244.
van Houte, J., Gibbons, R.J. and Pulkkinen, A.J. (1971), *Arch. Oral Biol.*, **16**, 1131–1141.
van Oss, C.J. (1971), *Infect. Immun.*, **4**, 54–59.
van Oss, C.J. and Gillman, C.F. (1972a), *J. Reticuloendothelial Soc.*, **12**, 283–292.
van Oss, C.J. and Gillman, C.F. (1972b), *J. Reticuloendothelial Soc.*, **12**, 497–502.
van Oss, C.J., Gillman, C.F. and Good, R.J. (1972), *Immun. Commun.*, **1**, 627–636.
Verwey, E.J.W. and Overbeek, J.T.G. (1948), *Theory of the Stability of Lyophobic Colloids*, Elsevier, London.
Waitkins, S. (1974), *Br. J. Vener, Dis.*, **50**, 272–278.
Ward, M.E., Robertson, J.N., Englefield, P.M. and Watt, P.J. (1975), In: *Microbiology 1975*, (Schlessinger, D., ed.), pp. 188–199, American Society for Microbiology, Washington, D.C.
Ward, M.E., and Watt, P.J. (1975), In: *Genital Infections and their Complications*, (Danielsson, L.J., Juhlin, L. and Mördh, P.-A., eds.), pp. 229–242, Almqvist and Wiksell, Stockholm.
Ward, M.E., Watt, P.J. and Robertson, J.N. (1974), *J. Infect. Dis.*, **129**, 650–659.
Weiss, L. (1970), *In Vitro* **5**, 48–78.

Weiss, L. (1972), In: *The Chemistry of Biosurfaces,* (Hair, M.L., ed.), Vol. 2, pp. 377–447, Marcel Dekker, Inc., New York.
Wickham, N. (1956), *J. comp. Path.,* **66**, 62–70.
Wickens, A.J. and Knox, K.W. (1975), *Science,* **187**, 1161–1167.
Williams, R.C. and Gibbons, R.J. (1975), *Infect. Immun.,* **11**, 711–718.
Wilson, M.H. and Collier, A.M. (1976), *J. Bact.,* **125**, 332–339.
Wilson, M.R. and Hohman, A.W. (1974), *Infect. Immun.,* **10**, 776–782.
Wistreich, G.A. and Baker, R.F. (1971), *J. gen. Microbiol.,* **65**, 167–173.
Wohlhieter, J.A., Brinton, C.C. and Baron, L.S. (1962), *J. Bact.,* **84**, 416–421.
Yanagawa, R. and Honda, E. (1976), *Infect. Immun.,* **13**, 1293–1295.
Yanagawa, R. and Otsuki, K. (1970), *J. Bact.,* **101**, 1063–1069.
Yanagawa, R., Otsuki, K. and Tokui, T. (1968), *Jap. J. vet. Res.,* **16**, 31–37.
ZoBell, C.E. (1972), In: *Marine Ecology,* (Kinne, O., ed.), Vol. 1, pp. 1251–1270, Wiley-Interscience, London, New York, Sydney and Toronto.
Zucker-Franklin, D., Davidson, M. and Thomas, L. (1966a), *J. exp. Med.,* **124**, 521–532.
Zucker-Franklin, D., Davidson, M. and Thomas, L. (1966b), *J. exp. Med.,* **124**, 533–542.

5 Binding and Entry of DNA in Bacterial Transformation

SANFORD A. LACKS

5.1	Introduction	page	179
	5.1.1 Significance		179
	5.1.2 Scope		179
	5.1.3 Approach		179
5.2	Basic methodology		180
	5.2.1 Genetic transformation		180
	5.2.2 Definition of binding and entry		181
	5.2.3 Molecular fate of DNA		181
	5.2.4 Physiological conditions		182
	5.2.5 Genetic dissection		182
	5.2.6 Biochemical analysis		182
5.3	DNA uptake by pneumococcal cells		183
	5.3.1 Binding of DNA		183
	5.3.2 Entry of DNA		185
	5.3.3 Competence		188
	5.3.4 Tentative model		191
5.4	Detailed consideration of various transformation systems		194
	5.4.1 Binding of DNA to the outer surface		194
	(a) *Specificity of donor DNA, 194*, (b) *Kinetics of the binding reaction, 195*, (c) *DNA breakage on binding, 197*, (d) *Analysis of binding sites, 198*, (e) *Distinguishing binding from entry, 200*		
	5.4.2 Entry of DNA into cells		202
	(a) *Ionic and energy requirements, 202*, (b) *Conversion to single strands, 203*, (c) *Donor DNA degradation, 205*, (d) *Role of nucleases, 207*, (e) *Polarity of DNA entry, 209*		
	5.4.3 Competence for DNA uptake		211
	(a) *Physiological variation, 211*, (b) *Changes in the cell surface, 212*, (c) *Competence as a hormonal control system, 212*		
5.5	Transfection with viral DNA		213
	5.5.1 Efficiency of transfection		213
	5.5.2 Mechanism of uptake		214

Contents

5.6	Non-physiological transformation	page	215
	5.6.1 Force-feeding of DNA		215
	5.6.2 Transformation of *Escherichia coli*		215
5.7	Comparative aspects of DNA transport		217
	5.7.1 Conjugative plasmid transfer		217
	5.7.2 Injection of viral DNA		217
	5.7.3 Packaging of viral DNA		218
5.8	Conclusions and forecast		218
	References		219

Acknowledgements

I would like to thank Dr A. Rosenthal for critically reading parts of the manuscript and Mr B. Greenberg for assistance with the references. Unpublished work from the author's laboratory was carried out under the auspices of the U.S. Atomic Energy Commission and the U.S. Energy and Research Administration.

Microbial Interactions
(*Receptors and Recognition,* series B, Volume 3)
Edited by J.L. Reissig
Published in 1977 by Chapman and Hall, 11 New Fetter Lane, London EC4P 4EE
© Chapman and Hall

5.1 INTRODUCTION

5.1.1 Significance

Genetic transformation of bacteria by DNA released from cells of a related strain is one of the intriguing examples of microbial interaction. Discovery of the transformation of capsular types in pneumococcus by Griffith (1928) led ultimately to a most convincing demonstration that the genetic information in a cell was codified in its DNA (Avery *et al.,* 1944). More recently, attention has turned to the mechanism by which the giant information-bearing molecules of DNA are transported into the bacterial cell. With growing interest in the prospect of 'genetic engineering', introduction of DNA into the cell takes on practical as well as theoretical significance.

5.1.2 Scope

The overall process of DNA uptake consists of two main steps — binding of donor DNA to the outside of the cell and entry of the bound DNA into the cell. Each step will be treated in detail. Inasmuch as these phenomena occur at the cell surface, they will be related to structures and functions of the cell wall and membrane. In addition, the development of competence, that is the formation of cell surface structures allowing DNA uptake, will be examined from both a physiological and evolutionary point of view. Genetic transfer mediated by free DNA is an obvious and important form of cellular interaction. The development of competence involves another, quite distinct system of interaction between bacterial cells.

5.1.3 Approach

This chapter has a two-fold purpose. Because it is the first review to devote itself entirely to the question of DNA uptake in transformation, it aims, on the one hand, to acquaint the general reader with progress in the area. On the other hand, it seeks to analyse critically the experimental basis of our present knowledge. To best accomplish this dual goal information of a more specialized nature has been relegated to Section 5.4. A reader desiring a quick overview of the subject will find it possible to skip that section without losing the continuity of the discourse.

Much of our understanding of bacterial transformation comes from studies of pneumococcus. Therefore, attention will focus first on DNA transport and competence in *Streptococcus pneumoniae* (also called *Diplococcus pneumoniae*). This species will serve as a model with which to compare transformation in other bacterial species, particularly *Bacillus subtilis* and *Haemophilus influenzae,* for there are many similarities as well as differences*.

The intention of this chapter is to provide a meaningful synthesis of available information. We are far, however, from a full understanding of the mechanisms involved in DNA uptake. Some speculation will be necessary to present a coherent picture.

5.2 BASIC METHODOLOGY

5.2.1 Genetic transformation

An essential tool for the study of transformation is the quantitative analysis of the frequency of transformation by genetic markers in donor DNA. This approach was developed largely by Hotchkiss (1957). Purified DNA is prepared from cells of a donor strain, which differs from the recipient strain by one or more genetic mutations that affect traits which can be readily selected. Measured amounts of DNA are added to recipient cells that have been made competent for transformation by growth of the culture according to a regime appropriate for the species. Transformation requires the intervention of several sequential processes. Donor DNA must first enter the cell. The genetic marker it carries must then be integrated into the host DNA. Integration must occur before the donor information can be expressed. Inasmuch as single-stranded segments of donor DNA are integrated (Sections 5.3.2 and 5.4.2(b)), replication and segregation of the DNA strands in the recipient chromosome must precede the appearance of a completely transformed cell. After allowing sufficient time for phenotypic expression, the frequency of transformed cells is determined by counting the number of colonies formed in selective medium.

Depending on the experimental variables, the frequency of transformed cells in the population can provide information on the competence of the recipient cells, on the potency of the donor DNA, or on the effect of environmental conditions on the interaction of cells and DNA. Under favorable circumstances, from one to ten per cent of the cells may be transformed for a particular marker. The extent of transformation is normally limited by the amount of DNA that can be taken up by the cells.

Once inside the cell the DNA generally transforms it with a high and uniform efficiency. (Exceptional cases of low integration efficiency are caused by deletions or by the *hex* system: see Lacks and Hotchkiss, 1960a; Ephrussi-Taylor *et al.*, 1965; Lacks, 1966, 1970.) Therefore, even though the frequency of genetically transformed

---

* For information on aspects of transformation not dealt with here the reader can turn to a recent general review (Notani and Setlow, 1974) or specialized reviews of competence (Tomasz, 1969), genetic recombination (Hotchkiss and Gabor, 1970), transfection (Trautner and Spatz, 1973) and transformation in *B. subtilis* (Young and Wilson, 1972; Dubnau, 1976).

cells measures the overall transformation reaction, it can be made to reflect DNA entry. For example, the kinetics of DNA entry into the cell can be followed by terminating the reaction by the addition of pancreatic deoxyribonuclease (a protein that cannot enter the cell) at successive times after mixing cells with DNA. Similarly, binding of marker DNA to the outside of the cell can be terminated either by dilution or by adding an excess of unmarked DNA to swamp out the marker DNA.

5.2.2 Definition of binding and entry

Labeling of donor DNA with radioactive isotopes allows it to be physically detected in small amounts and to be readily distinguished from recipient DNA. With radioactively labeled DNA the two main steps of DNA uptake — binding and entry — can be demonstrated. After incubation of labeled DNA with a suspension of cells, the cells are separated by centrifugation and *total DNA uptake* is measured as the radioactivity associated with the cells. Some of this radioactivity represents *binding** to the outside of the cell. This component is determined by treating the cells with pancreatic DNase, centrifuging again and measuring the radioactivity that remains with the cells. The loss of radioactivity represents the component that was bound on the outside, where it was susceptible to removal by the added enzyme. The component that was not removed represents *entry* of DNA into the cell. In the case of *S. pneumoniae* and *B. subtilis* it will be seen that an operational definition of bound DNA as susceptible to DNase, and entered DNA as resistant, corresponds well to the conceptually defined stages of binding and entry. In the case of *H. influenzae*, however, DNA at the stage that otherwise fits the conceptual definition of binding is resistant to added DNase (Section 5.4.1(e)).

5.2.3 Molecular fate of DNA

The molecular fate of isotopically labeled donor DNA can be traced through various stages of the transformation process. Intrinsic differences in the buoyant density of single-stranded and double-stranded DNA were exploited to show that DNA was converted to single strands on entry into pneumococcal cells (Lacks, 1962). In addition to radioactive labeling, density labeling of donor DNA with heavy atoms allows the tracking of large blocks of DNA, for example, in determining the size and strandedness of donor DNA segments integrated into the host DNA (Fox and Allen, 1964).

---

* The terms *binding* (or bound DNA) and *entry* (or entered DNA) are preferable to 'reversible' and 'irreversible' binding, which have sometimes been used, because current evidence indicates that DNA bound on the surface has already undergone irreversible reactions (Lacks 1977). Kinetic studies do suggest a transient, reversible binding prior to irreversible binding (Section 5.4.1(b)). In this chapter reversible binding refers to DNA that is removed by dilution or washing, and irreversible binding or, just, binding refers to DNA that cannot be so removed but, with the exception noted above, is still susceptible to DNase.

5.2.4 Physiological conditions

Controlled variation of the environment in which cells are grown or allowed to interact with DNA has given insight into the nature of competence and DNA uptake, respectively. Environmental variables such as temperature, the presence of particular nutrients, or population density during growth have been shown to affect the development of competence. Once cells have become competent they can be removed into simpler media to study the interaction with DNA. Conditions that bear on this latter process include the ionic composition of the medium and the availability of a source of energy, as well as temperature. Both competence and DNA uptake can be affected by specific inhibitory substances.

5.2.5 Genetic dissection

Bacterial mutations that prevent transformation are readily obtained. A rapid way to screen for non-transformable mutants depends on transformation occurring within colonies growing on agar plates (Ravin, 1954; Caster *et al.*, 1970; Tiraby *et al.*, 1973). In one such method (Lacks and Greenberg, 1973) DNA from a maltose-utilizing strain is plated in nutrient agar along with mutagen-treated cells of a maltose-negative mutant. The agar contains a small, limiting amount of sucrose and a large amount of maltose. All the cells can use sucrose and grow into small colonies. Clones that are transformable to maltose utilization will give rise to a large colony. Non-transformable clones are detected, therefore, as small colonies. Since these mutants may be blocked at any stage of transformation, their analysis can dissect out the individual steps in the process.

Another genetic approach seeks to relate transformation to particular cellular properties. Mutations that alter the property are obtained, and their effect on transformation is ascertained. For instance, nuclease-deficient mutants of pneumococcus were obtained by screening colonies with a DNase plate assay. Mutagen-treated cells were grown in nutrient agar containing DNA and methyl green, a dye that binds to DNA. DNases leak out of a normal colony and degrade the DNA in the vicinity, so that a colorless zone forms around the colony. Nuclease-deficient colonies do not form such a zone (Lacks, 1970). Pneumococcal mutants deficient in the major endonuclease and in the next most potent DNase in the cell, the major exonuclease, have thus been obtained. Mutants that lacked the major endonuclease were found to be defective in transformation.

5.2.6 Biochemical analysis

To explain the process of transformation in molecular terms, it is necessary to identify, isolate and characterize those cellular components that participate in the development of competence and the uptake of DNA, and to determine how they act.

Soluble factors that affect competence in species of *Streptococcus* have been identified and partially characterized with respect to chemical structure (Section 5.3.3). Investigation of surface-bound DNA was made feasible by its gentle removal on treatment with guanidine hydrochloride (Morrison and Guild, 1973b). Proteins and other cell components that bind to DNA are currently being examined for a role in DNA uptake. Attempts are also being made to relate cell wall lytic enzymes to competence.

Those cellular factors that are responsible for molecular alterations in DNA on binding and entry appear the most accessible to investigation because they may display predictable *in vitro* activities towards DNA. Once such a factor — e.g. the major endonuclease of pneumococcus — is identified, its structure and location in the cell, as well as its mode of enzymatic action, can be determined. Such information should clarify its role in the transformation process. Eventually, these pieces of information must be assembled into a coherent picture.

5.3 DNA UPTAKE BY PNEUMOCOCCAL CELLS

5.3.1 Binding of DNA

The first clearly defined step in the transformation of *S. pneumoniae* is the binding of donor DNA to the outside of the cell. In this location, the DNA is susceptible to removal by external agents, such as added deoxyribonuclease. It can also be partially removed by shear forces, as was found for DNA bound to *B. subtilis* (Dubnau and Cirigliano, 1972b), which suggests that the DNA molecules do not lie flat on the surface, but that parts of them extend into the medium where they are subject to the tug of shear.

The binding step has been demonstrated in several ways. It was originally observed that more donor DNA was associated with cells than the amount that had become resistant to added DNase (Lerman and Tolmach, 1957). When entry of DNA is specifically blocked by an inhibitor, such as the chelating agent EDTA, DNA remains bound on the outside (Seto and Tomasz, 1974). Mutations that block DNA entry similarly cause the accumulation of donor DNA at the cell surface (Lacks *et al.*, 1974). Such mutations have been found at a single genetic locus governing the major pneumococcal endonuclease; they have been called either *end*, indicating endonuclease deficiency, or *noz*, indicating no zone formed in the DNase plate assay (Lacks *et al.*, 1975).

That externally bound DNA is actually a precurosor of donor DNA inside the cell has been shown by briefly allowing ^{3}[H] DNA to bind to cells at 30°C, and then incubating at 0°C in medium devoid of DNA. After relatively long periods (80 min), even at 0°C external radioactivity was lost, and radioactivity inside the cell increased (Morrison and Guild, 1973b). Similarly, genetically marked DNA, bound to cells in the presence of EDTA, was shown to enter and give rise to transformants when divalent cations were restored, even if the cells were suspended in medium containing an excess of competitor DNA to rule out detachment before entry (Seto and Tomasz, 1974, 1976). The relevance to the transformation process of DNA bound by *end* and *noz* mutants is attested by the molecular specificity of such binding, and by its requirement of cellular competence (Sections 5.3.3 and 5.4.1(a)).

Environmental requirements for binding have been determined for the *end* mutant system. An energy source is essential; in the absence of sugar only 10% of the maximum level of binding is obtained (Lacks *et al.*, 1974). Consistent with the energy requirement is the necessity for incubation at temperatures around $30°C$. Maintenance of sulfhydryl reduction may be necessary. When competent pneumococci are suspended in medium lacking either serum albumin (which contains a free SH group) or a sulfhydryl compound like 2-mercaptoethanol, they often lose their ability to bind DNA. As was found for *Streptococcus sanguis* (Ranhand, 1974) and *B. subtilis* (Groves *et al.*, 1974), pneumococcal transformation is also inhibited by sulfhydryl reagents such as *N*-ethylmaleimide and mercuric ion (Seto and Tomasz, 1974; Lacks, unpublished). The most likely possibilities are that a sulfhydryl group is directly involved in the binding reaction or that it is needed for energy production.

DNA bound on the outside of the cell remains double-stranded. This has been shown for normal cells briefly treated with DNA (Morrison and Guild, 1973b), for cells inhibited with EDTA (Seto and Tomasz, 1974) and for mutants blocked in entry (Lacks and Greenberg, 1976). However, the single-strand molecular weight of bound DNA, measured by sedimentation in alkaline sucrose gradients, is much reduced. For example, on binding of coliphage T7 DNA to cells of the mutant *noz-48*, the single-strand molecular weight was reduced from 12.5×10^6 to a median, by weight, of 3.3×10^6 (Lacks and Greenberg, 1976). Single-strand breaks apparently occurred in both strands of bound DNA at random intervals averaging approximately 6000 nucleotides (number-average). This nicking of DNA appears concomitant with binding. It also occurs when DNA is bound by either normal or mutant cells in the presence of EDTA.

With normal cells in the absence of EDTA, bound donor DNA undergoes double-strand breakage. This was reported first for *B. subtilis* (Dubnau and Cirigliano, 1972a) where the molecular weight was reduced to 9×10^6; and for pneumococcus (Morrison and Guild, 1973b), where the double-strand molecular weight was reduced to an average (by weight) of 5×10^6. When DNA entry is blocked, as with non-leaky endonuclease mutants, or in the presence of EDTA, only single-strand breaks occur on binding, but when entry can follow binding, double-strand breaks are formed. Double-strand breaks may be produced in bound DNA by the initiation of the entry process. This view is supported by the behavior of a leaky *end* mutant, *noz-11*. Whereas the transformability of mutants lacking detectable endonuclease activity (e.g. *noz-48*) is reduced to one-thousandth of the normal level, *noz-11* gives about one-fiftieth of normal transformability (Lacks *et al.*, 1975). When *noz-11* cells bind DNA in the presence of EDTA only single-strand breaks occur. In the absence of EDTA, however, even though more than 90% of the cell-associated DNA is still on the outside of the cell, double-strand breakage is observed. The double-stranded fragments ranged in median molecular weight from 5 to 8×10^6 in different experiments (Lacks and Greenberg, 1976). Although DNA does not enter its cell efficiently, the leaky mutant seems to manage to initiate entry of most of the bound molecules, thereby producing double-strand breaks.

A reasonable hypothesis would be that donor DNA binds to the cell at sites where it is nicked. If this is so, the number of binding sites on the cell surface can be calculated from the maximum number of molecules that can be bound and the frequency of breaks. The maximum amount of DNA bound by non-leaky *end* mutants is 6.0×10^{-16} g per colony-forming unit (Lacks and Greenberg, 1976), or, since the cells come in chains of average length four, 1.5×10^{-16} g per cell. This is equivalent to 1×10^8 daltons (or 300 000 nucleotides), which would correspond to four molecules of a typical donor DNA. A nick frequency of 1 every 6000 nucleotides therefore indicates 50 binding sites per cell. Other estimates of the number of binding sites will be considered in Section 5.4.1(d). The mean length of double-stranded DNA between binding sites would be 3000 nucleotide pairs or 1.0 μm.

Mutations, called *ntr*, that prevent all binding of DNA to the cell have been obtained. The steps affected have not been identified: either the DNA-binding mechanism itself, or the antecedent development of competence could be altered. The binding reaction in pneumococcus is highly specific for duplex DNA. This specificity and other information bearing on the binding process in pneumococcus, such as the localization and isolation of binding proteins, will be discussed in Section 5.4.1.

5.3.2 Entry of DNA

On entry into the cell, donor DNA is converted to single strands. This was originally demonstrated in pneumococcal transformation by the increased buoyant density and affinity to methylated albumin columns of [^{32}P] DNA immediately after entry into cells (Lacks, 1962). The result has been confirmed in other laboratories (Morrison and Guild, 1972; Vovis, 1973). After a brief treatment with DNA (5 min at 30°C), removal of external material with pancreatic deoxyribonuclease and lysis of the cells, the only high molecular weight donor DNA present is single-stranded. Single strands, therefore, must carry the newly introduced genetic information. In the entire recipient population both complementary strands are found inside cells, inasmuch as it is possible to anneal single strands from such a population to form duplex molecules (Ghei and Lacks, 1967). However, for any one molecule that enters, it is likely that the strand complementary to the one that enters is destroyed. This fate on entry is the same for heterologous DNA as for homologous DNA (Lacks *et al.*, 1967).

During entry, an amount of donor DNA equal to the amount that has entered the cell is degraded to acid-soluble oligonucleotides which remain in the medium (Lacks and Greenberg, 1973; Morrison and Guild, 1973a). The appearance of acid-soluble products of donor DNA outside cells had been previously observed with *H. influenzae* (Stuy, 1965) and *B. subtilis* (Dubnau and Cirigliano, 1972a). Some of the donor DNA inside the cell is also degraded to acid-soluble fragments, which consist of 5'-deoxynucleotides and smaller metabolic products (Lacks *et al.*, 1967). Donor label also enters recipient DNA via normal synthesis from the mononucleotide products (Lacks, 1962; Fox and Allen; 1964; Lacks *et al.*, 1967). It had been thought that the internal fragments represent the degraded strand of donor DNA (Lacks *et al.*, 1967). However, the extent of internal degradation was shown to reflect the size of the strands in the donor DNA (Morrison and Guild, 1972). Using sheared DNA it was found that the smaller the donor strands were, the smaller the size of the single strands that entered the cell and the greater the proportion of degraded DNA inside. With unsheared donor DNA, single strands of high molecular weight account for more than 80% of the donor material in the cell. It was aptly pointed out by Morrison and Guild (1972) that the remaining 20%, which was degraded, could not account for the complementary strands. The currently prevailing hypothesis is that donor DNA enters the cell as a single strand (some of which may then be degraded) while the complementary strand is degraded to oligonucleotides that remain outside the cell.

Conversion of DNA to single strands explains the eclipse of donor marker-transforming activity immediately after uptake of DNA. If pneumococci are treated briefly with a genetically marked DNA and then the cells are broken open, the re-extracted DNA shows very little donor marker-transforming activity when tested on an appropriate recipient (Fox, 1960). When the cells are incubated before re-extraction, donor marker activity recovers rapidly. The kinetics of recovery correspond to the integration of introduced single strands into double-stranded host DNA as measured either by isotopic incorporation (Lacks, 1962) or genetic recombination (Ghei and Lacks, 1967). The eclipse results from the conversion to single strands because single-stranded DNA is not effective in pneumococcal transformation (see Section 5.4.2(b)). Conversion of DNA to single strands on entry may also explain the inability of newly introduced DNA to code for a new enzyme. After entry of equal amounts of wild-type DNA into recipients carrying various deletions of the structural gene for amylomaltase, the rate of synthesis of this enzyme reflected the extent of integration of the wild-type marker (Lacks and Hotchkiss, 1960b).

The conversion of DNA to single strands may have two-fold significance: it may be an essential part of the mechanism for entry and it may prepare the donor DNA for recombination with the host. Formation of at least partially single-stranded DNA could be a general requirement for genetic recombination (Hotchkiss, 1971). Fox and Allen (1964) showed, using density labeling, that a single-stranded segment of donor DNA is integrated into the host DNA to give a heteroduplex structure. The size of the integrated segment depends, at least in part, on the process of binding and entry. Morrison and Guild (1972) found that newly entered DNA strands were smaller than the single strands in the donor DNA. Donor strands that were 4×10^6 to 15×10^6

daltons in size gave rise to strands inside the cell of median molecular weight 2×10^6. Smaller donor strands gave rise to internal strands that were just about half the donor size. These reductions in size appear to result mainly from reactions at the surface — from the nicking of donor strands on binding and from the additional break that occurs on initiation of entry: analysis of DNA on the surface of the leaky *noz-11* mutant indicated a reduction in average molecular weight of single strands to 1.8×10^6 (Lacks and Greenberg, 1976); data on DNA bound to normal cells gave a median single strand size of 2.8×10^6 daltons (Morrison and Guild, 1973a). Cutting of DNA during binding and entry thus determines the weight-average size of DNA that enters the cell to be about 2×10^6 daltons and limits the size of pieces that can be integrated into host DNA.

The size of DNA that is integrated has been found by buoyant density measurements with heavy-atom donor DNA to average about 2×10^6 daltons (Gurney and Fox, 1968). Random breaks would produce pieces of number-average size just half the weight-average size (Tanford, 1961), or 3000 nucleotides long. Another estimate of the size of the integrated segment by a genetic approach gave a number-average length of 2000 nucleotides (Lacks, 1966). Refinement of the genetic calculation based on additional data for the amylomaltase gene locus gives an estimate closer to 3000 nucleotides (Lacks, unpublished). The reduction in strand size on entry is apparently sufficient to account for the size of integrated DNA. The randomness of the breaks is attested both by the size distribution of single-strand fragments on the surface (Lacks and Greenberg, 1976) and by the normal transforming activity for DNA of that size (Morrison and Guild, 1973b). Such random breakage could perhaps account for genetic recombination, that is, the separation of linked markers in transformation. It is possible, however, that an additional recombinatory process occurs after synapsis of the donor strand with host DNA.

There are several specific ionic requirements for entry of DNA. The divalent cations Mg^{2+} (Lacks and Greenberg, 1973; Lacks, 1977) and Ca^{2+} (Fox and Hotchkiss, 1957; Seto and Tomasz, 1976) are both absolutely essential. (However, Mn^{2+} can substitute for Mg^{2+}—Lacks, unpublished data.) Of the monovalent cations, entry of DNA requires K^+ but not Na^+ (Lacks, 1977). Parallel to the requirement of magnesium, calcium and potassium ions for entry of DNA, is the requirement of these ions for the simultaneous release of acid-soluble fragments from donor DNA into the medium. Entry of DNA may also require energy. When DNA was bound to cells in the presence of EDTA and the cells were then transferred to medium containing divalent cations with either glucose or a glucose analogue (2-deoxyglucose) present, the extent of DNA entry with the analogue present was only 25% of that obtained in the presence of glucose (Seto and Tomasz, 1974). Although there is no doubt that energy is required for DNA uptake, there is some uncertainty regarding the precise steps in the process that require energy. This matter will be considered further in Section 5.4.1(e).

The major endonuclease of pneumococcus appears to play an important role in DNA entry. Its surface location is consistent with such a role. Protoplast fractionation experiments show that the major endonuclease is located in the cell membrane, from which it can be released by mild detergents (Lacks and Neuberger, 1975). *In vitro* the enzyme attacks DNA endonucleolytically and gives rise to acid-soluble oligonucleotides (Lacks and Greenberg, 1967). The distribution in size of oligonucleotides formed by the endonuclease *in vitro* is very similar to that of the oligonucleotides that appear in the medium during entry of DNA into cells. In both cases the fragments contain from one to ten nucleotide residues with a mean length about five (Lacks *et al.*, 1974). Also parallel to the *in vivo* process is the requirement for Mg^{2+} that can be fulfilled equally well by Mn^{2+} (Lacks *et al.*, 1975).

Mutants that lack the major endonuclease can only bind DNA on the outside of the cell (Lacks *et al.*, 1974, 1975). Over a dozen mutations that affect the major endonuclease have been examined. All the mutations occurred at the same genetic locus, but they vary widely in their effect on the enzyme activity. There appears to be an excess of the enzyme normally present, for mutants that retain even a few per cent of wild-type activity are fully transformable. Most of the mutants that show no detectable enzyme activity are very severely inhibited in transformation. Levels of transformation obtained with these mutants are one-thousandth of those normally obtained. Some mutants do not show detectable endonuclease activity in extracts, but they may have some residual activity since they can be transformed to levels ranging from one to ten per cent of normal. Even with these leaky mutants most of the DNA associated with the cell is bound on the outside, and the extent of entry reflects the residual level of transformability.

The molecular fate of donor DNA, the properties of the major endonuclease and the analysis of the mutants point toward a role for the enzyme in facilitating DNA entry. By attacking one strand of duplex DNA bound to the surface and degrading it to oligonucleotides, the enzyme may enable the complementary strand to enter the cell. The alternation between attachment and hydrolysis exerted on phosphodiester linkages in one strand, sequentially, could even provide the motive force for entry of the other strand. Because the enzyme is fixed in position on the membrane, it would not be able to attack the entering strand, which is of opposite polarity, during the entry process.

5.3.3 Competence

The ability to take up DNA is not a permanent property of the bacterial cell. Physiological variation in this ability, which is called competence, depends on cultural conditions, such as temperature, pH, the presence of nutrients and cell density. Pneumococcal cells become most competent when they reach a particular density during the late logarithmic growth phase of a culture (Tomasz and Hotchkiss, 1964). This behavior results from the role of an extracellular activator of competence. The development of competence involves surface changes that enable one cell to bind DNA from another cell. In the case of the genus *Streptococcus*, activation of the cell to initiate this development itself constitutes a distinct phenomenon of cellular interaction, one which is preliminary to the genetic interaction.

Competence factors that appear in the medium of a competent culture and that will elicit competence in otherwise incompetent cultures were first reported in *S. sanguis* (Pakula and Walczak, 1963). They were found, independently, in *S. pneumoniae* by Tomasz and Hotchkiss (1964), who extracted an activator of competence in cell-free form by heating cells for 10 min at 60°C (Tomasz and Mosser, 1966). The pneumococcal competence factor has also been observed in cell-free culture filtrates (Kohoutová and Malek, 1966). Both the pneumococcal and streptococcal competence factors are small, basic proteins of molecular weight between 5000 and 10 000 (Tomasz and Mosser, 1966; Leonard and Cole, 1972). Their activity is rapidly destroyed by treatment with proteolytic enzymes (Tomasz and Mosser, 1966; Dobrzanski and Osowiecki, 1967).

Activator is a suitable term for the pneumococcal competence factor because it activates or initiates a sequence of cellular changes leading to competence for DNA uptake. Activator is released by cells into the medium where it can interact with other cells (Tomasz and Hotchkiss, 1964). Only when the cells grow to a certain density, usually about 10^7 cells ml^{-1}, does the concentration of activator accumulated in the medium become sufficiently high to activate the cells. Addition of proteolytic enzymes to the medium prevents activation without interfering with growth (Tomasz, 1966; Lacks and Greenberg, 1973). Interaction of a cell with external activator also increases the production of activator (Tomasz, 1966). This autocatalytic effect, in conjunction with the ever more rapid accumulation of activator due to exponential growth of the culture, assures that activation (and, subsequently, competence) will spread rapidly throughout the population. The rate of formation of competent cells is often much greater than the rate of cell proliferation (Tomasz and Hotchkiss, 1964). The activator thus serves as a messenger, a signal to the cells in the culture that they are present at a certain cell density.

Cells that have reacted with activator are not yet competent, but can develop competence and take up DNA even in the presence of trypsin (Tomasz, 1970; Lacks and Greenberg, 1973). The activated state is transient, however, for the cells lose this ability within a generation of growth after dilution of the culture or addition of trypsin (Lacks and Greenberg, 1973). Development of competence in the activated cell requires protein synthesis (Tomasz, 1970). With the R6 strain of pneumococcus, at least, the protein that must be synthesized is temperature sensitive either in its structure or location: competence is ten-fold greater when protein synthesis occurs at 30°C than at 37°C (Lacks and Greenberg, 1973). This explains why, contrary to purely kinetic expectations, cultures take up DNA more rapidly at 30°C than at 37°C (Fox and Hotchkiss, 1957; Lerman and Tolmach, 1957). Activator binds to specific proteins on the cell membrane (Pakula, 1967; Ziegler and Tomasz, 1970), but the manner in which it elicits new protein synthesis is not known.

Besides being able to take up DNA, competent pneumococci show other characteristic properties. They exhibit a novel surface antigen (Nava *et al.*, 1963) and they agglutinate at low pH (Tomasz and Zanati, 1971). These properties could reflect new protein synthesis but they could also come about secondarily, as a result of changes in the cell surface that unmask underlying properties. Surface changes that may be indicative of a fragility or porosity of the cell wall do occur. The appearance of DNA and nuclease activity in supernatant fluids from competent cultures (Ottolenghi and Hotchkiss, 1962) could result from lysis or leakage. Leakage from competent cells under certain conditions has been demonstrated (Seto and Tomasz, 1975a). Pneumococcal cells lose their cell walls spontaneously to form 'autoplasts' in hypertonic sucrose solutions, and competent cells form autoplasts much more rapidly, which suggests that increased cell wall lytic activity is associated with competence (Lacks and Neuberger, 1975; Seto and Tomasz, 1975a). The importance of the cell surface in competence is also emphasized by the requirement for choline in the cell wall. When choline, a normal component of teichoic acid in the wall, is substituted by ethanolamine, the cells cannot become competent (Tomasz, 1968). The presence of choline is also necessary for activity of the major pneumococcal autolysin (Mosser and Tomasz, 1970), which suggests a role for this enzyme. However, a mutant that is severely deficient in the autolysin is normally transformable (Lacks, 1970), even though it forms autoplasts more slowly than incompetent cells (Lacks and Neuberger, 1975). So the relevance of cell wall alteration to competence and the mechanism by which such alteration is brought about are still obscure.

The development of competence is essential for the initial binding of DNA to pneumococci (Lacks *et al.*, 1974; Seto *et al.*, 1975b). Mechanisms for subsequent steps may pre-exist. The endonuclease involved in entry, for example, is present in incompetent cells (Lacks *et al.*, 1974). Most of the non-transformable mutants which are not deficient in the endonuclease (*ntr* mutants) are unable to bind DNA (Lacks *et al.*, 1974). These mutations could either block steps in the activation of competence or alter components directly involved in DNA binding. Some *ntr* mutants show the cell wall fragility associated with competence and some do not (Lacks and Neuberger, 1975). Several genetically incompetent strains examined by Tomasz (1969) failed either to produce or respond to activator, but one strain did respond to added activator. This strain apparently failed to produce or release its own activator.

In a mutant called *trt* the activation step appears to be by-passed (Lacks and Greenberg, 1973). This strain will develop competence in the absence of activator, that is, at low cell densities or in the presence of proteolytic enzymes. The cells behave as if they are permanently activated, either because of a genetic structural change that dispenses with the need for activator, or because activator can act in the very cell that makes it, thereby short-circuiting the normal, extracellular route. Mutant *trt* cells still require protein synthesis at 30°C to develop competence. Thus, the *trt* mutation emphasizes the distinction between activation and subsequent steps in competence development. Existence of the mutation supports the idea that activation is part of a highly evolved genetic mechanism.

The hormone-like nature of the activator has been stressed by Tomasz (1965). Its three-fold action as a detector of cellular concentration, as a transmitter of this signal and as an effector of new developmental behavior allows the activator to serve as a precise control mechanism for timing genetic interaction. Only when the cells reach a density that makes genetic interaction feasible, or even advisable, do changes take place at the cell surface that release DNA from some cells and allow its reception by others. This bacterial system is analogous to hormonal mechanisms in eukaryotes and could conceivably be an evolutionary antecedent of some of them. As a monitor of cell growth it might bear some relationship to the mechanism for controlling development in higher organisms (see Section 5.4.3(c)).

5.3.4 Tentative model

Our current understanding of DNA binding and entry in pneumococcus can be summarized in the model shown in Fig. 5.1. Although it is based on the facts presented above, it contains some speculative features and should therefore be considered tentative.

In an incompetent cell, parts of the mechanism required for DNA uptake are already in place on the cell surface. The major endonuclease is located in the cell membrane. Receptor sites for activator are accessible from outside the cell. Small, ineffective amounts of activator are released into the medium. When the cell density increases and other conditions, such as pH, are favorable, the concentration of activator in the medium increases sufficiently to activate the cell. The activated cell then enters into a developmental path that culminates in a competent cell. Protein synthesis is required for this development, which under optimal conditions is completed in 10 min. During this time additional activator is elaborated and distinct changes in surface properties occur. The most significant change is emergence of the ability to bind DNA. Although it is possible that DNA binding sites are unmasked secondarily by other surface changes, the requirement for protein synthesis suggests that a newly synthesized surface protein participates directly in the binding reaction.

The binding reaction itself is more than a simple adsorption. It is highly specific for double-stranded DNA. Also, it requires a source of energy, and the cells must be incubated with the DNA at a physiological temperature. The bound DNA exhibits a number of single-strand breaks, which are distributed, apparently at random, over both strands of the duplex at a mean distance of about 6000 nucleotides. A reasonable assumption is that the nicking occurs at the site of binding. A single protein may be responsible for both binding and nicking. The protein itself may be attached to the DNA at the site of the nick. It could thus link the DNA to the cell surface. A single donor DNA molecule would generally be attached to several such sites on the cell. Portions of the molecule between binding sites may extend into the medium. If this is the case, then initiation of entry must occur at the binding site, where DNA is in contact with the cell.

Formation of a complex between the binding protein and the major endonuclease in the surface membrane would facilitate initiation of entry at the point of binding. Binding and nicking of DNA on the outside of the cell can proceed in the absence of

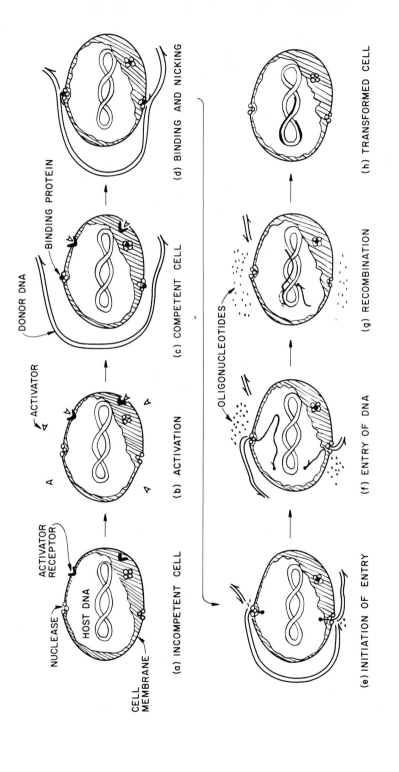

Fig. 5.1 Stages in the transformation of pneumococcus. Host cell surface (hatched area) is broken to reveal cellular DNA inside. Cell wall is not shown. (a) Incompetent cell, the cell membrane contains a nuclease in the form of an oligomeric complex and a receptor for the activator. (b) Activation, on accumulation of activator (A) in the medium. (c) Competent cell, develops as activation leads to the synthesis or unmasking of a binding protein, which is positioned over the pore formed by the nuclease complex. (d) Binding and nicking of donor DNA, binding proteins break one strand of the duplex and attach to the new 5'-ends (arrowheads). (e) Initiation of entry, the strand opposite a nick is broken by the nuclease as the bound strand enters, perhaps together the binding protein. (f) Entry of DNA, a single strand enters, facilitated by the nuclease degrading the complementary strand to oligonucleotides and possibly also by internal proteins (not shown) binding to single-stranded DNA. (g) Recombination, donor strand segments synapse with the homologous region of host DNA. (h) Transformed cell, formed as donor DNA substitutes for homologous strand segments of host DNA.

divalent cations, but initiation of entry requires the addition of magnesium ions. Single-strand breaks are first converted to double-strand breaks, presumably by action of the endonuclease on the strand of DNA opposite a nick. Entry of DNA follows, with the introduction of a single-stranded segment of DNA while the opposite strand is degraded to oligonucleotides by action of the endonuclease.

The major endonuclease is depicted as a doughnut-shaped oligomer located on the plane of the membrane. Such a structure could provide in its interior an aqueous channel for passage of the single strand of DNA. Because it is fixed in the membrane, the endonuclease would degrade only one of the two strands of opposite polarity in the DNA duplex. The complementary strand could be pulled in passively by alternate attachment and hydrolysis by the nuclease of the one strand. Such DNA translocation by the enzyme does not account for the observed requirements for energy and calcium and potassium ions, so the entry process may be more complex. Once entry has progressed appreciably the incoming single strand would anchor the complex in the cell. Initial insertion, however, may require internalization of the nicked end, perhaps with the binding protein attached as shown in Fig. 5.1. Quite possibly the incoming DNA then reacts with single-stranded-DNA-binding proteins within the cell as postulated for *B. subtilis* (see Section 5.4.2).

After it is inside the cell, the single-stranded DNA must interact with the homologous nucleotide sequence in the host DNA. The single-stranded status of the donor DNA at this point, of course, would facilitate base-specific interaction. If the surface binding protein is retained, it could aid in protecting the donor strand, in fostering synapsis, or in linking the donor and host material. The end result is a transformed cell in which a single-stranded segment of donor DNA replaces a homologous segment of host DNA. One consequence of the proposed model is that integrated segments, even in a single cell, may be derived from opposite strands of the donor molecule. Determination of which strand segments enter and which are degraded would depend on the (accidental) configuration of the DNA relative to the binding sites in the initial interaction.

5.4 DETAILED CONSIDERATION OF VARIOUS TRANSFORMATION SYSTEMS

5.4.1 Binding of DNA to the outer surface

(a) *Specificity of donor DNA*
Binding of nucleic acids to the competence-specific receptors of pneumococcal cells is restricted to unencumbered duplex DNA molecules. Recognition of the high level of specificity exerted at the binding step had to await the clear separation of this step from entry. Nucleic acids can now be tested directly for binding to cells blocked in entry by either a mutation or an inhibitor. Both direct binding and competitive ability were examined in the binding of denatured, single-stranded DNA by

pneumococcus (Lacks, 1977). Radioactively labeled single strands were bound by competent cells, but they were bound to the same extent by genetically incompetent cells. Furthermore, denatured DNA failed to reduce the binding of native DNA to competent cells. These results explain the inability of denatured DNA to compete effectively against native DNA in transformation (Roger and Hotchkiss, 1961). Competence-specific sites bind only duplex DNA; single-stranded DNA must be bound elsewhere on the surface.

Single-stranded RNA is not taken up by pneumococci (Lerman and Tolmach, 1957), and it does not compete with transforming DNA (Tomasz, 1973). Even RNA:DNA heteroduplex molecules do not compete with transforming DNA (Tomasz, 1973), and they are not bound by competent cells (Lacks, 1977). Glucosylation of DNA also interferes with binding. *B. subtilis* cells will not take up coliphage T6 in which the hydroxymethylcytosine bases are glucosylated, but they will take up T6 DNA that lacks the glucose residues (Soltyk *et al.*, 1975). Similarly, glucosylated coliphage DNA does not compete with transforming DNA in the *S. sanguis* (Ceglowski *et al.*, 1975) or *S. pneumoniae* (Lacks, 1977) systems. These quite specific structural requirements for binding must reflect a precise stereochemical fit between the DNA and its binding sites.

Most transformable species bind heterologous DNA as well as homologous DNA. Thus, *S. pneumoniae, S. sanguis* and *B. subtilis* indiscriminately bind and take up double-stranded DNA from many different sources. In *H. influenzae,* however, foreign DNA generally competes very poorly and only small amounts are taken up by the cells (Schaeffer *et al.*, 1960; Setlow and Boling, 1972; Scocca *et al.*, 1974), although more substantial uptake of DNA from *E. coli* or coliphage lambda has sometimes been reported (Steinhart and Herriott, 1968; Newman and Stuy, 1971). DNA from *Haemophilus parainfluenzae* or from *Haemophilus* phages is taken up as well as *H. influenzae* DNA. The basis of this discrimination in *H. influenzae* cannot be sequence homology with host DNA since binding occurs in mutants in which donor DNA fails to interact with host DNA at all (Sedgwick and Setlow, 1976). It is possible, however, that the binding sites in *H. influenzae* recognize a methylation pattern specific to *Haemophilus* DNA. This pattern could be present, also, in certain strains of *E. coli.* A common methylated sequence has been found in *H. influenzae* and *E. coli* DNA (Lacks and Greenberg, 1977). Discrimination by *H. influenzae* based on DNA methylation could explain the otherwise disparate results.

In studying binding, attention must be directed to the competence-connected component. At low ionic strength (less than 0.1 M NaCl) bacteria bind DNA in a non-specific manner (Barnhart and Herriott, 1963). Such non-specific binding is found with both physiologically and genetically incompetent cells. It is eliminated by the use of media of sufficient ionic strength.

(b) *Kinetics of the binding reaction*
Transformable bacteria show a characteristic dependence of DNA entry and transformation on DNA concentration. For *S. pneumoniae* (Hotchkiss, 1957),

H. influenzae (Stuy and Stern, 1964) and *B. subtilis* (Anagnostopoulos and Spizizen, 1961) transformants (obtained during a given time interval) increase linearly up to 0.1 µg ml^{-1} and level off above this DNA concentration. The saturation in rate of DNA entry must result from the limited number of binding sites on the cell surface. At each binding site DNA presumably enters at a fixed rate. When all the binding sites are filled, DNA enters at the saturating rate. The similarity in the DNA concentration required for saturation in the three species indicates that their binding sites have a similar affinity for DNA. This affinity is high. Per unit of interaction of DNA containing 6000 nucleotides (see Section 5.3.1), half saturation occurs at about 10^{-10} M. Although DNA is irreversibly bound on the outside of cells, the initial attachment of DNA may well be reversible, and the saturation observed could reflect the concentration dependence of the reversible step.

Irreversible binding of transforming DNA on the outside of cells of *B. subtilis* was demonstrated by Levine and Strauss (1965). When DNA uptake was terminated by dilution, transformants were obtained immediately after mixing the cells with DNA, but when uptake was terminated by the addition of DNase, no transformants appeared for uptake times of less than 2 minutes. Markers in the DNA, therefore, required a minimum of 2 min (at 30°C) to enter the cell. The subsequent rate of increase of transformants indicate that the average time for entry of a marker located in DNA that had been irreversibly bound to the cell was about 4 min. In addition to this estimate of the rate of DNA entry, an idea of the rate of DNA binding could also be gleaned from the experiment. When the reaction was stopped by dilution, the increase of transformants with time, even with saturating DNA concentrations, showed that the binding reaction is relatively slow. During the first four minutes new binding occurred while the initial binding sites were still occupied, so it must take at least this long for binding to progress to the irreversible step at the majority of sites.

Analysis of the fate of isotopically labeled donor DNA in *B. subtilis* gave similar kinetics (Dubnau and Cirigliano, 1972b; Davidoff-Abelson and Dubnau, 1973b). When donor DNA was continually present, binding proceeded with time so that externally bound DNA reached its peak at 5 min. When cells were given a 0.5 min pulse of donor DNA, terminated by swamping with unlabeled DNA, it took 2 min for half the externally bound DNA to be converted to single-stranded fragments inside the cells. These data correlate with the above-mentioned genetic results and show, moreover, that the rates of binding and entry are roughly similar. The overall uptake of DNA by *B. subtilis* requires energy, since it is inhibited by KCN and dinitrophenol (Young and Spizizen, 1963), and Dubnau and Cirigliano (1974) reported that energy is necessary for the binding step.

The time course of DNA uptake in *S. pneumoniae* is similar to that in *B. subtilis*. The availability in this species of mutants (*end*) blocked in entry, however, allows a direct examination of binding kinetics. At 30°C the amount of bound DNA increased rapidly for 5 to 10 minutes and then at a gradually slower rate for up to 30 min (Lacks *et al.,* 1974). The binding reaction thus appears to be rather slow.

In normal pneumococci the overall process of binding and entry results in a steady increase of donor DNA inside the cells for up to 30 min at 30°C (Kent and Hotchkiss, 1964). Termination of the reaction with DNase gives a linear increase of transformants with time (after a lag of $\simeq 1$ min) even at subsaturating DNA concentrations. Fox and Hotchkiss, (1957) carried out an ingenious kinetic analysis of this steady state by considering it as an enzymatic reaction. Their model should now be modified to include an irreversible step in binding, which corresponds to the nicking reaction, to give

$$DNA + B(s) \underset{k_2}{\overset{k_1}{\rightleftarrows}} B(s)DNA \overset{k_3}{\rightarrow} B(s)_{DNA} \overset{k_4}{\rightarrow} B \cdot DNA \ (s)$$

where DNA represents donor DNA; $B(s)$, bacterial binding sites; $B(s)DNA$, DNA reversibly bound to sites; $B(s)_{DNA}$, DNA irreversibly bound; $B \cdot DNA(s)$, DNA that has entered cells. Reaction rate constants are k_1 through k_4. The Lineweaver–Burk treatment can be applied to the data as before to determine the number of binding sites (see Section 5.4.1(d)). In the expanded model, with saturating DNA, $V_{max} = k_3 k_4 B(s)/k_3 + k_4$. The rate constant k_4 corresponds to the entry of bound DNA. Inasmuch as Gabor and Hotchkiss (1966) found a delay of 2 to 5 min between the time a DNA marker was bound and the time it became completely resistant to DNase, a reasonable estimate for k_4 would be 0.5 min^{-1}. Calculation from the data of Fox and Hotchkiss (1957) and values of 4×10^{-15} g for DNA uptake per transformant and 50 for the number of sites per bacterium, then gives $k_3 = 0.04$ min^{-1}. This rate corresponds to a half-time for binding equal to 17 min. The low value of k_3 originally calculated by Fox and Hotchkiss (1957) for the first irreversible step in uptake was puzzling in view of the rapid rate of entry of a bound molecule. It can now be ascribed to the slow rate of irreversible binding associated with nicking of donor DNA. The kinetic determination of this rate is consistent with direct measurements of binding to *end* mutant cells.

(c) DNA breakage on binding

DNA that is bound to pneumococcus contains single-strand breaks. Could it be that pre-existing nicked DNA is preferentially bound? This possibility is rendered unlikely by quantitative considerations (Lacks and Greenberg, 1976) and apparently precluded by the binding of covalently closed circular duplex DNA, which is also nicked in the process (Lacks, unpublished). The latter experiments show that double-stranded ends are not required for binding. Circular DNA can be taken up by pneumococcal cells (Morrison and Guild, 1973a) because the cell itself introduces a nick on binding and, eventually, a double-stranded end.

The nicks in DNA bound to pneumococcal cells are apparently distributed at random in both strands of the DNA to give single-stranded segments with a weight-average molecular weight of 3 to 4×10^6 (Lacks and Greenberg, 1976). If double-strand breaks are formed at the position of these nicks, their location will also be random. Double-stranded fragments removed from the surface of normal cells retain the transforming activity expected for donor DNA of this size (Morrison and Guild, 1973b). This finding is consistent with a random distribution of breaks. If binding and breakage occurred only at specific sites in the DNA, it would be expected that, with those sites broken after one round of binding, the DNA would not retain its original binding ability. As indicated in Section 5.3.2, the single-strand segments that result from double-strand breakage at the surface are $\simeq 2 \times 10^6$ in weight-average molecular weight and correspond in size to the single strands found within the cell after entry of high molecular-weight DNA. When DNA is sheared to a molecular weight less than 5×10^6 prior to binding, the mean size of donor strands inside the cell is always just half of the original donor strands. This was taken to indicate that donor DNA must be broken at least once during its uptake by the cell (Morrison and Guild, 1972). It is now evident that the break is required for binding.

Donor DNA that is bound to *B. subtilis* initially retains its original double-stranded molecular weight (Dubnau and Cirigliano, 1972b), but this material has not been tested for the presence of single-strand breaks. Soon after binding, however, double-strand breaks appear, which reduce the bound DNA to fragments of weight-average molecular weight $\simeq 9 \times 10^6$ (Dubnau and Cirigliano, 1972a; Arwert and Venema, 1973) or somewhat larger (Dubnau, 1976). These double-stranded fragments, when extracted from the cell, are deficient in transforming activity, more so than expected from their reduction in size (Dubnau and Cirigliano, 1972a; Arwert and Venema, 1973). For a discussion of this inordinately low activity see Section 5.4.2(d,e).

(d) *Analysis of binding sites*

The satisfactory description of DNA uptake as an enzyme—substrate reaction implies a reversible association of DNA with sites on the cell surface. Rate constants for the association and dissociation will determine the proportion of the sites that are filled at a given DNA concentration. From the dependence of DNA uptake on DNA and bacterial concentration Fox and Hotchkiss (1957) calculated that the maximum amount of DNA that could be occupying such sites, per bacterium, was 3.3 to 7.5×10^{-16} g or 33 to 75 molecules of molecular weight 6×10^6. If each molecule were reversibly bound at a single site, this would mean $\simeq 50$ sites per bacterial cell. If each 6000-nucleotide segment of DNA (the number-average size piece that undergoes irreversible binding) corresponded to a single site, there would have to be 100–230 binding sites per cell. Direct measurement of DNA segments bound on the surface of *end* mutants gave 50 sites (see Section 5.3.1). It may actually happen that a DNA molecule is reversibly bound at one point in its length to a single bacterial site. At a slow rate such a molecule is nicked and irreversibly bound. Segments of the bound molecule then displace other molecules from adjacent sites and eventually get nicked, so that the original molecule is now bound at several points along its length. Thus, 50 *molecules* could be bound reversibly, but only 50 *segments* could be irreversibly bound.

In *B. subtilis* the amount of DNA externally bound (irreversibly, by the current definition) at early times was used to calculate the number of binding sites. Only a small fraction of the cells in a competent culture of *B. subtilis* are competent (see Section 5.4.3(a)), so binding sites were computed per competent cell. Singh (1972) found 20 to 50 molecules of 17×10^6 daltons bound per cell. Dubnau and Cirigliano (1972b) reported 15 double-stranded fragments of 9×10^6 daltons per cell at 2 min after exposure to DNA. By this time half of the originally bound DNA was internalized (see Section 5.4.1(b)) so the original of binding corresponded to 30 such fragments. Each double-stranded fragment presumably corresponds to a segment of irreversibly bound DNA that has begun to enter a single binding/entry site. The number-average molecular weight of a segment would be half of 9×10^6, which is a weight-average (see Section 5.3.2), so that the number of binding sites would be 60 per cell. Singh (1972) made an independent estimate of entry sites by measuring the rate of entry of DNA per cell and dividing it by the rate of entry of a single molecule, 55 nucleotides s^{-1} at $28°C$ (Strauss, 1965; see Section 5.4.2(e)). These measurements indicated 35 to 65 uptake sites per competent cell. It appears that cells of *B. subtilis*, like those of *S. pneumoniae*, have $\simeq 50$ binding/entry sites. Binding in *H. influenzae* is not directly comparable and will be discussed in Section 5.4.1(e).

Much of the above discussion assumes that binding sites are re-used after a DNA molecule, or segment, has passed through them. It is possible, however, that a component of the binding site reacts stoichiometrically with the DNA and is used up in the process. This could explain the apparent limit to total DNA uptake, generally $\simeq 10\%$ of the DNA content of a cell, in all three highly transformable species.

Not much is known about the location of DNA binding sites on the cell surface. They may be simply disposed randomly over the surface, but it has been suggested that they are associated with zones of wall growth (Tomasz *et al.*, 1971) or with mesosomes (Miller and Landman, 1966; Akrigg *et al.*, 1969). Mesosomes are membranous structures that may be connected with septum formation (Greenawalt and Whiteside, 1975). Autoradiographic studies did indicate preferential binding of DNA to the ends or middle of *B. subtilis* cells (Javor and Tomasz, 1968) and in the vicinity of mesosomes (Wolstenholme *et al.*, 1966; Vermeulen and Venema, 1974), but so far such localization has been neither precise nor conclusive.

Circumstantial evidence suggests that the binding sites are composed of proteins, at least in part. The development of competence requires protein synthesis (see Section 5.4.3(b)), and inasmuch as the key feature of competence is the ability to bind DNA, the new protein could be directly involved in binding. Competent cells from several species carry an antigen that is not present in incompetent cells; this antigen provokes the production of an antibody that blocks transformation (Nava *et al.*, 1963; Pakula, 1965; Bingham and Barnhart, 1973). Since the antibody protein cannot enter the bacterial cell, it presumably acts at the surface and a likely target is the DNA binding site.

Attempts to isolate subcellular structures that could bind DNA have met with limited success. A cell wall—membrane complex prepared by disrupting competent pneumococci with glass beads showed much greater ability to bind DNA than either purified cell walls or protoplasts (Seto et al., 1975b). However, wall–membrane complexes from incompetent cells gave just as much binding, and the binding was not specific for double-stranded DNA. Pneumococcal protoplasts did show some DNA binding ability, as did, also, membrane vesicles from *B. subtilis* (Joenje et al., 1974), but the relevance of such binding to the transformation process has not yet been demonstrated. Membrane vesicles from both competent and incompetent cells bound DNA, although those from competent cells bound more at lower DNA concentrations (Joenje et al., 1975). Binding to vesicles was not inhibited by cyanide. Seto and Tomasz (1975b) have extracted binding factors, apparently proteins, that bind to DNA and cause the DNA to stick to nitrocellulose filters. The pH and ionic strength at which such binding can be revealed are both considerably lower than for cellular DNA binding. Under these conditions even bovine serum albumin can cause DNA to stick to filters (Lacks, unpublished). One pneumococcal protein, however, was obtained by gentle extraction from only competent cells, so its binding activity may be physiologically significant.

The binding reaction with whole cells requires energy. Quite possibly energy is required for linking a newly-formed end at the nick in bound DNA to a cell surface component. This linkage might even be covalent. Examples of proteins that appear to bind covalently after nicking DNA have been demonstrated in plasmid relaxation complexes (Blair and Helinski, 1975) and in the case of superhelical DNA untwisting enzymes (Champoux, 1976). Donor DNA bound on the surface of pneumococcal cells can be quantitatively removed by 5 M guanidine hydrochloride (Morrison and Guild, 1973b). This does not disprove a covalent linkage because the reagent may remove the presumed binding protein itself from the surface together with the DNA. Whether the linkage is covalent or not, the question arises whether the binding component functions stoichiometrically, by acting only once, or catalytically, by regenerating itself when the DNA enters the cell. Neither the questions of covalent binding or catalytic action can be resolved at present.

(e) *Distinguishing binding from entry*

This section will be somewhat controversial in that it aims, first, to explain divergent results reported by different laboratories studying pneumococcal transformation and, second, to reconcile the divergent behavior of the *H. influenzae* transforming system with the model derived from investigation of its gram-positive cousins.

Contradictory findings for the pneumococcal system have been reported for (1) energy requirement for binding, (2) excess binding by normal cells in the presence of EDTA and (3) excessive surface nuclease activity (cf. Lacks and Greenberg, 1973; Lacks et al., 1974; Seto and Tomasz, 1974; Seto et al., 1975a). Data of Seto et al. (1975b) indicate that DNA binds to cells in the presence of EDTA in two ways: a superficial binding from which DNA is released by removal of the cell wall, and a deep binding in which DNA remains with the protoplasts. Only deep binding required energy. One possible interpretation is that deep binding actually represents the initiation of entry and therefore requires energy, while the earlier step of superficial binding does not. A contrary point of view is that the superficial binding is not transformation-specific. According to this view certain treatments of cells (possibly shifts in pH or exposure to EDTA, which itself can eventually cause lysis) may uncover superficial binding sites. The superficial binding could be simply ionic, or else the sites might correspond to non-specific nucleases. In the presence of EDTA such sites would bind DNA in a non-specific manner and without a requirement for energy. The appearance of excess binding sites under conditions of low salt even in incompetent cells (Seto et al., 1975b) supports this view. The existence of such sites can explain superficial binding without expenditure of energy, excess binding in the presence of EDTA, and, if the sites correspond to nucleases, degradation of DNA in the medium far beyond the expectation for DNA uptake (see Section 5.4.2).

H. influenzae transformation differs from *S. pneumoniae* and *B. subtilis* in that donor DNA is still in a double-stranded form when it becomes protected from external DNase (Notani and Goodgal, 1966). However, this sequestered form may not represent entry, but rather the earlier binding reaction. This point of view is buttressed by the following arguments. Gram-negative bacteria are generally enclosed by an outer membrane as well as an inner cell membrane (Osborn, 1969; Braun and Hantke, this volume). In *H. influenzae* donor DNA may be bound within the outer membrane and therefore protected from external agents. The bound DNA is released by digitonin, which does not release nuclear DNA (Notani et al., 1972). This suggests a compartmentalization of the sort envisaged. The binding reaction is rapid; at saturating DNA concentrations it is completed in one minute (Stuy and Stern, 1964). It also requires energy — it is inhibited by dinitrophenol (Stuy, 1962; Barnhart and Herriott, 1963). Again, similarly to bound DNA in gram-positive bacteria, the bound DNA in *H. influenzae* undergoes double-strand breaks (Notani, 1971). In the KB6 mutant, which appears to be blocked in entry (see Section 5.4.2(b)), the double-stranded molecular weight of donor DNA is unaltered (Notani et al., 1972). Aside from the speed of the binding reaction and the protected situation of bound DNA, binding in *H. influenzae* may be otherwise similar to the other systems studied.

A kinetic analysis by Stuy and Stern (1964) suggested the existence of very few 'penetration' sites, perhaps one or two per cell. Their data can now be interpreted quite differently. It was calculated that, with saturating DNA, one DNA molecule was sequestered by a cell every 5 seconds. This interval was compared to the delay in 'entry' of a genetically marked DNA after adding it to cells previously treated with DNA bearing a different marker. The delay was 4 to 5 seconds and was assumed to represent the time needed by the first DNA molecule to 'enter' through the site. If DNA entry only took 5 seconds per molecule, then one site was sufficient to account for the overall rate of entry. The analysis rested on a belief that no reversible binding occurred. No DNA bound at $0°C$ could be shown to enter cells on dilution and transfer to $37°C$, but this result is by no means inconsistent with reversible binding since, on dilution, the DNA would come off the cell surface before it could be sequestered. In fact, the 4 to 5 second delay in binding of a second DNA molecule could represent the time for *reversible* release of the first.

Thus, irreversible binding in *H. influenzae* may take some tens of seconds and the number of binding sites may be appreciable.

5.4.2 Entry of DNA into cells

(a) *Ionic and energy requirements*

Entry of DNA in *S. pneumoniae* requires K^+, Ca^{2+} and either Mg^{2+} or Mn^{2+} (see Section 5.3.2). Only the need for Mg^{2+} or Mn^{2+} is understood inasmuch as either ion can serve as a co-factor for the nuclease implicated in entry. Entry of DNA in *B. subtilis* requires a divalent cation, and either Mg^{2+}, Ca^{2+}, Ba^{2+} or Sr^{2+} can fulfil the requirement (Young and Spizizen, 1963). Morrison (1971) suggested that Mg^{2+} was required only for the initiation of DNA entry, because transformants continued to appear (at a reduced rate) after addition of EDTA to cultures that had received DNA several minutes earlier. These results can be equally well explained, however, by a much reduced rate of entry in the presence of EDTA, if that rate is sufficient to allow entry of those marker segments on the verge of completing their penetration when the EDTA was added. It is possible that, in *B. subtilis,* the divalent cations are also needed for nuclease action during entry of DNA.

H. influenzae does not require divalent cations to sequester DNA from added DNase (Barnhart and Herriott, 1963). This is not surprising if the initial sequestration is equivalent to binding in other bacteria (see Section 5.4.1(e)). Such binding in *H. influenzae* does occur optimally at an ionic strength of 0.10–0.15, but no particular ions are essential. It takes 15–30 min for the bound DNA to enter the cell proper, as judged by the formation of genetically and physically recombined DNA (Voll and Goodgal, 1961; Notani and Goodgal, 1966). Entry in this system cannot be conveniently measured as conversion of donor DNA to a form resistant to added DNase. Therefore the ionic and energy requirements for entry in *H. influenzae* have not been defined.

In *S. pneumoniae,* energy may be necessary for entry as well as binding. When cells, loaded with DNA in the presence of EDTA, were transferred to medium containing divalent cations, with or without glucose, only one-quarter as much DNA entered the cells when glucose was absent from the medium (Seto and Tomasz, 1974). This inhibition is less severe than for the overall uptake process (Lacks and Greenberg, 1973). It had earlier been proposed that energy for entry may be derived from the hydrolysis of phosphodiester bonds in one strand of the DNA as the complementary strand enters the cell (Lacks, 1962).

In *B. subtilis,* the addition of cyanide abruptly stops DNA entry, so it appears that energy is essential for this step (Strauss, 1970). In fact, addition of cyanide after dilution away from DNA, but prior to DNase treatment, resulted in somewhat fewer transformants than addition of DNase alone. This led Strauss to suggest that some DNase-resistant DNA was sequestered outside the cell and that cyanide blocked its entry. An alternative explanation is that not enough DNase was added (the amount was not specified) to immediately destroy all of the external DNA.

In summary, the weight of evidence to date indicates that energy is probably required for both binding and entry in all of the highly transformable bacteria.

(b) *Conversion to single strands*

Donor DNA that is associated with cells of *B. subtilis* undergoes an eclipse of its biological activity similar to that found for *S. pneumoniae* (Venema et al., 1965). Even externally bound double-stranded fragments show lower biological activity, probably due to the nature of their termini (Dubnau and Cirigliano, 1972a; see Section 5.4.2(d,e)), but the biological eclipse of DNase-resistant donor DNA inside the cell must be due to its single-stranded nature. Although earlier studies reported some single-stranded donor DNA within cells (cf. Bodmer and Ganesan, 1964), it was Piechowska and Fox (1971) who showed that soon after uptake most of the donor DNA is single-stranded. To demonstrate this, the single strands had to be liberated from a protein complex by treating the extract with 4 M NaCl at 70°C. Davidoff-Abelson and Dubnau (1973a) confirmed this result and found the molecular weight of the single strands to be $\simeq 5 \times 10^6$. They reasoned that the procedure was effective because high concentrations of lysozyme, used to lyse the cells, protected the DNA on lysis; and that the high salt treatment was later needed to remove the DNA from the highly basic lysozyme protein. An alternative view (Piechowska and Fox, 1971) is that the single strands are tightly bound to a cellular protein when they enter the cells. As in the case of *S. pneumoniae* (Section 5.3.2), single-stranded donor DNA is integrated in *B. subtilis* as segments of mean molecular weight $\simeq 2.8 \times 10^6$ (Dubnau and Cirigliano, 1972c).

The fate of DNA in *S. sanguis* was examined by Raina and Ravin (1977). Donor DNA that had become resistant to external DNase was found to be in biological eclipse until the DNA became integrated. However, the donor material at this stage did not band as single-stranded DNA in CsCl density gradients, and was resistant both to enzymes that attack single-stranded DNA and those that attack double-stranded DNA. It could conceivably consist of single-stranded DNA bound up with protein, as was found for *B. subtilis*.

No eclipse can be demonstrated in the transformation of *H. influenzae* and free single strands have not been observed. At some stage, however, donor DNA may be converted to single strands because only single-stranded donor segments are integrated into the host DNA (Notani and Goodgal, 1966). According to one point of view (Section 5.4.1(e)), the DNA that is bound in *H. influenzae* lies outside a membrane barrier and entry through that barrier requires conversion to single strands as in the other highly transformable species. If the subsequent integration of single strands were rapid compared to entry, no single strands would accumulate. Another point of view (Sedgwick and Setlow, 1976) considers the bound DNA to be inside the cell, but in order for it to react with the host DNA, the latter itself must contain stretches of single-stranded DNA (LeClerc and Setlow, 1975), which then interact with donor DNA to ultimately integrate single-stranded donor segments.

Soon after binding, approximately 15% of the donor DNA in cells of *H. influenzae* is single-stranded, and this material appears to be located at the ends of the double-stranded molecules (Sedgwick and Setlow, 1976). After 30 min, when integration is complete, very little single-stranded material is left. Depending on the point of view, the partially single-stranded molecules can be considered to be in the process either of entry into the cell proper or of recombination with host DNA. Certain transformation-defective mutants (called KB), which do not interfere with bacteriophage recombination, bind DNA normally but fail to transform cells and do not even contribute atoms to the host DNA (Beattie and Setlow, 1971; Notani *et al.,* 1972). Donor DNA bound by the KB mutants does not become single-stranded at all (Sedgwick and Setlow, 1976), so these mutants may be blocked in entry of DNA similarly to the *end* mutants of *S. pneumoniae.* Another mutant, *rec-2,* which blocks recombination, also fails to form either single-stranded donor or recipient DNA. Unless the *rec-2* strain is a double mutant with separate blocks in entry and recombination, the observed dual effect supports the view that single-stranded donor material results from the recombination mechanism.

The tight binding of single strands to protein after DNA uptake by *B. subtilis* led Piechowska and Fox (1971) to suggest a role for the complex in transformation. Indeed, a protein that binds to single-stranded DNA was found only in competent cells of *B. subtilis* (Eisenstadt *et al.,* 1975). This protein was identified in extracts by its ability to protect denatured DNA from degradation by pancreatic DNase. It was absent in physiologically and genetically incompetent cells, but its appearance during the development of competence was not parallel to transformability. As pointed out by Eisenstadt *et al.* (1975), such a binding protein could facilitate entry of DNA strands, protect them from nucleases within the cell, and enable them to recombine with recipient DNA. An entry mechanism alternative to DNA translocation by a nuclease (see Section 5.3.4) could be based on the unwinding of donor DNA as the binding proteins bind to one (or both) complementary strand(s) inside the cell. It is quite possible that both mechanisms are combined in DNA entry. The unexplained K^+ and Ca^{2+} requirements for entry in *S. pneumoniae* could result from their role in binding DNA strands to a protein on entry. Systems other than *B. subtilis* may not synthesize a competence-specific single-stranded DNA binding protein, but may simply use those proteins, called unwinding proteins, that carry out this function in DNA replication (Kornberg, 1974).

Under normal physiological conditions, single-stranded donor DNA transforms cells very poorly. Observation of the eclipse in *S. pneumoniae* is predicated on this fact. In *S. pneumoniae* both the binding and entry reactions are specific for double-stranded DNA. Denatured DNA, however, can exhibit up to 5% of the native transforming activity in all three major transforming species. In each case the majority of this activity was shown to reside in a rapidly renaturing (or incompletely denatured) fraction of DNA containing, apparently, cross-linked strands (Roger and Hotchkiss, 1961; Barnhart, 1965; Chevallier and Bernardi, 1965, 1968; Rownd *et al.,* 1968; Mulder and Doty, 1968). Nevertheless, undisputably single-stranded DNA, isolated from CsCl density gradients, can also transform cells, albeit at a lower frequency.

The maximum frequency reported for transformation of *S. pneumoniae* by purified single strands is 0.5% of the equivalent native DNA (Miao and Guild, 1970).

Transformation of *H. influenzae* by single strands can be enhanced by absorbing DNA to cells at pH 4.8 in the presence of EDTA (Postel and Goodgal, 1966, 1967). On subsequent neutralization most of the DNA is released from the cell, but enough is retained and enters the cells to give $\simeq$ 5% of normal DNA transformation. In *B. subtilis* frequencies as high as 20% of normal transformation with native DNA have been reported with single-stranded DNA (Tevethia and Mandel, 1971) on addition of EDTA (Chilton, 1967) and lowering the pH.*

(c) *Donor DNA degradation*

The significance of the formation of donor DNA fragments in the medium during pneumococcal transformation rests on the following observations: (1) the cells must be competent, (2) the major endonuclease must be active, (3) sugar, as a source of energy, is required, (4) additional cations, Ca^{2+} and K^+, that are required for DNA entry must be present and (5) the amount of DNA in fragments appearing in the medium is equal to the amount of DNA taken into the cell. Although the last observation refers to the entire culture, a plausible assumption is that it reflects the fate of each donor molecule, so that for each segment of donor DNA taken into the cell, the complementary strand is degraded.

Equality between the amount of degraded DNA and that of entered DNA has not always been found. Excessive degradation of donor DNA in the medium (Seto *et al.*, 1975a) can perhaps be attributed to treatments of the pneumococcal cells that expose nucleases not functional in DNA uptake. Suspension of cells in a low-salt medium had precisely this effect (Seto *et al.*, 1975b). No change was observed in DNA entry, but three times as much acid-soluble DNA was found in the medium. Incompetent cells, which took up no DNA, produced no acid-soluble fragments in 150 mM NaCl, but in 15 mM NaCl they produced as much as competent cells. Low salt also increased the amount of DNA bound by competent cells in the presence of EDTA by a factor of six; the new binding sites may correspond to newly exposed enzymes not involved in DNA uptake (see Section 5.4.1(e)). A similar binding ability and nuclease activity that is not coupled to transformation has been reported for incompetent cells of *B. subtilis* (Haseltine and Fox, 1971). There have been two reports of deficient

---

* This relatively, high activity was attributed to the inhibition by EDTA and low pH of a nuclease active against single-stranded DNA (Tevethia and Mandel, 1971). In addition, however, the pH shift and EDTA may make the cells more porous so that denatured DNA enters the cell without the help of the normal binding and entry mechanisms, which are geared to native DNA. Nevertheless, uptake of single strands by both *H. influenzae* and *B. subtilis* under these special conditions required cellular competence (Postel and Goodgal, 1967; Tevethia and Caudill, 1971). If a protein that binds incoming single strands and aids their entry is, in fact, induced in competent cells, then this part of the uptake mechanism could also facilitate the uptake of single-stranded DNA.

external DNA degradation, where acid-soluble fragments corresponded to only half the donor DNA resistant to DNase. One is the case of the *noz-1* mutant of *S. pneumoniae*, in which transformation is reduced to $\simeq 10\%$ of the normal level (Lacks *et al.*, 1974). The other case occurs with normal pneumococci when Mg^{2+} is limiting (Lacks, 1977). In both instances, the slow entry of DNA may result in a partial sequestration of external DNA on the way in, which would render it resistant to added DNase and lead to a falsely high value for the amount of DNA that has entered the cell.

External donor DNA fragments were first observed by Stuy (1965) in *H. influenzae* transformation. These fragments consisted mainly of oligonucleotides (Stuy and van der Have, 1971). Since binding is rapid relative to entry in *H. influenzae*, it was possible to separate the cells from unbound donor DNA and show that fragments were released into the medium over the course of DNA integration. The amount of acid soluble fragments produced after 30 min ranged from 25 to 50% of the bound DNA. The lower values may result from incomplete entry since not all of the bound DNA becomes integrated in *H. influenzae* (Notani and Goodgal, 1965). The degradation does not require integration inasmuch as bacteriophage DNA is degraded even when the recipient cell is not lysogenic for the phage (Stuy, 1974). That the degradation accompanies entry of DNA is supported by its absence in the *rec-2* mutant (Stuy, 1974) if this mutant is indeed defective in entry (Section 5.4.2(b)). The data from *H. influenzae* are consistent, therefore, with a mode of entry in which at least part of the strand complementary to the entering strand is degraded to oligonucleotides.

In *B. subtilis*, acid-soluble donor DNA fragments are formed and released into the medium as bound DNA enters the cell (Dubnau and Cirigliano, 1972a; Joenje and Venema, 1975). The fragments contain some oligonucleotides but mostly mononucleotides, nucleosides and free bases; the latter possibly result from secondary decomposition. Ability to form external fragments was closely correlated with competence, as was the case for *S. pneumoniae* (Lacks and Greenberg, 1973); and, at least initially, fragment formation was approximately proportional to entry (Joenje and Venema, 1975). Loss from the cell of half the radioactivity bound in a short pulse suggests that one strand of DNA was degraded per molecule during entry (Davidoff-Abelson and Dubnau, 1973b). However, Bresler *et al.* (1964) have presented genetic evidence that both strands of the same DNA molecule can enter a cell. Hybrid DNA containing his^+, tyr^+ on one strand and his^+, tyr^- on the other was prepared by renaturation of the appropriate mixture of denatured DNA. Such hybrid DNA gave rise to mixed his^+, tyr^+ and his^+, tyr^- colonies at the rate of 6% of all his^+, tyr^+ transformants. The rate with the control native his^+, tyr^+ DNA was 2%. Although the cause of this background level was not explained, it is possible that the excess mixed colonies with hybrid DNA resulted from simultaneous transformation of a cell by both the his^+, tyr^+ strand and the complementary his^+, tyr^- strand and, then, segregation of the strands on cell division. If this explanation is correct, then DNA entry can occur, at least occasionally, without destruction of one strand. Perhaps such entry could be facilitated by the unwinding protein mechanism alone, as speculated for single-strand entry (see Section 5.4.2(b)).

Some donor DNA is converted to acid-soluble fragments inside cells of *S. pneumoniae* but, as explained earlier (Section 5.3.2), these fragments are no longer considered to result from the entry process. In general, donor nucleotides can be released within cells by exonuclease attack, especially when the donor DNA is not integrated either because of absence of homology with the recipient or a genetic block in recombination. In *B. subtilis*, such internally produced nucleotides appear to be excreted by the cell (Dubnau and Cirigliano, 1972a), but in *S. pneumoniae* and *H. influenzae* they are incorporated into recipient DNA (Lacks, 1962; Fox and Allen, 1964; Lacks *et al.*, 1967; Notani and Goodgal, 1966; Notani *et al.*, 1972). In the latter cases they create the problem of distinguishing resynthesis from integration.

(d) *Role of nucleases*

The most definitive evidence implicating a nuclease in transformation is the defective transformation of mutants of *S. pneumoniae* that lack the major endonuclease (see Section 5.3.2). Kohoutová (1967) had suggested a role for a nuclease on the basis of a common inhibition by tRNA of transformation and of a pneumococcal nuclease activity. In *S. sanguis* no mutant strains have been analysed, but two surveys of various wild strains showed that all transformable strains exhibited a nuclease activity either in gentle extracts (Nalecz and Dobrzanski, 1972) or in culture filtrates (Pakula *et al.*, 1972). Most of the non-transformable strains tested lacked the nuclease activity.

A number of properties of the pneumococcal nuclease implicated in DNA entry have been investigated. This enzyme, which is similar in many respects to *E. coli* Endo I (Lehman *et al.*, 1962), is also the major endonuclease of the cell. It can be called Endo *Dpn* I to distinguish it from endonucleases of other species and from Endo R·*Dpn* I, a restriction type endonuclease specific for methylated DNA (Lacks and Greenberg, 1975). Endo *Dpn* I appears to be a generalized nuclease since it attacks native DNA, denatured DNA and RNA equally well (Rosenthal and Lacks, 1977). Its affinity for RNA may account for the inhibition of its DNase activity in crude extracts (Lacks, 1970; Lacks *et al.*, 1975) and for the inhibition of transformation by tRNA (Kohoutová, 1967). The enzyme is a fairly small, basic protein composed of a single polypeptide chain of 25 000 molecular weight (Rosenthal and Lacks, 1977). It is located in the cell membrane (Lacks and Neuberger, 1975). Its spontaneous release into the medium in aging cultures (Ottolenghi and Hotchkiss, 1962) or colonies in agar (Lacks, 1970) suggests that it may also serve a nutritive function. Normally, it can only be released from membranes by detergents, which indicates that it is an integral membrane protein.

The endonuclease has been purified to homogeneity (Rosenthal and Lacks, unpublished), and from the yield it appears that each cell contains $\simeq 1000$ molecules. This number is considerably greater than the calculated number of binding/entry sites (see Section 5.4.1(d)), which may explain the normal transformability of partially deficient *end* mutants (Lacks *et al.*, 1975). The correlation between level of enzyme activity in the more deficient mutants and the tranformation phenotype (Section 5.3.1) suggests that the endonuclease itself functions in entry. It has not been definitely shown that these mutations affect the structure of the enzyme. If the lack of endonuclease resulted from regulatory mutations, then another protein could be missing and that protein might be the one essential for entry. Although this possibility seems unlikely, it cannot yet be excluded.

In the pneumococcal membrane the endonuclease molecules may form an oligomeric complex among themselves or with other proteins to create a doughnut-shaped structure with an aqueous channel through which the donor DNA strand could pass (Fig. 5.1, Section 5.3.4). Some of the enzyme in extracts is associated with a complex of large size (Lacks *et al.,* 1975). *In vivo* the complex may be associated with a binding protein that nicks the donor DNA and binds it to the entry site. If the binding protein were positioned over the hole in the 'doughnut', its attachment to the donor DNA could trigger its passage through the hole to the interior and allow the previously masked nuclease to degrade the unattached strand. *In vivo* the enzyme appears to release oligonucleotide fragments from an end of a DNA strand. But *in vitro* it acts purely endonucleolytically, making approximately equal numbers of double- and single-strand breaks (Lacks, unpublished), so that it reduces the sedimentation constant or viscosity of DNA before any acid-soluble product appears (Lacks and Greenberg, 1967). Presumably the *in vivo* context of the enzyme, particularly its anisotropy and immobility relative to the substrate, results in the exonucleolytic mode of action by which it functions as a translocase to bring DNA into the cell.

Cells of *B. subtilis* contain perhaps ten nucleases capable of acting on DNA (Rosenthal and Lacks, 1977). Several nuclease activities have been related in one way or another to transformation. One enzyme, which requires Ca^{2+}, has been found in the cell membrane (Birnboim, 1966) and extracellularly (Kerr *et al.,* 1965). This enzyme acts preferentially on single-stranded DNA, and its only connection with transformation is that inhibition of its activity enhances transformation by single-stranded DNA (Tevethia and Mandel, 1970). Another enzyme, with a periplasmic location, is an endonuclease that requires Ca^{2+} or Mn^{2+} and makes just a few double-strand breaks in DNA (Scher and Dubnau, 1976). It was, suggested, therefore, that this enzyme might produce the double-stranded fragments that are observed in DNA uptake. These fragments, which are bound to the outside of the cell, appear to be intermediates in transformation although they themselves have inordinately low transforming activity (see Sections 5.4.1(c) and 5.4.2(e)). Interestingly, exposure of DNA to incompetent cells of *B. subtilis* also produces fragments that are inordinately poor in their ability to transform or to compete with transforming DNA for binding by competent cells (Haseltine and Fox, 1971). The relationship of this nuclease activity to that described by Scher and Dubnau (1976) is not known.

Competent cells of *B. subtilis*, after separation from incompetent cells in the culture (Section 5.4.3(a)), do not exhibit the surface endonuclease of incompetent cells, unless they are heated at 50°C for 10 min (Haseltine and Fox, 1971). Such heating destroys competence (McCarthy and Nester, 1969a). McCarthy and Nester (1969b) extracted a heat-activated endonuclease from *B. subtilis*. The enzyme requires either Mg^{2+} or Mn^{2+}, which is reminiscent of the pneumococcal endonuclease, and is lacking in the genetically incompetent *B. subtilis* strain W23.

An intriguing hypothesis can be developed by applying the model for DNA uptake in *S. pneumoniae* to *B. subtilis* as follows. A nuclease on the cell surface functions in DNA entry, but only when the cell is competent; that is, when a binding protein is present at the entry site. In competent cells the binding protein blocks access to DNA in the medium, so that the nuclease cannot degrade the DNA. Removal of the binding protein by heating at 50°C simultaneously destroys competence and unmasks the nuclease. This hypothesis fits very well with the observation of Williams and Green (1972) that for a brief period (less than 5 min) after being bound, DNA can be prevented from ever entering the cell by heating to 47°C. In this interval, the DNA is presumably removed along with the binding protein. Once the DNA has begun to enter it is no longer sensitive to heating, although much of it can still be removed by DNase.

A mutant of *B. subtilis* has been found to be blocked in DNA entry in a manner similar to the *end* mutants of *S. pneumoniae* (Venema et al., 1977). Analysis of the nucleases in this mutant may reveal which enzyme is involved in transformation of *B. subtilis*.

(e) *Polarity of DNA entry*

Either strand of a DNA molecule can enter the cell. This was shown for *S. pneumoniae* by the recovery of biological activity on *in vitro* annealing of single strands isolated from cells that had just taken up DNA (Ghei and Lacks, 1967). Although only a single strand of a DNA molecule enters one cell, both complementary strands are present in the entire population. Also attesting to this fact is the finding that either strand can be integrated into the host genome. Hybrid density DNA, derived from integration of a heavy-labeled donor strand into light recipient DNA, contains recipient marker transforming activity (Notani and Goodgal, 1966; Gurney and Fox, 1968). The original integration of the heavy strand, and the potential for integration of the recipient marker on the complementary strand, combine to show that either strand can be integrated.

The decision as to which strand enters the cell probably depends on the orientation of the DNA when it encounters the binding site, which presumably selects (by nicking) the strand of polarity appropriately oriented to its active site. More precisely, this selection acts at the level of strand segments (cf. Fig. 5.1). Different segments of the same donor molecule are nicked and, presumably, bound at different cellular sites. Pairwise orientation between DNA and site at the different points appears to be random inasmuch as nicks occur randomly on both donor strands. Thus, one entering segment may be derived from one strand while another segment may be derived from the adjacent complementary strand. This could account for the entry of markers on complementary strands of artificial heteroduplex donor molecules (Roger, 1972).

Bound DNA molecules appear to enter the cell in a linear fashion. Evidence for this assertion comes from the delayed entry of linked marker pairs relative to single markers in *B. subtilis* (Strauss, 1965, 1966) and in *S. pneumoniae* (Gabor and Hotchkiss, 1966) and from the size of entered DNA.*

Investigation of the uptake into *B. subtilis* of the lengthy (103×10^6 daltons) DNA of bacteriophage SP82G indicated that entry occurred linearly after binding of the DNA molecule at either (or both) end(s), since internal markers (measured by marker rescue of mutant phage) entered later than those located near the ends (Williams and Green, 1972). Complete entry of the phage DNA at $33°C$ required 15 min. The implication, here, is that with *B. subtilis* entry occurs only at bound ends of the original donor molecule and not at the end of each double-stranded fragment (of mean length 9 to 20×10^6 daltons). Certain peculiarities of DNA binding to cells of *B. subtilis*, such as the inordinately low transforming activity of double-stranded fragments isolated from the surface of competent cells or produced in the medium by incompetent cells (Section 5.4.2(d)), could result from a requirement for binding to an end having a particular structure (Dubnau and Cirigliano 1972a; Williams and Green, 1972). Ends produced by shearing would have the appropriate structure. It is significant that shearing double-stranded donor DNA fragments, after removing them from the cell surface, restored their biological activity (Arwert and Venema, 1973). Whatever the requirements for ends may be in *B. subtilis*, binding and entry in *S. pneumoniae* do not require pre-existing ends because, as indicated in Section 5.4.1(c) circular duplex donor DNA is taken up very well (Lacks, unpublished).

A model for entry of DNA concomitant with integration was proposed by Erickson and Braun (1968). In this model, integration of the donor strand into the host chromosome would provide the motive force for entry, and integration would occur at the replicating fork, where single-stranded host DNA would be available to interact with the donor strand. A number of facts argue against this model. Heterologous DNA enters the cell just as well as homologous DNA in most systems (Lerman and Tolmach, 1957; Lacks *et al.*, 1967; Soltyk *et al.*, 1975). Integration of donor DNA does not occur at the replication fork (Laird *et al.*, 1968), and arrest of

* In *B. subtilis* the minimum delay for a single marker was 2.5 min at $28°C$, this presumably represents the time needed for the smaller segments of DNA to enter; for a linked pair the delay was 4.0 min, reflecting the longer length of segments carrying both markers. The rate of entry was calculated by Strauss (1965) to be 55 nucleotides per second. In *S. pneumoniae,*, Gabor and Hotchkiss (1966) found that most bound markers entered in less than 3 min at $30°C$, while a loosely linked marker pair took as long as 10 min. In a similar experiment, but with a physical rather than a genetic measurement, Morrison and Guild (1973a) found that 2 min at $25°C$ were necessary before internalized single-strand segments reached their maximum size — addition of DNase at earlier times clipped off the strands. Taking a mean length (weight average) of donor segment to be 6000 nucleotides and an entry time of 2 min, the rate of entry for pneumococcus is 50 nucleotides per second.

chromosome replication does not block uptake (Archer and Landman, 1969b). In no system is DNA synthesis required either for the development of competence or during the uptake process (Fox, 1960; Lacks et al., 1967; Tomasz, 1970; Bodmer, 1965; Archer and Landman, 1969a; Levin and Landman, 1973; Dubnau and Cirigliano, 1973; LeClerc and Setlow, 1974). A variant of the model that couples entry with integration, but does not require DNA synthesis, may be applicable to *H. influenzae* (Sedgwick and Setlow, 1976). When this bacterium becomes competent, the host DNA appear to develop single-stranded regions (LeClerc and Setlow, 1975) which may interact with donor strands and pull them into the nuclear compartment of the cell. (See also Section 5.4.2(b)).

In conclusion, the conversion of donor DNA to single strands in the entry process could serve a two-fold purpose. First, it would allow the DNA to squeeze through a small pore into the cell. Second, by uncovering the base sequence specificity, it would enable donor DNA to interact with host DNA. Furthermore, the enzyme that facilitates this conversion may, under other circumstances, function in a nutritive or digestive capacity. Such economy is typical of the efficient use of resources that we have come to admire in Nature.

5.4.3 Competence for DNA uptake

(a) *Physiological variation*

Competence in *S. pneumoniae* was covered in some detail in Section 5.3.3. No attempt will be made here to describe competence in other bacterial species exhaustively. The purpose of this section is mainly to compare certain features of the different systems. In *H. influenzae* and *B. subtilis* competence develops when cells enter a stationary phase, as growth of the culture ceases (Goodgal and Herriott, 1961; Anagnostopoulos and Spizizen, 1961), rather than during late stages of exponential growth as in *S. pneumoniae* and *S. sanguis* (Section 5.3.3). *Neisseria* species are exceptional in that they remain competent throughout the culture cycle (Catlin, 1960; Lie, 1965; Sparling, 1966). Appearance of competence in most systems, however, occurs only in dense populations of cells — a condition favorable for both intercellular transmission of genes and selection of better adapted variants.

The fraction of cells that are competent in a given population can be determined by genetic (Goodgal and Herriott, 1961; Balassa and Prévost, 1962; Porter and Guild, 1969) or autoradiographic (Young, 1967; Javor and Tomasz, 1968) methods. With most species all the cells in a culture are competent, but in *B. subtilis* cultures only a fraction of the cells, which varies from 1–25%, become competent (Nester and Stocker, 1963; Young, 1967; Somma and Polsinelli, 1970; Vermeulen and Venema, 1971). Competent *B. subtilis* cells differ from their incompetent sisters in exhibiting a lower buoyant density (Cahn and Fox, 1968; Hadden and Nester, 1968; Singh and Pitale, 1967) and biosynthetic latency (Nester and Stocker, 1963). These properties of competent cells may reflect differentiation along a presporulation pathway, since mutants blocked at an early stage of sporulation do not become competent (Schaeffer, 1969).

(b) *Changes in the cell surface*

An essential feature of competence is the appearance of specific DNA binding sites on the cell surface. Such sites could arise in either of two ways: (1) by synthesis of a new component or (2) by unmasking of an existing component. New protein synthesis is required for the development of competence in several systems (Fox and Hotchkiss, 1957; Tomasz, 1970; Pakula and Walczak, 1963; Nester, 1964; Stuy, 1962; Leidy *et al.*, 1962; Spencer and Herriott, 1965). Specific protein differences in the membranes of competent and incompetent cells were observed by Ranhand *et al.* (1970) in *S. sanguis* and by Zoon and Scocca (1975) in *H. influenzae*. The nature of the proteins involved and their significance for transformation remain to be established.

Possible support for the unmasking hypothesis comes from an observed correlation between competence and cell wall lytic activity (see Section 5.3.3 for evidence from the pneumococcal system). In *S. sanguis,* an autolytic activity associated with competence is absent in physiologically or genetically incompetent cells (Ranhand *et al.,* 1971; Ranhand, 1973, 1974). The problem, here, is to show that autolytic activity generates DNA binding activity and that the two activities are not simply parallel developments. A possible correlation between lytic activity and transformation in *B. subtilis* does not appear to be completely consistent (Young and Wilson, 1972). Partial removal of the cell wall of *B. subtilis* did enhance transformation, however, but complete conversion of the cells to protoplasts abolished DNA uptake (Prozorov, 1965; Tichy and Landman, 1969; Wilson and Bott, 1970). A role for an enzyme in opening up a hole in the cell wall for DNA passage is attractive but not yet established. It is quite possible that both novel membrane proteins and cell wall lytic activity are necessary for competence.

(c) *Competence as a hormonal control system*

Competence may possibly exhibit more facets. In addition to a hole in the cell wall and a duplex DNA binding protein on the membrane, the development of competence may entail synthesis of a single-stranded DNA binding protein (Eisenstadt *et al.,* 1975) and may involve changes in the structure of recipient cell DNA (Harris and Barr, 1971; LeClerc and Setlow, 1975). Co-ordination of these multiple events could involve a hormone-like control.

Possible candidates for control of competence are the 'competence factors'. As a matter of definition, the term competence factor may be used in general for any substance that affects the competence of a cell. The term activator, however, should be reserved for the type of fairly well characterized control substance observed in the *S. pneumoniae* and *S. sanguis* systems. Thus, competence factor would be a suitable term for the cell-derived material that enhances competence of *B. subtilis,* possibly by dint of its cell wall lytic activity (Akrigg and Ayad, 1970). The activator, on the other hand, resembles in many respects a mammalian hormone (Tomasz, 1965; Section 5.3.3). It consists of a relatively small polypeptide. It is secreted by cells into the medium. It interacts with surface receptors on target cells to produce multiple effects, including the synthesis of new proteins. Cyclic adenosine monophosphate (cyclic AMP) often serves as an intermediary of hormonal action in mammals. Although an activation system has not been demonstrated in *H. influenzae,* it is interesting that cyclic AMP can increase the competence of incompetent cells of *H. influenzae* a thousand-fold (Wise *et al.,* 1973). The induced level, however, is still only 10^{-2} that of a maximally competent culture, so the effect of cyclic AMP may have little bearing on the normal development of competence.

If the analogy of activation to hormone action is valid, the activator could even be a prokaryotic homologue of a mammalian hormone. In this respect it may be worthwhile to examine the function performed by the bacterial activator. Because it is secreted by each cell at a constant rate but acts only when it reaches a certain concentration in the medium, it serves as a sensor and signal of cell density. Analogous mechanisms have been proposed for the control of normal growth in animal tissues (see Burch, 1976). These theories envisage that cells of a particular type produce a specific substance. Accumulation of a certain concentration of the substance, which would indicate that the cells have sufficiently proliferated, would then either inhibit growth directly, trigger synthesis of a specific growth inhibitor or turn off synthesis of a growth stimulator. Except for the end result which would be growth inhibition rather than the development of competence, the system is analogous to activation of competence in *Streptococcus* species.

5.5 TRANSFECTION WITH VIRAL DNA

5.5.1 Efficiency of transfection

The ability of bacteriophage DNA to enter a competent cell of *B. subtilis* and give rise to an infective center was first demonstrated by Romig (1962). This process has been differentiated from transformation by bacterial DNA and from whole virus infection by the useful term transfection (Földes and Trautner, 1964). It can now be demonstrated in practically all transformable species. Cells become competent simultaneously for transformation and transfection.

A priori one might expect very high frequencies of transfection because the added DNA is homogeneous. If a bacterial marker present on only one of a hundred DNA molecules gives a transformation frequency of 1%, the expectation for transfection would approach 100%. Actually, phage DNA never infects more than 1% of the cells and fewer than one infected cell is produced per 1000 molecules of phage DNA added. These low efficiencies apparently result from fragmentation of the DNA, and from the need to reconstitute a whole virus genome. The fragmentation itself results from the processes of binding and entry.*

For most phage DNAs, each infective center requires the interaction of two or

---

* Another indication of fragmentation and reconstitution during transfection is the requirement for recombination. Mutants defective in genetic recombination are poorly transfected (Okubo and Romig, 1966).

more DNA molecules with one cell. This is indicated by the multiple-hit kinetics of transfection (see review by Trautner and Spatz, 1973). On the other hand, marker rescue, which requires only a small fragment of DNA, displays single-hit kinetics (Green, 1966).

5.5.2 Mechanism of uptake

At least in *B. subtilis* and *S. pneumoniae,* phage DNA is presumably converted to single strands on entry. The difficulty in observing the single strands is partly technical in that double-stranded fragments on the surface must be removed with adequate DNase treatment, and single strands must be released from protein complexes (Section 5.4.2(b). In addition, however, the homogeneity of the phage DNA will allow annealing of complementary strands that enter the cell (from different donor molecules) to give at least partly double-stranded phage DNA within the cell. Such annealing does not occur with bacterial donor DNA because complements rarely enter a single cell. Transfection of *S. pneumoniae* does not occur in the transformation defective *end* mutants (Porter and Guild, 1976). These mutants are — as we have seen — blocked in the conversion of donor DNA to single strands. Their block in transfection, as well, indicates that in this species transfecting DNA enters by the same path as transforming DNA. In addition to conversion to single strands, phage DNA molecules undergo strand breaks on binding to the cell. For example, *B. subtilis* phage H1 DNA (mol. wt. 83×10^6) suffers an average of six double-strand breaks on binding (Arwert and Venema, 1974).

For most phage DNAs the formation of an intact genome will require piecing together strand segments from two or more donor molecules. In the simplest case, two complementary strand segments each covering more than half of the genome from opposite ends wi'l anneal in their overlapping region. Replicative extension would follow. More generally, several segments would be pieced together in a process requiring host recombination functions.

Some phage DNAs are exceptional in that they do not show the multiple hit concentration dependence for transfection (Rutberg *et al.,* 1969; Spatz and Trautner, 1971). These phage DNA molecules are generally small, but, what is more important, they have cohesive ends, which enable them to circularize during the course of infection. Such a molecule would be broken in half by a double-strand break after binding, but if the opposite ends of the complementary strands both entered the cell, they could anneal at the cohesive ends to give a full-length molecule, which could be completed by replication and then circularized by a double-strand ligase (Sgaramella, 1972). An interesting example is a circularizing phage DNA of *H. influenzae* that transfects normal cells efficiently with a second power dependence on concentration; but transfects recombination-deficient mutants much less efficiently and with a first order dependence on concentration (Boling *et al.,* 1972).

It should be pointed out that these considerations hold only for DNA that enters through the specially evolved mechanisms of highly competent cells. When transfecting

DNA must be otherwise introduced, as in *E. coli* for example (Section 5.6), it probably does not undergo the same kind of fragmentation.

5.6 NON-PHYSIOLOGICAL TRANSFORMATION

5.6.1 Force-feeding of DNA

High frequencies of transformation, on the order of 10^{-2}, have been obtained only in several bacterial genera: *Bacillus, Haemophilus, Neisseria* and *Streptococcus*. Certain species in these genera evolved mechanisms for the introduction of free DNA, presumably as a quasi-sexual means of achieving genetic variability. Transformation at lower frequencies, on the order of 10^{-5}, has been reported for other species, for example, *Agrobacterium tumefaciens* (Klein and Klein, 1953), *Rhizobium lupini* (Balassa, 1960) and *Micrococcus radiodurans* (Moseley and Setlow, 1968). The Enterobacteriaceae, however, have been refractory to transformation under normal physiological conditions.

Several special procedures, nevertheless, were devised to force DNA into *E. coli* cells. These techniques were developed using transfection, which can be a more sensitive assay than chromosomal transformation. Helper phage particles will introduce λ DNA or fragments thereof into the cell (Kaiser and Hogness, 1960). The λ DNA must contain a single-stranded cohesive end (Kaiser and Inman, 1965), so entry probably occurs by attachment of the DNA to the cohesive end of the helper phage DNA molecule that is injected into the cell. Phage DNA by itself can transfect spheroplasts (Meyer *et al.*, 1961). Removal of the cell wall apparently allows adsorption of DNA molecules, which occasionally penetrate the membrane. Entry of double-stranded DNA is facilitated by the addition of protamine (Benzinger *et al.*, 1971), which may coat the spheroplast surface and enable more DNA to adhere to it. Single-stranded DNA transfection is not so facilitated, perhaps because single-stranded DNA readily adheres to the surface, anyway (compare with pneumococcal cells, Section 5.4.1(a)). Treatment of intact *E. coli* cells with very high concentrations of Ca^{2+} (0.03 M) before addition of DNA also allows some DNA to enter (Mandel and Higa, 1970). Although the harsh treatment kills some of the cells (Oishi and Irbe, 1977), enough survive so that they can be examined with respect to bacterial characters. With the calcium technique, bacterial transformation could be demonstrated in *E. coli*. Similar procedures were used to transform *Staphylococcus aureus* (Nomura *et al.*, 1971; Lindberg, *et al.*, 1972).

5.6.2 Transformation of *Escherichia coli*

Transformation by bacterial DNA markers occurs with a frequency of the order of 10^{-5} under the conditions described above (Cosloy and Oishi, 1973a). Use of DNA enriched for the marker, such as *trp*⁺ DNA isolated from the specialized transducing phage $\phi 80$ *ptrp,* can increase the frequency 30-fold (Oishi *et al.,* 1974). These frequencies of chromosomal transformation are attained only when the recipient bacteria carry certain mutations (Oishi and Cosloy, 1972; Cosloy and Oishi, 1973b; Wackernagel, 1973). The presence of either *sbcA* or *sbcB* provides a recombination pathway (Clark, 1973) necessary for the integration of donor DNA. Elimination of the ATP-dependent DNase by a *recBC* mutation increases transformation five-fold. The *recBC* mutation probably prevents enzymatic attack on the linear double-stranded donor DNA after its introduction into the cell (Oishi and Irbe, 1977). A similar enhancement is observed in transfection of *recBC* spheroplasts, but not with single-stranded or closed circular phage DNA, which are not substrates of the *recBC* enzyme (Benzinger *et al.,* 1975). Double-stranded DNA is required for transformation (Cosloy and Oishi, 1973a), probably to enable recombination with the chromosome rather than entry, inasmuch as single-stranded phage DNA can transfect cells (Taketo, 1975).

The foregoing observations suggest that double-stranded DNA enters substantially intact in this system. The absence of donor DNA fragmentation is attested by the linear dependence of transfection on phage DNA concentration (Mandel and Higa, 1970). Unlike transformable species, which take up much DNA but show low efficiencies of transfection because they fragment it, the low transfection efficiency of *E. coli* probably represents limited entry of DNA. Absence of a specific binding mechanism, which might fragment the DNA, is also indicated by the high concentrations of DNA needed to saturate the *E. coli* system ($> 10\ \mu g\ ml^{-1}$). In keeping with the difference in mechanism between DNA uptake in *S. pneumoniae* and *E. coli*, *end* mutants of *E. coli* that lack the major endonuclease are not impaired in transformation (Oishi and Cosloy, 1972). It should be pointed out here that the most deficient *end* mutant of *S. pneumoniae* can still be transformed at approximately the frequency of *E. coli* transformation, 10^{-5}. Direct physical studies on DNA taken up by the cells have not yet revealed much, probably because very little DNA actually enters the cell (Cosloy and Oishi, 1973b). A considerably amount of donor DNA adheres to the cells, reportedly in a form resistant to added DNase (Sabelnikov *et al.*, 1975). It remains to be seen whether this material has entered the cell or has simply been precipitated on the cell surface as a calcium salt.

Transformation of *E. coli* by plasmid DNA (Cohen *et al.*, 1972) should prove most useful for 'genetic engineering'. Because this bacterium, its phages and its plasmids have been so well studied genetically, it is a convenient host for artificially produced 'recombinant' plasmid DNA containing genes from other organisms. The relative ease of introduction of the 'recombinant' DNA into calcium-treated *E. coli* will facilitate the selection and propagation, that is, 'cloning', of potentially useful eukaryotic genes (e.g. Morrow *et al.,* 1974).

5.7 COMPARATIVE ASPECTS OF DNA TRANSPORT

5.7.1 Conjugative plasmid transfer

Plasmids are extrachromosomal DNA molecules capable of independent replication. Transmissible plasmids are those that can implement their own transfer from cell to cell by cellular conjugation. One such much-studied plasmid is the *E. coli* sex factor, F, which when inserted into the chromosome mobilizes it for conjugative transfer (see Chapter 6, Achtmann and Skurray). Certain facets of DNA transfer in F-mediated conjugation are similar to DNA uptake in bacterial transformation.

Only a single strand of donor DNA is transferred in conjugation (Cohen *et al.*, 1968). Synthesis of DNA, beginning at the origin of transfer of the F factor, normally accompanies conjugation (Sarathy and Siddiqi, 1973a). It has been supposed that this synthesis is asymmetric and produces only one new strand, which replaces the strand that is transferred (Vapnek and Rupp, 1971). However, when synthesis of DNA in the donor cell was blocked completely by deprivation of thymine, conjugative transfer still occurred (Sarathy and Siddiqi, 1973b). In this case no single-stranded DNA was found in the donor cell. It is therefore possible that, just as in pneumococcal transformation, the strand complementary to the entering strand is digested (as proposed by Kunicki-Goldfinger, 1968) to facilitate entry of the other strand. Persistence of pre-existing strands of F factor that are complementary to the transferred strands (Vapnek and Rupp, 1970), under conditions allowing DNA synthesis, can be explained by degradation of complementary strands that are synthesized just before transfer.

A curious property of many plasmids, including F, is their coupling with protein to form 'relaxation complexes' (Helinski, 1973). Following an appropriate stimulus, a protein that is bound to the twisted covalently closed circular DNA duplex molecule nicks the DNA at a specific site in one strand, thereby allowing the molecule to 'relax' and untwist. After it nicks the DNA the protein appears to be bound covalently to the $5'$-end formed at the nick (Guiney and Helinski, 1975). It is known that in conjugation the $5'$-end of the DNA strand enters the cell first (Ohki and Tomizawa, 1968; Rupp and Ihler, 1968). If the protein bound to the $5'$-end helps the DNA adhere to and penetrate the recipient cell membrane, it is functioning just like the hypothetical nicking-binding protein in transformation.

5.7.2 Injection of viral DNA

Another instance of DNA transport is the introduction of DNA into cells in viral infection. Bacteriophages like T4 or λ, after specifically attaching themselves by their tail fibers to receptors on the cell surface, seem to physically inject their DNA. In the case of T4 interaction of the base plate of the tail with the cell wall triggers contraction of the sheath so that the needle-like core of the tail penetrates the cell envelope (Simon and Anderson, 1967). The DNA, packed into the head and poised to spring forth on first opportunity, is released into the cell. Bacterial DNA that is transferred by transduction would also enter by this mechanism. Available evidence indicates that transducing DNA does appear to be double-stranded after entry (Ebel-Tsipis *et al.*, 1972).

Viruses without tails must introduce their DNA by other means. In phage M13 infection a minor coat protein, the product of gene 3, appears to be essential for uncoating the viral DNA and transporting it through the cell membrane (Jazwinski *et al.*, 1973). The protein is bound to the entering DNA and remains associated with the replicative form. Similar 'pilot' proteins have been found with other phages and it has been suggested that, in addition, they serve to lead incoming nucleic acids to appropriate internal sites (Kornberg, 1974). Their ability to bind to DNA and guide it through the membrane are both properties attributed to the hypothetical binding protein in transformation.

5.7.3 Packaging of viral DNA

Insertion of virus DNA into the viral capsid can also be considered a transport process, in the reverse direction to injection. This process, which also condenses the DNA, is complex and requires energy, but in some cases, at least, it shares one feature with DNA uptake in transformation. Encapsidation of DNA in phages λ (Mackinlay and Kaiser, 1969), T4 (Frankel, 1968) and P22 (Botstein *et al.*, 1973) involves cutting of concatemeric DNA into unit phage lengths. In the case of λ, the *A* gene product, which cuts the DNA at specific sites, also binds to the DNA, and such binding is essential for encapsidation (Hohn, 1975). This dual function of binding and nicking in DNA transport was also postulated for the surface receptor in transformation.

5.8 CONCLUSIONS AND FORECAST

Bacterial transformation is no accidental phenomenon! On the contrary, the ability to take up DNA represents a highly evolved mechanism for gene transfer, an important form of microbial interaction. Witness to this assertion are (1) the loss of components of this mechanism by mutations, which do not affect other cellular functions, and (2) the elaborate provisions for the induction of competence at a stage of growth most propitious for genetic interchange.

It is the mechanism of DNA uptake that evolution has granted to the highly transformable bacterial species. Once inside the cell the DNA can interact with host DNA by mechanisms of genetic recombination that are probably universally present in cells. These recombination systems may have evolved for quite another purpose, such as the repair of damage to DNA. Thus, when by special means DNA is made to enter bacteria that are not normally transformable, the bacteria can be genetically transformed.

From numerous studies on the highly transformable species an outline of a possibly general mechanism for DNA uptake is beginning to emerge. At successive stages donor DNA is reversibly and then irreversibly bound on the outside of the cell membrane, nicked to give single-strand breaks, nicked again on initiation of entry to give double-strand breaks, and, finally, one strand is degraded by a surface nuclease while the other strand enters the cell, perhaps aided by one or more DNA binding proteins. From this sequence a model of the molecular mechanism for DNA binding and entry can be proposed (see Section 5.3.4 and Fig. 5.1). Some solid facts support this model. Much of its support, however, is still preliminary, suggestive or hypothetical. Future work should flesh out the model or, perhaps, disprove certain aspects of it. In particular, the protein or proteins that bind and nick the DNA on the surface of the cell, and those that bind the donor strands within the cell, need to be isolated and characterized. Nucleases that participate in the entry process must be identified and their mode of action determined. Energy requirements at each stage should be elucidated. And the similarities and differences among the various bacterial systems must be clarified.

Elucidation of the molecular basis of DNA transport in bacterial transformation is essential to understanding a major pathway of genetic transfer in microbial systems. It should contribute in general to our knowledge of the structure and function of cell membranes. Insight may also be gained into the transport of macromolecules by eukaryotic cells. Possible applications of such knowledge could include the introduction of DNA into cells of higher organisms, an application that would facilitate 'genetic engineering'.

REFERENCES

Akrigg, A. and Ayad, S.R. (1970), Studies on competence inducing factor of *Bacillus subtilis. Biochem. J.,* **117**, 397–403.

Akrigg, A., Ayad, S.R. and Blamire, J. (1969), Uptake of DNA by competent bacteria – a possible mechanism. *J. theor. Biol.,* **24**, 266–272.

Anagnostopoulos, C. and Spizizen, J. (1961), Requirements for transformation in *Bacillus subtilis. J. Bact.,* **81**, 741–746.

Archer, L.J. and Landman, O.E. (1969a), Development of competence in thymine-starved *Bacillus subtilis* with chromosomes arrested at the terminus. *J. Bact.,* **97**, 166–173.

Archer, L.J. and Landman, O.E. (1969b), Transport of donor deoxyribonucleic acid into the cell interior of thymine-starved *Bacillus subtilis* with chromosomes arrested at the terminus. *J. Bact.,* **97**, 174–181.

Arwert, F. and Venema, G. (1973), Transformation in *Bacillus subtilis.* Fate of newly introduced transforming DNA. *Mol. gen. Genet.,* **123**, 185–198.

Arwert, F. and Venema, G. (1974), Transfection of *Bacillus subtilis* with bacteriophage H1DNA: fate of transfecting DNA and transfection enhancement in *B. subtilis* uvr^+ and uvr^- strains. *Molec. gen. Genet.,* **128**, 55–72.

Avery, O.T., MacLeod, C.M. and McCarty, M. (1944), Studies on the chemical nature of the substance inducing transformation of pneumococcal types. Induction of transformation by a desoxyribonucleic acid fraction isolated from pneumococcus type III. *J. exp. Med.*, **89**, 137–158.

Balassa, R. (1960), Transformation of a strain of *Rhizobium lupini*. *Nature*, **188**, 246–247.

Balassa, G. and Prévost, G. (1962), Etude théorique de la transformation bactérienne. *J. theor. Biol.*, **3**, 315–334.

Barnhart, B.J. (1965), Residual activity of thermally denatured transforming deoxyribonucleic acid from *Haemophilus influenzae*. *J. Bact.*, **89**, 1271–1279.

Barnhart, B.J. and Herriott, R.M. (1963), Penetration of deoxyribonucleic acid into *Haemophilus influenzae*. *Biochim. biophys. Acta*, **76**, 25–39.

Beattie, K.L. and Setlow, J.K. (1971), Transformation defective strains of *Haemophilus influenzae*. *Nature New Biol.*, **231**, 177–179.

Benzinger, R., Enquist, L.W. and Skalka, A. (1975), Transfection of *Escherichia coli* spheroplasts. V. Activity of *rec*BC nuclease in rec^+ and rec^- spheroplasts measured with different forms of bacteriophage DNA. *J. Virol.*, **15**, 861–871.

Benzinger, R., Kleber, I. and Huskey, R. (1971), Transfection of *Escherichia coli* spheroplasts. I. General facilitation of double-stranded deoxyribonucleic acid infectivity by protamine sulfate. *J. Virol.*, **7**, 646–650.

Bingham, D.P. and Barnhart, B.J. (1973), Inhibition of transformation by antibodies against competent *Haemophilus influenzae*. *J. gen. Microbiol.*, **75**, 249–258.

Birnboim, H.C. (1966), Cellular site in *Bacillus subtilis* of a nuclease which preferentially degrades single-stranded nucleic acids. *J. Bact.*, **91**, 1004–1011.

Blair, D.G. and Helinski, D.R. (1975), Relaxation complexes of plasmid DNA and protein. *J. biol. Chem.*, **250**, 8785–8789.

Bodmer, W.F. (1965), Recombination and integration in *Bacillus subtilis* transformation: involvement of DNA synthesis. *J. mol. Biol.*, **14**, 534–557.

Bodmer, W.R. and Ganesan, A.T. (1964), Biochemical and genetic studies of integration and recombination in *Bacillus subtilis* transformation. *Genetics*, **50**, 717–738.

Boling, M.E., Setlow, J.K. and Allison, D.P. (1972), Bacteriophage of *Haemophilus influenzae*. I. Differences between infection by whole phage, extracted phage DNA and prophage DNA extracted from lysogenic cells. *J. mol. Biol.*, **63**, 335–348.

Botstein, D., Waddell, D.H. and King, J. (1973), Mechanism of head assembly and DNA encapsulation in *Salmonella* phage P22. I. Genes, proteins, structures and DNA maturation. *J. mol. Biol.*, **80**, 669–695.

Bresler, S.E., Kreneva, R.A. Kushev, V.V. and Mosevitskii, M.I. (1964), Molecular mechanism of genetic recombination in bacterial transformation. *Z. Vererbungsl.*, **95**, 288–297.

Burch, P.R.J. (1976), *The Biology of Cancer: a new approach*, University Park Press, Baltimore, pp. 11–16.

Cahn, F.H. and Fox, M.S. (1968), Fractionation of transformable bacteria from competent cultures of *Bacillus subtilis* on renografin gradients. *J. Bact.*, **95**, 867–875.

Caster, J.H., Postel, E.H. and Goodgal, S.H. (1970), Competence mutants: isolation of transformation deficient strains of *Haemophilus influenzae*. *Nature* **227**, 515–517.

Catlin, B.W. (1960), Transformation of *Neisseria meningitidis* by deoxyribonucleates from cells and from culture slime. *J. Bact.*, **79**, 579–590.

Ceglowski, P., Fuchs, P.F. and Soltyk, A. (1975), Competitive inhibition of transformation in group H *Streptococcus* strain Challis by heterologous deoxyribonucleic acid. *J. Bact.*, **124**, 1621–1623.

Champoux, J.J. (1976), Evidence for an intermediate with a single-strand break in the reaction catalyzed by the DNA untwisting enzyme. *Proc. natn. Acad. Sci., U.S.A.*, **73**, 3488–3491.

Chevallier, M.R. and Bernardi, G. (1965), Transformation by heat-denatured deoxyribonucleic acid. *J. mol. Biol.*, **11**, 658–660.

Chevallier, R.M. and Bernardi, G. (1968), Residual transforming activity of denatured *Haemophilus influenzae* DNA. *J. mol. Biol.*, **32**, 437–452.

Chilton, M.-D. (1967), Transforming activity in both complementary strands of *Bacillus subtilis* DNA. *Science*, **157**, 817–819.

Clark, A.J. (1973), Recombination deficient mutants of *E. coli* and other bacteria. *Ann. Rev. Genet.*, **7**, 67–86.

Cohen, S.N., Chang, A.C.Y. and Hsu, L. (1972), Nonchromosomal antibiotic resistance in bacteria: genetic transformation of *Escherichia coli* by R-factor DNA. *Proc. natn. Acad. Sci. U.S.A.*, **69**, 2110–2114.

Cohen, A., Fisher, W.D., Curtiss, R., III and Adler, H.I. (1968), DNA isolated from *Escherichia coli* minicells mated with F^+ cells. *Proc. natn. Acad. Sci. U.S.A.*, **61**, 61–68.

Cosloy, S.D. and Oishi, M. (1973a), Genetic transformation in *Escherichia coli* K12. *Proc. natn. Acad. Sci. U.S.A.*, **70**, 84–87.

Cosloy, S.D. and Oishi, M. (1973b), The nature of the transformation process in *Escherichia coli* K12. *Molec. gen. Genet.*, **124**, 1–10.

Davidoff-Abelson, R. and Dubnau, D. (1973a), Conditions affecting the isolation from transformed cells of *Bacillus subtilis* of high-molecular-weight single-stranded deoxyribonucleic acid of donor origin. *J. Bact.*, **116**, 146–153.

Davidoff-Abelson, R. and Dubnau, D. (1973b), Kinetic analysis of the products of donor deoxyribonucleate in transformed cells of *Bacillus subtilis*. *J. Bact.*, **116**, 154–162.

Dobrzanski, W.T. and Osowiecki, H. (1967), Isolation and some properties of the competence factor from Group H *Streptococcus* strain Challis. *J. gen. Microbiol.*, **48**, 299–304.

Dubnau, D. (1976), Genetic transformation of *Bacillus subtilis:* A review with emphasis on the recombination mechanism. In: *Microbiology–1976*, (Schlessinger, D., ed.), American Society for Microbiology, Washington, D.C.

Dubnau, D. and Cirigliano, C. (1972a), Fate of transforming DNA following uptake by competent *Bacillus subtilis*. III. Formation and properties of products isolated from transformed cells which are derived entirely from donor DNA. *J. mol. Biol.*, **64**, 9–29.

Dubnau, D. and Cirigliano, C. (1972b), Fate of transforming DNA following uptake by competent *Bacillus subtilis*. IV. The endwise attachment and uptake of transforming DNA. *J. mol. Biol.*, **64**, 31–46.

Dubnau, D. and Cirigliano, C. (1972c), Fate of transforming deoxyribonucleic acid after uptake by competent *Bacillus subtilis*: Size and distribution of the integrated donor segments. *J. Bact.*, **111**, 488–494.

Dubnau, D. and Cirigliano, C. (1973), Fate of transforming deoxyribonucleic acid after uptake by competent *Bacillus subtilis*: Non requirement of deoxyribonucleic acid replication for uptake and integration of transforming deoxyribonucleic acid. *J. Bact.*, **113**, 1512–1514.

Dubnau, D. and Cirigliano, C. (1974), Uptake and integration of tranforming DNA in *Bacillus subtilis*. In *Mechanisms in Recombination*, (Grell, R.F., ed.), p. 167–178, Plenum Press, New York.

Ebel-Tsipis, J., Fox, M.S. and Botstein, D. (1972), Generalized transduction of bacteriophage P22 in *Salmonella typhimurium*. II. Mechanism of integration of transducing DNA. *J. mol. Biol.*, **71**, 449–469.

Eisenstadt, E., Lange, R. and Willecke, K. (1975), Competent *Bacillus subtilis* cultures synthesize a denatured DNA binding activity. *Proc. natn. Acad. Sci. U.S.A.*, **72**, 323–327.

Ephrussi-Taylor, H., Sicard, A.M. and Kamen, R. (1965), Genetic recombination in DNA-induced transformation of pneumococcus. I. The problem of relative efficiency of transforming factors. *Genetics*, **51**, 455–475.

Erickson, R.J. and Braun, W. (1968), Apparent dependence of transformation on the stage of deoxyribonucleic acid replication of recipient cells. *Bact. Rev.*, **32**, 291–296.

Földes, J. and Trautner, R.A. (1964), Infectious DNA from a newly isolated *B. subtilis* phage. *Z. Vererbungsl.*, **95**, 57–65.

Fox, M.S. (1960), Fate of transforming deoxyribonucleate following fixation by transforming bacteria. II. *Nature*, **187**, 1004–1006.

Fox, M.S. and Allen, M.K. (1964), On the mechanism of deoxyribonucleate integration in pneumococcal transformation. *Proc. natn. Acad. Sci. U.S.A.*, **52**, 412–419.

Fox, M.S. and Hotchkiss, R.D. (1957), Initiation of bacterial transformation. *Nature* **179**, 1322–1325.

Frankel, F.R. (1968), DNA replication after T4 infection. *Cold Spring Harbor Symp. Quant. Biol.*, **33**, 485–493.

Gabor, M. and Hotchkiss, R.D. (1966), Manifestation of linear organization in molecules of pneumococcal transforming DNA. *Proc. natn. Acad. Sci. U.S.A.*, **56**, 1441–1448.

Ghei, O.K. and Lacks, S.A. (1967), Recovery of donor deoxyribonucleic acid marker activity from eclipse in pneumococcal transformation. *J. Bact.*, **93**, 816–829.

Goodgal, S.H. and Herriott, R.M. (1961), Studies of transformation of *Haemophilus influenzae*. I. Competence. *J. gen. Physiol.*, **44**, 1201–1227.

Green, D.M. (1966), Intracellular inactivation of infective SP82 bacteriophage DNA. *J. mol. Biol.*, **22**, 1–13.

Greenawalt, J.W. and Whiteside, T.L. (1975), Mesosomes: membranous bacterial organelles. *Bact. Rev.*, **39**, 405–463.

Griffith, F. (1928), The significance of pneumococcal types. *J. Hyg.*, **27**, 113–159.

Groves, D.J., Wilson, G.A. and Young, F.E. (1974), Inhibition of transformation of *Bacillus subtilis* by heavy metals. *J. Bact.*, **120**, 219–226.

Guiney, D.G. and Helinski, D.R. (1975), Relaxation complexes of plasmid DNA and protein. III. Association of protein with the $5'$ terminus of the broken DNA strand in the relaxed complex of plasmid ColEl. *J. biol. Chem.*, **250**, 8796–8803.

Gurney, T., Jr. and Fox, M.S. (1968), Physical and genetic hybrids formed in bacterial transformation. *J. mol. Biol.*, **32**, 83–100.

Hadden, C. and Nester, E.W. (1968), Purification of competent cells in *Bacillus subtilis* transformation system. *J. Bact.*, **95**, 876–885.

Harris, W.J. and Barr, G.C. (1971), Structural features of DNA in competent *Bacillus subtilis*. *Mol. gen. Genet.*, **113**, 316–330.

Haseltine, F.P. and Fox, M.S. (1971), Bacterial inactivation of transforming deoxyribonucleate. *J. Bact.*, **107**, 889–899.

Helinski, D.R. (1973), Plasmid determined resistance to antibiotics: molecular properties of R factors. *Ann. Rev. Microbiol.*, **27**, 437–470.

Hohn, B. (1975), DNA as substrate for packaging into bacteriophage lambda, *in vitro*. *J. mol. Biol.*, **98**, 93–106.

Hotchkiss, R.D. (1957), Criteria for quantitative genetic transformation of bacteria. In: *The Chemical Basis of Heredity*, (McElroy, W.D. and Glass, B., eds.), p. 321–335, Johns Hopkins Press, Baltimore.

Hotchkiss, R.D. (1971), Toward a general theory of genetic recombination in DNA. *Adv. Genet.*, **16**, 325–348.

Hotchkiss, R.D. and Gabor, M. (1970), Bacterial transformation, with special reference to recombination process, *Ann. Rev. Genet.*, **4**, 193–224.

Javor, G.T. and Tomasz, A. (1968), An autoradiographic study of genetic transformation. *Proc. natn. Acad. Sci. U.S.A.*, **60**, 1216–1222.

Jazwinski, S.M., Marco, R. and Kornberg, A. (1973), A coat protein of the bacteriophage M13 virion participates in membrane-oriented synthesis of DNA. *Proc. natn. Acad. Sci. U.S.A.*, **70**, 205–209.

Joenje, H., Konings, W.N. and Venema, G. (1974), Interactions between exogenous deoxyribonucleic acid and membrane vesicles isolated from *Bacillus subtilis* 168. *J. Bact.*, **119**, 784–794.

Joenje, H., Konings, W.N. and Venema, G. (1975), Interactions between exogenous deoxyribonucleic acid and membrane vesicles isolated from competent and non-competent *Bacillus subtilis*. *J. Bact.*, **121**, 771–776.

Joenje, H. and Venema, G. (1975), Different nuclease activities in competent and non-competent *Bacillus subtilis*. *J. Bact.*, **122**, 25–33.

Kaiser, A.D. and Hogness, D.S. (1960), The transformation of *Escherichia coli* with deoxyribonucleic acid isolated from bacteriophage λdg. *J. mol. Biol.*, **2**, 392–415.

Kaiser, A.D. and Inman, R.B. (1965), Cohesion and the biological activity of bacteriophage lambda DNA. *J. mol. Biol.*, **13**, 78–91.

Kent, J.L. and Hotchkiss, R.D. (1964), Kinetic analysis of multiple, linked recombinations in pneumococcal transformation. *J. mol. Biol.*, **9**, 308–322.

Kerr, I.M., Pratt, E.A. and Lehman, I.R. (1965), Exonucleolytic degradation of high-molecular-weight DNA and RNA to nucleoside 3'-phosphates by a nuclease from *B. subtilis*. *Biochem. biophys. Res. Comm.*, **20**, 154–162.

Klein, D.T. and Klein, R.M. (1953), Transmittance of tumor-inducing ability to avirulent crown-gall and related bacteria. *J. Bact.*, **66**, 220–228.

Kohoutová, M. (1967), Role of deoxyribonuclease in genetic transformation II. RNA as an inhibitor of deoxyribonuclease activity in pneumococcal strains. *Folia Microbiol.* (Prague), **12**, 316–322.

Kohoutová, M. and Malek, I. (1966), Stimulation of transformation frequency by sterile filtrates from pneumococcus, In: *The Physiology of Gene and Mutation Expression,* (Kohoutová, M. and Hubacek, J., eds), p. 195–200, Academia, Prague.

Kornberg, A. (1974), *DNA synthesis,* W. H. Freeman and Co., San Francisco.

Kunicki-Goldfinger, W1. (1968), Mechanism of bacterial conjugation and recombination; a tentative model. *Acta Microbiol. Polon.,* **17,** 147–180.

Lacks, S. (1962), Molecular fate of DNA in genetic transformation of pneumococcus. *J. mol. Biol.,* **5,** 119–131.

Lacks, S. (1966), Integration efficiency and genetic recombination in pneumococcal transformation. *Genetics,* **53,** 207–235.

Lacks, S. (1970), Mutants of *Diplococcus pneumoniae* that lack deoxyribonucleases and other activities possibly pertinent to genetic transformation. *J. Bact.,* **101,** 373–383.

Lacks, S. (1977), Binding and entry of DNA in pneumococcal transformation. In: *Modern Trends in Bacterial Transformation and Transfection,* (Portolés, A., López, R. and Espinosa, M., eds), p. 35–44, Elsevier/North-Holland, Amsterdam.

Lacks, S. and Greenberg, B. (1967), Deoxyribonucleases of *Pneumococcus, J. biol. Chem.,* **242,** 3108–3120.

Lacks, S. and Greenberg, B. (1973), Competence for deoxyribonucleic acid uptake and deoxyribonuclease action external to cells in the genetic transformation of *Diplococcus pneumoniae, J. Bact.,* **114,** 152–163.

Lacks, S. and Greenberg, B. (1975), A deoxyribonuclease of *Diplococcus pneumoniae* specific for methylated DNA. *J. biol. Chem.,* **250,** 4060–4066.

Lacks, S. and Greenberg, B. (1976), Single-strand breakage on binding of DNA to cells in the genetic transformation of *Diplococcus pneumoniae. J. mol. Biol.,* **101,** 255–275.

Lacks, S. and Greenberg, B. (1977), Complementary specificity of restriction endonucleases of *Diplococcus pneumoniae* with respect to DNA methylation. *J. mol. Biol.,* (In press).

Lacks, S., Greenberg, B. and Carlson, K. (1967), Fate of donor DNA in pneumococcal transformation. *J. mol. Biol.,* **29,** 327–347.

Lacks, S., Greenberg, B. and Neuberger, M. (1974), Role of a deoxyribonuclease in the genetic transformation of *Diplococcus pneumoniae. Proc. natn. Acad. Sci. U.S.A.,* **71,** 2305–2309.

Lacks, S., Greenberg, B. and Neuberger, M. (1975), Identification of a deoxyribonuclease implicated in genetic transformation of *Diplococcus pneumoniae. J. Bact.,* **123,** 222–232.

Lacks, S. and Hotchkiss, R.D. (1960a), A study of the genetic material determining an enzyme activity in pneumococcus. *Biochim. biophys. Acta,* **39,** 508–518.

Lacks, S. and Hotchkiss, R.D. (1960b), Formation of amylomaltase after genetic transformation of pneumococcus. *Biochim. biophys. Acta,* **45,** 155–163.

Lacks, S. and Neuberger, M. (1975), Membrane location of a deoxyribonuclease implicated in the genetic transformation of *Diplococcus pneumoniae. J. Bact.,* **124,** 1321–1329.

Laird, C.D., Wang, L. and Bodmer, W.F. (1968), Recombination and DNA replication in *Bacillus subtilis* transformation. *Mutation Res.,* **6,** 205–209.

LeClerc, J.E. and Setlow, J.K. (1974), Transformation in *Haemophilus influenzae.* In: *Mechanisms in Recombination,* (Grell, R.F., ed.), p. 187–207, Plenum Press, New York.

LeClerc, J.E. and Setlow, J.K. (1975), Single-strand regions in the deoxyribonucleic acid of competent *Haemophilus influenzae*. *J. Bact.*, **122**, 1091–1102.

Lehman, I.R., Roussos, G.G. and Pratt, E.A. (1962), The deoxyribonucleases of *Escherichia coli*. II. Purification and properties of a ribonucleic acid-inhibitable endonuclease. *J. biol. Chem.*, **237**, 819–828.

Leidy, G., Jaffee, I. and Alexander, H.E. (1962), Emergence of competence (for transformation) of three *Haemophilus* species in a chemically defined environment. *Proc. Soc. exp. Biol. Med.*, **111**, 725–731.

Leonard, C.G. and Cole, R.M. (1972), Purification and properties of streptococcal competence factor isolated from chemically defined medium. *J. Bact.*, **110**, 273–280.

Lerman, L.S. and Tolmach, L.J. (1957), Genetic transformation. I. Cellular incorporation of DNA accompanying transformation in pneumococcus. *Biochim. biophys. Acta*, **26**, 68–82.

Levin, B.C. and Landman, O.E. (1973), DNA synthesis inhibition by 6-(p-hydroxyphenylazo)-uracil in relation to uptake and integration of transforming DNA in *Bacillus subtilis*, In: *Bacterial Transformation*, (Archer, L.J., ed.), p. 217–240, Academic Press, New York.

Levine, J.S. and Strauss, N. (1965), Lag period characterizing the entry of transforming deoxyribonucleic acid into *Bacillus subtilis*. *J. Bact.*, **89**, 281–287.

Lie, S. (1965), Studies on the phenotypic expression of competence in transformation of *Neisseria meningitidis*. *Acta path. microbiol. scand.*, **64**, 119–129.

Lindberg, M., Sjostrom, J.-E., and Johansson, T. (1972), Transformation of chromosomal and plasmid characters in *Staphylococcus aureus*. *J. Bact.*, **109**, 844–847.

Mackinlay, A.G. and Kaiser, A.D. (1969), DNA replication in head mutants of bacteriophage λ. *J. mol. Biol.*, **39**, 679–683.

McCarthy, C. and Nester, E.W. (1969a), Heat-sensitive step in deoxyribonucleic acid-mediated transformation of *Bacillus subtilis*. *J. Bact.*, **97**, 162–165.

McCarthy, C. and Nester, E.W. (1969b), Heat-activated endonuclease in *Bacillus subtilis*, *J. Bact.*, **97**, 1426–1430.

Mandel, M. and Higa, A. (1970), Calcium-dependent bacteriophage DNA infection. *J. mol. Biol.*, **53**, 159–162.

Meyer, F., Mackal, R.P., Tao, M., and Evans, E.A. (1961), Infectious deoxyribonucleic acid from lambda bacteriophage *J. biol. Chem.*, **236**, 1141–1151.

Miao, R. and Guild, W.R. (1970), Competent *Diplococcus pneumoniae* accept both single- and double-stranded deoxyribonucleic acid. *J. Bact.*, **101**, 361–364.

Miller, I.L. and Landman, O.E. (1966), On the mode of entry of transforming DNA into *Bacillus subtilis*. In: *Physiology of Gene and Mutation Expression*, (Kohoutová, M. and Hubacek, J., eds.), p. 187–194, Academia, Prague.

Morrison, D.A. (1971), Early intermediate state of transforming deoxyribonucleic acid during uptake by *Bacillus subtilis*. *J. Bact.*, **108**, 38–44.

Morrison, D.A. and Guild, W.R. (1972), Transformation and deoxyribonucleic acid size: extent of degradation on entry varies with size. *J. Bact.*, **112**, 1157–1168.

Morrison, D.A. and Guild, W.R. (1973a), Breakage prior to entry of donor DNA in pneumococcus transformation. *Biochim. biophys. Acta*, **299**, 545–556.

Morrison, D.A. and Guild, W.R. (1973b), Structure of deoxyribonucleic acid on the cell surface during uptake by pneumococcus. *J. Bact.*, **115**, 1055–1062.

Morrow, J.F., Cohen, S.N., Chang, A.C.Y., Boyer, H.W., Goodman, H.M., Helling, R.B. (1974), Replication and transcription of eukaryotic DNA in *Escherichia coli*. *Proc. natn. Acad. Sci. U.S.A.*, **71**, 1743–1747.

Moseley, B.E.B. and Setlow, J.K. (1968), Transformation in *Micrococcus radiodurans* and the ultraviolet sensitivity of its transforming DNA. *Proc. natn. Acad. Sci. U.S.A.*, **61**, 176–183.

Mosser, J.L. and Tomasz, A. (1970), Choline-containing teichoic-acid as a structural component of pneumococcal cell wall and its role in sensitivity to lysis by an autolytic enzyme. *J. biol. Chem.*, **245**, 237–298.

Mulder, C. and Doty, P. (1968), Residual activity of denatured transforming DNA of *Haemophilus influenzae*: A naturally occurring cross-linked DNA. *J. mol. Biol.*, **32**, 423–435.

Nalecz, J. and Dobrzanski, W.T. (1972), Correlation between the occurrence of competence in the transformation of group H streptococci in the presence of the competence factor and the *in vitro* DNA inactivating factor. *Mol. gen. Genet.*, **114**, 249–260.

Nava, G., Galis, A. and Beiser, S.M. (1963), Bacterial transformation: an antigen for competent pneumococci. *Nature*, **197**, 903–904.

Nester, E.W. (1964), Penicillin resistance of competent cells in deoxyribonucleic acid transformation in *Bacillus subtilis*. *J. Bact.*, **87**, 867–875.

Nester, E.W. and Stocker, B.A.D. (1963), Biosynthetic latency in early stages of deoxyribonucleic acid transformation in *Bacillus subtilis*. *J. Bact.*, **86**, 785–796.

Newman, C.M. and Stuy, J.H. (1971), Fate of bacteriophage λ DNA after adsorption by *Haemophilus influenzae*. *J. gen. Microbiol.*, **65**, 153–159.

Nomura, H., Udou, T., Yoshida, K., Ichikawa, Y., Naito, Y. and Smith, M.R. (1971), Induction of hemolysin synthesis by transformation in *Staphylococcus aureus*. *J. Bact.*, **105**, 673–675.

Notani, N.K. (1971), Genetic and physical properties of unintegrated donor DNA molecules during *Haemophilus* transformation. *J. mol. Biol.*, **59**, 223–226.

Notani, N.K. and Goodgal, S.H. (1965), Decrease in integration of transforming DNA of *Haemophilus influenzae* following ultraviolet irradiation. *J. mol. Biol.*, **13**, 611–613.

Notani, N. and Goodgal, S.H. (1966), On the nature of recombinants formed during transformation in *Haemophilus influenzae*. *J. gen. Physiol.*, **49**, Part 2, 197–209.

Notani, N.K. and Setlow, J.K. (1974), Mechanism of bacterial transformation and transfection. *Prog. Nucl. Acid Res. Mol. Biol.*, **14**, 39–100.

Notani, N.K., Setlow, J.K., Joshi, U.R. and Allison, D.P. (1972), Molecular basis for the transformation defects in mutants of *Haemophilus influenzae*. *J. Bact.*, **110**, 1171–1180.

Ohki, M. and Tomizawa, J-I. (1968), Asymmetric transfer of DNA strands in bacterial conjugation. *Cold Spring Harbor Symp. Quant. Biol.*, **33**, 651–657.

Oishi, M. and Cosloy, S.D. (1972), The genetic and biochemical basis of the transformability of *Escherichia coli* K12. *Biochem. biophys. Res. Commun.*, **49**, 1568–1572.

Oishi, M., Cosloy, S.D. and Basu, S.R. (1974), Transformation-mediated recombination in *Escherichia coli*. In: *Mechanisms in Recombination*, (Grell, R.F., ed.), p. 145–154, Plenum Press, New York.

Oishi, M. and Irbe, R.M. (1977), Circular chromosomes and genetic transformation in *Escherichia coli*. In: *Modern Trends in Bacterial Transformation and Transfection*, (Portolés, A., Lopéz, R. and Espinosa, M., eds.), p. 121–134, Elsevier/North-Holland, Amsterdam.

Okubo, S. and Romig, W.R. (1966), Impaired transformability of *Bacillus subtilis* mutant sensitive to mitomycin C and ultraviolet radiation. *J. mol. Biol.*, **15**, 440–454.

Osborn, M.J. (1969), Structure and biosynthesis of the bacterial cell wall. *Ann. Rev. Biochem.*, **38**, 501–538.

Ottolenghi, E. and Hotchkiss, R.D. (1962), Release of genetic transforming agent from pneumococcal cultures during growth and disintegration. *J. exp. Med.*, **116**, 491–519.

Pakula, R. (1965), Inhibition of transformation of streptococci by antibodies. *J. Bact.*, **90**, 1501–1502.

Pakula, R. (1967), Cellular sites for the competence-provoking factor of streptococci. *J. Bact.*, **94**, 75–79.

Pakula, R., Spencer, L.R. and Goldstein, P.A. (1972), Deoxyribonuclease and competence factor activities of transformable and non-transformable group H streptococci. *Can. J. Microbiol.*, **18**, 111–119.

Pakula, R. and Walczak, W. (1963), On the nature of competence of transformable streptococci. *J. gen. Microbiol.*, **31**, 125–133.

Piechowska, M. and Fox, M.S. (1971), Fate of transforming deoxyribonucleate in *Bacillus subtilis*. *J. Bact.*, **108**, 680–689.

Porter, R.D. and Guild, W.R. (1969), Number of transformable units per cell in *Diplococcus pneumoniae*. *J. Bact.*, **97**, 1033–1035.

Porter, R.D. and Guild, W.R. (1976), Evidence favoring single-strand entry of transfecting DNA in pneumococcus. *Fedn. Proc. fedn. Am. Socs. exp. Biol.*, **35**, 1595.

Postel, E.H. and Goodgal, S.H. (1966), Uptake of 'single-stranded DNA in *Haemophilus influenzae* and its ability to transform. *J. mol. Biol.*, **16**, 317–327.

Postel, E.H. and Goodgal, S.H. (1967), Further studies on transformation with single-stranded DNA of *Haemophilus influenzae*. *J. mol. Biol.*, **28**, 247–259.

Prozorov, A.A. (1965), The effect of egg-white lysozyme on the permeability of cells of *B. subtilis* to transforming DNA. *Dokl. Akad. Nauk SSR.*, **160**, 472–474.

Raina, J.L. and Ravin A.W. (1977), The fate of transforming DNA bound to competent *Streptococcus sanguis*. In: *Modern Trends in Bacterial Transformation and Transfection*, (Portolés, A., Lopéz, R. and Espinosa, M., eds.), p. 143–148, Elsevier/North-Holland, Amsterdam.

Ranhand, J.M. (1973), Autolytic activity and its association with the development of competence in group H streptococci. *J. Bact.*, **115**, 607–614.

Ranhand, J.M. (1974), Inhibition of the development of competence in *Streptococcus sanguis* (Wicky) by reagents that interact with sulfhydryl groups. *J. Bact.*, **118**, 1041–1050.

Ranhand, J.M., Leonard, C.G. and Cole, R.M. (1971), Autolytic activity associated with competent group H streptococci. *J. Bact.*, **106**, 257–268.

Ranhand, J.M., Theodore, T.S. and Cole, R.M. (1970), Protein difference between competent and noncompetent cultures of a group H streptococcus. *J. Bact.*, **104**, 360–362.

Ravin, A.W. (1954), A quantitative study of autogenic and allogenic transformations in *Pneumococcus*. *Exp. Cell Res.*, **7**, 58–82.

Roger, M. (1972), Evidence for conversion of heteroduplex transforming DNAs to homoduplexes by recipient pneumococcal cells. *Proc. natn. Acad. Sci. U.S.A.*, **69**, 466–470.

Roger, M. and Hotchkiss, R.D. (1961), Selective heat inactivation of pneumococcal transforming deoxyribonucleate. *Proc. natn. Acad. Sci. U.S.A.*, **47**, 653–669.

Romig, W.R. (1962), Infection of *Bacillus subtilis* with phenol-extracted bacteriophages. *Virology*, **16**, 452–459.

Rosenthal, A.L. and Lacks, S.A. (1977), Nuclease detection in SDS-polyacrylamide gel electrophoresis. *Analyt. Biochem.*, (in press).

Rownd, R., Green, D.M., Sternglanz, R. and Doty, P. (1968), Origin of the residual transforming activity of denatured *B. subtilis* DNA. *J. mol. Biol.*, **32**, 369–377.

Rupp, W.D. and Ihler, G. (1968), Strand selection during bacterial mating. *Cold Spring Harbor Symp. Quant. Biol.*, **33**, 647–650.

Rutberg, L., Hoch, J.A. and Spizizen, J. (1969), Mechanism of transfection with deoxyribonucleic acid from the temperate bacillus bacteriophage $\phi 105$. *J. Virol.*, **4**, 50–57.

Sabelnikov, A.G., Avdeeva, A.V. and Ilyashenko, B.N. (1975), Enhanced uptake of donor DNA by Ca^{2+}-treated *Escherichia coli* cells. *Mol. gen. Genet.*, **138**, 351–358.

Sarathy, P.V. and Siddiqi, O. (1973a), DNA synthesis during bacterial conjugation. I. Effect of mating on DNA replication in *Escherichia coli* Hfr. *J. mol. Biol.*, **78**, 427–441.

Sarathy, P.V. and Siddiqi, O. (1973b), DNA synthesis during bacterial conjugation. II. Is DNA replication in the Hfr obligatory for chromosome transfer? *J. mol. Biol.*, **78**, 443–451.

Schaeffer, P. (1969), Sporulation and the production of antibiotics, exoenzymes and exotoxins. *Bact. Rev.*, **33**, 48–71.

Schaeffer, P., Edgar, R.S. and Rolfe, R. (1960), Sur l'inhibition de la transformation bactérienne par des désoxyribonucléates de compositions variées. *Compt. Rend. Soc. Biol.*, **154**, 1978–1983.

Scher, B. and Dubnau, D. (1976), Purification and properties of a manganese stimulated endonuclease from *Bacillus subtilis*. *J. Bact.*, **126**, 429–438.

Scocca, J.J., Poland, R.L. and Zoon, K.C. (1974), Specificity in deoxyribonucleic acid uptake by transformable *Haemophilus influenzae*. *J. Bact.*, **118**, 369–373.

Sedgwick, B. and Setlow, J.K. (1976), Single-stranded regions in transforming deoxyribonucleic acid after uptake by competent *Haemophilus influenzae*. *J. Bact.*, **125**, 588–594.

Setlow, J.K. and Boling, M.E. (1972), Bacteriophage of *Haemophilus influenzae*. II. Repair of ultraviolet-irradiated phage DNA and the capacity of irradiated cells to make phage. *J. mol. Biol.*, **63**, 349–362.

Seto, H., Lopez, R., Garrigan, O. and Tomasz, A. (1975a), Nucleolytic degradation of homologous and heterologous deoxyribonucleic acid molecules at the surface of competent pneumococci. *J. Bact.*, **122**, 676–685.

Seto, H., Lopez, R. and Tomasz, A. (1975b), Cell surface located deoxyribonucleic acid receptors in transformable pneumococci. *J. Bact.*, **122**, 1339–1350.

Seto, H. and Tomasz, A. (1974), Early stages in DNA binding and uptake during genetic transformation of pneumococci. *Proc. natn. Acad. Sci. U.S.A.*, **71** 1493–1498.

Seto, H. and Tomasz, A. (1975a), Protoplast formation and leakage of intramembrane cell components: induction by the competence activator substance of pneumococci. *J. Bact.*, **121**, 344–353.

Seto, H. and Tomasz, A. (1975b), Selective release of a deoxyribonucleic acid-binding factor from the surface of competent pneumococci. *J. Bact.*, **124**, 969–976.

Seto, H. and Tomasz, A. (1976), Calcium-requiring step in the uptake of deoxyribonucleic acid molecules through the surface of competent pneumococci. *J. Bact.*, **126**, 1113–1118.

Sgaramella, V. (1972), Enzymatic oligomerization of bacteriophage P22 DNA and of linear Simian Virus 40 DNA. *Proc. natn. Acad. Sci. U.S.A.*, **69**, 3389–3393.

Simon, L.D. and Anderson, T.F. (1967), The infection of *Escherichia coli* by T2 and T4 bacteriophages as seen in the electron microscope I. Attachment and penetration. *Virology*, **32**, 279–297.

Singh, R.N. (1972), Number of deoxyribonucleic acid uptake sites in competent cells of *Bacillus subtilis*. *J. Bact.*, **110**, 266–272.

Singh, R.N. and Pitale, M.P. (1967), Enrichment of *Bacillus subtilis* transformants by zonal centrifugation. *Nature*, **213**, 1262–1263.

Soltyk, A., Shugar, D. and Piechowska, M. (1975), Heterologous deoxyribonucleic acid uptake and complexing with cellular constituents in competent *Bacillus subtilis*. *J. Bact.*, **124**, 1429–1438.

Somma, S. and Polsinelli, M. (1970), Quantitative autoradiographic study of competence and deoxyribonucleic acid incorporation in *Bacillus subtilis*. *J. Bact.*, **101**, 851–855.

Sparling, P.V. (1966), Genetic transformation of *Neisseria gonorrhoeae* to streptomycin resistance. *J. Bact.*, **92**, 1364–1371.

Spatz, H. Ch. and Trautner, T.A. (1971), The role of recombination in transfection of *B. subtilis*. *Molec. gen. Genet.*, **113**, 174–190.

Spencer, H.T. and Herriott, R.M. (1965), Development of competence of *Haemophilus influenzae*. *J. Bact.*, **90**, 911–920.

Steinhart, W.L. and Herriott, R.M. (1968), Genetic integration in the heterospecific transformation of *Haemophilus influenzae* cells by *Haemophilus parainfluenzae* deoxyribonucleic acid. *J. Bact.*, **96**, 1725–1731.

Strauss, N. (1965), Configuration of transforming deoxyribonucleic acid during entry into *Bacillus subtilis*. *J. Bact.*, **89**, 288–293.

Strauss, N. (1966), Further evidence concerning the configuration of transforming deoxyribonucleic acid during entry into *Bacillus subtilis*. *J. Bact.*, **91**, 702–708.

Strauss, N. (1970), Early energy-dependent step in the entry of transforming deoxyribonucleic acid. *J. Bact.*, **101**, 35–37.

Stuy, J.H. (1962), Transformability of *Haemophilus influenzae* *J. gen. Microbiol.*, **29**, 537–549.

Stuy, J.H. (1965), Fate of transforming DNA in the *Haemophilus influenzae* transformation system. *J. mol. Biol.*, **13**, 554–570.

Stuy, J.H. (1974), Acid-soluble breakdown of homologous deoxyribonucleic acid adsorbed by *Haemophilus influenzae*: Its biological significance. *J. Bact.*, **120**, 917–922.

Stuy, J.H. and Stern, D. (1964), The kinetics of DNA uptake by *Haemophilus influenzae. J. gen. Microbiol.*, **35**, 391–400.

Stuy, J.H. and van der Have, B. (1971), Degradation of adsorbed transforming DNA by *Haemophilus influenzae. J. gen. Microbiol.*, **65**, 147–152.

Taketo, A. (1975), Ba^{2+}-induced competence for transfecting DNA. *Z. Naturf.*, **30c**, 520–522.

Tanford, C. (1961), *Physical Chemistry of Macromolecules*. John Wiley and Sons, New York.

Tevethia, M.J. and Caudill, C.P. (1971), Relationship between competence for transformation of *Bacillus subtilis* with native and single-stranded deoxyribonucleic acid. *J. Bact.*, **106**, 808–811.

Tevethia, M.J. and Mandel, M. (1970), Nature of the ethylenediaminetetraacetic acid requirement for transformation of *Bacillus subtilis* with single-stranded deoxyribonucleic acid. *J. Bact.*, **101**, 844–850.

Tevethia, M.J. and Mandel, M. (1971), Effects of pH on transformation of *Bacillus subtilis* with single-stranded deoxyribonucleic acid. *J. Bact.*, **106**, 802–807.

Tichy, P. and Landman, O.E. (1969), Transformation in quasispheroplasts of *Bacillus subtilis. J. Bact.*, **97**, 42–51.

Tiraby, G., Claverys, J.-P. and Sicard, M.A. (1973), Integration efficiency in DNA-induced transformation of pneumococcus. I. A method of transformation in solid medium and its use for isolation of transformation-deficient and recombination-modified mutants. *Genetics*, **75**, 23–33.

Tomasz, A. (1965), Control of the competent state in *Pneumococcus* by a hormone-like cell product: and example for a new type of regulatory mechanism in bacteria. *Nature*, **208**, 155–159.

Tomasz, A. (1966), Model for the mechanism controlling the expression of competent state in pneumococcus cultures. *J. Bact.*, **91**, 1050–1061.

Tomasz, A. (1968), Biological consequences of the replacement of choline by ethanolamine in the cell wall of pneumococcus: chain formation, loss of transformability, and loss of autolysis. *Proc. natn. Acad. Sci. U.S.A.*, **59**, 86–93.

Tomasz, A. (1969), Some aspects of the competent state in genetic transformation. *Ann. Rev. Genet.*, **3**, 217–232.

Tomasz, A. (1970), Cellular metabolism in genetic transformation of pneumococci: requirement for protein synthesis during induction of competence. *J. Bact.*, **101**, 860–871.

Tomasz, A. (1973), The binding of polydeoxynucleotides to the surface of competent pneumococci. In: *Bacterial Transformation*, (Archer, L.J. ed.), p. 81–88, Academic Press, New York.

Tomasz, A. and Hotchkiss, R.D. (1964), Regulation of the transformability of pneumococcal cultures by macromolecular cell products. *Proc. natn. Acad. Sci. U.S.A.*, **51**, 480–487.

Tomasz, A. and Mosser, J.L. (1966), On the nature of the pneumococcal activator substance. *Proc. natn. Acad. Sci. U.S.A.*, **55**, 58–66.

Tomasz, A. and Zanati, E. (1971), Appearance of a protein 'agglutinin' on the spheroplast membrane of pneumococci during induction of competence. *J. Bact.*, **105**, 1213–1215.

Tomasz, A., Zanati, E. and Ziegler, R. (1971), DNA uptake during genetic transformation and the growing zone of the cell envelope. *Proc. natn. Acad. Sci. U.S.A.*, **68**, 1848–1852.

Trautner, T.A. and Spatz, H.Ch. (1973), Tranfection in *B. subtilis. Curr. Top. Microbiol. Immunol.*, **62**, 61–88.

Vapnek, D. and Rupp, W.D. (1970), Asymmetric segregation of the complementary sex-factor DNA strands during conjugation in *Escherichia coli. J. mol. Biol.*, **53**, 287–303.

Vapnek, D. and Rupp, W.D. (1971), Identification of individual sex-factor DNA strands and their replication during conjugation in thermosensitive DNA mutants of *Escherichia coli. J. mol. Biol.*, **60**, 413–424.

Venema, G., Joenje, H. and Vermeulen, C.A. (1977), Differences between competent and non-competent cells of *Bacillus subtilis* and their possible significance for competence. In: *Modern Trends in Bacterial Transformation and Transfection*, (Portolés, A., López, R. and Espinosa, M., eds), p. 69–84, Elsevier/North-Holland, Amsterdam.

Venema, G., Prichard, R.H. and Venema-Schröder, T. (1965), Fate of transforming deoxyribonucleic acid in *Bacillus subtilis. J. Bact.*, **89**, 1250–1255.

Vermeulen, C.A. and Venema, G. (1971), Autoradiographic estimation of competence and the relationship between competence and transformability in cultures of *Bacillus subtilis. J. gen. Microbiol.*, **69**, 239–252.

Vermeulen, C.A. and Venema, G. (1974), Electron microscope and autoradiographic study of ultrastructural aspects of competence and deoxyribonucleic acid absorption in *Bacillus subtilis. J. Bact.*, **118**, 342–350.

Voll, M.J. and Goodgal, S.H. (1961), Recombination during transformation in *Haemophilus influenzae. Proc. natn. Acad. Sci. U.S.A.*, **47**, 505–512.

Vovis, G. (1973), Adenosine triphosphate-dependent deoxyribonuclease from *Diplococcus pneumoniae*: Fate of transforming deoxyribonucleic acid in a strain deficient in the enzymatic activity. *J. Bact.*, **113**, 718–723.

Wackernagel, W. (1973), Genetic transformation in *E. coli*: The inhibitory role of the recBC DNAse. *Biochem. biophys. Res. Commun.*, **51**, 306–311.

Williams, G.L. and Green, D.M. (1972), Early extracellular events in infection of competent *Bacillus subtilis* by DNA of bacteriophage SP82G. *Proc. natn. Acad. Sci. U.S.A.*, **69**, 1545–1549.

Wilson, G.A. and Bott, K.R. (1970), Effects of lysozyme on competence for *Bacillus subtilis* transfection. *Biochim. biophys. Acta*, **199**, 464–475.

Wise, E.M., Alexander, S.P. and Powers, M. (1973), Adenosine $3':5'$-cyclic monophosphate as a regulator of bactcrial transformation. *Proc. natn. Acad. Sci. U.S.A.*, **70**, 471–474.

Wolstenholme, D.R., Vermeulen, C.A. and Venema, G. (1966), Evidence for the involvement of membranous bodies in the processes leading to genetic transformation in *Bacillus subtilis. J. Bact.*, **92**, 1111–1121.

Young, F.E. (1967), Competentce in *Bacillus subtilis* transformation system. *Nature*, **213**, 773–775.

Young, F.E. and Spizizen, J. (1963), Incorporation of deoxyribonucleic acid in the *Bacillus subtilis* transformation system. *J. Bact.*, **86**, 392–400.

Young, F.E. and Wilson, G.A. (1972), Genetics of *Bacillus subtilis* and other grampositive sporulating bacilli. In: *Spores V,* (Halvorson, H.D., Hanson, R. and Campbell, L.L., eds.), p. 77–106, American Society for Microbiology, Ann Arbor, Michigan.

Ziegler, R. and Tomasz, A. (1970), Binding of the competence factor to receptors in the spheroplast membrane of pneumococci. *Biochem. biophys. Res. Commun.*, **41**, 1342–3149.

Zoon, K.C. and Scocca, J.J. (1975), Constitution of the cell envelope of *Haemophilus influenzae* in relation to competence for genetic transformation. *J. Bact.*, **123**, 666–677.

6 A Redefinition of the Mating Phenomenon in Bacteria

MARK ACHTMAN and RON SKURRAY

6.1	Introduction	page	235
6.2	Some generalities		236
	6.2.1 Plasmids and their ecology		236
	6.2.2 Sex factor taxonomy		237
	6.2.3 F factor genetics		239
6.3	The mating phenomenon		243
	6.3.1 Requirements for mating		243
	(a) *F pili*, 243, (b) *Recipient's cell surface*, 249, (c) *DNA transfer requirements*, 255, (d) *DNA transport across cell membranes*, 257		
	6.3.2 Cell–cell interactions during mating		258
	(a) *Mating aggregates instead of mating pairs*, 258, (b) *Stable mating aggregates*, 262, (c) *Formation of mating aggregates*, 263, (d) *Active disaggregation*, 264, (e) *Surface exclusion: a barrier*, 264, (f) *Lethal zygosis: a pathological interaction*, 266, (g) *Purified components*, 267		
6.4	Summary		268
	6.4.1 Stages in the mating cycle		268
	6.4.2 Overview		268
	References		269

Acknowledgements

We wish to thank all our colleagues who have provided us with unpublished information. We are also particularly grateful to Naomi Datta and Y.A. Chabbert for supplying us with their current compilations of compatibility groupings and for allowing us to incorporate the data in this review. Similarly, C.M. To has been exceedingly generous in supplying us with the photographs in Figs. 6.3–6.5 especially as this material is still unpublished. Our thanks for critically reviewing the manuscript are extended to Russell Thompson, Thomas Trautner, and John Womack. We have not attempted to present a review of the literature beyond October, 1976. RS was supported by a California Division American Cancer Society Fellowship (no. J-275).

Microbial Interactions
(*Receptors and Recognition,* Series B, Volume 3)
Edited by J.L. Reissig
Published in 1977 by Chapman and Hall, 11 New Fetter Lane, London EC4P 4EE
© Chapman and Hall

6.1 INTRODUCTION

The phenomenon of bacterial conjugation has been known for 30 years, and a broad outline of the process with emphasis on DNA transfer and subsequent recombination has become dogma. The dogma focuses on a rolling circle mechanism of transfer of chromosomal DNA between cells via the F pilus. Conjugation is thought to occur in two successive stages; that of 'specific pair formation' in which donor and recipient cells first encounter each other, followed by that of 'effective contact formation' in which DNA transfer is stimulated (de Haan and Gross, 1962). Thereafter, the only phenomena considered to be of relevance are those that involve the transferred DNA, such as recombination. This dogma does reflect some aspects of conjugation seen with *Escherichia coli* Hfr x F$^-$ matings in the laboratory, but it has also become clear that certain of its features are unproven, that others are irrelevant to the ecological role of conjugation, and that important aspects of conjugation at the level of cell—cell interactions have largely been ignored. The biochemical mechanism of conjugation is not known, the role of F pili is uncertain, and there are probably serious errors in the prevailing concepts of some stages in the process. This lack of understanding is all the more remarkable when one considers that hundreds of publications have been addressed to the topic of conjugation (for reviews of these see Curtiss, 1969; Brinton, 1971).

It is our intention to present here a speculative and biased analysis of the information available on conjugation and to attempt to put it in context using the current knowledge of the genetics and biochemistry of plasmids. Since an adequate review of plasmids is beyond the scope of this publication, the interested reader is referred elsewhere (Falkow, 1975; Meynell, 1972; Schlessinger, 1975). Attention will be concentrated on conjugation mediated by F and F-like sex factors. The major points of the redefinition of conjugation offered here are:

(a) the rolling circle model is shown to be unproven for DNA transfer,
(b) the concept of mating aggregates, containing variable numbers and proportions of donor and recipeint cells, replaces that of mating pairs,
(c) the role of F pili is questioned, and the importance of wall—wall contact in mating is explored, and
(d) conjugation is subdivided into five experimentally observable stages: binding of F pili, formation of wall to wall contacts, stabilization of mating aggregates, transfer of DNA and active disaggregation.

6.2 SOME GENERALITIES

6.2.1 Plasmids and their ecology

Plasmids are supercoiled DNA molecules which are commonplace in numerous bacterial species. Despite their diversity, it is possible to make some generalizations that have only few exceptions. A plasmid is classed as a sex factor if it is self-transmissible from one cell to another. In addition, sex factors can promote the transfer of DNA from co-resident plasmids which are non-sex factors, by a process called mobilization. Sometimes they can also bring about the inefficient transfer of chromosomal DNA. Sex factors are, in general, large plasmids of over 17 Mdal (Megadalton or millions of daltons) in size (Crosa et al., 1975) and usually between 40 and 70 Mdal (Grindley et al., 1973). Most are present in relatively few copies per cell (1–3 per chromosome; Grindley et al., 1973) although exceptions, such as R6K with a copy number of 11 (Crosa et al., 1975) and mutants with elevated copy numbers (Uhlin and Nordstrom, 1975), have been found. Sex factors have been found primarily within gram-negative and gram-variable species. Two notable exceptions are *Streptococcus faecalis* (Jacob and Hobbs, 1974; Tomura et al., 1973) and *Streptomyces coelicolor* (Hopwood and Wright, 1976).

The fact that chromosomal DNA can be transferred by conjugation from one bacterium to another has provided an invaluable tool for bacterial geneticists to unravel the genetic and regulatory complexities of enteric bacteria. However, it has become clear that the ecological relevance of conjugation lies in the dissemination of bacterial sex factors and plasmids rather than in the transfer of chromosomal DNA: transfer of the bacterial chromosome, followed by recombination into the recipient's chromosome, has never to our knowledge been reported as a natural event.

It was formerly believed that the mechanism of conjugation in Hfr x F$^-$ matings was representative of bacterial conjugation in general. However, the diversity of sex factors and the lack of DNA homology between different groups of sex factors makes it likely that no single mechanism of conjugation is true for all bacteria. Rather, a number of different mechanisms may be present whose only similarity is that cell contact is necessary for the transfer of DNA.

Non-sex factor plasmids are prevalent in the same bacterial species that harbour sex factors, but they are also found in species (including gram-positive bacteria) not known to conjugate. In general, they are smaller and range from 1 to 10 Mdal (Smith et al., 1974). Often the number of DNA copies per cell is high relative to that of sex factors, and plasmids with 10 to 20 copies per chromosome are not unusual (Smith et al., 1974). The distinction between large and small plasmids is blurred by the fact that the two can interact to form a composite plasmid exhibiting some of the properties of both (Guerry et al., 1974). Furthermore, DNA from one can insert into or recombine with the other under certain conditions (Guerry et al., 1974). The reader interested in more details is referred to recent review articles (Cohen and Kopecko, 1976; Starlinger and Saedler, 1976).

A comprehensive compilation of properties for which plasmids may code has been published (Novick, 1974). All plasmids tested, with the exceptions of special mutants, control their DNA replication and the number of DNA copies per cell, and exhibit incompatibility with related plasmids. As mentioned above, sex factors code for DNA transfer as well. These phenomena will be discussed in more detail below. However, many plasmids also code for supernumerary properties that are not so obviously related to maintenance or transfer of the plasmid itself. These include resistance to or the production of antibiotics, production of and immunity to bacteriocins, synthesis of whole enzyme pathways, and virulence. The expression of these supernumerary properties is assumed to justify the widespread occurrence of plasmids in nature. DNA transfer probably occurs rarely enough in nature that the existence of conjugation alone would not be sufficient to have resulted in the prevalence of sex factors. However, selection for the supernumerary properties can easily account for the overgrowth of a previously sex factor-free population by the few clones which have acquired the appropriate sex factor. This being so, it is all the more puzzling that as many as one third of the sex factors isolated from enteric bacteria possess no detectable supernumerary properties (Tschaepe, 1977). Hence the selective pressures responsible for their common occurrence are not fully understood.

6.2.2 Sex factor taxonomy

The instability observed when two plasmids are co-resident in a host cell (giving segregation of clones carrying only one or the other plasmid) is termed incompatibility, and our understanding of the taxonomy of sex factors has been greatly increased by the analysis of compatibility groups. Sex factors *incompatible* with each other are assigned to the same compatibility group. Thus, since any two genetically distinguishable F factors are incompatible, all F factors have been assigned to compatibility group FI (Datta, 1975a) together with the sex factors ColV2 and R386. All 3 sex factors are thought to have a different origin. The closely related sex factors R100 and R1 can stably co-exist with F but not with each other and thus define a second compatibility group, FII (Datta, 1975a). On the basis of similar reasoning, at least 16 compatibility groups can be defined among sex factors of the enteric bacteria which are transmissible to *Escherichia coli* (Datta, 1975a and 1975b; Le Minor et al., 1976). The clearly distinguishable groups are listed in Table 6.1 (Datta and Chabbert, personal communication); still others not as well defined have also been described (Datta, 1975a). Similarly, although the analysis is not as complete, at least 8 compatibility groups (of which three are transmissible to *E. coli* and two of which are included in Table 6.1) can be defined among *Pseudomonas* sex factors (Jacoby, 1975; Sagai et al., 1976) and it is reasonable to expect that a similar plasmid diversity will be found in other species of bacteria.

If compatibility groupings can be used to easily and reliably identify related sex factors, then conclusions regarding the evolution and dissemination of sex factors may be facilitated. The number of compatibility groups would then yield a minimal

Table 6.1 Compatibility groups of sex factors transmissible to *E. coli*

Compatibility		Sex pili	Representatives
Overgroup F			
	FI	F	F42, R386, ColV2
	FII	3 F-like groups	R1, R100, R538-1, R136
	FIII	F-like	ColB-K98, MIP240 Hly
Overgroup I			
	I alpha	I	R64, R144, ColIb-P9
	I2	I-like	TP114, MIP241 Hly
Other groups			
	C		R40a, R576-1
	H		R27, R726-1
	J		R391, R391·3b-1
	L		R472, R831
	M		R446b, R930
	N	N	N3, R447b
	O		R7, R724
	P	RP4	RP4, R702
	S		R478, R477-1
	T	E	Rts1, R401
	W	W	S-a, R388

The compatibility groups summarized here are those with more than one representative sex factor. Spaces left blank indicate that no information is available. The remaining F-like groups described elsewhere (Datta, 1975a), possess only a single representative; the remaining I-like groups described (Datta, 1975a) have become somewhat uncertain (Datta and Barth, 1976). The groupings listed here are those currently considered to be certain (N. Datta and Y.A. Chabbert, personal communication). Most of the data is according to Datta (1975a). Group I2 was first defined by Grindley *et al.* (1973), and Le Minor *et al.* (1976) have found the second representatives of groups FII and I2 listed. Photographs of F, N, RP4, and E pili are to be seen in Figs. 6.2—6.5. F pili are described in the text; I pili were described by Lawn *et al.* (1967); N and E pili have been detected by Sam C.M. To (personal communication); and P and W pili by Bradley (1974, 1976) and Olsen and Thomas (1973). Pilus specific phages are known for F pili (Brinton, 1965), I pili (Lawn *et al.*, 1967), E pili (To, personal communication), W pili (Bradley, 1976) and P pili (Bradley, 1974). Phage Ike (Khatoon *et al.*, 1972) is specific for N sex factor-carrying cells.

estimate for the number of unrelated plasmid 'species'. This hope has been partially substantiated by the use of independent criteria to estimate genetic relatedness, such as contour length measurements and DNA-DNA hybridization. Experiments both with sex factors (Grindley *et al.*, 1973; Falkow *et al.*, 1974; Anderson *et al.*, 1975) and with non-sex factors (Smith *et al.*, 1974 and 1976) have shown that plasmids of different origin and supernumerary properties, which have been assigned to the same compatibility group, are also similar in size and share extensive common DNA sequences. Plasmids from different compatibility groups are much less related with regard to both size and DNA sequence. One exception to this is that plasmids belonging to the three compatibility groups in the F overgroup (Table 6.1), so defined because they all produce F-like pili, display considerable DNA homology. A comparison of representatives of the FI and FII groups by electron microscopic analysis of DNA heteroduplices (Sharp et al., 1973) has shown that the primary homology is in that part of the sex factor coding for pilus production and conjugation. Plasmids of the I overgroup are characterized by production of I-like pili.

Given the DNA homology within the conjugation region of the F-like plasmids, it is likely that they use basically similar mechanisms of conjugation. In contrast, the fact that at least 10 compatibility groups exist among enteric sex factors with no known DNA homologies or any other similarities (Grindley et al., 1973; Falkow et al., 1974; Anderson et al., 1975) implies that as many totally different mechanisms of conjugation may exist. Sex factors in other non-enteric bacteria may code for still different mechanisms of conjugation. Given these arguments it is unrealistic, though not uncommon, to extrapolate findings obtained with one sex factor to unrelated sex factors. Partially for historical reasons, the conjugation system most extensively investigated has been that coded by F and F-like plasmids in *E. coli*; the rest of this review will therefore concentrate on results obtained with this archetypal system.

6.2.3 F factor genetics

F factor genetics have been recently reviewed (Achtman, 1973a) and only a brief summary will be presented here. The current map of the F factor is presented in Fig. 6.1. Co-ordinates are in kb (kilobases or 1000s of nucleotide bases (single-stranded DNA) or base pairs (double-stranded DNA)) units, and the whole F factor is 94.5 kb (equals 62 Mdal) in size (Sharp et al., 1972). All cistrons assigned physical locations to date are clustered according to function. The physical locations of the cistrons shown are based on two sources of reliable information: (a) plasmid chimeras in which defined segments of F DNA are cloned on a non-sex factor plasmid and which code for the same properties as the corresponding DNA on the intact F factor (Skurray et al., 1976b; Clark et al., 1976); and (b) mutants with deletions of known regions of the F factor which have been shown to have lost certain functions (Sharp et al., 1972). On this basis, the *pif* cistrons (Morrison and Malamy, 1971), which lead to poor growth of T7 and related bacteriophages on F-carrying bacteria, map between approximately 33 and 43 kb (Anthony et al., 1974; Skurray et al., 1976b). All the cistons necessary for F factor DNA replication (*frp*) and for incompatibility (*inc*) are located between 40.3 and 49.3 kb, the region spanned by *Eco*RI fragment 5, since chimeric plasmids carrying this fragment behave identically to F with respect to these two functions (Guyer et al., 1976; Ohtsubo and Ohtsubo, personal communication; Timmis et al., 1975; Lovett and Helinski, 1976;

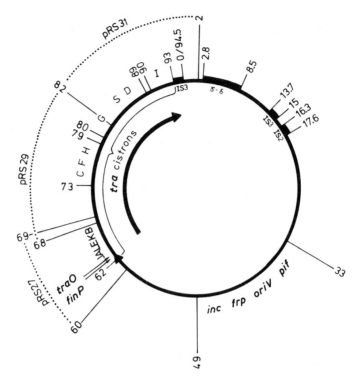

Fig. 6.1 The F factor map. Mnemonics: *inc* for *inc*ompatibility; *frp* for
*F r*e*p*lication; *oriV* for *ori*gin of *v*egetative replication; *pif* for *p*hage
*i*nfection inhibition; IS2, IS3, γ-δ are insertion sequences; *tra* for DNA
*tra*nsfer-related cistrons; *traO* is the site defined by cis-dominant mutants
resistant to repression by the FinOP repression system; *finP* is the F-coded,
F-specific component of the FinOP repression system. The location of the
cistrons shown were assigned according to the arguments in the text. Changes
from previously published maps (Davidson *et al.*, 1975; Helmuth and Achtman,
1975; Skurray *et al.*, 1976b) are on the basis of our unpublished experiments
with plasmid chimeras. Three of these chimeras are also depicted: all three are
pSC101 derivatives carrying *Eco*RI F DNA fragments. pRS27 and pRS29
overlap by approximately 1 kb whereas pRS29 and pRS31 end at the same
*Eco*RI site (Skurray *et al.*, 1976b). That *Eco*RI site, at 82 kb, splits *traG*
(unpublished observations). The exact kilobase assignments of the F DNA
carried by the chimeras may be in error by as much as 2 kb. Additional *tra*
cistrons are known which are not as precisely mapped yet. *traM* maps
between 60 and 68 kb whereas *traT*, a surface exclusion cistron, maps near
traS. ilzA, mapping near *traS*, and *ilzB*, mapping clockwise of the *tra* region,
are not shown. The location of *oriT* is indicated by an arrowhead at 62 kb.
Transcription of the *tra* operon is indicated by the heavy arrow.

Skurray et al., 1976a). Within this region, the origin of vegetative replication, designated *oriV* (Clark et al., 1976) has been located at 42.6 kb (Helinski, personal communication), whereas the *inc* genes lie between 46.5 and 48.5 kb (Santos et al., 1975). These results support the idea that plasmid replication and incompatibility may be functionally independent (Novick and Schweisinger, 1976). F also carries various insertion sequences (indicated as IS or|γ−δ in Fig. 6.1 which are involved in integration of F into the bacterial chromosome to form Hfr strains (Davidson et al., 1975). In the original F factor all these map in the region 93−0−18 kb, but insertion sequences have been found at still other locations in certain F derivatives (Davidson et al., 1975), and it is known that such insertion sequences can insert at numerous sites within DNA. The DNA region from 0 to 60 kb presumably codes for other as yet unknown functions since only few cistrons have been assigned to it. The alternative is that much of this DNA is genetically silent.

The cistrons for the functions just described vary in their location among representatives of the FI and FII groups. The *pif* cistrons are not even present on some F-like plasmids (Morrison and Malamy, 1970; Willetts and Finnegan, 1970). The DNA regions assigned to these functions also show minimal homology between these different sex factors. In contrast, the region between 63 and 93 kb − the transfer or *tra* region − is conserved with minor differences among F-like plasmids (Sharp et al., 1973). Nine *tra* cistrons (*traA, L, E, K, B, C, F, H,* and part of *traG*) are necessary for F pilus synthesis (Achtman et al., 1971, 1972) and these all map between approximately 65 and 82 kb (Sharp et al., 1972). The *traA* cistron codes for F pilin, the structural subunit of the F pilus (Minkley et al., 1976). The role of the 8 other cistrons is unknown, but the possibility exists that they code for F pilin processing enzymes or possibly for an as yet speculative F pilus basal structure. Mutants in these nine cistrons were isolated as being Tra⁻ (incapable of promoting DNA transfer). In contrast, Brinton's group has isolated transfer-proficient *traA* mutants with altered adsorption of F pilus-specific bacteriophages (Brinton, C.C., Jr., personal communication).

Four other cistrons have been identified by the isolation of Tra⁻ mutants (Achtman et al., 1971, 1972; Willetts and Achtman, 1972; Achtman, unpublished observation): Some *traG* mutants synthesize F pili whereas others do not, although all are Tra⁻. All *traD* mutants synthesize F pili, but f2 and related RNA pilus-specific bacteriophages do not infect *traD* cells. *traI* and *traM* mutants have no known effects on F pilus structure. *traG, traD* and *traI* are clustered, and all map between approximately 80 and 93 kb (Sharp et al., 1972). *traM* maps between 60 and 68 kb (our unpublished observations). The mapping of all these *tra* cistrons has been confirmed by the demonstration that chimeric plasmids carrying the appropriate fragments of F DNA express the individual gene functions (Achtman, Skurray, Thompson, Helmuth, Hall, Beutin and Clark, manuscript submitted; Willetts, personal communication). Other than *traA*, the detailed function of these cistrons is not known.

In contrast to the cistrons directly involved in donor ability, cistrons which reduce the recipient ability of F⁺ bacteria have been also assigned to the *tra* region. These are the cistrons *traS* and *traT* which map between *traG* and *traD* (Willetts, 1974a; our unpublished observations) and code for surface exclusion. Surface exclusion is a property of sex factor-carrying cells which converts the host into a poor recipient of DNA. Mutants in these surface exclusion cistrons affect neither F pilus function nor DNA transfer, but instead turn the hosts carrying these sex factors into better recipients of DNA than the wild type F⁺ cell (Achtman and Helmuth, 1975; our unpublished observations). Minkley and Ippen-Ihler (1977) have detected a cell membrane protein of 24 000 daltons associated with expression of this very region of F DNA. This protein has now been demonstrated to be the gene product of the *traT* cistron and to be an outer membrane protein (Achtman, Kennedy and Skurray, manuscript submitted).

Lethal zygosis is a phenomenon whereby a high ratio of Hfr to F⁻ cells kills many of the F⁻ cells. Surface exclusion leads to the protection of F⁺ cells when used as recipients under these conditions. However, a second line of defence exists since even surface-exclusion deficient F⁺ cells are also immune to lethal zygosis. By using deletion analysis, Skurray *et al.* (1976c) were able to map a region on F called *ilzA* encoding this immunity. *ilzA* maps near *traS* and *traT*.

All the *tra* cistrons just described have been assigned to one operon, the *tra* operon, of more than 15 Mdal in size (Achtman and Helmuth, 1975; Helmuth and Achtman, 1975) with its beginning counterclockwise of *traA* and its end clockwise of *traI*. The operon's function is controlled by a positive control gene, *traJ*, mapping just outside an counterclockwise of the *tra* operon (Willetts and Achtman, 1972; Achtman, 1973b; R. Thompson, personal communication). *traJ* mutants are incapable of DNA transfer or surface exclusion and do not synthesise any F pili (Achtman *et al.*, 1972). *traJ* has recently been shown to be needed for the transcription of the *tra* operon (Willetts, 1977).

Many of the F-like sex factors isolated from nature are repressed for the functions of the *tra* operon, whereas the F factor is constitutive (Meynell *et al.*, 1968). Mutants of such repressed F-like sex factors have been used to define two cistrons, *finO* and *finP*, both necessary for self repression (Willetts, 1972b; Gasson and Willetts, 1975). The *finO* product seems to be non-specific and can interact with *finP* products of different specificities. Thus a *finO*⁺ R factor such as R100 will also repress F DNA transfer from a cell carrying both R100 and F, due to an interaction between the R100 *finO* product and the F *finP* product. The repressor formed by this interaction can act on F (Finnegan and Willetts, 1971). The F factor does not make an active *finO* product and is therefore naturally derepressed. The R100 *finP* product does not act on F, as shown by the fact that *finP* mutants of F are not repressed by R100 (Finnegan and Willetts, 1971). The complete *finO-finP* repressor seems to act at a site, *traO*, to prevent synthesis of the *traJ* control protein (Finnegan and Willetts, 1973; Achtman, 1973b), and thus results in lack of transcription from the *tra* operon (Davis and Vapnek, 1976, Willetts, 1977). The order of these cistrons and sites is *finP, traO* and *traJ* (Willetts *et al.*, 1976) as indicated in Fig. 6.1.

The last genetic site to be discussed is *oriT*, the origin of DNA transfer, indicated by an arrowhead in Fig. 6.1. It has long been known that DNA transfer begins at a specific region on Hfr chromosomes located within or at the end of the chromosomally integrated F factor. A *cis*-dominant F site which is essential for DNA transfer has been mapped by deletion analysis as lying counterclockwise of *finP* (Willetts, 1972a; Willetts *et al.*, 1976); and even more accurately, by DNA heteroduplex analysis, at 62 kb or slightly clockwise thereof (Guyer and Clark, 1976; Guyer, Davidson and Clark, in preparation). DNA is transferred starting from this site, *oriT*, in the counterclockwise direction (Willetts, 1972a). Confirmation of this mapping is supplied by studies with chimeras carrying this region of F DNA and which are mobilized efficiently by F-like sex factors (Clark *et al.*, 1976). The location and orientation of *oriT* results in the *frp* and *inc* cistrons being transferred first in conjugation, followed by the *pif* cistrons, and finally by the *tra* region. *oriT* is plasmid-specific since very few other sex factors can recognize F's *oriT* and bring about mobilization of the DNA carrying it (Reeves and Willetts, 1974; Clark *et al.*, 1976).

6.3 THE MATING PHENOMENON

Genetic experiments in the 1950's and 1960's demonstrated that F factor-carrying cells quickly formed specific contacts with F^- cells, and that the contact stage could be distinguished from a subsequent DNA transfer stage (deHaan and Gross, 1962; Curtiss, 1969). A requirement for F pili on the donor cells, and for a suitable recipient cell surface, were also demonstrated. Very recently, it has proven possible to define stages in conjugation by physical techniques which measure aggregation directly rather than by means of indirect genetic measurements. These experiments demonstrated that conjugation involves cells aggregated in 'mating aggregates' of variable size and composition. It has also been possible to purify and partially characterize F pili and the outer membrane protein pOmpA, both of which function as mating receptors on cell surfaces. These receptors and DNA transfer will be discussed first, followed by the formation and properties of mating aggregates.

6.3.1 Requirements for mating

(a) *F pili*
The properties of sex pili have been the topics of several reviews (Brinton, 1971; Paranchych, 1975; Tomoeda *et al.*, 1975; Achtman, 1973a) which served primarily to expose our ignorance about their function. However, recent advances have lent hope that the full role of F pili in conjugation may become clear in the near future. F pili are protein organelles of about 90 Å diameter and up to several microns in length. Mid-exponential phase F^+ cells typically possess 1 to 3 F pili per cell. F pili

serve as the adsorption sites for male-specific bacteriophages, with RNA phages adsorbing at the sides and DNA phages at the tip. Any further role F pili may have in bacteriophage infection, such as nucleic acid transport to the cell surface, is still not proven. However, there is evidence that during infection with RNA male-specific phages the A protein and the RNA leave the phage capsid and are transported to the cell; the process occurs on the extracellular pili and the A protein seems to promote attachment to the F pili and then pull the RNA into the host cell (Krahn et al., 1972; Paranchych, 1975; Wong and Paranchych, 1976; Shiba and Miyake, 1975; Leipold and Hofschneider, 1975). Distant or unrelated sex factors code for sex pili with different properties. The production of sex pili has been demonstrated for overgroup F and I sex factors and for sex factors of some other compatibility groups (see Table 6.1). Figures 6.2–6.5 show photographs of some of these pili. F-like and I-like pili are not known to show any immunological cross-reaction to each other or to any of the other sex pilus types. Even F and F-like pili are sufficiently different that 4 serological groups can be defined among them (Lawn and Meynell, 1970). They are also similar enough that cells carrying two sex factors coding for serologically distinguishable F-like pili synthesize mixed pili containing subunits of both types (Lawn et al., 1971). The ultrastructure of these pili is not yet clear. Originally, Brinton (1965) suggested that F pili were hollow tubes and that nucleic acids could be transported through an empty 25 Å core. Later, he (Brinton, 1971) presented evidence that F pili consist of two parallel rods and suggested alternative speculative models for how they could transport nucleic acids.

Evidence for retraction of F pili upon infection with DNA male-specific bacteriophages has been presented (Jacobson, 1972). It has also been reported that F pili retract when the cells are incubated at a temperature of 25°C (Novotny and Lavin, 1971) or 50°C (Fives-Taylor and Novotny, 1976 and personal communication), or in the presence of CN^- (Novotny and Fives-Taylor, 1974). All these experiments used only electron microscopy, and are subject to the limitations of that technique. Either degradation of pili into submicroscopic structures, or a change in pilus properties such that discarded pili no longer adsorbed to electron microscope grids, would have yielded a false conclusion. Thus it is only certain that the extended sex pili seen on normal male cells change somehow upon phage infection or after certain treatments.

Somewhat more success has been obtained with the chemical purification of isolated sex pili, although much of this work is still not published in any detail. Sex pili coded for by the F factor (Brinton, 1971; Tomoeda, personal communication; Helmuth and Achtman, unpublished observations) or by the F-like sex factor R1*drd*19 (Beard et al., 1972) have been successfully purified in yields of several milligrams. The primary subunit, pilin, has a molecular weight of about 11 000 daltons (Brinton, 1971; Beard et al., 1972; Helmuth and Achtman, unpublished observations) and contains two phosphate and one glucose residue per molecule. The question of whether there are additional minor molecular species within the purified pili is not clear and the possibility of an attachment site or a basal structure at the cell surface is

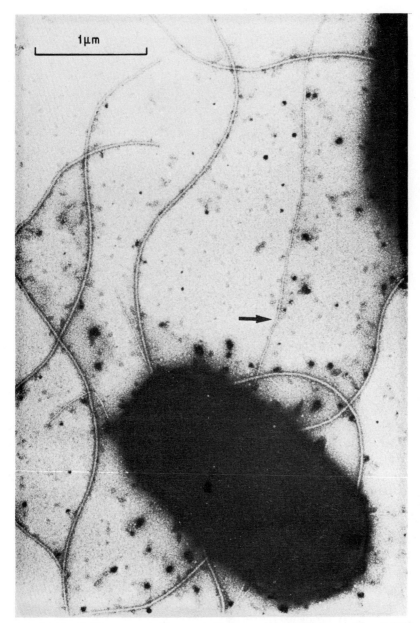

Fig. 6.2 F pili. The arrow indicates an F pilus projecting from an *E. coli* K12 (F*lac*)[+] cell. The thicker filaments are flagellae, whereas the thinner filaments are common type I pili. Photograph supplied by R. Helmuth.

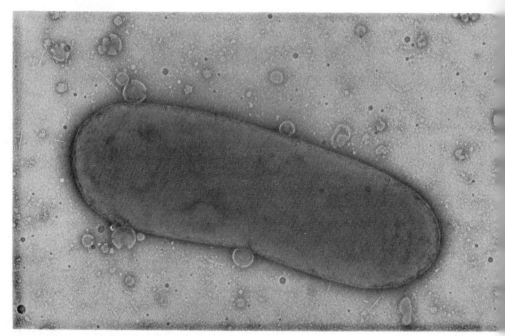

Fig. 6.3 N Pili. *Salmonella typhimurium* LT2 carrying sex factor N3. These cells synthesized no pili when plasmid-free. Photograph supplied by C.M. To.

unresolved. Beard and Connolly (1975) have shown that R1*drd*19 cells have a pool of pilin subunits within the outer membrane equal to approximately 50% of the mass of the assembled pili. The assembly mechanism is unknown.

At least two of these purified pilus preparations have been shown to be active in adsorption of male-specific RNA bacteriophages (Beard *et al.,* 1972; Helmuth and Achtman, unpublished observations) and one preparation has been shown to bind to *E. coli* cells (Helmuth and Achtman, unpublished observations; see Fig. 6.6). Furthermore, the binding to *E. coli* cells is strong enough that the cells can be washed several times and still retain at least some of their bound pili. Finally, purified pili can act at biological concentrations to agglutinate F^- cells when purified anti-F pilus antiserum is added (Helmuth and Achtman, unpublished observations). These purified pilus preparations still represent a difficult system to analyze. All the preparations contain crystallized rather than single pili, and there is no convincing evidence to indicate that these pili express all the original functions of the cell-bound pili. No suitable quantitative assays for pilus number or function are available, and electron microscopy has remained the primary tool for their analysis.

The evidence is convincing that sex pili are at least necessary at some stage in conjugation before DNA transfer. Brinton (1965) reported that he could see motile

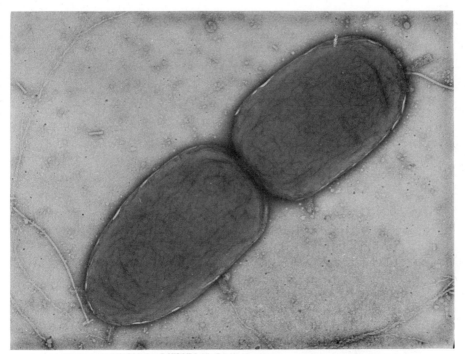

Fig. 6.4 RP4 pili. *Ps. aeruginosa* **PAT904 carrying sex factor RP1. PRR1** bacteriophages have been added to identify the sex pili. Photograph supplied by C.M. To.

donor cells towing non-motile recipient cells within mating cultures; the connection between the cells was unresolved under the light microscope, but presumed to be the F pilus. In other experiments he was able to show that blending of F piliated-cells removed F pili and mating ability with the same kinetics and that mating ability and F piliation returned to the blended cells with identical kinetics (Brinton, 1965). DNA transfer between donor and recipient cells can be prevented by treatment with male-specific bacteriophages (Knolle, 1967; Novotny *et al.*, 1968; Ippen and Valentine, 1967). No mutant is known which does not possess F pili and can still donate DNA whereas 100 different Tra⁻ mutants are known which are resistant to male-specific phages and are thus presumed not to produce any F pili or at least only drastically altered ones (Achtman *et al.*, 1971); the lack of F pilus production has been confirmed by electron microscopy in selected cases. Finally, a number of chemical treatments are known which both prevent mating and reduce the number of F pili on the treated cells. These treatments include incubation at 25°C (Novotny and Lavin, 1971; Walmsley, 1976), in 5 mM CN⁻ (Novotny and Fives-Taylor, 1974; Achtman, 1975), in 0.01% SDS (sodium dodecyl sulfate) (Tomoeda *et al.*, 1975; Achtman, 1975) or

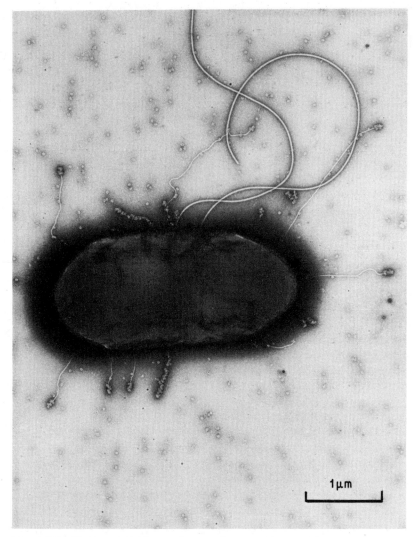

Fig. 6.5 E pili. *E. coli* K-12 carrying sex factor R402. ER1 bacteriophages have been added to help identify the sex pili. Photograph supplied by C.M. To.

in 1 mM Zn^{2+} (Ou and Anderson, 1972; Achtman, 1975). Treatment at 25°C or with CN^- causes 'retraction', i.e., no F pili can be seen with the electron microscope either on the cell surface or free in the cell supernatant (Novotny and Lavin, 1971; Novotny and Fives-Taylor, 1974); Zn^{2+} causes shedding of full-length F pili into the culture medium (Novotny, personal communication); and sodium dodecyl sulfate seems to dissolve F pili (Tomoeda *et al.*, 1975). The observation (Fig. 6.6) that F pili can adsorb to F⁻ cells also argues that they can recognize suitable cell surfaces.

Although the evidence for the involvement of F pili at an early stage in conjugation is convincing, their function in DNA transfer is controversial. The only evidence for a role in DNA transport is the observation by Ou and Anderson (1970) that DNA transfer between mating bacteria occurred in some cases between cells that did not appear to come into wall to wall contact. Since no extracellular appendage obligatory for DNA transfer other than the sex pilus is known, this observation has been interpreted as meaning that F pili are at least occasionally involved in DNA transport. Theoretically it is equally possible that another structure may be formed during conjugation which can stretch up to a length of a few microns, forming a real conjugation bridge†. An alternative role for F pili in conjugation has been proposed independently by Curtiss (1969) and by Marvin and Hohn (1969), according to which F pili serve firstly as recognition organelles for suitable recipient cells and then serve to bring the cells into wall to wall contact by their retraction into the donor cell. A retraction model of some sort is suggested by the observation that most mating cells are found in relatively close contact by the time DNA transfer occurs (Ou and Anderson, 1970; Ou, personal communication; Achtman, Schwuchow, Morelli and Helmuth, unpublished observations). However, these results could equally well be explained by pili binding along their sides to two cells at once.

(b) *Recipient's cell surface*

E. coli cell envelopes have been analyzed by a combination of biochemical, physical and genetic techniques. Among a wealth of other cell envelope proteins, a major outer membrane protein variously called 3a (Schnaitman, 1974), G (Chai and Foulds, 1974), II* (Henning *et al.*, 1973) or d (Lugtenberg *et al.*, 1975) has been identified. Mutants in a cistron now called *ompA* (Datta *et al.*, 1976) and previously termed *con* (Skurray *et al.*, 1974), *tolG* (Chai and Foulds, 1974) or *tut* (Henning *et al.*, 1976) can be isolated which do not contain that protein in their outer membrane. We will use the convention of Datta *et al.* (1976) and refer to the cistron as *ompA*. In the absence of a published neutral designation for the protein, we will refer to it as pOmpA. *ompA* mutants lacking pOmpA can be isolated on the basis of resistance to phages K3 (Skurray *et al.*, 1974; Manning *et al.*, 1976) or TuII* (Henning and Haller, 1975; Henning *et al.*, 1976), or on the basis of tolerance to colicins L or K (Foulds and Barrett, 1973; Davies and Reeves, 1975). Regardless of how they were isolated, most of the *ompA* mutants tested are resistant to K3 and TuII*, tolerant to colicins L and K, and poor recipients in conjugation with F factor-carrying donors (Manning and Reeves, 1976b, Manning *et al.*, 1976). They are also sensitive to

† The idea of a conjugation bridge was first suggested on the basis of an electron microscopic analysis of mating cells (Anderson *et al.*, 1957). The original observations are generally accepted as being artefactual, but the possibility of an extendable conjugation bridge has not been ruled out.

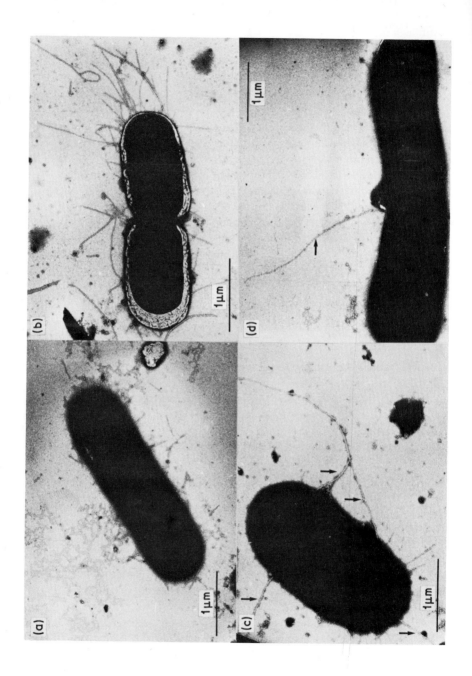

Fig. 6.6 Cell-bound purified F pili. A presents a cell from an *E. coli* F$^-$ culture before adding F pili. Type I pili can be seen. B and C show cells from the same culture after addition of purified F pili followed, respectively, by 1 and 2 cycles of washing by centrifugation and resuspension. D shows a cell from a mixture to which F pili were added at a concentration of 2 F pili per cell. Most of the extraneous filaments in B and all those indicated by arrows in C and D are F pili which have bound to the cells. Photograph supplied by R. Helmuth.

ethylene diamine tetraacetic acid, phenethyl alcohol, eosin yellow and novobiocin (Foulds and Barrett, 1973; Manning and Reeves, 1976b) and in some cases to sodium citrate (Achtman, unpublished observations); pOmpA is therefore functionally pleiotropic. *ompA* maps at 21.5 min on the *E. coli* map (Foulds, 1974; Henning *et al.*, 1976; Manning *et al.*, 1976) and has also been demonstrated to be the structural gene for pOmpA (Henning *et al.*, 1976).

Protein pOmpA has been purified from *E. coli* membrane preparations in several laboratories (Schnaitman, 1974; Reithmeier and Bragg, 1974; Hindennach and Henning, 1975). pOmpA is largely insoluble in water but can be solubilized at low concentrations with detergents such as SDS. When solubilized at temperatures under 50°C, the protein migrates in SDS polyacrylamide gel electrophoresis with an apparent molecular weight of 28 000; after solubilization at higher temperatures, the protein migrates as if it were 33 000 daltons in size (Henning, personal communication). Purified preparations of the 28 000 form in 100 mg amounts have been obtained (Hindennach and Henning, 1975) and the amino acid composition as well as the cyanogen bromide fragmentation pattern determined (Garten *et al.*, 1975). The molecular weight of the cyanogen bromide fragments indicates that the native protein has a molecular weight of approximately 28 000 daltons and that the 33 000 dalton form migrates aberrantly on SDS gel electrophoresis (Garten *et al.*, 1975). Only the 28 000 form has been shown to express any biological function (see below). It is still unclear whether the *E. coli* K-12 pOmpA is glycosylated or not (Garten *et al.*, 1975) but it certainly contains very few sugar residues if any (Schnaitman, 1974). In *E. coli* and related bacterial species, pOmpA is present in approximately 10^5 copies per cell (Garten *et al.*, 1975).

The primary interest of pOmpA for this review is that *ompA* mutants (prototype P460) are usually poor recipients in conjugation with F$^+$ donors (Skurray *et al.*, 1974; Manning *et al.*, 1976). Such mutants are referred to as Con$^-$ mutants (for *con*jugation-defective). The most extensively analyzed Con$^-$ mutants in this respect are those isolated by Reeves's group on the basis of resistance to phage K3 (Skurray *et al.*, 1974; Manning *et al.*, 1976). Most of their *ompA* mutants manifest between 50 and 5000-fold lower recipient ability than the wild type parent. (Manning *et al.*, 1976) and possess much less pOmpA in the outer membrane (Manning *et al.*, 1976; Manning and Reeves, 1976a). It has been suggested that protein pOmpA is the receptor for mating aggregation (Skurray *et al.*, 1974). This question is discussed in further detail in Section 6.3.2f.

Eight mutations mapping at *ompA*, but which allow the synthesis of pOmpA, have been described by Henning *et al.*, (1976). They were able to show a change in the electrophoretic mobility of protein pOmpA in two of these mutants, one of which demonstrated an altered fragment pattern after cyanogen bromide treatment of the isolated protein. Thus it seems clear that different mutations in *ompA* can result in loss of the protein whereas others are best interpreted as affecting its function but not its incorporation into the outer membrane. A Con$^-$ mutant isolated directly on the basis of poor recipient ability, with the mutation mapping at *ompA*,

Table 6.2 Conjugation-defective mutants

Cistron	Map position	Resistance to	Prototype strain	Structural defects	References	Aggregation with F+	Aggregation with I+
ompA	21.5	K3, TuII*, Col L and K	P460	Lacks pOmpA	1, 2, 3, 4, 5	–	+
tfrA(?)	5.5–6.5	T3, T4, T7	PC2040	Heptose-deficient LPS	6	–	+
?	79–82	T3, T4, T7, P1, ST-1	FR19B	Heptose-deficient LPS	6, 7	–	+
?	16.7	χ, T3, T4, T7, T2	AM4023	Galactose- and phosphate-deficient LPS	8	+	–

The genetic assignments listed here are based primarily on chromosomal location. Genetic locations have been listed in terms of the current modified *E. coli* map (Bachman *et al.*, 1976). No complementation analyses have yet been reported and for most of these mutants no cistron designations have been suggested. Over one hundred mutants of *ompA* have been described. PC2040 shares some properties with *tfrA* mutants but the assignment is tentative. With the exception of FR19B, where another mutant (PC2041) has been described [7], the other classes are based on the properties of single mutants. The conclusions on whether mating aggregates are formed or not are derived from references 1, 5, 6, and 8 and from unpublished observations by M.A.
References: [1] Skurray *et al.*, 1974; [2] Chai and Foulds, 1974; [3] Henning *et al.*, 1976; [4] Manning and Reeves, 1976b; [5] Havekes and Hoekstra, 1976; [6] Havekes *et al.*, 1976; [7] Reiner, 1974; [8] Havekes *et al.*, 1977.

has recently been described (Havekes and Hoekstra, 1976). No changes have been detected in this mutant's cell membrane (Havekes and Hoekstra, 1976; Havekes et al., 1976) and this may again reflect a functional mutation at *ompA*.

A variety of Con⁻ mutants with other biochemical lesions have been isolated and a summary of the properties of the various Con⁻ mutants is presented in Table 6.2. 5% of mutants isolated as resistant to bacteriophage ST-1 were Con⁻ (Reiner, 1974). One subclass, type B mutants (prototype FR19B), is temperature-sensitive for growth, shows increased susceptibility to antibiotics and deoxycholate (Reiner, 1974) and has been shown to produce a defective lipopolysaccharide which is heptose-deficient (Lugtenberg, cited in Havekes et al., 1976). Two other mutants, similar in phenotype, but isolated as resistant to bacteriophages T3, T4 and T7, were also shown to produce a heptose-deficient lipopolysaccharide (Havekes et al., 1976a). One of these mutants, PC2040, defines a different cistron from that defined by FR19B; the other one maps near FR19B. Cell envelope preparations of these mutants display a markedly altered protein composition including a partial deficiency of pOmpA. The effect of these mutations on recipient ability may be indirect and they may affect the components of a receptor complex, possibly including pOmpA.

The Con⁻ mutant types described above are deficient in forming stable mating aggregates in liquid but can form mating aggregates stable enough to allow DNA transfer on solid surfaces (Havekes and Hoekstra, 1976; Havekes et al., 1976; Achtman, unpublished observations).

Still other Con⁻ mutants isolated on the basis of resistance to ampicillin or bacteriophages have been shown to contain deficient lipopolysaccharides (Monner et al., 1971; Hancock and Reeves, 1975, 1976; Manning and Reeves, 1975). Con⁻ mutants which are not yet biochemically defined have been isolated as resistant to fosfomycin, requiring alanine, or being unable to ferment several carbohydrates (Falkinham and Curtiss, 1976). As yet these various mutants have not been tested for their ability to form stable mating aggregates or to act as recipients on solid surfaces.

The availability of the Con⁻ mutants has shown that different mating systems involve different cell—cell interactions at the molecular level. All the Con⁻ mutants just mentioned were isolated as poor recipients for DNA transfer from an F factor carrying donor culture. The *ompA* mutants lacking pOmpA, isolated by Reeve's group, were also poor recipients in matings with the FI sex factors ColV2, ColVBtrp, R386, or with the FII sex factors R538-1 or R1 (Skurray et al., 1974; Manning and Reeves, 1976b). However, they were perfectly good recipients in matings with donor cultures carrying the FII sex factors R100 or R136, or any of the 4 I group sex factors tested (Skurray et al., 1974; Manning and Reeves, 1976b). Similarly, the *ompA* mutant isolated by Havekes and Hoekstra (1976) which possesses pOmpA was also a poor recipient with F^+ donors, but a good recipient with donors carrying the FII sex factor R100 or the I sex factor R144. Thus, when analyzed at the molecular level, conjugation coded by these sex factors can be classified in at least two categories: one represented by FI and certain FII sex factors and the other defined by R100. The sex pili synthesized by F and R100 are similar enough to co-polymerize on cells carrying both plasmids (Lawn et al., 1971). The different transferability of the two sex factors to Con⁻ mutants is an indication that these Con⁻ mutants do not affect the primary receptor for F pilus binding. pOmpA is shown below to play a direct role in the cell–cell interactions of F-coded matings, and we interpret the observations described above as indicating that R100-coded matings use a different cell protein to achieve the analogous result.

A third specificity is indicated by the isolation of Con⁻ mutants (Havekes et al., 1977) which do not allow transfer from the I group sex factor R144. One of these mutants, AM4023, is unable to form mating aggregates with R144 (Achtman, unpublished observation). These mutants are good recipients for both F and R100 and were shown to be affected in their lipopolysaccharide. Since lipopolysaccharide extracts of F⁻ cells are efficient at inhibiting mating with an I sex factor but not with an F sex factor, it was proposed that lipopolysaccharides are involved in the aggregation stages in I-coded matings, whereas only proteins are directly involved in F mating aggregate formation (Havekes et al., 1977).

(c) DNA transfer requirements

Much of the conjugation literature has been addressed to the question of whether DNA transfer is dependent on concomitant DNA synthesis (see reviews by Brinton, 1971 and Curtiss, 1969). Autoradiographic experiments showed that only one of the DNA strands synthesized in F⁺ or Hfr cells before the beginning of conjugation, was transferred to recipient cells (Gross and Caro, 1966; Herman and Forro, 1964). The observation that the transferred DNA was largely single-stranded in minicell recipients (Cohen et al., 1968a and b) led to the realization that single-stranded rather than double-stranded DNA was transferred in conjugation. The transferred DNA enters 5' end first (Ohki and Tomizawa, 1968; Rupp and Ihler, 1968) whereas its complementary strand remains behind in the donor cell (Vapnek and Rupp, 1970). When autonomous sex factor-carrying cells are used as donors to F⁻ cells, both single strands are immediately synthesized into double-stranded DNA and eventually into covalently closed circles (Vapnek and Rupp, 1970, 1971). Thus a coupled process of DNA transfer and DNA synthesis normally occurs. However, this coupling between transfer and synthesis can be disrupted by thymine starvation of Hfr donor cells (Sarathy and Siddiqi, 1973), or by using *dnaE* recipients at a non-permissive temperature (Wilkins and Hollom, 1974); the former treatment preventing DNA synthesis of the complementary strand in the donor and the latter treatment in the recipient cells. In neither case is DNA transfer impaired. Thus it now seems clear that DNA replication is not needed for DNA transfer.

Given an F-piliated donor cell and a suitable recipient, what actually triggers the DNA transfer event? It is clear that transfer replication *is* triggered during mating since it does not occur in the absence of recipient cells (Bresler et al., 1968; Marinus and Adelberg, 1970; Sarathy and Siddiqi, 1973; Wilkins and Hollom, 1974). Ou (1975) has confirmed earlier observations (Fenwick and Curtiss, 1973) that DNA transfer replication can also be stimulated by addition of minicells. He found, however, that the replication triggered by addition of minicells need not be accompanied by DNA transfer. Even minicells derived from F⁺ parent cells, which exert surface exclusion and hence prevent DNA transfer, nevertheless stimulated DNA synthesis in the donor cells. Thus he concluded (Ou, 1975) that a mating signal occurs which acts as a trigger for the initiation of DNA transfer and replication, regardless whether such transfer is successful or not. We can conclude that, although conjugational DNA replication and DNA transfer are normally coupled, neither is dependent on the other.

The enzymes responsible for the conjugation-stimulated DNA synthesis or for the DNA transfer are not known, and none of the numerous mutants affecting chromosomal DNA replication which have been tested, prevents DNA transfer (Hirota, personal communication). It is possible that the sex factor codes for the enzymes responsible, and the products of *traM*, *traD* and *traI* must be considered for this role. Recently, Wilkins (1975) has shown that I group sex factors code for an enzyme activity capable of replacing the chromosomal *dnaG* product, and that this activity is only expressed by de-repressed mutants as if it were located in the *tra* region of these sex factors. Such an enzyme may well play a role in DNA synthesis during conjugation. Since the genetics of the I group sex factors has not been developed and furthermore since an analogous activity was not found with F-like sex factors, it is not currently possible to further analyze the genetics of that *dnaG*-like activity.

As described above, the site called *oriT* is necessary for transfer of a DNA molecule. It has been speculated that *oriT* defines a nucleotide sequence which is recognized and nicked or cut by a sex factor-coded endonuclease (Willetts, 1972b); *traI* may, in fact, code for this hypothetical endonuclease since F's *traI* product is not replaceable by that of R100 (Willetts, 1971), suggesting (Alfaro and Willetts, 1972) that the *traI* product recognizes a specific DNA site which is different in the two sex factors. This idea is supported by the observation that cells carrying a *traI* mutant of F and the non-sex factor ColE1 can efficiently transfer the ColE1 plasmid. Thus nothing may be wrong with the transfer machinery of *traI* cells except the absence of a DNA-specific enzyme (Reeves and Willetts, 1974). This interpretation assumes however, that ColE1 synthesizes its own specific endonuclease or that it can use a cellular enzyme. Since we have speculated that F and R100 use different cell surface proteins for conjugation, an alternative is that the *traI* product is involved in the specific recognition of recipient cell surface structures. This alternative hypothesis leaves the problem of ColE1 mobilization from *traI*⁻ cells unresolved.

The mode of DNA replication during transfer is not clear, although it has been generally assumed that a rolling circle mechanism without a defined terminus is involved (Gilbert and Dressler, 1968; Ohki and Tomizawa, 1968). The primary evidence for a rolling circle mechanism is two-fold. Genetic evidence has been presented that late in mating there is increased recombinational linkage within the recipient of genes straddling *oriT* (Fulton, 1965). It has also been observed that transferred sex factor DNA isolated from recipient cells sediments faster than unit length DNA molecules (Matsubara, 1968; Ohki and Tomizawa, 1968). The former result can easily be explained by multiple DNA transfer events from more than one donor to a single recipient (see below). The latter result may be criticised on the basis that experimental techniques to distinguish monomer length supercoiled or open circular molecules from linear multimers were not employed (see Achtman, 1973a). Thus, the evidence for a rolling circle mechanism of DNA transfer in F-coded conjugation is not convincing.

In a different conjugation system, Fenwick and Curtiss (1973) have presented data showing that new RNA and protein synthesis are needed for transfer of more than one unit length of DNA. Therefore, Curtiss and Fenwick (1975) have proposed two alternative models for R64 (I group) or R1 (FII group) transfer. Both these models assume a defined terminus for DNA transfer but are different in their details. One model invokes transfer from a linear duplex molecule, derived from the covalently closed circle by an endonuclease yielding sticky ends; the other invokes transfer of DNA from a circular molecule, whose transfer terminus is defined by an RNA transcript synthesized near *oriT* at the same time as DNA transfer begins. Both models also envisage nuclease action after DNA transfer and before the single strand can be converted to a covalently closed circle. Evidence for active disaggregation of mating aggregates (see below) also suggests that there is a natural limit to how much DNA is transferred in conjugation.

(d) *DNA transport across cell membranes*

In the absence of any direct data on DNA transport across cell membranes during conjugation, we shall briefly speculate on possible mechanisms. Regardless of the transfer model invoked, the DNA must at least pass across the inner cytoplasmic membrane and this membrane, normally a barrier to the entry of most molecules, must somehow change to accomodate the transferred DNA.

Information on DNA transport across the cell membrane in other systems may be relevant to conjugation. For pneumococcal transformation (see Lacks, Chapter 5), DNA entry is dependent on the conversion of double-stranded to single-stranded DNA (Lacks and Greenberg, 1976) through intermediates with single- then double-stranded breaks. Mutants of *Diplococcus pneumoniae* that lack a membrane-localized endonuclease bind DNA but do not allow its entry (Lacks and Neuberger, 1975). Similar but less well defined mechanisms may operate with *Bacillus subtilis* (Davidoff-Abelson and Dubnau, 1973) and *Haemophilus influenzae* (Sedgwick and Setlow, 1976). Conjugation may involve single-stranded DNA transfer for similar reasons. But how is the transferred DNA protected from nucleolytic attack?

Lacks and Greenberg (1976) suggested that a nuclease in *D. pneumoniae* not only nicks the DNA, but also acts as a 'leader' molecule by binding to breaks in the 5'-ended strands. These bound strands are postulated to enter the cells via a multimeric protein channel that spans the membrane while another nuclease progressively degrades the opposite strand. The 'pilot protein' hypothesized by Kornberg (1974) to account for the multi-functional protein involved in the adsorption, penetration and early multiplication and expression stages of viral infection may be similar. Pilot proteins (gene 3 protein of M13, or gene H protein of ϕx174 (Jazwinski *et al.*, 1973, 1975)) are postulated to initially bind the phage to the bacterial cell, but then to be released into the cell envelope to form a protein pore lined with DNA binding sites through which DNA penetration occurs. Protein A of the RNA phage R17 may well be the counterpart pilot protein for the RNA penetration which follows binding of the phage to the F pilus. The idea that a pilot protein may be involved in conjugational DNA transfer has been suggested by Kornberg (1974).

Transmembrane channels or pores have been proposed by Singer (1974) to account for the transport of small molecules, and Inouye (1974) has proposed a model in which the *E. coli* lipoprotein forms a hydrophilic channel across the outer membrane (see also Braun and Hantke, Chapter 3). It should be noted in this regard that the cell envelope of competent *Haemophilus* cells differs from that of non-competent cells both in protein composition and in its organization (Zoon *et al.*, 1976).

Possible candidates for DNA transport channels include the membrane adhesion sites described by Bayer (1968). Approximately 50 adhesions between outer and inner membranes can be detected per plasmolyzed cell and they have been implicated in bacteriophage adsorption and penetration as well as being the sites at which F pili are extruded (Bayer, 1968 and 1976). As proposed previously (Skurray, 1974), one of the stages in conjugation might well involve the apposition of such pores on the donor and recipient cell surfaces. The exchange of outer membrane material during conjugation observed by Goldschmidt and Curtiss (personal communication) might well occur during the envelope modification involved in forming or apposing membrane pores.

On the basis of the above discussion, there might be some features of DNA penetration common to conjugation and transformation. Skurray, Nagaishi and Clark (in preparation) have used surface exclusion-proficient chimeric plasmids carrying the F factor's surface exclusion cistrons (Skurray *et al.*, 1976b) to show that these plasmids not only render the host surface exclusion-proficient, but also reduce its competence in transformation. T7 bacteriophage infection of these cells is also reduced, although the plasmids do not carry the *pif* cistrons. These results suggest that the surface exclusion phenomenon imposes a barrier on DNA penetration across the cell envelope. This brief exposé serves to highlight our ignorance on this topic and is presented in the hope of stimulating some badly needed research.

6.3.2 Cell–cell interactions during mating

(a) *Mating aggregates instead of mating pairs*
The information given above summarizes the topics investigated in most analyses of conjugation. Remarkable by its absence until recently is an analysis of conjugation as a cell–cell interaction. What actually interacts with what and how?

Early analyses of conjugation reported that shortly after mixing donor and recipient cultures, some mating cells were to be found in pairs, whereas a short time

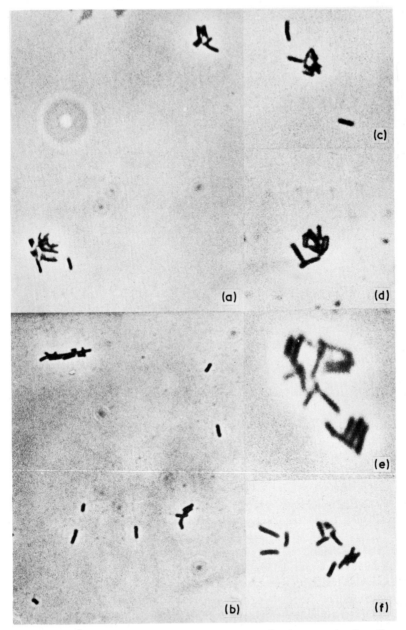

Fig. 6.7 Larger mating aggregates. Light microscopic photograph of mating aggregates from an *E. coli* K-12 F*lac*$^+$ × F$^-$ mating. Reprinted from Achtman (1975) with permission from ASM Press.

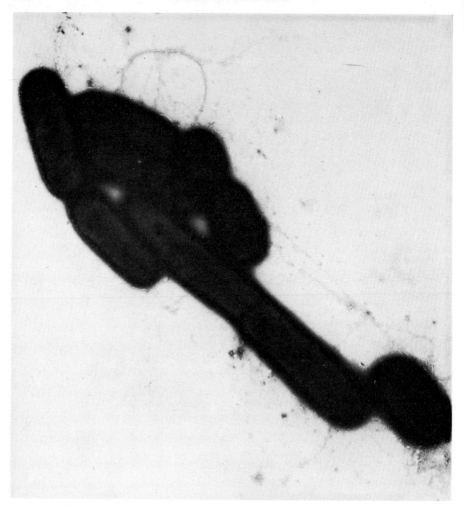

Fig. 6.8 Mating aggregates. A typical mating aggregate from an *E. coli* K-12 Hfr x *E. coli* C F$^-$ mating. Photograph supplied by G. Morelli.

later most of the mating cells were in large aggregates (Lederberg, 1956; Anderson and Maze, 1957; Anderson, 1958). These analyses were purely descriptive and not quantitative; they were, however, unbiased by preconceptions of what mating should be. Later descriptions focussed on 'mating pairs' (Brinton, 1965; deHaan and Gross, 1962) and this terminology has become accepted. In fact, many workers who have examined mating mixtures directly by microscopy have noticed the existence of clumps and large aggregates without ever commenting on this fact in print. This long-neglected observation has recently been re-analyzed quantitatively with the help of microscopy, sucrose gradient centrifugation and a modified Coulter Counter

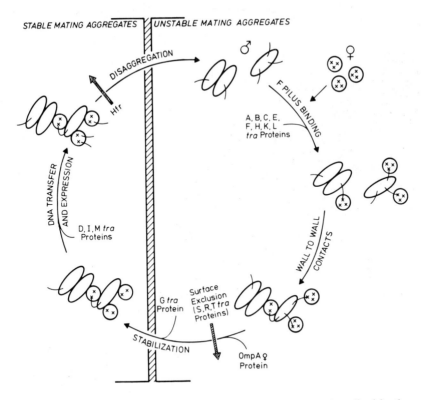

Fig. 6.9 A model for the mating cycle. The various stages are described in the text. The male cells are depicted as rod-shaped with slightly curved F pilus projections. The female cells are depicted as round: X represents F pilus receptors. The hatched arrows indicate inhibition of the mating cycle at the stabilization stage by surface exclusion, and at the disaggregation stage when Hfr donors are used.

(Achtman, 1975). Matings between donor cells carrying F-like or I group sex factors, and suitable recipients, may contain more pairs than larger aggregates but usually contain more *cells* in larger aggregates than in pairs (Achtman, 1975). The visual impact is primarily one of pairs, but a quantitative analysis showed that there were both small (2–5 cells) and large (6–20 cells) aggregates in every mating mixture, and that under optimal mating conditions most of the mating cells were in the larger aggregates.

Figures 6.7 and 6.8 shows light and electron micrographs, respectively, of selected larger aggregates of mating cells. The term mating aggregates was devised to represent the entire spectrum of mating cells in the population, including pairs (Achtman, 1975). Of the methods tested, electronic particle analysis proved to be the most useful for quantitating the number of cells in stable mating aggregates, and electron microscopy the most useful for quantitating unstable interactions. Most the following is based upon experiments using these techniques.

An analysis of mating aggregation kinetics is rendered complicated because the population of cells is not synchronous in its aggregation, and even more asynchronous in transfer and subsequent disaggregation. The population may simultaneously contain some cells that are just aggregating, others that are in stable and transferring aggregates and still others that are disaggregating. Furthermore, both the donor and recipient populations show some spontaneous aggregation prior to mixing. Despite these problems, it has proven possible to analyze the course of bacterial conjugation between F-like sex factor donors and F^- recipients. Fig. 6.9 presents a preliminary model for bacterial conjugation, the individual features of which are discussed in detail below. We shall first discuss the relevant characteristics of stable mating aggregates, and then the stages of the mating cycle.

(b) *Stable mating aggregates*

A common belief has been that the interaction between conjugating cells is sensitive to mild shear forces and that only very few conjugating cells stay together long enough to transfer large amounts of DNA. However, a number of observations demonstrated that mating aggregates are really quite sturdy (Ou and Anderson, 1970; Achtman, 1975; Achtman, 1977). The prototype population chosen to define stable mating aggregates consists of HfrC or HfrH x F^- matings approximately 15 to 20 min after mixing at $37°C$ (Achtman, 1977). These populations do not disaggregate (see Section 6.3.2) and most of the cells will have been aggregated for at least 5 or 10 min. Moderate shear forces such as are caused by pipetting, centrifugation, or passage through the orifice of a Coulter counter, have only minimal disruptive effects on the mating aggregates. These are totally resistant (Achtman, 1977) to concentrations of SDS which dissolve F pili (Tomoeda *et al.*, 1975) and which also prevent mating aggregate formation (Achtman, 1975). Electron microscopic observations (Achtman and Morelli, unpublished observations) demonstrated that such treatment also dissolved all F pili within the mating aggregates. Such SDS-treated aggregates can transfer DNA at least 50% as well as untreated mating aggregates demonstrating that extended F pili are not the primary connection or conjugation bridge. An F^+ or $F' \times F^-$ mating also contains such stable mating aggregates 15 to 20 min after mixing the parental cultures, but a few of these cells are already beginning to disaggregate.

Stable mating aggregates do not have a defined size or proportion of donor cells (Achtman, 1975). Depending upon the input ratios and concentrations of the parental cells and upon their growth history, one can obtain mating populations in which most of the mating cells are either in small or in larger aggregates and one can shift the donor: recipient ratio within the aggregates by a factor of up to three- or four-fold (Achtman, 1975). However, these variations in aggregate size and makeup do not affect the transfer efficiency per aggregated cell. Under optimal conditions, every aggregated donor cell can transfer DNA, and every aggregated recipient cell can receive it (Achtman, 1975, 1977). Electron microscopic analysis (Fig. 6.8) indicated that cells in mating aggregates often establish contacts with more than one cell at a time, and genetic experiments show that multiple transfer events per aggregated cell are common (Achtman, 1977).

A feature evident from Figs. 6.7 and 6.8 is that many of the mating cells are in wall to wall contact. This feature has been noted previously (Ou and Anderson, 1970) and these authors were able to show that mating cells in wall to wall contact were both more common, and also more efficient at DNA transfer, than those (presumably) connected by an F pilus. There is preliminary evidence for exchange of outer membrane proteins between mating cells (Goldschmidt and Curtiss, personal communication), indicating real contact between some mating cells and possibly even partial membrane fusion. Another feature evident from Fig. 6.8 is that, although F pilus connections are visible between some of the mating cells, this is not usually the case. The model presented in Fig. 6.9 includes the unproven assumption that the F pilus connections seen in mating aggregates represent the initial contacts between the mating cells, and that wall to wall contact represents a subsequent stage in conjugation.

(c) *Formation of mating aggregates*

Shortly after mixing donor and recipient cultures, mating aggregates are formed which can survive mild shear forces and can transfer DNA. DeHaan and Gross (1962) proposed the terms specific and effective pairs to distinguish dilution-sensitive pairs from dilution-resistant pairs. These terms seem inappropriate in the light of our current and more sophisticated knowledge of the interactions, and we have adopted the terminology of unstable and stable mating aggregates.

The availability of genetically defined sex factor and F^- mutants has permitted an analysis of the requirements for the formation of stable mating aggregates (Achtman, Schwuchow, Morelli and Helmuth, unpublished observations). Sex factor mutants which do not produce F pili do not form any detectable stable mating aggregates (Achtman, Schwuchow, Morelli and Helmuth unpublished observations; Achtman et al., 1971). In contrast, donor cells carrying *traD, traM* or *traI* mutants of F*lac* do aggregate specifically with F^- cells and form stable mating aggregates. Thus aggregation can be measured in the absence of DNA transfer and therefore matings with mutants in which less stable forms of mating aggregates are detectable can be used to define intermediate stages in conjugation.

Most of the Con^- mutants tested (Table 6.2) do not allow the formation of stable mating aggregates. It can be speculated that their whole effect on conjugation is that they form shear-sensitive mating aggregates, since mating efficiency is normal on solid surfaces (Achtman, unpublished observations; Havekes and Hoekstra, 1976). In agreement, electron microscopic analysis demonstrated the presence of normal numbers of presumably unstable mating aggregates in Hfr x Con^- matings (Achtman, Schwuchow, Morelli and Helmuth, unpublished observations). Mating mixtures with piliated *traG* donors contain few mating aggregates although those detected included cells in wall to wall contact (Achtman, Schwuchow, Morelli and Helmuth, unpublished observations). These few *traG* x F⁻ mating aggregates cannot even transfer DNA on solid surfaces but are detectable with both electron microscopic and Coulter counter analysis. Since no detectable mating aggregates are formed with non-piliated donor cells, these results imply that the Con⁻ and the *traG* mutants are blocked in conjugation at a stage subsequent to F pilus binding. These results allow the conclusion that F pilus binding and wall to wall contacts are not sufficient for the formation of normal numbers of dilution-resistant mating aggregates and that an as yet unknown process must occur for stabilization of the mating aggregates.

Suggestive evidence for preliminary stages in conjugation has also been supplied by Ou and Anderson (1972) who showed that conversion of mating aggregates to a Zn^{2+}-resistant form took several minutes and was slower than the formation of dilution stable mating aggregates. Ou (1973) reported similar results for the formation of MS2 phage-resistant mating aggregates. A further subdivision is indicated by the finding that 1, 10-phenanthroline acts at a later stage than Zn^{2+} to inhibit formation of mating aggregates capable of DNA transfer (Ou and Reim, 1976).

(d) *Active disaggregation*

Mating mixtures containing F-like or I-like donors and F⁻ recipients can be obtained which contain up to 80% of all cells in mating aggregates. However, when the fate of the mating aggregates was analyzed, it became clear that individual mating cells disaggregated soon after DNA transfer was complete (Achtman, 1977). The observed time sequence in which DNA transfer was followed by disaggregation suggested that disaggregation is an active process triggered after completion of DNA transfer (Achtman, 1977). Since mating aggregates are stable structures in the case of Hfr x F⁻ matings, disaggregation in sex factor matings cannot be due to an inherent instability of the intercellular connection. The disaggregation phenomenon may be related to an observation by Wilkins (personal communication) that if either RNA or protein synthesis in the recipient were inhibited, DNA transfer in a ColI x F⁻ mating was enhanced. A model (Achtman, 1977) to explain these results postulates that, after transfer, the sex factor expresses enzymes which act in the recipient to specifically change the cell surface rendering the mating aggregates unstable and leading to disaggregation. It should be stressed that active disaggregation is different from surface exclusion: active disaggregation can be observed in situations where the F factor does not code for surface exclusion in the reciepient (Achtman, 1977).

(e) *Surface exclusion: a barrier*

Sex factor-carrying cells are usually poor recipients in conjugation with other cells carrying the same sex factor due to surface exclusion (Lederberg *et al.,* 1952). The *traS* and *traT* cistrons are responsible: cells carrying *traS* or *traT* mutants of F are better recipients than F⁺ cells.

Initial observations indicate that few mating aggregates were detectable in pure F⁺ cultures. Furthermore, in mating mixtures in which the recipient culture was a non-piliated but surface exclusion-positive mutant, a similar lack of stable mating aggregates was found (Achtman *et al.,* 1971; Achtman, 1977). This lessened mating aggregate formation is at least part of the basis for the surface exclusion phenomenon since F mutants which are totally surface-exclusion deficient are as capable as F⁻ cells of forming normal numbers of mating aggregates when used as recipients. Closer examination of both Coulter counter and electron microscopic data indicated, however, that similar to *traG* × F⁻ matings, the primary effect of surface exclusion was a reduction of mating aggregate formation rather than a total lack of formation (Achtman, Schwuchow, Morelli and Helmuth, unpublished observations). Furthermore, the few mating aggregates formed were extremely sensitive to SDS concentrations which had no effect on stable mating aggregates. The mating aggregates formed between F⁺ cells did resemble stable mating aggregates in regard to size and the presence of both F pilus and wall to wall connections (Achtman, Schwuchow, Morelli and Helmuth unpublished observations). A final point of resemblance between surface exclusion and the defect in *traG* × F⁻ matings is that neither is abolished on millipore membranes. Thus we can conclude that surface exclusion acts at that part of the stabilization stage for which the *traG* protein is necessary, and that surface exclusion may even act by inhibiting the *traG* protein's interaction with a cell surface component.

Although polar insertions, deletions, or *traJ* mutants completely abolish surface exclusion, neither *traS* nor *traT* mutations do so. It is possible that surface exclusion consists of two independent mechanisms. In this event, the second level of action of surface exclusion may be directly at the DNA transfer level. As mentioned in Section 6.3.1(d), surface exclusion is correlated with the inhibition of transformation (Skurray, Nagaishi and Clark, in preparation). Furthermore, *traT* mutants form more mating aggregates as recipients than do either *traT*⁺ *traS*⁺ or *traT*⁺ *traS*⁻ cells. Despite the almost normal number of mating aggregates formed with *traT* mutants, DNA transfer is still relatively inefficient (Achtman, Kennedy and Skurray, manuscript submitted). A further indication for the existence of two independent surface exclusion mechanisms comes from our interpretation of the results presented by Ou (1975): he found that F⁺ cells used as recipients could not trigger DNA replication in the donors, whereas minicells from F⁺ parents could do so. No DNA transfer occurred to the minicells, indicating that at least one surface exclusion mechanism was still operative. We speculate that the *traS* gene product is either not segregated to or is not functional in minicells, and interpret these data as further support for our subdivision of surface exclusion into two distinct mechanisms.

Beard and Bishop (1975) have shown that surface exclusion can be circumvented in R1 × R1 matings by converting the donor cells (but not the recipient cells) to sphaeroplasts. These results suggest to us that mating aggregate formation normally involves major changes in the cell surface and the peptidoglycan, which are prevented by surface exclusion. We suggest further that these changes are triggered after the conjugating cells have reached a substage in mating aggregate formation and that surface exclusion acts before this stage is reached. Since surface exclusion first acts at the stabilization stage, the postulated changes must first occur during or after stabilization.

In trying to explain surface exclusion at the molecular level, models such as a chemical modification of cell surface proteins necessary for conjugation are attractive. FI and FII sex factors were assigned to four categories on the basis of their surface exclusion specificity (Willetts and Maule, 1974) such that surface exclusion coded by any one of them only prevented DNA transfer from other sex factors in the same category. These categories only correlate to some extent with serological grouping on the basis of F pilus specificity (Lawn and Meynell, 1970), an observation which agrees with the conclusion just drawn that surface exclusion acts at a relatively late stage in mating aggregation and probably beyond the initial F pilus-cell wall interaction. Three of the surface exclusion groups (SfxI, SfxII and SfxIII) include FI and FII sex factors (F, ColV2 and R538-1, ColVB*trp* and R1-19, respectively) known to use pOmpA in conjugation whereas the fourth group (SfxIV) includes two FII sex factors (R100 and R136) which can transfer DNA to *ompA* mutants (Willetts and Maule, 1974; Manning *et al.*, 1976). We have argued above that the Con$^-$ mutants may subdivide conjugation by F-like sex factors into at least two specificity groups. These results on surface exclusion specificities would then subdivide one group further although they are also explainable by different modification of the same protein.

(f) *Lethal zygosis: a pathological interaction*

These considerations now allow us to place the phenomenon of lethal zygosis in context. When an excess of Hfr cells is mixed with F$^-$ cells, up to 99% of the F$^-$ cells die within one to two hours (Alfoldi *et al.*, 1957; Nagel de Zwaig *et al.*, 1962; Clowes, 1963; Gross, 1963; Skurray and Reeves, 1973a, b). The physiological changes in the F$^-$ cells include membrane permeability and transport defects (Skurray and Reeves, 1973b). In addition, enzyme induction and DNA synthesis are inhibited although these may be indirect effects of the membrane dysfunction. Since similar although less dramatic effects were observed even at Hfr : F$^-$ ratios of 1:1 (Skurray and Reeves, 1973a, b), it is likely that lethal zygosis is an extreme manifestation of events normally occurring during conjugation.

F$^+$ donors do not cause lethal zygosis and F$^+$ recipients are not susceptible (Skurray and Reeves, 1974). We speculate that active disaggregation in F$^+$ x F$^-$ matings disrupts otherwise potentially lethal cell contacts. The surface exclusion phenomenon presents a barrier to lethal zygosis when F$^+$ cells are used as recipients. However, even F$^+$ recipients which are surface exclusion-deficient are resistant to lethal zygosis (Skurray *et al.*, 1976c) indicating a second line of defence expressed by F$^+$ cells and termed Ilz (*i*mmunity to *l*ethal *z*ygosis). A genetic analysis mapped a cistron on F outside the transfer region between 90–0–33 kb, termed *ilzB* (Skurray *et al.*, 1976c). This may represent a disaggregation cistron. A second cistron, called *ilzA* with a similar role was mapped as being between *traD* and *traG*. This may correspond to a surface exclusion cistron or may define a new protein. No point mutations are known in either *ilz* locus.

Since DNA penetration may involve gross membrane rearrangements as discussed in Section 6.3.1(d), lethal zygosis could be due to the prolonged existence of these membrane changes and the resultant excessive membrane permeability. This model then demands that these very membrane changes are prevented by surface exclusion and reversed by disaggregation. Surface exclusion and disaggregation represent at least two membrane changes induced by conjugation proteins; if *ilzA* is unrelated to surface exclusion and/or *ilzB* is unrelated to disaggregation, then even more membrane alterations are determined by the F factor.

(g) *Purified components*

The evidence presented so far indicates that mating aggregation is a complicated process. Experiments with two purified components have confirmed this viewpoint and have provided a preliminary definition of two of these stages at a molecular level.

Purified active F pili have been analyzed for their binding specificity. As predictable from the observations above, F pili bind to F^- cells, to F^+ (surface-exclusion positive) cells and to Con^- mutants either lacking pOmpA or making a heptose-deficient lipopolysaccharide (Helmuth and Achtman, unpublished observations). The F pili did not bind to either *Bacillus* or *Pseudomonas* cells. Thus we can clearly define a stage in aggregation at which F pilus binding to the recipient cell surface is involved but at which pOmpA, surface exclusion and the *traG* protein are not.

Purified pOmpA in the 28 000 dalton form is active in inhibiting conjugation (Schweizer and Henning, 1977; von Alphen *et al.*, 1977) at a concentration of 10^{-8} M in the presence of lipopolysaccharide. This mating inhibition can be fully accounted for by the inhibition of the formation of stable mating aggregates (Achtman, Morelli and Helmuth, unpublished observations). The formation of unstable mating aggregates is only inhibited at higher protein concentrations (Achtman, Morelli and Helmuth, unpublished observations). However, F pilus binding to cells is not inhibited even at the higher concentrations of pOmpA (Helmuth and Achtman, unpublished observations). Neither pOmpA inhibition of mating aggregation, nor mating aggregation *per se*, nor F pilus binding to cells can be prevented by the addition of low concentrations of purified lipopolysaccharide; these low concentrations are sufficient to allow full action of the purified pOmpA (Achtman, Helmuth and Henning, unpublished observations). As might be expected, F^+ x F^- matings are not affected by low concentrations of purified lipopolysaccharide (Havekes *et al.*, 1977). Thus the components on the cell surface with which F pili and pOmpA interact are presumably protein in nature although interactions with a lipopolysaccharide matrix may also be involved. The results to date do not indicate whether the inhibition of conjugation by purified pOmpA acts on the donor or recipient cells. In our model the interpretation is simpler if we assume that an unidentified donor component (sex factor-coded?) must bind to the recipient's pOmpA for stabilization to be successful. Thus two of the components implicated in mating aggregation by experiments with whole cells have been purified in active form and found to interact with whole cells but not with one another.

6.4 SUMMARY

6.4.1 Stages in the mating cycle

The model in Fig. 6.9 summarizes our interpretation of the data discussed here (Achtman, Schwuchow, Morelli and Helmuth, in preparation). The requirement for F pili in order to obtain even unstable mating aggregates, and the properties of the purified F pili, demand that F pili participate in mating aggregate formation at an early stage (the 'F pilus binding stage'). The common occurrence of cells in wall to wall contact in mating aggregates suggests, but does not prove, that a second stage is involved, called the 'wall to wall contact stage'. No mutants are known which are blocked in this stage. A subsequent 'stabilization' stage is demanded by the observations that pOmpA is essential before any stable mating aggregates are formed. The lessened formation of mating aggregates in *traG* x F^- matings or in matings with an F^+ recipient has led us to assign the function of the *traG* gene product and the inhibitory effects of surface exclusion to the stabilization stage as well. This assignment is supported by the SDS-sensitive nature of the mating aggregates formed between F^+ cells. It may well be that at the molecular level, the *traG* gene product and surface exclusion define a stage different from that defined by pOmpA. As discussed above, surface exclusion may very well also act at a later stage.

The lack of transfer by stable mating aggregates with *traD*, *traI* or *traM* donors clearly indicates that DNA transfer occurs at a subsequent stage. Finally, there is the 'disaggregation stage' in autonomous sex factor matings but not in Hfr matings. It should be stressed that the definition of the five stages is based on a combination of physical and genetic techniques, and that mutants are available which form mating aggregates blocked specifically at each stage except wall to wall contact formation. Since surface exclusion prevents lethal zygosis and the initiation of DNA transfer replication, these must both occur beyond the stabilization stage; we presume that the peptidoglycan changes, which we have postulated to be stimulated by conjugation, also occur at or beyond the stabilization stage.

6.4.2 Overview

One can summarize the information given above in terms of a series of events and requirements which together define a mating cycle. In cells carrying the F factor or other constitutive F-like sex factors, at least 16 *tra* proteins are synthesized continuously which result in the production of F or F-like pili, stabilization, DNA transfer enzyme and surface exclusion. Successful mating within the donor population is prevented by surface exclusion. When these sex factor-carrying populations encounter a suitable recipient population, the piliated donor cells interact with the recipient cells to form mating aggregates whose size and composition reflects the concentrations of the two fertile subpopulations. A series of events occurs in which F pili bind to the recipient cell surface, the *traG* protein exerts an unknown role, and in which the recipient's pOmpA is also involved, such that the mating aggregates attain a dilution- and SDS-stable form. Transfer is initiated by an endonuclease cut at *oriT*, and single-stranded DNA is transferred while the complementary strand remains in the donor and is returned to the double-stranded circular form. The nature of the 'mating bridge' between the cells is unknown but does not involve extended F pili. The single strand arriving in the recipient is also converted into double stranded circular DNA, and is used for transcription and translation of sex factor-coded proteins. Among the proteins made in the recipient, a disaggregating protein(s) is postulated to exist which prevents the initiation of further DNA transfer from the same donor and also causes disaggregation of the mating cells. At the same time, the former recipient synthesizes the *traJ* protein, leading to synthesis of sex pili and expression of surface exclusion. The former recipient is now a donor cell and is essentially unable to act as a recipient for the formation of stable mating aggregates.

In general, however, F-like sex factors are repressed by the FinOP system, and the majority of cells carrying them are infertile as donors (Meynell et al., 1968). A small proportion of these cells will, however, have somehow synthesized enough *traJ* product so that the *tra* operon is transcribed and translated. These few cells are capable of acting as donors. In the repressed sex factor population, mating probably occurs continuously from the small proportion of cells which are fertile and take advantage of the fact that the majority of the population is repressed for surface exclusion (Willetts and Finnegan, 1970). However, transferred sex factor DNA is immediately repressed and further DNA transfer is prevented. On the other hand, upon contact with a suitable recipient population — i.e., a population not harboring an analogous plasmid — these few donors are immediately able to undergo the mating cycle described above such that sex factor DNA is quickly transferred (albeit from only few cells) to the new recipients. However, the FinOP proteins are either not active or not synthesized in the former recipient quickly enough after DNA transfer to prevent formation of the *traJ* product (Willetts, 1974b). The former recipients (now donors) are thus highly fertile. The new donors can then efficiently transfer the sex factor further to other recipient cells, thus repeating the mating cycle and causing the exponential spread of the sex factor through the recipient population. When all the former recipients have received a sex factor, surface exclusion prevents further mating. The FinOP system prevents new synthesis of the *traJ* product and the existing *traJ* product will eventually become either inactivated or diluted too low a concentration for further activity; the cells become repressed and infertile. The former recipient population now resembles the former donor population in its properties.

REFERENCES

Achtman, M. (1973a), Genetics of the F sex factor in *Enterobacteriaceae*, Current Topics Microb. Immun., **60**, 79–123.

Achtman, M. (1973b), Transfer-positive J-independent revertants of the F factor in *Escherichia coli* K12, *Genet. Res.*, **21**, 67–77.

Achtman, M. (1975), Mating aggregates in *Escherichia coli* conjugation, *J. Bact.*, **123**, 505–515.

Achtman, M. (1977), A physical analysis of mating in *Escherichia coli*, In: *3rd International symposium on Antibiotic Resistance*, (Rosival, L. and Krcmery, V., eds.), Avicenum, Prague, (in press).

Achtman, M. and Helmuth, R. (1975), The F factor carries an operon of more that 15×10^6 daltons coding for deoxyribonucleic acid transfer and surface exclusion, *Microbiology-1974*, (Schlessinger, D., ed.), ASM Press, Washington, DC, p. 95–103.

Achtman, M., Willetts, N. and Clark, A.J. (1971), Beginning a genetic analysis of conjugational transfer determined by the F factor in *Escherichia coli* by isolation and characterization of transfer-deficient mutants, *J. Bact.*, **106**, 529–538.

Achtman, M., Willetts, N. and Clark, A.J. (1972), A conjugational complementation analysis of transfer-deficient mutants of F*lac* in *Escherichia coli*, *J. Bact.*, **110**, 831–851.

Alfaro, G. and Willetts, N. (1972), The relationship between the transfer systems of some bacterial plasmids, *Genet. Res.*, **20**, 279–289.

Alfoldi, L., Jacob, F. and Wollman, E.L. (1957), Zygose létal dans les croisements entre souches coliconogènes et non colicinogènes d'*Escherichia coli*. *C. r. Acad. Sci. (Paris)*, **244**, 2974–2976.

von Alphen, L., Havekes, L. and Lugtenberg, B. (1977), Major outer membrane protein *d* of *Escherichia coli* K12, purification and *in vitro* activity on bateriophage K3 and F-pilus mediated conjugation, *FEBS Letters*, **75**, 285–290.

Anderson, E.S., Humphreys, G.O. and Willshaw, G.A. (1975), The molecular relatedness of R factors in *Enterobacteria* of human and animal origin, *J. gen. Microbiol.*, **91**, 376–382.

Anderson, T.F. (1958), Recombination and segregation in *Escherichia coli*, *Cold Spring Harbor Symp. Quant. Biol.*, **23**, 47–58.

Anderson, T.F. and Maze, R. (1957), Analyse de la descendance de zygotes formé par conjugasion chez *Escherichia coli* K12, *Ann. Inst. Past. (Paris)*, **93**, 194–198.

Anderson, T.F., Wollman, E.L. and Jacob, F. (1957), Sur les processus de conjugation et de recombinaison chez *Escherichia coli* III. Aspects morphologiques en microscopie électronique, *Ann. Inst. Pasteur (Paris)*, **93**, 450–455.

Anthony, W.M., Deonier, R.C., Lee, H.J., Hu, S., Ohtsubo, E., Davidson, N. and Broda, P. (1974), Electron microscope heteroduplex studies of sequence relations among plasmids of *Escherichia coli*. IX. Note on the deletion mutant of F, F_{del} (33–43), *J. mol. Biol.*, **89**, 647–650.

Bachmann, B.J., Low, K.B. and Taylor, A.L. (1976), Recalibrated linkage map of *Escherichia coli* K12, *Bact. Rev.*, **40**, 116–167.

Bayer, M.E. (1968), Adsorption of bacteriophages to adhesions between wall and membrane of *Escherichia coli*, *J. Virol.*, **2**, 346–356.

Bayer, M.E. (1976), Role of adhesion zones in bacterial cell-surface function and biogenesis, In: *Membrane Biogenesis*, (Tzagoloff, A., ed.), Plenum, New York, p. 393–427.

Beard, J.P. and Bishop, S.F. (1975), Role of the cell surface in bacterial mating: requirement for intact mucopeptide in donors for the expression of surface exclusion in R^+ strains of *Escherichia coli, J. Bact.*, **123**, 916–920.

Beard, J.P. and Connolly, J.C. (1975), Detection of a protein, similar to the sex pilus subunit in the outer membrane of *Escherichia coli* carrying a derepressed F-like R factor, *J. Bact.*, **122**, 59–65.

Beard, J.P., Howe, T.G.B. and Richmond, M.H. (1972), Purification of sex pili from *Escherichia coli* carrying a derepressed F-like R factor, *J. Bact.*, **111**, 814–820.

Bradley, D.E. (1974), Adsorption of bacteriophages specific for *Pseudomonas aeruginosa* R. factors, *Biochem. biophys. Res. Comm.*, **57**, 893–900.

Bradley, D.E. (1976), Adsorption of the R-specific bacteriophage PR4 to pili determined by a drug resistance plasmid of the W compatibility group, *J. gen. Microbiol.*, **95**, 181–185.

Bresler, S.E., Lanzov, V.A. and Lukjaniec-Blinkova, A.A. (1968), On the mechanism of conjugation in *Escherichia coli* K12, *Mol. gen. Genet.*, **102**, 269–284.

Brinton, C.C., Jr. (1965), The structure, function, synthesis and genetic control of bacterial pili and a molecular model for DNA and RNA transport in Gram negative bacteria, *Trans. N.Y. Acad. Sci.*, **277**, 1003–1054.

Brinton, C.C., Jr. (1971), The properties of sex pili, the viral nature of 'conjugal' genetic transfer systems, and some possible approaches to the control of bacterial drug resistance, *Critical Rev. Microbiol.*, **1**, 105–160. CRC Press, Cleveland, Ohio.

Chai, T. and Foulds, J. (1974), Demonstration of a missing outer membrane protein in *tolG* mutants of *Escherichia coli, J. mol. Biol.*, **85**, 465–474.

Clark, A.J., Skurray, R.A., Crisona, N.J. and Nagaishi, H. (1976), Use of molecular cloning in the genetic and functional analysis of the plasmid F, In: *Molecular Mechanism In The Control of Gene Expression,* (Nierlich, D.P., Rutter, W.J. and Fox, C.F. eds.), Academic Press, New York, pp. 565–579.

Clowes, R.C. (1963), Colicin factors and episomes, *Genet. Res.*, **4**, 162–165.

Cohen, A., Fisher, W.D., Curtiss, R., III and Adler, H.I. (1968a), DNA isolated from *Escherichia coli* minicells mated with F^- cells, *Proc. natn. Acad. Sci. U.S.A.*, **68**, 2826–2829.

Cohen, A., Fisher, W.D., Curtiss, R., III and Adler, H.I. (1968b), The properties of DNA transferred to minicells during conjugation, *Cold Spring Harbor Symp. Quant. Biol.*, **33**, 635–641.

Cohen, S.N. and Kopecko, D.J. (1976), Structural evolution of bacterial plasmids: role of translocating genetic elements and DNA sequence insertions, *Fedn Proc. fedn Socs. exp. Biol.*, **35**, 2031–2036.

Crosa, J.H., Luttrop, L.K., Heffron, F. and Falkow, S. (1975), Two replication initiation sites on R-plasmid DNA, *Mol. gen. Genet.*, **140**, 39–50.

Curtiss, R., III (1969), Bacterial conjugation, *A. Rev. Microbiol.*, **23**, 69–136.

Curtiss, R., III and Fenwick, R.G., Jr. (1975), Mechanism of conjugal plasmid transfer, *Microbiology-1974,* (Schlessinger, D., ed.), ASM Press, Washington, DC, p. 156–165.

Datta, D.B., Kramer, C. and Henning, U. (1976), Diploidy for a structural gene specifying a major protein of the outer cell envelope membrane from *Escherichia coli* K12, *J. Bact.*, **128**, 834–841.

Datta, N. (1975a), Epidemiology and classification of plasmids, in *Microbiology-1974*, (Schlessinger, D., ed.), ASM Press, Washington, DC, p. 9–15.

Datta, N. (1975b), The diversity of the plasmids that determine bacterial conjugation, *Proc. Soc. gen. Microbiol.*, **3**, 26–27.

Datta, N. and Barth, P.T.H. (1976), Compatibility properties of R483, a member of the I plasmid complex, *J. Bact.*, **125**, 796–799.

Davidoff-Abelson, R. and Dubnau, D. (1973), Kinetic analysis of the products of donor deoxyribonucleate in transformed cells of *Bacillus subtilis, J. Bact.*, **116**, 154–162.

Davidson, N., Deonier, R.C., Hu, S. and Ohtsubo, E. (1975), Electron microscope heteroduplex studies of sequence relations among plasmids of *Escherichia coli*. X. Deoxyribonucleic acid sequence organization of F and F-primes, and the sequences involved in Hfr formation, *Microbiology-1974*, (Schlessinger, D., ed.), ASM Press, Washington, DC, p. 56–65.

Davies, J.K. and Reeves, P. (1975), Colicin tolerance and map location of conjugation-deficient mutants, *J. Bact.*, **123**, 372–373.

Davis, R. and Vapnek, D. (1976), *In vivo* transcription of R-plasmid deoxyribonucleic acid in *Escherichia coli* strains with altered antibiotic resistance levels and/or conjugal proficiency, *J. Bact.*, **125**, 1148–1155.

Falkinham, J.O. and Curtiss, R., III (1976), Isolation and characterization of conjugation-deficient mutants of *Escherichia coli* K12, *J. Bact.*, **126**, 1194–1206.

Falkow, S. (1975), *Infectious multiple drug resistance*, Pion Ltd., London.

Falkow, S., Guerry, P., Hedges, R.W. and Datta, N. (1974), Polynucleotide sequence relationships among plasmids of the I compatibility complex, *J. gen. Microb.*, **85**, 65–76.

Fenwick, R.G., Jr. and Curtiss, R., III (1973), Conjugal deoxyribonucleic acid replication by *Escherichia coli* K12: effect of chloramphenicol and rifampin, *J. Bact.*, **116**, 1224–1235.

Finnegan, D.J. and Willetts, N.S. (1971), Two classes of F*lac* mutants insensitive to transfer inhibition by an F-like R factor, *Mol. gen. Genet.*, **111**, 256–264.

Finnegan, D.J. and Willetts, N.S. (1973), The site of action of the F transfer inhibitor, *Molec. gen. Genet.*, **127**, 307–316.

Fives-Taylor, P. and Novotny, C.P. (1976), Evidence for the involvement of ribonucleic acid in the production of F pili, *J. Bact.*, **125**, 540–544.

Foulds, J. (1974), Chromosomal location of the *tolG* locus for tolerance to bacteriocin JF246 in *Escherichia coli* K32, *J. Bact.*, **117**, 1354–1355.

Foulds, J. and Barrett, C. (1973), Characterization of *Escherichia coli* mutants tolerant to bacteriocin JF246: two new classes of tolerant mutants, *J. Bact.*, **116**, 885–892.

Fulton, C. (1965), Continuous chromosome transfer in *Escherichia coli, Genetics*, **52**, 55–74.

Garten, W., Hindennach, I. and Henning, U. (1975), The major proteins of the *Escherichia coli* outer cell envelope. Characterization of proteins II* and III, comparison of all proteins, *Eur. J. Biochem.*, **59**, 215–221.

Gasson, M.J. and Willetts, N.S. (1975), Five control systems preventing transfer of *Escherichia coli,* K12 sex factor F, *J. Bact.*, **122**, 518–525.

Gilbert, W. and Dressler, D. (1968), DNA replication: the rolling circle model, *Cold Spring Harbor Symp. Quant. Biol.*, **33**, 473–484.

Grindley, N.D.F., Humphreys, G.O. and Anderson, E.S. (1973), Molecular studies of R factor compatibility groups, *J. Bact.*, **115**, 387–398.

Gross, J.D. (1963), Cellular damage associated with multiple mating in *Escherichia coli, Genet, Res.*, **4**, 463–469.

Gross, J.D. and Caro, L.G. (1966), DNA transfer in bacterial conjugation, *J. mol. Biol.*, **16**, 269–284.

Guerry, P., vanEmbden, J. and Falkow, S. (1974), Molecular nature of two non-conjugative plasmids carrying drug resistance genes, *J. Bact.*, **117**, 619–630.

Guyer, M.S. and Clark, A.J. (1976), cis-Dominant, transfer-deficient mutants of the *Escherichia coli* K12 F sex factor, *J. Bact.*, **125**, 233–247.

Guyer, M.S., Figurski, D. and Davidson, N. (1976), Electron microscope study of a plasmid chimera containing the replication region of the *Escherichia coli* F plasmid, *J. Bact.*, **127**, 988–997.

deHaan, P.G. and Gross, J.D. (1962), Transfer delay and chromosome withdrawal during conjugation in *Escherichia coli, Genet, Res.*, **3**, 251–272.

Hancock, R.E. and Reeves, P. (1975), Bacteriophage resistance in *Escherichia coli* K12: General pattern of resistance, *J. Bact.*, **121**, 983–993.

Hancock, R.E.W. and Reeves, P. (1976), Lipopolysaccharide-deficient, bacteriophage-resistant mutants of *Escherichia coli* K12, *J. Bact.*, **127**, 98–108.

Havekes, L.M. and Hoekstra, W.P.M. (1976), Characterization of an *Escherichia coli* K12 F$^-$ Con$^-$ mutant, *J. Bact.*, **126**, 593–600.

Havekes, L.M., Lugtenberg, B.J.J. and Hoekstra, W.P.M. (1976), Conjugation deficient *Escherichia coli* K12 F$^-$ mutants with heptose-less lipopolysaccharides, *Mol. gen. Genet.*, **146**, 43–50.

Havekes, L., Tommassen, J., Hoekstra, W. and Lugtenberg, B. (1977), Isolation and characterization of *Escherichia coli* K12 F$^-$ mutants defective in conjugation with an I-type donor, *J. Bact.*, **129**, 1–8.

Helmuth, R. and Achtman, M. (1975), Operon structure of DNA transfer cistrons on the F sex factor, *Nature*, **257**, 652–656.

Henning, U. and Haller, I. (1975), Mutants of *Escherichia coli* K12 lacking all 'major' proteins of the outer cell envelope membrane, *FEBS Letters*, **55**, 161–164.

Henning, U., Hindennach, I. and Haller, I. (1976), The major proteins of the *Escherichia coli* outer cell envelope membrane: evidence for the structural gene of protein II*, *FEBS Letters*, **61**, 46–48.

Henning, U., Hoehn, B. and Sonntag, I. (1973), Cell envelope and shape of *Escherichia coli* K12, *Eur. J. Biochem.*, **39**, 27–36.

Herman, R.K. and Forro, F. (1964), Autoradiographic study of transfer of DNA during bacterial conjugation, *Biophys. J.*, **4**, 335–353.

Hindennach, I. and Henning, U. (1975), The major proteins of the *Escherichia coli* outer cell envelope membrane, *Eur. J. Biochem.*, **59**, 207–213.

Hopwood, D.A. and Wright, H.M. (1976), Interactions of the plasmid SCP1 with the chromosome of *Streptomyces coelicolor* A3(2), in *Second International Symposium on the Genetics of Industrial Microorganisms*, (Macdonald, K.D., ed.), Academic Press, New York, pp. 607–619.

Inouye, M. (1974), A three-dimensional molecular assembly model for a lipoprotein from the *Escherichia coli* outer membrane, *Proc. natn. Acad. Sci., U.S.A.*, **71**, 2396–2400.

Ippen, K.A. and Valentine, R.C. (1967), The sex hair of *Escherichia coli* as a sensory fiber, conjugation tube, or mating arm?, *Biochem. biophys. Res. Comm.*, **27** 674–680.

Jacob, A.E. and Hobbs, S. (1974), Conjugal transfer of plasmid-borne multiple antibiotic resistance in *Streptococcus faecalis* var. *zymogenes, J. Bact.*, **121**, 863–872.

Jacobson, A. (1972), Role of F pili in the penetration of bacteriophage f1, *J. Virol.*, **10**, 835–843.

Jacoby, G.A. (1975), Properties of R plasmids in *Pseudomonas aeruginosa, Microbiology-1974,* (Schlessinger, D., ed.), ASM Press, Washington, DC, p. 36–42.

Jazwinski, S.M., Marco, R. and Kornberg, A. (1973), A coat protein of the bacteriophage M13 virion participates in membrane-orientated synthesis of DNA, *Proc. natn. Acad. Sci., U.S.A.*, **70**, 205–209.

Jazwinski, S.M., Marco, R. and Kornberg, A. (1975), The gene H spike protein of bacteriophages ϕx174 and S13. II. Relation to the synthesis of the parental replication form, *J. Virol.*, **66**, 294–305.

Khatoon, H., Iyer, R.V. and Iyer, V.N. (1972), A new filamentous bacteriophage with sex-factor specificity, *Virology*, **48**, 145–155.

Knolle, P. (1967), Evidence for the identity of the mating-specific site of male cells of *Escherichia coli* with the receptor site of an RNA phage, *Biochem. biophys. Res. Comm.*, **27**, 81–87.

Kornberg, A. (1974), *DNA Synthesis,* Freeman, San Francisco.

Krahn, P.M., O'Callaghan, R.J. and Paranchych, W. (1972), Stages in R17 infection. VI. Injection of A protein and RNA into the host cell, *Virology*, **47**, 628–637.

Lacks, S. and Greenberg, B. (1976), Single-strand breakage on binding of DNA to cells in the genetic transformation of *Diplococcus pneumoniae, J. mol. Biol.*, **101**, 255–275.

Lacks, S. and Neuberger, M. (1975), Membrane location of a deoxyribonuclease implicated in the genetic transformation of *Diplococcus pneumoniae, J. Bact.*, **124**, 222–232.

Lawn, A.M. and Meynell, E. (1970), Serotypes of sex pili, *J. Hyg.*, **68**, 683–694.

Lawn, A.M., Meynell, E. and Cooke, M. (1971), Mixed infections with bacterial sex factors: sex pili of pure and mixed phenotypes, *Ann. Inst. Pasteur (Paris),* **120**, 3–8.

Lawn, A.M., Meynell, E., Meynell, G.G. and Datta, N. (1967), Sex pili and the classification of sex factors in the *Enterobacteriaceae, Nature,* **216**, 343–346.

Lederberg, J. (1956), Conjugal pairing in *Escherichia coli, J. Bact.*, **71**, 497–198.

Lederberg, J., Cavalli, L.L. and Lederberg, E.M. (1952), Sex compatibility in *Escherichia coli, Genetics,* **37**, 720–730.

Leipold, B. and Hofschneider, P.H. (1975), Isolation of an infectious RNA-A-protein complex from bacteriophage M12, *FEBS Letters,* **55**, 50–52.

Le Minor, L., Coynault, C., Chabbert, Y., Gerbaud, G., and Le Minor, S. (1976), Groupes de compatibilité de plasmides métaboliques, *Ann. Microbiol. (Inst. Pasteur),* **127B**, 31–40.

Lovett, M.A. and Helinski, D.R. (1976), Method for the isolation of the replication region of a bacterial replicon: construction of a mini-F'*km* plasmid, *J. Bact.*, **127**, 982–987.

Lugtenberg, B., Meijers, J., Peters, R., v.d. Hoek, P. and v. Alphens, L. (1975), Electrophoretic resolution of the major outer membrane protein of *Escherichia coli* K12 into four bands, *FEBS Letters*, **58**, 254–258.

Manning, P.A. and Reeves, P. (1975), Recipient ability of bacteriophage-resistant mutants of *Escherichia coli* K12, *J. Bact.*, **124**, 576–577.

Manning, P.A. and Reeves, P. (1976a), Outer membrane of *Escherichia coli*: demonstration of the temperature sensitivity of a mutant in one of the major outer membrane proteins, *Biochem. biophys. Res. Comm.*, **72**, 694–700.

Manning, P.A. and Reeves, P. (1976b), Outer membrane of *Escherichia coli* K12: differentiation of proteins 3A and 3B on acrylamide gels and further characterization of *con* (*tolG*) mutants, *J. Bact.*, **127**, 1070–1079.

Manning, P.A., Puspurs, A. and Reeves, P. (1976), Outer membrane of *Escherichia coli* K12, isolation of mutants with altered protein 3a using host-range mutants of bacteriophage K3, *J. Bact.*, **127**, 1080–1084.

Marinus, M.G. and Adelberg, E.A. (1970), Vegetative replication and transfer replication of deoxyribonucleic acid in temperature-sensitive mutants of *Escherichia coli* K12, *J. Bact.*, **104**, 1266–1272.

Marvin, D. and Hohn, B. (1969), Filamentous bacterial viruses, *Bact. Rev.*, **33**, 172–209.

Matsubara, K. (1968), Properties of sex factor and related episomes isolated from purified *Escherichia coli* zygote cells, *J. mol. Biol.*, **38**, 89–108.

Meynell, E., Meynell, G.G. and Datta, N. (1968), Phylogenetic relationships of drug-resistance factors and other transmissible bacterial plasmids, *Bact. Rev.*, **32**, 55–83.

Meynell, G.G. (1972), *Bacterial plasmids. Conjugation, colicinogeny and transmissible drug-resistance*, MacMillan Press Ltd., London.

Minkley, E.G., Jr. and Ippen-Ihler, K. (1977), Identification of a membrane protein associated with expression of the surface exclusion region of the F transfer operon, *J. Bact.*, **129**, 1613–1622.

Minkley, E.G., Jr., Polen, S., Brinton, C.C., Jr. and Ippen-Ihler, K. (1976), Identification of the structural gene for F-pilin, *J. mol. Biol.*, **108**, 111–121.

Monner, D.A., Jonsson, S. and Boman, H.G. (1971), Ampicillin-resistant mutants of *Escherichia coli* K12 with lipopolysaccharide alterations affecting mating ability and susceptibility to sex-specific bacteriophages, *J. Bact.*, **107**, 420–432.

Morrison, T.G. and Malamy, M.H. (1970), Comparisons of F factors and R factors: existence of independent regulation groups in F factors, *J. Bact.*, **103**, 81–88.

Morrison, T.G. and Malamy, M.H. (1971), T7 translational control mechanisms and their inhibition by F factors, *Nature New Biol.*, **231**, 37–41.

Nagel de Zwaig, R., Anton, D.N. and Puig, J. (1962), The genetic control of colicinogenic factors E2, I and V, *J. gen. Microbiol.*, **29**, 473–484.

Novick, R.P. (1974), Bacterial plasmids, In: *Handbook of Microbiology*, Vol. IV, p. 537–586, (Laskin, A.I. and Lechevalier, H.A., eds.), CRC Press, Cleveland, Ohio.

Novick, R.P. and Schweisinger, M. (1976), Independence of plasmid incompatibility and replication control functions in *Staphylococcus aureus*, *Nature*, **262**, 623–626.

Novotny, C. and Fives-Taylor, P. (1974), Retraction of F pili, *J. Bact.*, **117**, 1306–1311.

Novotny, C., Knight, W.S. and Brinton, C.C., Jr. (1968), Inhibition of bacterial conjugation by ribonucleic acid and deoxyribonucleic acid male-specific bacteriophages, *J. Bact.,* **95**, 314–326.

Novotny, C. and Lavin, K. (1971), Some effects of temperature on the growth of F pili, *J. Bact.,* **107**, 671–682.

Ohki, M. and Tomizawa, J. (1968), Asymmetric transfer of DNA strands in bacterial conjugation, *Cold Spring Harbor Symp. Quant. Biol.,* **33**, 651–658.

Olsen, R.H. and Thomas, D.D. (1973), Characteristics and purification of PRR1, an RNA phage specific for the broad host range *Pseudomonas* R1882 drug resistance plasmid, *J. Virol.,* **12**, 1560–1567.

Ou, J.T. (1973), Inhibition of formation of *Escherichia coli* mating pairs by f1 and MS2 bacteriophages as determined with a Coulter Counter, *J. Bact.,* **114**, 1108–1115.

Ou, J.T. (1975), Mating signal and DNA penetration deficiency in conjugation between male *Escherichia coli* and minicells, *Proc. natn. Acad. Sci. U.S.A.,* **72**, 3721–3725.

Ou, J.T. and Anderson, T.F. (1970), Role of pili in bacterial conjugation, *J. Bact.,* **102**, 648–654.

Ou, J.T. and Anderson, T.F. (1972), Effect of Zn^{2+} on bacterial conjugation: inhibition of mating pair formation, *J. Bact.,* **111**, 177–185.

Ou, J.T. and Reim, R. (1976), Effect of 1,10-phenanthroline on bacterial conjugation in *Escherichia coli* K12: inhibition of maturation from preliminary mates into effective mates, *J. Bact.,* **128**, 363–371.

Paranchych, W. (1975), Attachment, ejection and penetration stages of the RNA phage infectious process, in *RNA Phages,* (Zinder, N.D., ed.), Cold Spring Harbor Laboratory U.S.A., p. 85–111.

Reeves, P. and Willetts, N. (1974), Plasmid specificity of the origin of transfer of sex factor F, *J. Bact.,* **120**, 125–130.

Reiner, A.M. (1974), *Escherichia coli* females defective in conjugation and in adsorption of a single-stranded deoxyribonucleic acid phage, *J. Bact.,* **119**, 183–191.

Reithmeier, R.A.F. and Bragg, P.D. (1974), Purification and characterization of a heat-modifiable protein from the outer membrane of *Escherichia coli,* *FEBS Letters,* **41**, 195–198.

Rupp, W.D. and Ihler, G. (1968), Strand selection during bacterial matings, *Cold Spring Harbor Symp. Quant. Biol.,* **33**, 647–650.

Sagai, H., Hasuda, K., Iyobe, S., Bryan, L.E., Holloway, B.W. and Mitsuhashi, S. (1976), Classification of R plasmids by incompatibility in *Pseudomonas aeruginosa, Antimicrob. Agents Chemother.,* **10**, 573–578.

Santos, D.S., Palchaudhuri, S. and Maas, W.K. (1975), Genetic and physical characteristics of an enterotoxin plasmid, *J. Bact.,* **124**, 1240–1247.

Sarathy, P.V. and Siddiqi, O. (1973), DNA synthesis during bacterial conjugation. III. Is DNA replication in the Hfr obligatory for chromosome transfer? *J. mol. Biol.,* **78**, 443–451.

Schlessinger, D. (1975), *Microbiology-1974,* American Society of Microbiology, Washington, DC.

Schnaitman, C.A. (1974), Outer membrane proteins of *Escherichia coli*. III. Evidence that the major protein of *Escherichia coli* 0111 outer membrane consists of four distinct polypeptide species, *J. Bact.*, **118**, 442–453.

Schweizer, M. and Henning, U. (1977), Action of a major outer cell envelope membrane protein in conjugation of *E. coli* K12, *J. Bact.*, **129**, 1651–1652.

Sedgwick, B. and Setlow, J.K. (1976), Single-stranded regions in transforming deoxyribonucleic acid after uptake by competent *Haemophilis influenzae*, *J. Bact.*, **125**, 588–594.

Sharp, P.A., Hsu, M., Ohtsubo, E. and Davidson, N. (1972), Electron microscope heteroduplex studies of sequence relations among plasmids of *Escherichia coli*. I. Structure of F prime factors, *J. mol. Biol.*, **71**, 471–497.

Sharp, P.A., Cohen, S.N. and Davidson, N. (1973), Electron microscope heteroduplex studies of sequence relations among plasmids of *Escherichia coli*. II. Structure of drug resistance (R) and F factors, *J. mol. Biol.*, **75**, 235–255.

Shiba, T. and Miyake, T. (1975), New type of infectious complex of *Escherichia coli* RNA phage, *Nature*, **254**, 157–158.

Singer, S.J. (1974), The molecular organization of membranes, *A. Rev. Biochem.*, **43**, 805–833.

Skurray, R.A. (1974), Physiology of *Escherichia coli* K12 during conjugation, Ph. D. thesis, University of Adelaide, Australia.

Skurray, R.A., Guyer, M.S., Timmis, K., Cabello, F., Cohen, S.N., Davidson, N. and Clark, A.J. (1976a), Replication region fragments cloned from Flac^+ are identical to *Eco*RI fragment f5 of F, *J. Bact.*, **127**, 1571–1575.

Skurray, R.A., Hancock, R.E.W. and Reeves, P. (1974), Con$^-$ mutants: Class of mutants in *Escherichia coli* K12 lacking a major cell wall protein and defective in conjugation and adsorption of a bacteriophage, *J. Bact.*, **119**, 726–735.

Skurray, R.A., Nagaishi, H. and Clark, A.J. (1976b), Molecular cloning of DNA from F sex factor of *Escherichia coli* K12, *Proc. natn. Acad. Sci. U.S.A.*, **73**, 64–68.

Skurray, R.A. and Reeves, P. (1973a), Characterization of lethal zygosis associated with conjugation in *Escherichia coli* K12, *J. Bact.*, **113**, 58–70.

Skurray, R.A. and Reeves, P. (1973b), Physiology of *Escherichia coli* K12 during conjugation: altered recipient functions associated with lethal zygosis, *J. Bact.*, **114**, 11–17.

Skurray, R.A. and Reeves, P. (1974), F factor-mediated immunity to lethal zygosis in *Escherichia coli* K12, *J. Bact.*, **117**, 100–106.

Skurray, R.A., Willetts, N. and Reeves, P. (1976c), Effect of *tra* mutations on F factor-specified immunity to lethal zygosis, *Mol. gen. Genet.*, **146**, 161–165.

Smith, H.R., Humphreys, G.O. and Anderson, E.S. (1974), Genetic and molecular characterization of some non-transferring plasmids, *Mol. gen. Genet.*, **129**, 229–242.

Smith, H.R., Humphreys, G.O., Willshaw, G.A. and Anderson, E.S. (1976), Characterization of plasmids coding for the restriction endonuclease *Eco*RI, *Mol. gen. Genet.*, **143**, 319–325.

Starlinger, P. and Saedler, H. (1976), IS-elements in microorganisms, *Current Topics Microbiol. Immunol.*, **75**, 111–152.

Timmis, K., Cabello, F. and Cohen, S.N. (1975), Cloning, isolation and characterization of replication regions of complex genomes, *Proc. natn. Acad. Sci., U.S.A.*, **72**, 2242–2246.

Tomoeda, M., Inuzuka, M. and Date, T. (1975), Bacterial sex pili, *Prog. Biophys., mol. Biol.*, **30**, 23–56.

Tomura, T., Hirano, T., Ito, T. and Yoshioka, M. (1973), Transmission of bacteriogenicity by conjugation in group D *Streptococci*, *Jap. J. Microbiol.*, **17**, 445–452.

Tschaepe, H. (1977), Genetic characterization of transferable plasmids presumed to be unloaded, In: *3rd International Symposium on Antibiotic Resistance*, (Rosival, L. and Krcmery, V., eds.), Avicenum, Prague, (in press).

Uhlin, B.E. and Nordstrom, K. (1975), Plasmid incompatibility and control of replication: Copy mutants of the R factor R1 in *Escherichia coli* K12, *J. Bact.*, **124**, 641–649.

Vapnek, D. and Rupp, W.D. (1970), Asymmetric segregation of the complementary sex-factor DNA strands during conjugation in *Escherichia coli*, *J. mol. Biol.*, **53**, 287–303.

Vapnek, D. and Rupp, W.D. (1971), Identification of individual sex-factor DNA strands and their replication during conjugation in thermosensitive DNA mutants of *Escherichia coli*, *J. mol. Biol.*, **60**, 413–424.

Walmsley, R.H. (1976), Temperature dependence of mating-pair formation in *Escherichia coli*, *J. Bact.*, **126**, 222–224.

Wilkins, B.M. (1975), Partial suppression of the phenotype of *Escherichia coli* K12 *dnaG* mutants by some I-like conjugative plasmids, *J. Bact.*, **122**, 889–904.

Wilkins, B.M. and Hollom, S.E. (1974), Conjugational synthesis of F*lac*$^+$ and ColI DNA in the presence of rifampicin and in *Escherichia coli* K12 mutants defective in DNA synthesis, *Mol. gen. Genet.*, **134**, 143–156.

Willetts, N.S. (1971), Plasmid specificity of two proteins required for conjugation in *Escherichia coli* K12, *Nature New Biol.*, **230**, 183–185.

Willetts, N. (1972a), Location of the origin of transfer of the sex factor F, *J. Bact.*, **112**, 773–778.

Willetts, N. (1972b), The genetics of transmissible plasmids, *A. Rev. Genet.*, **6**, 257–268.

Willetts, N. (1974a), Mapping loci for surface exclusion and incompatibility on the F factor of *Escherichia coli* K12, *J. Bact.*, **118**, 778–782.

Willetts, N.S. (1974b), The kinetics of inhibition of F*lac* transfer by R100 in *Escherichia coli*, *Mol. gen. Genet.*, **129**, 123–130.

Willetts, N. (1977), The transcriptional control of fertility in F-like plasmids, *J. mol. Biol.*, **112**, 141–148.

Willetts, N.S. and Achtman, M. (1972), A genetic analysis of transfer by the *Escherichia coli* sex factor F, using P1 transductional complementation, *J. Bact.*, **110**, 843–851.

Willetts, N.S. and Finnegan, D.J. (1970), Characteristics of *Escherichia coli* K12 strains carrying both an F prime and an R factor, *Genet. Res.*, **16**, 113–122.

Willetts, N. and Maule, J. (1974), Interactions between the surface exclusion systems of some F-like plasmids, *Genet. Res.*, **24**, 81–89.

Willetts, N. Maule, J. and McIntire, S. (1976), The genetic locations of *traO*, *finP* and *tra-4* on the *E. coli* sex factor F, *Genet. Res.*, **26**, 255–263.

Wong, K. and Paranchych, W. (1976), The effect of ribonuclease on the penetration of R17 phage A-protein and RNA, *Can. J. Microbiol.*, **22**, 826–831.

Zoon, K.C., Habersat, M. and Scocca, J.J. (1976), Synthesis of envelope polypeptides by *Haemophilus influenzae* during development of competence for genetic transformation, *J. Bact.*, **127**, 545–554.

7 Cell–Cell Interactions during Mating in *Saccharomyces cerevisiae*

THOMAS R. MANNEY and JAMES H. MEADE

7.1	Introduction *page*	283
7.2	A popular picture of the events leading to zygote formation	284
	7.2.1 The life cycle of *Saccharomyces cerevisiae*	284
	7.2.2 An overview of cell–cell interactions during mating	285
	7.2.3 Evidence from studies on mixed cultures	285
	(a) *Cell division inhibition, 285*, (b) *Sex-specific aggregation, 286*, (c) *Zygote formation, 288*	
	7.2.4 Evidence from analysis of sterile mutants	288
	7.2.5 Evidence from studies with isolated components of mixed cultures	289
	(a) *Extracellular factors, 289*, (b) *Recovery from α-factor inhibition, 290*, (c) *Cell surface changes during mating, 290*	
7.3	Yeast sex factors: a hard-nosed look	293
	7.3.1 An appeal for order	293
	7.3.2 Mating type specificity	293
	7.3.3 Multiple-factors versus multiple-functions	294
	7.3.4 α-Factors	295
	(a) *Partially purified α-factor, 295*	
	7.3.5 *a*-Factor	296
	(a) *Isolation of a-factor, 297*, (b) *Other diffusible a-factors, 297*	
	7.3.6 Agglutination-inducing factors	298
7.4	A comprehensive approach to mating studies	299
	7.4.1 Quantitative assays for *Saccharomyces* sex factors	300
	7.4.2 Methods for purification of factors	303
	7.4.3 Mating type-specific aggregation	304
	(a) *Methods for observing agglutination, 304*, (b) *Quantitation of aggregation, 304*, (c) *Strain differences, 306*, (d) *Do only unbudded cells aggregate?, 308*, (e) *Agglutination and agglutinability, 309*, (f) *Initiation of aggregation, 310*	

			page	
7.4.4	Studies with α-factor-deficient mutants			310
	(a) *Isolation of mutants, 311*, (b) *Genetics, 312*, (c) *Production of α-factor, 312*, (d) *Mating, 312*, (e) *Agglutination, 315*, (f) *Summary, 316*			
7.4.5	Courting revisited			316
7.5	Summary			318
	References			319

... 'Do you believe in ghosts?'

'No,' I say.

'Why not?'

'Because they are *un*-sci-en-*ti*fic.'

... 'They contain no matter,' I continue, 'and have no energy and therefore, according to the laws of science, do not exist except in people's minds.'

... 'Of course,' I add, 'the laws of science contain no matter and have no energy either and therefore do not exist except in people's minds. It's best to be completely scientific about the whole thing and refuse to believe in either ghosts or the laws of science. That way you're safe. That doesn't leave you very much to believe in, but that's scientific too.'

Robert M. Pirsig, Zen and The Art of Motorcycle Maintenance, Bantam Books

Acknowledgements

We gratefully acknowledge and thank the many people who have helped put this chapter together, especially the following: Patricia Jackson, Dan Paretsky, Lela Riley, and Verna Woods, who carried out most of the original experiments described; Dr Ken Conrow, who made the computer analysis of our data work; Drs Vivian MacKay, Wolfgang Duntze, and Mike Fox, who read the manuscript and made many suggestions that helped us say what we mean; Gloria Manney, who put our commas in the proper places; and many others, mentioned in the text, who shared their unpublished data with us.

Microbial Interactions
(*Receptors and Recognition*, Series B, Volume 3)
Edited by J.L. Reissig
Published in 1977 by Chapman and Hall, New Fetter Lane, London EC4P 4EE
© Chapman and Hall

7.1 INTRODUCTION

Among the interactions that can occur between a yeast cell and its environment, the most interesting are those that occur with other yeast cells, and among those, the most interesting are those that occur between cells of opposite mating type. This subset includes all of the interactions that are involved in mating, the process by which two cells fuse to form a zygote. At present, it seems clear that some of the interactions are mediated by extracellular sex factors, while others require direct cell–cell contact. There are many aspects of these processes that we do not understand, but the tools are now available for a concerted genetic, physiological, and biochemical assault. We should be able to manipulate this system so as to elucidate the nature of the interactions on the cell surface and determine how these cell-surface interactions are translated into changes in gene regulation. Indeed, the ease with which the yeast, *Saccharomyces cerevisiae,* can be manipulated provides the primary value for its use as a model system for studies of cell–cell interactions.

The literature describing mating in *Saccharomyces* has accumulated for the past 20 to 30 years, but only in recent years has much excitement been generated. In 1956, Levi reported that a morphological change, the development of 'copulatory processes' in mating-type *a* cells of *S. cerevisiae,* occurred only after an interaction between *a* and α cells. This interaction did not require cell–cell contact but appeared to be the result of a diffusible factor. In 1970, Duntze, MacKay and Manney isolated and partially purified this factor, now called α-factor, from cultures of α cells and determined it to be an oligopeptide. The main interest in this factor, however, came after 1973 (Bucking-Throm *et al.,* 1973), when it was reported that α-factor not only induced a morphological change in *a* cells (now called 'shmooing' because of the resemblance of the altered cells to Al Capp's comic characters), but that it also arrested *a* cells in the unbudded stage of the cell cycle and inhibited initiation of DNA synthesis. Since that time a number of other factors or responses have been interpreted as being involved in the mating interaction. Several recent reviews (Crandall *et al.,* 1976; Crandall, 1977; MacKay, 1976) have summarized the literature. This chapter will not be another recapitulation of the literature, but rather a critical appraisal of what is known, what is believed, what is assumed, and some important directions we think the research should take from here.

7.2 A POPULAR PICTURE OF THE EVENTS LEADING TO ZYGOTE FORMATION

7.2.1 The life cycle of *Saccharomyces cerevisiae*

Saccharomyces cerevisiae has a simple life cycle which can be considered as four developmental phases (Fig. 7.1; Hartwell 1974). Changes from one phase to another are regulated by the culture conditions, the genotype of the cells, and by interactions

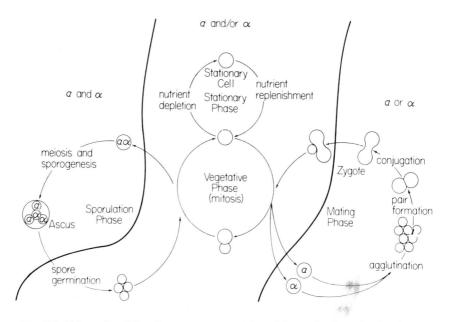

Fig. 7.1 Life cycle of *Saccharomyces cerevisiae* with emphasis on its developmental phases. The relationships among the four phases are adapted from Hartwell (1974). The influence of the mating-type alleles, a and α, in the regulation of these phases is illustrated by division of the diagram into three regions. The functions associated with sporulation (left side of diagram) are normally found only in strains that are heterozygous (e.g. a/α). Functions associated with mating (right side of diagram) are normally found only in strains that are hemizygous or homozygous for one of the mating types. The functions that are associated with the cell division cycle and stationary phase (center) are not restricted to any particular mating genotype.

with other cells in the culture. The genetic control of the life cycle is primarily determined by the mating-type alleles, either a or α; mating interactions are initiated when cells of a and α mating type are mixed in the same culture.

7.2.2 An overview of cell–cell interactions during mating

From the literature that has been amassed to date, a popular view of the events that occur during mating has emerged. We begin this chapter by reviewing what is believed, and the data on which this is based.

Extracellular sex factors are produced both by a and α cells. These factors are made during the growth of the cells in pure culture. In the case of a-factor, there is some evidence that increased levels of activity are obtained from mixed cultures and from cultures of a cells that have been treated with α-factor. When cells of opposite mating type are mixed in culture, each produces a factor in the medium that causes an arrest of the opposite mating type, synchronizing them as unbudded cells. Cell surface agglutination factors are believed to be induced by extracellular factors or to be produced constitutively, depending on the strain. These events, synchronization by cell division arrest, and agglutination, are seen as a prelude to cell fusion. This picture is a composite built from many fragments of mostly indirect evidence.

7.2.3 Evidence from studies on mixed cultures

When a and α cells are mixed, zygotes are not formed immediately. This lag period suggests that there are interactions between a and α cells that are a necessary prelude to mating. These events have been termed 'courtship' by L.H. Hartwell.

Hartwell demonstrated the requirement for such a courtship by experimenting with a 'controlled orgy' (Hartwell, 1973). He mixed a and α cells and incubated this mixture for two hours. He then sonicated the mixture and added a second a and a second α strain to the mating mixture. The strains he used carried different auxotrophic markers, so he could identify the four possible diploids by plating on different media. The time course of appearance of the four possible diploid strains was then followed (Fig. 7.2). A lag was observed between the appearance of the diploid formed between the strains that had been mixed for the first 2 hours and the other three possible diploids. He concluded that in the initial mixture the a and α cells courted each other and hence were ready to mate, but that the lag in formation of the other three diploids demonstrated a requirement for courting of both partners in a mating. His experiment does not indicate the nature of the required events. There are some conspicuous changes, however, that are reasonable candidates.

(a) *Cell division inhibition*

When exponentially growing cultures of a and α cells are mixed, a temporary inhibition in cell division in the culture can be noted within a few hours, depending upon the culture conditions (Fig. 7.3a; see also Hartwell, 1973). This inhibition is paralleled by a decrease in budded cells in the mixture (Fig. 7.3b; see also Hartwell, 1973; Bilinski *et al.*, 1974) and a depression in DNA synthesis (Bilinski *et al.*, 1974; Zuk *et al.*, 1975). These observations indicate that one of the first events in the interaction between the a and α cells is the synchronization of the cultures in the

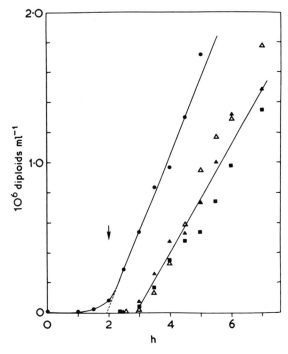

Fig. 7.2 Kinetics of zygote formation in Hartwell's yeast orgy (Hartwell, 1973). From exponentially growing cultures of A364A (*a*) and H136RS141 (α), aliquots of 2×10^6 cells were mixed and collected onto Millipore filters. After 2 h of incubation on complex medium at 23°C the cells were resuspended in liquid medium and 2×10^6 cells each of strains H138RS56 (*a*) and H137RS64 (α) were added to the suspension. The mixture containing all four strains was sonicated and again collected onto Millipore filters; the filters were incubated at 23°C on complex medium. At various times filters were removed; the cells were resuspended in distilled water, sonicated, and plated for the various diploid colony-forming units: A364A x H136RS141 (closed circles), A364A x H137RS64 (closed squares), H138RS56 x H136RS141 (closed triangles), and H138RS56 x H137RS64 (open triangles). After Hartwell (1973).

G1 phase of the cell division cycle. The cell cycle arrest is not permanent, and mating is not required for recovery from arrest, as can be noted by the increase in budded cells after a short period of inhibition (Fig. 7.3b).

(b) *Sex-specific aggregation*

When *a* and α cells are cultured together the formation of aggregates can be detected by a number of different methods (see below). In Fig. 7.3c, the formation of aggregates was determined by simply plating sonicated and unsonicated samples on complete medium. It has been suggested (Sakai and Yanagishima, 1972) that

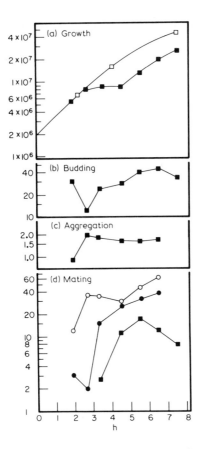

Fig. 7.3 Indices of cell–cell interactions in a mating mixture of exponentially growing haploid yeast cells of opposite mating types. Growing cultures of XP300-26C (α) and XP300-29B (a) in synthetic medium were mixed at $t = 0$. Cultures were shaken at 200 rev/min at 30°C. (a) Cell number. Sonicated aliquots were counted with a Coulter Counter; mixture (closed squares), average values of unmixed haploids (open squares). (b) % budded cells in the mixture. Sonicated aliquots were scored under microscope for budded and unbudded cells. (c) Aggregation in the mixture. Ratio of viable counts on sonicated and unsonicated aliquots is taken as an index of agglutination. (d) Mating. The fraction of cells in the mixture that have the morphology of zygotes was determined by microscopic examination (closed squares). Sonicated (closed circles) and unsonicated (open circles) aliquots from the mixture were plated on a minimal medium selective for diploids and on complete medium. The results are expressed as percent of colony forming units on selective medium compared to complete medium.

aggregation consists of two stages. The first stage, which consists of the formation of weak bonds between cells, does not appear to require metabolism. Campbell (1973) demonstrated a well-defined sex-specific aggregation occuring in buffer by gently plating cells in soft agar. He also presented indirect evidence that only unbudded cells participated in this aggregation. It has been reported (Sakai and Yanagishima, 1972) that a weak association can even occur between boiled cells and live cells of the opposite mating type as well as in cell mixtures incubated with cycloheximide. The second stage is the development of larger, visible clumps in mating mixtures. Formation of these clumps is inhibited by cycloheximide (Sakai and Yanagishima, 1972; D. Radin, personal communication).

It has been reported (Sakai and Yanagishima, 1972) that a cells can be pretreated with cell-free medium in which α cells have been grown, and thereby be rendered agglutinable. The factor that induces agglutinability in mating-type a cells has been reported to be identical with α-factor (Sakurai et al., 1976b). These results have been interpreted as induction of new cell wall components by extracellular factors produced by cells of the opposite mating type. In addition, there have been reports of cell wall composition changes induced by the same treatment (Lipke et al., 1976). These are important findings, because they suggest an additional function for the sex factors in the mating process.

(c) *Zygote formation*

Following agglutination of a culture, the formation of diploid zygotes can be detected (Fig. 7.3d). Different methods of analysis measure different stages of zygote formation. The aggregation of mating cells into clumps can be measured by plating samples from sonicated and unsonicated cultures. Later stages are indicated by the formation of sonication-resistant unions and finally the appearance of microscopically observable zygotes. Hartwell has indicated that the formation of zygotes is the result of mating between unbudded cells (Hartwell, 1973). In addition, studies by Reid (cited in Hartwell, 1974) showed that a temperature-sensitive cell cycle mutant, blocked in cell division as unbudded cells (cdc28), retained the ability to mate while mutants blocked in later stages of cell division showed drastically reduced mating abilities. Mortimer, however, has reported a series of studies of dominant lethality with X-rays, which involved observing a large number of individual mating pairs microscopically (Mortimer, 1959). He reported that approximately 50% of the budded cells in such pairs mated while they were budded. Since these cells had been irradiated, and similar observations have not been made with un-irradiated cells, it is not known whether the radiation played a role in the mating behavior. The question of whether budded cells can mate without being irradiated remains to be tested by Mortimer's direct pairing method.

7.2.4 Evidence from analysis of sterile mutants

At this time, we cannot state whether the ability to produce extracellular sex factors or the ability to respond to them are essential for mating. Some attempts to answer

Table 7.1 Comparison of sterile strains isolated by different techniques

Technique	Number of mutants	Number that produce their own sex factor	Number that respond to the opposite sex factor
a mutants:			
Selected as non-mating (MacKay and Manney, 1974)	66	7/66	0/66
Selected as α-factor-resistant (Manney and Woods, 1976)	283	40/120	0/283
No selection (J. Rine, personal communication)	20	NT	2/20
α mutants:			
Selected as non-mating (MacKay and Manney, 1974; MacKay, 1976)	383	196/383	1/107

these questions have been made by isolating and characterizing mutants affected in these processes. Table 7.1 presents a comparison of *a* and α mutants isolated by different techniques. All but three of the mutants isolated as steriles fail to form shmoos in response to the sex factor produced by cells of the opposite mating type, while all *a* mutants in this sample that were isolated as resistant to division inhibition by α-factor also fail to form shmoos and are sterile. However, Duntze (personal communication), in an independently isolated sample, found α-factor resistant mutants that retained the ability to shmoo and to mate. The invariable occurrence of sterility in mutants that have lost the ability to shmoo provides circumstantial evidence that shmooing reflects some function that is essential for mating.

If the ability to shmoo in response to a sex factor is essential for mating, then production of that factor should also be essential, and one would expect to find a major class of sterile mutants whose only defect is lack of sex-factor production. Only three sterile mutants are possible candidates for this class (MacKay and Manney, 1974a; J. Rine, personal communication). Accordingly, as discussed by MacKay and Manney (1974a), one must consider the sampling methods used in isolating these mutants to be suspect. It is likely that there are other possible classes of mutants that are not isolated by these screening procedures.

7.2.5 Evidence from studies with isolated components of mixed cultures

(a) *Extracellular factors*
As we have indicated, several of the events that have been observed in mixed cultures appear to involve extracellular sex factors. Bucking-Throm *et al.,* (1973)

have shown that the addition of α-factor to cultures of *a* cells inhibits growth in the culture, decreases the percentage of budded cells, and inhibits initiation of DNA synthesis. Wilkinson and Pringle (1974) have described an *a*-factor which arrests α cells in G1 phase. Another report of an *a*-factor which arrests α cells in G1 and also has some additional properties has been described by Betz, MacKay and Duntze (personal communication). (Analysis of the evidence concerning *a*-factor will be presented below.) Sakurai *et al.* (1976a,b) and Radin (personal communication) have reported that α-factor induces, in *a* cells, the ability to agglutinate with α cells. Duntze (personal communication) has reported that purified α-factor can induce agglutination in cultures of *a* cells by themselves.

(b) *Recovery from α-factor inhibition*

Cells arrested in G1 after exposure to α-factor are not permanently arrested, and recovery is not dependent upon mating. The response of *a* cells to α-factor, and their subsequent recovery, is shown in Fig. 7.4. As the concentration of α-factor is increased the time required for recovery from α-factor inhibition also increases. Several workers (unpublished observations from this laboratory and others) indicate that a decline in α-factor activity begins immediately after addition of α-factor to cultures of *a* cells (see Section 7.3.5(b)). This decline was noted when α-factor was added to cultures of two normal mating-type *a* strains (X2180-1A and XT1177-S47c) but not when α-factor was added to cultures of two sterile mutants (these sterile mutants had been isolated by MacKay and Manney (1974a) from XT1177-S47c) or to a culture of an *a*/α diploid (Chan, personal communication). This appears to indicate a mating type-specific activity which causes a decline in α-factor in the medium, rather than a non-specific effect of normal yeast metabolism. Duntze (personal communication) has found that if small amounts of α-factor are periodically added to *a* cells, recovery is blocked. Therefore, the recovery of *a* cells appears to be due solely to the decline in α-factor in the medium, and this decline is mediated by the *a* cells.

(c) *Cell surface changes during mating*

Upon exposure to α-factor, *a* cells enlarge and elongate asymmetrically, i.e. they shmoo (Duntze, MacKay and Manney, 1970). Lipke, Taylor and Ballou (1976) have studied changes in mating type *a* cells after treatment with partially purified α-factor. (For a discussion of experiments with partially purified α-factor, see below). They have shown that exogenous D-glucose, nitrogen, and phosphate are also required for these changes. One to two hours after the addition of α-factor, *a* cells become more susceptible to lysis by glusulase, a crude mixture of snail gut enzymes which includes glucanases. This change in susceptibility to lysis appears to indicate that there is a structural change in the cell wall. Analysis of chemical changes in the cell wall composition during this time shows that α-factor-treated cells contain more glucan and less mannan than untreated cells, and that the mannan of treated cells contains an increased proportion of shorter side chains and unsubstituted mannose units. The

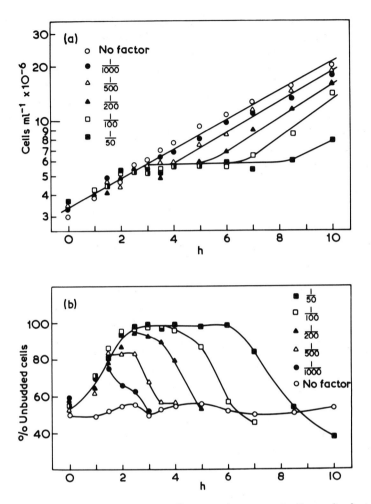

Fig. 7.4 Response of mating type *a* cells to varying concentrations of α-factor. X2180-1A was grown with shaking in synthetic complete medium at 23°C to 3×10^6 cells ml^{-1} and divided into six parallel cultures. α-factor (from a concentrated α-factor preparation of unspecified activity) was added to five of the cultures at 0 h to give final α-factor dilutions of 1/1000, 1/500, 1/200, 1/100, and 1/50. The cultures were shaken at 23°C. At intervals, samples were taken from each culture, diluted in 0.15M NaCl–3.7% formaldehyde. At the end of the experiment samples were counted in a particle counter (Model B; Coulter Electronics Inc.) to determine cell concentration (a) and were observed through a Zeiss phase-contrast microscope to determine the percentage of cells that were unbudded (b). (Courtesy of Russell Chan, personal communication).

glucans are believed to provide the main structural elements of the yeast cell wall, while the manno-proteins serve to immobilize enzymes, agglutination factors, and other cell-surface components. Tkacz and MacKay (personal communication) have detected some specific changes in the surface composition of cells in mating mixtures, which seem to be mediated by diffusible factors. Shmoos that are formed either in mating mixtures or by addition of partially purified a- or α-factor have a greater affinity for fluorescent-labeled concanavalin A. This increased binding is not uniformly distributed over the surface of the shmoo, but is localized at one end of the elongated cell. Asymmetric binding patterns were also seen in stained zygotes, with the increased fluorescence in the middle where the two cells have fused.

There are no biochemical studies on changes in the cell surface of *Saccharomyces cerevisiae* during mating, but some changes have been observed in a cells treated with α-factor, as was mentioned previoulsy. To illustrate the directions in which studies on cell-surface agglutinins may proceed, we will describe the surface-bound agglutinins of another yeast, *Hansenula wingei*.

Hansenula wingei is the only yeast for which a detailed biochemical analysis of cell-bound mating-type agglutinins is available. That the agglutinins are so well characterized is not surprising since the agglutination in mating mixtures of type 5 and type 21 cells (the two mating types) is a rapid, massive reaction. Production of the agglutinins is constitutive in haploid cells but repressed in heterozygous diploids (Crandall and Brock, 1968). The surface agglutinin isolated from type 5 cells is a glycoprotein composed of approximately 85% carbohydrate (mostly mannose), 10% protein and 5% phosphate (Yen and Ballou, 1973; Yen and Ballou, 1974). The phosphate does not appear to be essential since the agglutination reaction of a mutant whose type 5 agglutinin lacks phosphate is not different from a wild-type reaction (Ballou, personnal communication). The type 5 agglutinin appears to be multivalent rather than univalent, in that addition of isolated type 5 agglutinin to type 21 cells produces a strong agglutination reaction. No comparable agglutinin could be isolated from type 21 cells, but a glycoprotein could be isolated from type 21 cells that inhibited the activity of the type 5 factor (Crandall and Brock, 1968). This blocking factor may well be a univalent agglutinin.

No experiments comparable to the work with *Hansenula wingei* have been reported with *Saccharomyces cerevisiae*. One laboratory (Shimoda and Yanagishima, 1975; Shimoda, *et al.*, 1975) has attempted similar studies, but we believe that some of the methods used make them difficult to interpret. In *Hansenula wingei,* since the agglutinins are always present on the surface of the cells, the use of dead cells in an agglutination assay may be permissible. In *Saccharomyces,* where such factors appear to be induced in some strains, and where at least two different mechanisms of aggregation seem to operate, the use of dead cells leads to ambiguity, since it is difficult to determine what is really being measured.

7.3 YEAST SEX FACTORS: A HARD-NOSED LOOK

Comprehensive summaries of the evidence for the existence and the properties of *Saccharomyces* mating factors have been published (Crandall *et al.*, 1976; Crandall, 1977). These summaries have included a number of unpublished results in addition to an admirable survey of the published evidence. The picture that emerges from these studies (sketched above), when taken at face value, is a complex array of diverse physiological responses seemingly being mediated by an equally diverse array of diffusible hormone-like substances. However, a critical look at the experiments which are the basis of this picture dictates that much caution must be excercised in interpreting and summarizing these results.

7.3.1 An appeal for order

It is characteristic of developing areas of investigation, such as this one, that different laboratories will employ different methods, different organisms, and different criteria in an attempt to sketch the general outlines of the subject. One reviewer has suggested that in this case the situation is analogous to blind men describing different parts of the same elephant (Crandall, 1977). A less charitable critic might even question whether we are indeed examining the same elephant. This confusion is in part a direct and natural result of the excitement that has been generated in this area. But it is clearly time for a hard-nosed, critical appraisal of the evidence and of the assumptions on which it has been interpreted. The purpose of this section, therefore, is to scrutinize the evidence that has been presented to date for the purpose of defining and clarifying the important questions that still need to be answered and, most importantly, to suggest a set of rigorous criteria for establishing the existence and biological role of the factors that mediate sexual conjugation in yeast.

In Table 7.2 we have summarized the variety of yeast sex factors, their chemical characteristics, and their functions as reported to date. Taken at face value, these results indicate that there may be as many as nine different molecular species, produced by the two mating types of yeasts, that mediate changes in cell morphology, inhibition of budding and DNA synthesis, agglutinability, changes in the cell wall, and both the induction of and antagonism to the sex factor produced by the opposite mating type. At the same time, one of these factors — α-factor — has been attributed as many as five different activities. We have the ironic situation of proliferating new factors to explain new phenomena, on one hand, and proliferating new activities to be attributed to the same factors on the other hand. Table 7.2 clearly shows the need to distinguish between the biological activities and the molecules which possess those activities.

7.3.2 Mating type specificity

For any physiological effect to be attributed to a sex factor it must, by definition, be mating type-specific. The effect must be observed only when cells of one mating

Table 7.2 Sex factors

Sex factor	Chemistry	Function
1. α-Factor	Four oligopeptides (see Fig. 7.5)	(a) Arrest a cells in G1 (b) Cause a cells to shmoo
2. α-Substance I	Two different oligopeptides; one is the same as α-factor α2	Induce cell-surface agglutinins in a cells
3. α-Hormone	Octanoic acid	
4. a-Factor I	None	Arrest of α cells in G1
5. a-Factor II	Protein	(a) Arrest α cells in G1 (b) Cause α cells to shmoo
6. a-Hormone	None	Expansion of α cells
7. Barrier factor	None	Makes a cells immune to α-factor either by modifying the a cells or by inactivating α-factor

type are exposed to some factor produced only by cells of the opposite mating type. However, to rigorously establish this, it is necessary to employ strains of opposite mating type that differ only at mating-type locus; the strains must be as nearly isogenic as possible. Fortunately many of the investigators working in this area have appreciated this necessity and have employed the nearly isogenic strains X2180-1A(a) and X2180-1B(α), which have been developed and widely distributed by R.K. Mortimer of the University of California, Berkeley.

7.3.3 Multiple-factors versus multiple-functions

There is a diverse array of genetic and physiological evidence — all indirect and circumstantial — for an important, if not essential, role for diffusible sex factors in conjugation. On the strength of this evidence it is reasonable to conclude that there is at least one extracellular sex factor produced by each mating type. Beyond that, the situation becomes more treacherous. When faced with an additional biological activity, one cannot conclude that it represents a new molecular species until one has established the existence of that factor as a new chemical entity. However, on the other hand, one cannot assume that a new activity is associated with a previously characterized molecule until that association has been established chemically. These rigorous criteria have only been satisfied in a limited number of cases. The remainder of the evidence is much softer and mostly circumstantial.

7.3.4 α-Factor

The only yeast sex factor that has been characterized in chemically pure form is α-factor (Duntze et al., 1973; Stotzler and Duntze, 1976; Stotzler et al., 1976). In this work, a group of closely related oligopeptides have been purified to a high degree of homogeneity and their composition and primary structure have been determined (Fig. 7.5). The dilution assay, based on the biological activity of α-factor (Duntze et al., 1973), was used to monitor the activity throughout the purification.

α1: NH_2-Trp-His-Trp-Leu-Gln-Leu-Lys-Pro-Gly-Gln-Pro-Met-Tyr-COOH

α2: NH_2-His-Trp-Leu-Gln-Leu-Lys-Pro-Gly-Gln-Pro-Met-Tyr-COOH

α3: NH_2-Trp-His-Trp-Leu-Gln-Leu-Lys-Pro-Gly-Gln-Pro-Met(SO)-Tyr-COOH

α4: NH_2-His-Trp-Leu-Gln-Leu-Lys-Pro-Gly-Gln-Pro-Met(SO)-Tyr-COOH

Fig. 7.5 Primary structure of four α factors (Stotzler et al., 1976).

These purified oligopeptides produce the characteristic morphological changes in cells of X2180-1A (shmooing) and inhibit their cell division. Accordingly, these two activities can be unequivocally associated with these molecules. However, none of the other characteristics which have been attributed to α-factor from studies with partially purified preparations have been studied in this purified material.

(a) *Partially purified α-factor*

Many of the properties that have been attributed to α-factor, particularly cell surface changes, are based on experiments with preparations that have employed only the first one or two steps in the purification scheme (Duntze et al., 1973). The first step employs adsorption chromatography on Amberlite CG50 ion exchange resin, and the second is a methanol extraction. These steps produce specific activities of 200 and 360 units mg^{-1} dry weight respectively. Compared with specific activities of 70 000 units mg^{-1} dry weight that have been achieved with highly purified α-factor preparations (Duntze et al., 1973), these partially purified preparations are still highly impure, containing appreciably less than 1% α-factor by dry weight (Duntze et al., 1973; Stolzer and Duntze, 1976). Clearly, it is not valid to conclude that the activities found in such preparations are necessarily associated with the peptides in Fig. 7.5.

Procedures for purifying α-factor, however, are sufficiently well established that there is no longer any reason for continuing these ambiguities. Any additional activity that can be assayed can be rigorously associated with these molecules or it can be established to be the result of some other chemical species.

7.3.5 *a*-Factor

No molecular species that can be identified as an *a*-factor has been purified. However, from the indirect evidence we can conclude that there is at least one such factor.

The existence of an *a*-factor analogous to α-factor described above was first inferred from observations on mixed cultures of *a* and α mating-type cells (Bucking-Throm *et al.*, 1973). X2180-1A and X2180-1B were vigorously shaken in a mixed culture in minimal medium. At intervals, shmoos were isolated from the mixture by micromanipulation, and their mating types were determined after they were grown into colonies. In Fig. 7.6 we have plotted the percent of shmoos of each mating

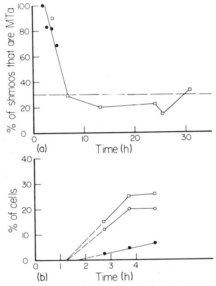

Fig. 7.6 Mating types of shmoos in a vigorously shaken mating mixture. X2180-1A(*a*) and X2180-1B(α) cells were mixed at a ratio of 1:3 in minimal medium. At intervals, samples were streaked on an agar slab and shmoo-shaped cells isolated by micromanipulation. Isolated clones were tested for mating type with standard test strains. (a) Fraction of shmoos isolated that were of mating type *a*. Data from two experiments are represented by different symbols. (b) Fraction of shmoos of either mating type (squares); fraction of shmoos of mating type *a* (open circles); fraction of shmoos of mating type α (closed circles). Note different time scales in (a) and (b).

type against the time at which the samples were taken. The occurence of shmoos of mating type α indicated that the *a* cells must make a substance analogous to α-factor.

Evidence that the production of α shmoos is mediated by a diffusible factor and does not require direct cell—cell contact has been provided by the observation of α shmoos in confrontation tests (see Duntze et al., 1970; MacKay, 1972; MacKay and Manney, 1974a). However, the response of α cells to a-factor by this test is appreciably weaker than in the reciprocal case, which constitutes a routine test for α-factor production.

(a) *Isolation of a-factor*

It has been shown that growth medium from X2180-1A contains stable activities analogous to those associated with α-factor; they produce shmooing and inhibit cell division in X2180-1B (MacKay, 1976).

Limited success in the isolation of this factor has been reported by Wilkinson and Pringle (1974) who were able to demonstrate an activity in culture medium from X2180-1A that specifically inhibits division in mating-type α cells. The inhibition of cell division is not as pronounced as observed in the case of α-factor and is always transitory, lasting no more than about two hours. They were also able to show that both budding and DNA synthesis were depressed in these inhibited cells. However, they failed to observe any morphological changes in the α cells. The activity that produced these effects could be concentrated in the culture filtrates by ultrafiltration but no attempts to characterize it chemically were reported.

More encouraging success in isolating an *a*-factor analogous to α-factor was achieved recently by Betz, MacKay, and Duntze (personal communication). They have succeeded in isolating stable activity in culture medium from X2180-1A, and other *a* strains, that produces both shmooing and inhibition of cell division. This activity is not present in cultures of α cells or a/α diploids. It has been partially purified by ion exchange chromatography and some of its characteristics have been examined. In approximately 200-fold purified preparations the activity appears to be associated with protein that may contain carbohydrate moieties and behaves in gel filtration as either a very large molecule, greater than 600 000 daltons, or an aggregate. The two activities, for morphological change and for cell division inhibition, have not been resolved by the purification method used to date, and could reside on the same molecule. Both activities are resistant to boiling.

The parallels between these activities and those of the well-characterized α-factor preparations, taken together with the evidence from mixed cultures for the existence of two analogous factors, encourage the optimist to conjecture the existence of a simple reciprocal system of extracellular sex factors, perhaps only two, that account for the behavior observed in mating mixtures. However, some of the striking contrasts between *a*-factor and α-factor, in particular the exasperating, elusive behavior of the former, encourages the realist toward more cautious conclusions.

(b) *Other diffusible a-factors*

Hicks and Herskowitz (1976) have reported evidence which suggests the possible existence of another type of diffusible sex factor in *Saccharomyces*. This is an activity which they call the Barrier Effect and is specifically produced by cells of mating-type *a*. It is an antagonist or an inhibitor of α-factor. It appears to be a diffusible activity that irreversibly inactivates α-factor in agar diffusion tests. They have attempted to infer something of the role of this activity in mating by studying a variety of mutants affected in their ability to mate. One of the most intriguing observations that they report is the existence of two mutants that produced this barrier factor but that had been previously shown by MacKay and Manney (1974a) to be defective for production of *a*-factor. It could be that the mutants make reduced levels of *a*-factor, below the detection level of the assay, but sufficient to produce the barrier effect. An alternative interpretation of these results is that the barrier factor and *a*-factor are two different molecular entities. Since these mutants show pleiotropic phenotypes for several mating-type-associated characteristics, however, this explanation is certainly not the only one possible. The final resolution of the question awaits purification of one or both of these activities. The possible existence of an analogous antagonist to *a*-factor by some α cell product is more difficult to examine because of the slow diffusion of *a*-factor in agar.

An additional question raised by Hicks and Herskowitz (1976), which may relate to the regulation of these factors and responses, concerned the ability of *a*/α diploids to produce a barrier effect. They reported that they could not detect a definite barrier effect with *a*/α diploids, but that the assay cells recovered more quickly than controls in these tests. We have taken another approach to this question (Manney and Riley, unpublished). In our experiments, test cells (wild-type *a* cells, sterile *a* cells, or *a*/α diploids) from overnight cultures were collected on Millipore filters and placed on a complex solid medium containing α-factor. After incubation overnight the filters were removed and X2180-1A cells were streaked where the filter had been and then were examined after 6–8 hours for the presence of shmoos. By this method we were able to detect an α-factor removal activity by mating-type *a* haploids, *a*/α diploids, and some of the *a* sterile mutants. Comparable results were obtained between the two studies as far as sterile mutants are concerned. This test clearly showed that *a*/α cells can remove α-factor activity from the medium.

7.3.6 Agglutination-inducing factors

An inordinate amount of the confusion in the literature concerning the multiplicity of the mating-type factors has been generated by papers reporting induction of agglutinability by diffusible factors produced by cells of the opposite mating type (Shimoda and Yanagishima, 1975; Shimoda *et al.*, 1975; Sakurai *et al.*, 1976a, 1976b). With one possible exception however (Sakurai *et al.*, 1976b), none of these studies have employed chemically pure preparations, so it is impossible to conclude whether or not these agglutinability-inducing activities are associated with the same molecules that produce shmooing and cell division inhibition. Sakurai *et al.*, (1976b) have reported that they have purified an oligopeptide having the composition of α-factor, and that it induces agglutinability in *a* cells. However, they do not report whether or not this material produces the morphological changes or G1 arrest of α-factor. Furthermore, the same authors earlier reported an oligopeptide from a different strain of *Saccharomyces* with the same agglutination activity, but with different chemical properties. These contradictions are yet to be resolved.

Some indirect evidence bearing on this question, based on studies of mutants, will be discussed below. At this stage, however, it is only possible to say with confidence that when cells of opposite mating type are incubated in mixed cultures they undergo changes that cause them to agglutinate.

7.4 A COMPREHENSIVE APPROACH TO MATING STUDIES

The genetic and biochemical studies of functions related to mating-type that have proliferated during the past few years have opened a door into a fascinating area of biological phenomena. We know of no other comparable aspect of biological control in a eukaryotic organism that appears so readily accessible to the tools of genetic and biochemical analysis that are available in *Saccharomyces*.

The general progression of events to be explained are those processes that occur in mating mixtures and lead to the formation of a zygote. Recognition of those events is facilitated by the fact that they are a subset of the events that are unique to mixtures of a and α cells. Unfortunately, however, it cannot be assumed that all such events play important roles in the mating process, and even in the case of those that do, the exact nature of the role is not self-evident; the task is to identify the important events and relate them to the overall process. At present we know of some specific events that are clearly mediated by extracellular factors — the cell-division arrest and morphological response to α-factor, for instance — but the biological significance of these events is still speculative. On the other hand, we can describe, at least in operational terms, some biological events — such as sexual agglutination and the courting changes — but the evidence that these are mediated by extracellular factors is still not firm.

We have attempted, during the past several years, to develop a comprehensive approach to study the process of mating. The feasibility of this endeavor derives from considerable innovation and development of ideas, methods, and approaches that have emerged from many different laboratories working in this area. We have tried to collect the best that has been developed by others, and to augment it whenever we can. The key elements of the system that is evolving from these efforts are the following:

(1) A precise, quantitative method for assaying sex factor activity.
(2) Chemical purification to establish the chemical identity of biologically active factors.
(3) A methodology and experimental protocol for defining the acquisition of agglutinability and the process of sexual agglutination under mating conditions.
(4) Genetic and physiological analysis of mating-associated functions in α-factor-deficient mutants in a controlled genetic background. Special emphasis has been placed on deriving genetically marked strains that give strong, well-defined mating-type responses.

We hasten to emphasize that the above is a list of objectives, not necessarily accomplishments.

In the following sections we will present a partial progress report on these efforts. Many of the experiments need further refinement, and much of the genetic and biochemical analysis is still at a relatively primitive stage. However, we feel that these studies already provide an integrated view of the events leading to mating.

7.4.1 Quantitative assays for *Saccharomyces* sex factors

The lack of a simple, precise, and objective assay for sex factor activity appears to remain a major impediment to resolving many questions. The failure of some workers in the field to agree to use even a standard definition of activity units, which can at least be reproduced within a factor of about two, makes it impossible to fully evaluate the significance of some apparently important work. Agreement on the unit of the activity would at least facilitate comparing results reported from different laboratories on the different physiological effects that can be attributed to one or another diffusible sex factor. But perhaps an even more important consideration is the fact that the nature of the assay which is used determines what molecular species is isolated in a purification system.

An indication of the amount of variation that exists is indicated in Table 7.3. This table summarizes the different types of assay that have been used by workers in different laboratories for assaying mating type-specific activities associated with diffusible a and α sex factors. The only assay, and associated definition of the unit of activity, that has had any general use is the one described by Duntze and his co-workers (Duntze *et al.*, 1970; Duntze *et al.*, 1973). This test is based on the morphological response (shmooing) and defines a unit of activity as the lowest concentration that will cause a detectable response under a prescribed set of conditions. Quantitation within a factor of about two is achieved by serial two-fold dilutions of the active material being tested. An analogous test system has been adopted by Betz, MacKay and Duntze for quantitating a-factor activity (personal communication). In this case, however, the factor does not diffuse freely through agar, so the entire procedure must be carried out in liquid suspension. Aside from the details of medium, the general principle of the assay is the same and the unit of activity is defined in exactly the same way.

The virtues of dilution-end-point assays based on shmooing are that they are reasonably objective and reproducible. Particularly in the case of α-factor, an assay based on the specific morphological transformation of a cells, shmooing, (which incidentally is always accompanied by cell division arrest) is the criterion of choice, for these are the activities that have been associated with purified α-factor. It should be noted that assays based solely on G1 arrest can be artifactual, since many different events can produce cell division inhibition. The principal disadvantage of these assays is that they do not provide sufficiently precise quantitation of the activity to answer many experimental questions.

Table 7.3 Assays for sex factor activity

Assay	Reference
α-Factor	
1. G1 arrest (% unbudded cells)	Wilkinson and Pringle (1974); Bucking-Throm et al. (1973)
2. Recovery of cells from G1 arrest	Wilkinson and Pringle (1974)
3. Change in cell morphology (shmoo)	Duntze et al. (1970); Duntze et al. (1973)
4. Binding of fluorescent concanavalin	MacKay (1976)
5. Half-time for lysis by glusulase	Lipke and Ballou (1976)
6. Induction of agglutinability	Sakurai et al. (1975)
a-Factor	
1. G1 arrest (% unbudded cells)	Wilkinson and Pringle (1974)
2. Recovery of α cells from G1 arrest	Wilkinson and Pringle (1974)
3. Change in cell morphology (shmoo)	Bucking-Throm et al. (1973); MacKay and Manney (1974); Manney and Woods (1976)
4. Binding of fluorescent concanavalin	MacKay (1976)

We have developed an extension of this method which adds precision, at the expense of tedium (Fox et al., 1976). This assay makes use of the diffusion of α-factor in agar to achieve a continuous dilution. It was made possible by the recognition of a well-defined end point that can, with a little practice, be ascertained with considerable sensitivity and objectivity. The details of the method are described in the legend to Fig. 7.7. The geometry of the test is designed to permit a simple mathematical solution to the physical diffusion equation so that a predictable relationship between the diffusion distance and the initial concentration of α-factor can be calculated. Fig. 7.7 is a calibration curve for this method. In this figure the units of activity were determined by dilution of a preparation of α-factor which was assayed by the standard dilution-end-point well test method (Duntze et al., 1973). The logarithm of the relative activity, determined by dilution, is plotted against the square of the diffusion distance. The linear relationship that is observed is predicted by the diffusion equation. In addition to providing a more precise method for quantitating the relative α-factor activity in different preparations, this method also provides a way to determine the diffusion coefficient of the α-factor molecule. The principal disadvantage of the method, however, is that it is tedious and slow. In order to satisfy the conditions of the diffusion equation it is necessary to allow the α-factor to diffuse in the agar for 24 hours before the test cells are applied so that the 3-hour period required for the end-point determination is a small portion of the diffusion time. If one shortens the total diffusion time to three hours, (Fig. 7.8) a different but

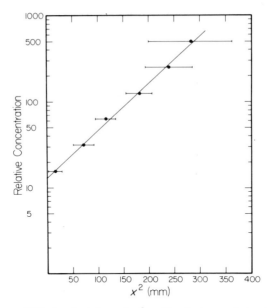

Fig. 7.7 24-hour diffusion test for α-factor. The test sample is placed near the end of a thin slab of synthetic agar medium on a microscope slide and allowed to diffuse into the agar for 21 h. Test cells of mating type *a* are then applied to the surface of the agar, incubated for 3 h, and then examined for morphological response. The relative concentration of α-factor is plotted on a log scale against the square of the diffusion distance, x^2, on a linear scale. The error bars represent the square of the standard deviations of means of at least 12 independent readings.

also useful result is obtained. In this case, a plot of the logarithm of the relative α-factor concentration against x can be fitted to the data. This of course does not satisfy the simple solution to the diffusion equation, as the diffusion is continuing during the time that the cells are responding to the α-factor. However, it does give a precise empirical relationship between the measurable diffusion distance and the α-factor concentration. Consequently, it provides a more rapid quantitative method for assaying α-factor.

Several workers have employed or suggested assay methods based on different physiological effects (Table 7.3). However, at this stage, no assay that is based on an end-point other than the morphological effect or perhaps cell division arrest can be interpreted as an assay for α-factor. They are instead assays for activities that may or may not be related to α-factor. Some of these approaches, however, offer promise of being highly sensitive and useful assays when their validity has been established.

A particularly promising approach has been reported by MacKay (MacKay, 1976). The idea for this assay is based on the observation that shmoos produced in either *a* or α mating type exhibit a greater binding affinity for fluorescent concanavalin A than

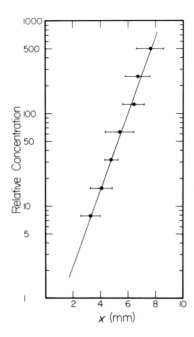

Fig. 7.8 3-hour diffusion test for α-factor. The method is the same as in Fig. 7.7, except that the 24 h diffusion period is omitted. The relative concentration of α-factor is plotted on a log scale against the diffusion distance, x, on a linear scale. End points were read 3–4 hours after cells were applied. The error bars represent standard deviations of means of at least 6 independent readings.

do vegetative cells. A quantitative method based on this observation is still being calibrated (MacKay, personal communication). If, in fact, it proves to be a direct measure of the concentration of the factor to which the test cells were exposed, it could well become the most sensitive, versatile, and the least subjective method available. This is particularly true since the diffusion assay described above would seem to be unsuitable for assaying a-factor, which shows little diffusion in agar.

7.4.2 Methods for purification of factors

The original purification procedures have been published and are reproducible (Duntze *et al.*, 1973; Stotzler and Duntze, 1976). They have been further refined in several laboratories (Duntze, MacKay, Chan and Lipke, personal communication). These methods have proved satisfactory for isolating α-factor from several different α strains, including the mutants (described below) that make reduced levels of α-factor (Fox, personal communication).

7.4.3 Mating type-specific aggregation

Aggregation or agglutination that is specific to mixtures of cells of opposite mating type has been reported to occur under a remarkably diverse variety of conditions. In our own experience, we have observed that there is also a striking degree of variety among laboratory strains of *Saccharomyces* in the extent to which they aggregate. Aggregation brings about intimate contact between cells of opposite mating type even in relatively dilute suspensions, and accordingly seems likely to play an important role in the mating process. For this reason, it has attracted the attention of a number of workers interested in describing the events that lead to sexual conjugation. However, the large volume of experimental data and interpretations that have been published (see Crandall *et al.*, 1976) do not provide a clear, coherent picture of the factors controlling agglutination, but rather leave us confused, with difficult to interpret, conflicting reports. The confusion can be traced to two major sources. First, there is little agreement between the workers in this field on the methods to be used to measure agglutination or aggregation; no two workers employ the same conditions of growth medium, incubation conditions, or criteria for judging the amount of agglutination that has occurred. Second, there is little consistency in the choice of standard strains for these studies, sometimes not even within the same laboratory.

(a) *Methods for observing agglutination*

Methods which have been used by others to estimate degree of aggregation of cells in mating mixtures include visual observation, sedimentation judged by changes in optical density, changes in optical density produced by sonication, changes in viable counts produced by sonication, and changes in the mean particle volume estimated visually. Media in which such studies have been carried out include water, buffer, chemically defined media, and complex media. The effects of these differences have seldom been established by controlled experiments.

We have developed procedures for studying agglutination with strains and experimental conditions compatible with the study of other aspects of mating. Consequently, some of the published methods, especially those which employ boiled cells, arbitrarily chosen strains and undefined media, could not be adopted.

(b) *Quantitation of aggregation*

We have employed three different methods for estimating the extent of aggregation, in addition to visual observation. In all three methods the culture conditions and handling of controls is the same. All experiments are conducted on cultures that have been growing exponentially for at least 12 hours. Cultures are mixed, without any adjustment of cell concentration, when they have reached $1-2 \times 10^6$ cells ml^{-1}. Samples are carefully removed with large-bore pipettes and tested by one of the procedures described below. A control mixture is prepared and tested at the same time. The controls are important because the degree of aggregation of cells growing in pure culture changes, especially after sonication. The controls consist of freshly

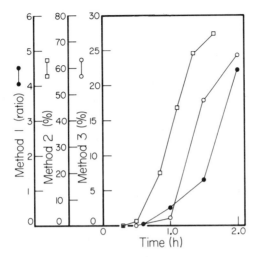

Fig. 7.9 Comparison of methods for estimating aggregation of yeast cells. In each case an exponentially growing mixture of XP300-29B(a) and XP300-26C(α) in synthetic medium (1–2 × 10^6 cells ml^{-1}) was followed. Cells were mixed at time 0. Each point has been corrected for aggregation in a control mixture prepared at the time of sampling.

Method 1 (closed circles): ratio of Coulter Counter counts on sonicated and unsonicated samples; Method 2 (open squares): fraction of cells in aggregates, determined by particle volume analysis (see Fig. 7.11); Method 3 (open circles): sedimentation index, calculated as % cells that sediment into a 5% dextran step gradient in a 20 minute period.

Visible agglutination is first detected in these cultures at approximately 30 minutes.

made mixtures of samples taken from the original pure cultures. These control mixtures are treated in a manner identical to the experimental mixture, and their agglutination values are subtracted.

The three methods for measuring agglutination are:

(1) Particle counts before and after sonication. A quick, simple estimate of agglutination is obtained by counting the number of particles with the Coulter Counter using sonicated and unsonicated samples.

(2) Volume distributions. A more sensitive test for aggregation is obtained by measuring particle size distributions with a Coulter Counter as described in Fig. 7.10.

(3) Sedimentation. Samples are carefully layered on top of a dextran solution. After 20 min of sedimentation the two layers are separated. The lower layer contains the large aggregates, which sediment faster than single cells and small aggregates. Control experiments with unaggregated cells demonstrated that the dextran layer contained less than 2% unaggregated cells.

The kinetics of agglutination in mixtures of a and α mating-type strains that show

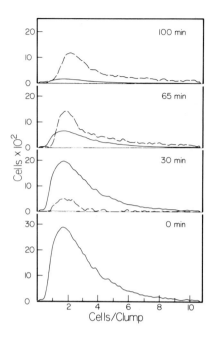

Fig. 7.10 Particle size distribution showing growth of sex-specific cell aggregates. Particle size distributions determined with Coulter Counter and Canberra 8100/e pulse height analyzer. Broken lines show particle size distribution of sex-specific aggregates; solid lines show distributions of control cells. The latter was estimated by normalizing the distribution obtained from a control mixture. The distribution of sex-specific aggregates was then obtained by subtracting the normalized control distribution from the experimental mixture distribution. The volume distribution for a sonicated control sample was used to calibrate the volume scale in units of mean cell volume.

a very strong agglutination reaction was followed by each of these methods. The results are shown in Figs. 7.9 and 7.10. Each of these methods yields a different result, although Methods 1 (particle counts) and 3 (sedimentation) give qualitatively similar results. Method 2 (volume distribution), however, is far more sensitive, and consequently detects aggregation at appreciably earlier times.

(c) *Strain differences*

Another obviously important variable that must be controlled is the genetic background of the yeast strains used. Fig. 7.11 shows the kinetics of agglutination reactions for three *a* strains and three α strains in all possible *a* plus α combinations. Each combination of *a* and α strains followed the same qualitative pattern. After a lag of approximately 30 min there was a linear increase in the ratio of sonicated to unsonicated counts (mean number of cells per aggregate). The lag period appears to be

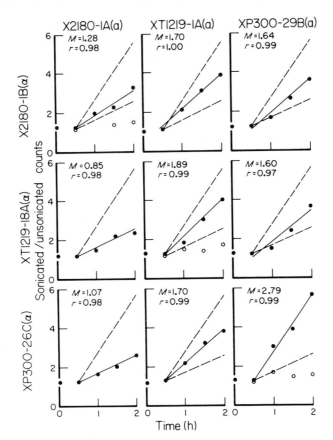

Fig. 7.11 Kinetics of agglutination of some standard strains. Agglutination was measured by counting sonicated and unsonicated samples. Closed symbols show results for each experimental mixture; open symbols are for control mixtures. Broken lines indicate the extreme values observed among all of these experimental mixtures. The highest values came from the mixture of XP300-26C and XP300-29B, while the lowest value came from the mixture of X2180-1A and XT1219-18A. They are included in each graph as references. The slope of the linear regression fit to each set of values (M) and the correlation coefficient (r) are given on the graphs. Genotypes of strains: X2180-1A: *a gal2*, X2180-1B: α *gal2*, XP300-29B: *a his6 lys1 trp5 ade2 gal2*, XP300-26C: α *thr4 his6 lys1 gal2*, XT1219-1A: *a trp1 his2 ade1 gal1*, and XT1219-18A: α *trp1 his2 ade1 gal1*.

essentially the same for all combinations. The slope of the linear increase after the lag period, however, varied considerably. All combinations which include the widely used strain X2180-1A show significantly lower agglutination rates; the combination of XP300-26C and -29B shows a strikingly higher rate than any other combination. These results are confirmed by visual observations.

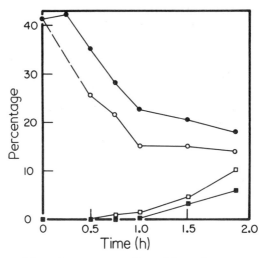

Fig. 7.12 Composition of aggregates collected by sedimentation into 5% dextran. Sonicated samples from sedimentation experiment (Fig. 7.9) were scored by microscopic examination for frequencies of zygotes and budded cells. Closed circles: % budded cells in total mixture; open circles: % budded cells in aggregates; closed squares: % zygotes in total mixture; open squares: % zygotes in aggregates.

(d) *Do only unbudded cells aggregate?*

The conclusion that only unbudded cells participate in aggregation is based on rather indirect evidence (Campbell, 1973). To get a more direct answer to this question, we examined the cells in isolated aggregates produced by agglutination of XP300-26C and XP300-29B. Aggregates were collected, free of unaggregated cells, by the sedimentation method described above. In Fig. 7.12, the composition of the aggregates collected is compared with the overall composition of the cell mixture, as a function of time of incubation of the mixture. Clearly, there is a slight enrichment for unbudded cells in the aggregates but certainly there are many budded cells that are able to form aggregates. In the actual analysis of the samples, we distinguished between the cells having large buds and those having small buds, but no significant segregation on this basis was observed. The results in this figure also show that while there is a slightly higher frequency of mating in the aggregates there is still a significant amount of mating in the unaggregated fraction of the population. We must however, exercise some caution in generalizing on these results. Fig. 7.9 shows that this method for estimating agglutination is relatively insensitive. Evidently small aggregates of cells that are formed early in the process do not sediment at a sufficient rate to be collected by this method. Consequently, we cannot conclude from these results that the very early stages of aggregation do not show a stronger preference for unbudded cells (Campbell, 1973).

(e) *Agglutination and agglutinability*

A distinction can be made between the agglutination reaction — the actual aggregation of cells of opposite mating type — and the acquisition of agglutinability — which implies a change in the surface of a cell that makes it stick to cells of the opposite mating type. The case for this distinction has been argued (Sakai and Yanagishima, 1972, Radin, personal communication) from a variety of evidence. The experiment shown in Fig. 7.13 was designed to verify these results in our own system by answering

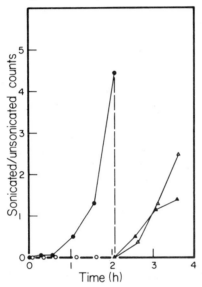

Fig. 7.13 Induction of agglutinability. Agglutination was followed in mixtures of XP300-29B and XP300-26C with cycloheximide, 10 mg l^{-1}, (open circles) and without cycloheximide (closed circles). After two hours, the mixture without cycloheximide was sonicated and divided into two portions; cycloheximide was added to one (open triangles) but not to the other (closed triangles).

two questions. First, is the lag period for the onset of agglutination a reflections of a period during which the cells are undergoing changes that are prerequisite for agglutination? Second, do these changes require protein synthesis? Agglutination was followed in two cell mixtures, one that contained cycloheximide and one that did not. After two hours the mixture without cycloheximide was sonicated to disrupt the aggregates that had formed, and divided into two samples. Cycloheximide was added to one, and agglutination was again followed in both. Similar results were obtained when cryptopleurine or canavanine was used to inhibit normal protein synthesis.

This experiment can be interpreted to show that agglutinability of at least one of these strains must be acquired during the time that they are incubated together, and that the process by which this agglutinability is acquired is sensitive to inhibition by cycloheximide and, therefore, most likely involves protein synthesis. Once this change has occurred, however, the actual formation of aggregates does not apparently require protein synthesis.

This type of experiment defines these two processes in relatively straightforward operational terms. The process that leads to cells becoming agglutinable is inhibited by cycloheximide (or cryptopleurine, or canavanine). It requires active cell metabolism. If cells are taken from growing cultures and mixed, they agglutinate only if they are under conditions that would permit growth. However, after being mixed for about two hours under metabolically favorable conditions, the ability of cells in mixed culture to aggregate becomes independent of active metabolism; they agglutinate in the presence of inhibitors and even in buffer. It has been reported that they will aggregate even after being boiled (Sakai and Yanagishima, 1972) but under such extreme conditions it is difficult to be confident that the same process is being observed.

(f) *Initiation of aggregation*

The agglutination experiment described above employed a relatively insensitive method based on particle counts for detecting and measuring cell aggregates. The comparison of methods in Fig. 7.9 shows that this method does not detect small aggregates formed during the initial stages of the process. It appears likely, as discussed in Section 7.2.2(b), that two distinct stages in the agglutination process can be distinguished. Experiments to be discussed below, using mutants with altered agglutination behavior, suggest that these stages are under independent genetic control. Ambiguity with respect to which of these stages is measured by a particular method may account for some apparently conflicting reports in the literature.

7.4.4 Studies with α-factor-deficient mutants

From the earliest observations that some diffusible factors were detectable in mating mixtures, it has been a virtually implicit assumption that these factors play an important, if not essential, role in the mating process (Levi, 1956). Yet, in excess of two dozen papers have been published on the subject in the past 20 years, and there is still no direct evidence, that we know of, establishing an essential role for these factors. There is abundant indirect evidence, but the absence of direct evidence is becoming an embarrassment. Furthermore, it has led to an excessive amount of speculation and extrapolation from indirect evidence, which may not be justified.

The definitive experiments that need to be done are obvious; they are the mainstay of biochemical genetics. One first isolates a mutant whose sole defect is the inability to make α-factor (or a-factor). The phenotype of such a mutant then defines the biological role of the missing factor. The validity of the characterization would be rigorously established if, when α-factor was supplied from another source, all the mutant phenotypic characteristics were reversed. Specifically, if α-factor is essential for mating, it should be possible to find among sterile mutants some that are not sterile in the presence of added α-factor. But such a mutant has not been reported. A candidate for this type of mutant would be expected to have only a single defect. In fact, we know of only three sterile mutants that don't have at least one other phenotypic defect in addition to the inability to make a sex factor. We have tried unsucessfully to mate one of these mutants in the presence of a normal mating mixture where all possible mating factors would be present at near-optimal concentrations.

It is possible, of course, that the role of these factors in mating is not essential. They may only increase the efficiency of mating by promoting agglutination and/or synchrony. In that case, mutants that specifically lack the ability to make a sex factor would not be sterile.

(a) *Isolation of mutants*

As an approach to the general question of the biological role of α-factor, we used the inability to make α-factor as the primary criterion for selecting mutants. The system that we used is based on a method devised by Fink and Styles (1972) and modified by L. Hartwell (personal communication) to look for mutants that produce high levels of α-factor. Mutagenized cells of mating-type α are plated (about 100 per plate) with a large excess (about 100 000) of mating-type *a* cells that are temperature-sensitive for protein synthesis ($ts185$). The plates are incubated at the restrictive temperature of the *ts* mutant (36°C) to allow the α cells to form colonies, and then are incubated at the permissive temperature (22°C) to allow the *a* cells to form a confluent background growth. The α-factor in the medium around the α colonies inhibits this background growth, leaving a clear zone. Mutants that are deficient in α-factor lack this clear zone. The method is quite efficient, and has allowed us to isolate and characterize a number of such mutants.

These mutants have been the source of a number of surprises. The first surprise was their ability to mate. Approximately 75% of them had no obvious defect in mating as judged by the standard nutritional complementation test. However, it soon became apparent that these non-sterile mutants were not altogether defective in α-factor. While they are negative by the relatively insensitive zone test, they still make reduced levels of α-factor that are detectable by other methods. We decided to concentrate a large portion of our efforts on these mutants as they seemed to offer a chance to observe the role of α-factor in mating.

Further characterization led to further surprises. Most of these mutants have additional phenotypic abnormalities. Some have abnormal morphologies, indistinguishable in appearance from the shmoos normally found only in the presence of opposite mating-type factors. Some, but not all are deficient in the ability to agglutinate with cells of mating type *a*. Many of them frequently sired spores which were inviable. This poor spore viability effect could not be reversed by mating the mutants in the presence of normal α-factor producers.

Among the mutants that gave good spore viability there was further attrition because the phenotype (presence or absence of a zone of inhibition on a background of *a* cells) was not transmitted with fidelity. While some produced α progeny with either no zone or a normal wild-type zone, others segregated α progeny that produce zones of varying sizes. Only the former type were studied further. But even some of these are subsequently proving to be the result of more than one mutation, when the segregation of different phenotypic characteristics is monitored.

(b) *Genetics*

We have data on the behaviour of three α-factor deficient mutants that stili appear to be single-gene mutations. They are designated PD6, PD7, and PD23, and they are derived from the normal α strain XP300-26C. Table 7.4 summarizes the results of

Table 7.4 Segregation of α-factor production and mating-type in tetrads from three α-factor deficient mutants

Mutant	Parental ditype	Non-parental ditype	Tetratype
PD6	6	7	17
PD7	1	3	18
PD23	7	4	18

tetrad analysis of these mutants. The values are the parental ditype (non-cross-over), non-parental ditype (double cross-over), and tetratype (single cross-over) classes, relative to the mating-type locus. The presence of non-parental ditype classes demonstrates that none of these mutations is linked to the mating-type locus. In these tests the mutant phenotype was scored by a semiquantitative estimate of α-factor (confrontation test, see Duntze, MacKay and Manney, 1970). Although the phenotype can only be determined in the α mating-type segregants, further genetic crosses to confirm the presence of the mutant or normal allele in mating-type *a* segregants were done with selected *a* segregants. Two of the mutants, PD6 and PD7, grow with abnormal cell morphologies. In every case the abnormal morphology segregates with the α-factor deficiency in mating-type α cells, and is not expressed in mating-type *a* cells.

(c) *Production of α-factor*

Our plan was to correlate altered mating characteristics with reduced α-factor levels. Production of α-factor was estimated by two methods, both of which indicate that these mutants make no more than 15% as much as their parent α strain (XP300-26C). These methods and results are summarized in Table 7.5. We do not judge the differences among the three mutants to be significant.

(d) *Mating*

The standard complementation test used to detect mating is extremely sensitive. It does not distinguish between the full wild-type frequency of mating and frequencies as low as 1% of wild-type. Consequently a quantitative estimate of mating efficiency was necessary to judge whether the mutants had any mating defect at all. This was also

Table 7.5 α-factor activity of deficient mutants

Strain	Diffusion distance (mm)*		Dilution endpoint†
	P.J.	T.M.	
XP300-26C(α)	5.3	5.4	64
XP300-29B(a)	0	0	0
PD6	0.4	0.4	
PD7	0.8	0.8	8
PD23	0.3	0.3	8

* Diffusion distance is a relative measure of α-factor production in growing cultures. A streak of cells is grown on synthetic agar medium for 24 hours. A thin streak of mating-type a test cells is then streaked at right angles to this streak. After 4 hours the diffusion endpoint is measured. The two sets of values shown are independent readings by different technicians.
† Dilution endpoint was determined on 10-fold concentrated culture medium. Cells were grown in 50 ml of synthetic medium for 49 hours. The medium was collected by centrifugation and concentrated 10-fold in a flash evaporator at 40°C. The α-factor activity in 2-fold serial dilutions was determined by well tests.

done by two methods. One, in which the mating mixture is collected and incubated on the surface of a Millipore filter, approximates the mating conditions in the standard complementation test, where the cells are mixed on the surface of an agar plate. The second method, in which the mixture is shaken in liquid medium, was chosen to provide a more sensitive test for small differences in mating ability, and to emphasize any role of agglutination. The results using both methods are shown for the parent strain and the three mutants in Table 7.6.

Once again, the picture is not simple. If we look at the tests on the filters we see no effect of the mutations. Yet if we look at the results from the shake cultures, we see a definite effect, but it does not seem to correlate with the α-factor assays. All three mutations have approximately the same reduction of α-factor activity, and yet PD7 shows a substantial reduction in mating efficiency, PD6 only a slight reduction, and PD23 shows no reduction at all. Apparently the effects must be explained by qualitative differences in the α-factor, or by other changes in these strains associated with these mutations.

To distinguish among several possible explanations, we conducted the experiment summarized in Table 7.7. In this experiment a three-way mixture containing one a strain and two α strains is made, with the a and one α carrying the recessive mutation for canavanine resistance. This marker is unlinked to mating-type and has no effect on mating by itself; it merely allows us to distinguish which of the two α strains was incorporated into each diploid. With this system we can accurately measure the relative mating efficiencies of two α strains in the same culture. When

Table 7.6 Mating efficiency of α-factor deficient mutants

α Strain	% diploids with XP300-29B	
	On filters	Shake cultures
XP300-26C	67	70
PD6	73	49
PD7	68	3
PD23	87	68

Mixtures of a and α strains prepared in synthetic medium. One portion was collected on a Millipore filter (0.45 μm) and the filter was incubated, face up, on synthetic agar medium. After 24 hours, filters were resuspended in water. The percent of diploids in each mixture was determined by plating on selective and non-selective media.

Table 7.7 Mating competition between normal and α-factor deficient mutants

Strain	% WT Diploids	No. expts.
WT	51 ± 9	3
PD6	88 ± 15	1
PD7	86 ± 3	3
PD23	40 ± 5	3

Mixtures contained equal numbers of wild-type a, wild-type α and mutant α cells at a total concentration of approximately 1×10^6 cells ml^{-1}. The a strain and one of the α strains carry the recessive mutation conferring resistance to canavanine. The mixture, in synthetic medium, was sonicated and then shaken at 200 rev/min, 30°C, for 2 hours. Aliquots were sonicated and plated on non-selective medium, on medium selective for all diploids and on medium selective for canavanine-resistant diploids. From the colony counts the fraction of diploids that contain the wild-type α strain ('WT Diploids') could be calculated. In the WT control experiment equal numbers of canavanine-sensitive and resistant wild-type α cells were used.

one of them is the normal strain and one is the mutant, the experiment measures their relative, and competitive, mating efficiencies with the same a cells, which have been exposed to the α-factor produced by both strains together. If the only effect of the mutation is to reduce the amount of α-factor produced, there should be no distinction between the mutant and the normal α strain; 50% of the diploids should be wild-type. However, Table 7.7 shows that PD6 and PD7 do not mate as well as the normal strain under these conditions, while PD23 consistently mates slightly better. It becomes increasingly difficult to explain these mutants in terms of a simple reduction in the amount of α-factor produced. This experiment seems to indicate

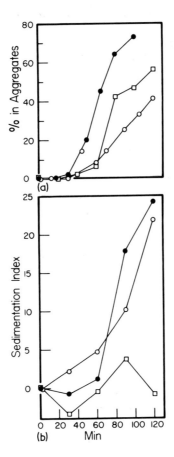

Fig. 7.14 Aggregation and sedimentation of α-factor-deficient mutants and wild-type α cells with XP300-29B. Closed circles: XP300-26C; open circles: PD23; open squares: PD7. (a) Growth of aggregates estimated by particle volume analysis (Method 2). (b) Change in sedimentation index (Method 3).

that their ability to mate with *a* cells that have been exposed to normal levels of α-factor is altered.

(e) Agglutination
Another phenotypic difference that could be related to mating efficiency is the ability to agglutinate. The normal α strain, XP300-26C, from which these mutants were derived, and the strain XP300-29B produce an unusually strong, visible agglutination reaction in mixed culture. This strong visible reaction was apparent in mixtures containing either PD6 or PD23, with the *a* strain, but not with PD7. Accordingly, we have studied cell aggregation quantitatively in these strains by several methods. The most informative results are summarized in Fig. 7.14.

Fig. 7.14a represents the fraction of cells in a culture that are aggregated into multicell particles that can be counted in a Coulter Counter with a 100 μm aperture. As noted above, this method measures small aggregates formed at the onset of agglutination. Fig. 7.14b summarizes the changes in the sedimentation of mixed cultures measured by particle counts. While this method is also controlled so that it only measures sex-specific changes, it is relatively insensitive to small aggregates. For the most part it gives information about the growth of large aggregates that occur later.

The results in Fig. 7.14 strongly suggest that the two methods measure different stages of the agglutination process that are under independent genetic control. The formation of large, rapidly sedimenting aggregates (Fig. 7.14b) correlates with the visible agglutination seen in the cultures, but the initial formation of small aggregates (Fig. 7.14a) does not.

(f) *Summary*

We have isolated three mutants whose mutations segregate as single-gene differences from the parent strain, and that were isolated by the same criterion. Each retains 10–15% of the parent activity for α-factor. One of the three has lost the ability to agglutinate with a normal *a* cell (or perhaps has lost the ability to induce the *a* cell to agglutinate), but all are able to initiate small sex-specific aggregates. All three retain the normal efficiency for mating in mixtures where close cell–cell contact is provided, but two show different degrees of impaired mating ability in dilute shake cultures. However, this impairment not only fails to correlate with α-factor activity, but rather, correlates with an apparent impairment of ability to mate with normal *a* cells even in the presence of normal α cells. Finally, two of the mutants have abnormal morphologies during growth that make them appear as though they were responding to *a*-factor. In short, these mutants, selected only for impaired α-factor production, display a variety of pleiotropic characteristics that appear to be related to mating type, and yet none of these abnormalities can be unambiguously correlated with the deficiency in α-factor.

A pessimistic moral to this still incomplete story could be borrowed from Gandalf, a wizard in *The Lord of the Rings*: 'He that breaks a thing to find out what it is has left the path of wisdom'. (Tolkien, 1965). But we remain optimists, and prefer to believe that there is something to be learned from this puzzle.

7.4.5 Courting revisited

While much of the indirect evidence, summarized above, suggests that α-factor production and response play a role in the events that lead to zygote formation, the same evidence can also be interpreted to demonstrate an important role of direct cell–cell contact in the courting process. Hartwell's mating orgy did not distinguish these two roles, as the pair of cells that did the courting were the same cells which mated first. Therefore, we have experimented with a slightly different orgy (Fig. 7.15),

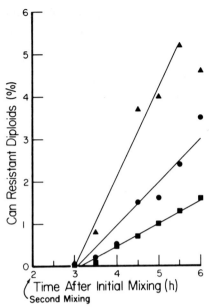

Fig. 7.15 Kinetics of zygote formation in our own orgy. Two mating mixtures were prepared in liquid complex medium from exponentially growing cultures. One mixture contained canavanine-resistant (recessive) a cells and canavanine-sensitive (dominant) α cells and the second was the reciprocal combination. The strains of the same mating type were isogenic except at the $can1$ locus. The mixtures were incubated on a shaker at 23°C. After 2 hours, we sonicated the mixtures and prepared three new mixtures. Mixture a (triangles) was prepared from part of the first mixture plus some fresh canavanine resistant α cells, Mixture b (circles) was prepared from part of the second mixture plus some fresh canavanine resistant a cells, while Mixture c (squares) was prepared by mixing aliquots from each of the initial cultures. The new mixtures were incubated on a shaker at 23°C and samples were removed at various intervals, sonicated, and plated on minimal medium containing canavanine, to count the canavanine-resistant diploids, and on non-selective medium, to count the total number of viable cells.

in which we observe the mating of cells courted by surrogates. We set up two reciprocal mating mixtures, each having one canavanine-resistant (recessive) and one canavanine-sensitive (dominant) strain. In each of these mixtures a and α cells could interact and even mate, but no canavanine-resistant pairs could be formed. After two hours we sonicated these mating mixtures and made three new mixtures. One was made by mixing uncourted canavanine-resistant α cells with part of the mating mixture that had sensitive α cells; the second was made with uncourted resistant a cells added to the reciprocal mixture; the third was a mixture of the two initial mating mixtures. The formation of canavanine-resistant diploids, which could occur in all three mixtures, is shown in Fig. 7.15. The production of resistant diploids begins at approximately the same time in all three cultures. The different rates merely reflect the differences in relative cell numbers. It does not appear that the canavanine-resistant a and α cells which had been courted for two hours (mixture c), had any advantage over the uncourted resistant a and α cells in the other two mixtures. The results of this experiment and Hartwell's are not necessarily contradictory, as there is an important difference in the procedures. In his experiment, the courted cell pairs were together during the courtship whereas in ours they were courted by surrogates, so there was no opportunity for direct cell—cell interaction. There is, in fact, one explanation that is consistent with both results. The interactions that are essential for mating may not be cell cycle synchronization, which may only be an incidental event. The courting that is essential may be direct cell—cell interactions between pairs of cells resulting from formation of a sonication-resistant union that precedes the appearance of visible zygotes. This would mean that courting is not merely a mass interaction of cells with extracellular factors in a mating culture, but also involves an intimate interaction between pairs of cells. In support of this thesis, Fig. 7.3d shows that, in mating mixtures, sonication-resistant plating units that ultimately become diploid cells precede the formation of visible zygotes.

7.5 SUMMARY

There are some obvious unanswered questions, including, but not limited to, the possible allelism of these mutations, and possible qualitative differences in the α-factor that they produce. We are addressing these questions, and others, by further genetic analysis and by attempting to purify the α-factors produced by these strains. But these are long and tedious jobs, and to relieve the monotony we indulge in an occasional speculation.

We find it instructive to play devil's advocate and consider the hypothesis that α-factor plays no essential role in mating under the usual laboratory conditions, but that direct cell—cell contact does.

The possible role of α-factor as an inducer of agglutinability in a cells would not appear to be essential to mating. The initial phase of agglutination seems to be sufficient to establish intimate a—α cell-pair formation, and apparently does not depend on any diffusible factor. The behavior of the α-factor-deficient mutants is quite consistent with this notion. While there is some correlation between the ability to mate and the ability to form large aggregates, neither of these is apparently correlated with the amount of α-factor produced by these mutants.

Consideration of all the characteristics of these mutants suggests that the changes in mating and agglutination are reflections of changes in the cell surface. However, the fact that these mutants were selected solely on their impaired ability to produce α-factor, strongly implies an association — either regulatory or structural — among all of these characteristics. The occurrence of abnormal morphology as one of them strongly suggests a structural association.

This leads to a model of a structural complex, determined by several genes, which is an integral part of the cell envelope. This complex would mediate many aspects of mating. By this model, α-factor would be considered as one functional subunit which is released into the medium. The pleiotropic phenotypes of the mutants could then be understood as the consequences of alteration of different genes whose products contribute to this complex.

Another possible prediction of this scheme would be that α-factor, and other sex factors, exists as a surface-bound activity as well as the extracellular form that has been characterized. Consequently, it is possible that all of the effects on cells of the opposite mating type attributable to different sex factors are also mediated by surface-bound factors whenever direct cell–cell contact is established. In this case, the diffusible factors would not be essential. Their natural biological function could well be to promote greater aggregation under conditions where direct cell–cell contact was infrequent.

REFERENCES

Bilinski, T., Jachymczyk, W., Litwinska, J., Zuk, J. and Gajewski, W. (1974), Mutual inhibition of DNA synthesis in a- and α-cells of *Saccharomyces cerevisiae* during conjugation. *J. gen. Microbiol.*, **82**, 97–101.

Bucking-Throm, E., Duntze, W., Hartwell, L.H. and Manney, T.R. (1973), Reversible arrest of haploid yeast cells at the initiation of DNA synthesis by a diffusible sex factor. *Exp. Cell Res.*, **76**, 99–110.

Campbell, D.A. (1973), Kinetics of the mating-specific aggregation in *Saccharomyces cerevisiae*. *J. Bact.*, **116**, 323–330.

Crandall, M. (1977), Mating-type interactions in micro-organisms. In: *Receptors and Recognition* (Cuatrecasas, P. and Greaves, M.F., eds), (Series A) Vol. 3, Chapman and Hall, London.

Crandall, M.A. and Brock, T.D. (1968), Molecular basis of mating in the yeast *Hansenula wingei*. *Bact. Rev.*, **32**, 139–163.

Crandall, M., Egel, R. and MacKay, V.L. (1976), Physiology of mating in three yeasts. In: *Advances in Microbial Physiology* (Rose, A.H. and Tempest, D.W. eds), Academic Press, London and New York (in press).

Duntze, W., MacKay, V. and Manney, T.R. (1970), *Saccharomyces cerevisiae*: A diffusible sex factor. *Science*, **168**, 1472–1473.

Duntze, W., Stotzler, D., Bucking-Throm, E. and Kalbitzer, S. (1973), Purification and partial characterization of α-factor, a mating-type specific inhibitor of cell reproduction from *Saccharomyces cerevisiae*. *Eur. J. Biochem.*, **35**, 357–365.

Fink, G.R. and Styles, C.A. (1972), Curing of a killer factor in *Saccharomyces cerevisia*. *Proc. natn. Acad. Sci., U.S.A.*, **69**, 2846–2849.

Fox, M.H., Meade, J.H. and Manney, T.R. (1976), Diffusion bioassay for α-factor (abstract) *Genetics*, **83**, 523.

Hartwell, L.H. (1973), Synchronization of haploid yeast cell cycles, a prelude to conjugation. *Exp. Cell Res.*, **76**, 111–117.

Hartwell, L.H. (1974), *Saccharomyces cerevisiae* cell cycle. *Bact. Rev.*, **38**, 164–198.

Hicks, J.B. and Herskowitz, I. (1976), Evidence for a new diffusible element of mating pheromones in yeast. *Nature*, **260**, 246–248.

Levi, J.D. (1956), Mating reaction in yeast. *Nature*, **177**, 753–754.

Lipke, P.N., Taylor, A. and Ballou, C.E. (1976), Morphogenic effects of α-factor on *Saccharomyces cerevisiae a* cells. *J. Bact.*, **127**, 610–618.

MacKay, V.L. (1972), Genetic and functional analysis of mutations affecting sexual conjugation and related process in *Saccharomyces cerevisiae*. Ph. D. Thesis, Case Western Reserve University.

MacKay, V.L. (1976), Mating-type specific pheromones as mediators of sexual conjugation in yeast. *Proc. Soc. Dev. Biol.*, **35**, (in press).

MacKay, V. and Manney, T.R. (1974a), Mutations affecting sexual conjugation and related processes in *Saccharomyces cerevisiae*. I. Isolation and phenotypic characterization of non-mating mutants. *Genetics*, **76**, 255–271.

MacKay, V. and Manney, T.R. (1974b), Mutations affecting sexual conjugation and related processes in *Saccharomyces cerevisiae*. II. Genetic analysis of non-mating mutants. *Genetics*, **76**, 273–288.

Manney, T.R. and Woods, V. (1976), Mutants of *Saccharomyces cerevisiae* resistant to the α mating type factor. *Genetics*, **82**, 639–644.

Mortimer, R.K. (1959), Invited discussion. *Rad. Res.*, Sup. **1**, 394–402.

Sakai, K. and Yanagishima, N. (1972), Mating reaction in *Saccharomyces cerevisiae* II. Hormonal regulation of agglutinability of *a* type cells. *Arch. Mikrobiol.* **84**, 191–198.

Sakurai, A., Tamura, S., Yanagishima, N., Shimoda, C., Hagiya, M. and Takao, N. (1974), Isolation and identification of a sexual hormone in yeast. *Agric. biol. Chem.*, **38**, 231–232.

Sakurai, A., Tamura, S., Yanagishima, N. and Shimoda, C. (1975), Isolation of a peptidyl factor controlling sexual agglutination in *Saccharomyces cerevisiae*. *Proc. Jap. Acad.*, **51**, 291–294.

Sakurai, A., Tamura, S., Yanagishima, N. and Shimoda, C. (1976a), Isolation and chemical characterization of the peptidyl factor inducing sexual agglutination in *Saccharomyces cerevisiae*, *Agric. biol. Chem.*, **40**, 255–256.

Sakurai, A., Tamura, S., Yanagishima, N. and Shimoda, C. (1976b) Structure of the peptidyl factor inducing sexual agglutination in *Saccharomyces cerevisiae*. *Agric. biol. Chem.*, **40**, 1057–1058.

Shimoda, C., Kitano, S. and Yanagishima, N. (1975), Mating reaction in *Saccharomyces cerevisiae*. VII. Effect of proteolytic enzymes on sexual agglutinability and isolation of crude sex-specific substances responsible for sexual agglutination. *Antonie van Leeuwenhoek J. Microbiol. Serol.*, **41**, 513–519.

Shimoda, C. and Yanagishima, N. (1975), Mating reaction in *Saccharomyces cerevisiae*. VIII. Mating-type-specific substances responsible for sexual cell aglutination. *Antonie van Leeuwenhoek, J. Microbiol. Serol.*, **41**, 521–532.

Stotzler, D., and Duntze, W. (1976), Isolation and characterization of four related peptides exhibiting α-factor activity from *Saccharomyces cerevisiae*. *Eur. J. Biochem.*, **65**, 257–262.

Stotzler, D., Klitz, H. and Duntze, W. (1976), Primary structure of α-factor peptides from *Saccharomyces cerevisiae*. *Eur. J. Biochem.*, **69**, 397–400.

Tolkien, J.R.R. (1965), *The Fellowship of the Ring,* Ballantine Books, New York.
Wilkinson, L.E. and Pringle, J.R. (1974), Transient G1 arrest of *S. cerevisiae* cells of mating type *alpha* by a factor produced by cells of mating type *a*. *Exp. Cell Res.,* **89,** 175–187.
Yanagishima, N. (1969), Sexual hormones in yeast. *Planta,* **87,** 110–118.
Yen, P.H. and Ballou, C.E. (1973), Composition of a specific intercellular agglutination factor. *J. biol. Chem.,* **248,** 8316–8318.
Yen, P.H. and Ballou, C.E. (1974), Partial characterization of the sexual agglutination factor from *Hansenula wingei* Y-2340 type 5 cells. *Biochemistry,* **13,** 2428–2437.
Zuk, J., Zaborowska, D., Litwinska, J., Chlebowicz, E. and Bilinski, T. (1975), Macromolecular synthesis during conjugation in yeast. *Acta microbiol. Polon.,* **7,** 67–75.

8 Mating Interactions in *Chlamydomonas*

URSULA W. GOODENOUGH

8.1	Introduction	page	325
8.2	Flagellar agglutination		326
	8.2.1 Morphology of the process		326
	8.2.2 Membrane participation in agglutination		327
	8.2.3 Polypeptide composition of membranes and mastigonemes		328
	8.2.4 Protein participation in *C. reinhardi* agglutination		329
	8.2.5 Carbohydrate role in *C. reinhardi* agglutination		330
	8.2.6 Protein and carbohydrate participation in *C. moewusii* agglutination		332
	8.2.7 The glycosyltransferase hypothesis		334
	8.2.8 Acquisition of agglutinability		335
	8.2.9 Loss of agglutinability by mutation		336
	8.2.10 Lectin agglutinability in *C. reinhardi*		337
	8.2.11 Lectin agglutinability in *C moewusii*		338
	8.2.12 Antibodies against the flagellar surface		339
	8.2.13 The requirement for a living cell		340
	8.2.14 Overview and speculations		341
8.3	Cell fusion		342
	8.3.1 The fusion apparati		342
	8.3.2 Cell wall lysis		342
	8.3.3 Mating structure activation		343
	8.3.4 Zygotic cell fusion		344
	8.3.5 Mutations affecting zygotic cell fusion		345
8.4	The flagellum-to-gamete and zygote-to-flagellum signals		345
	8.4.1 The flagellum-to-gamete signal		345
	8.4.2 The zygote-to-flagellum signal		346
	References		347

Acknowledgements

Critical readings of this review by Drs Kenneth Bergman, Charlene Forest, Donald Weeks, and Lutz Wiese are gratefully acknowledged, as are the grants from the N.I.H. and the Maria Moors Cabot Foundation for Botanical Research at Harvard University which supported the research reported from my laboratory.

Microbial Interactions
(*Receptors and Recognition,* Series B, Volume 3)
Edited by J.L. Reissig
Published in 1977 by Chapman and Hall, 11 New Fetter Lane, London EC4P 4EE
© Chapman and Hall

8.1 INTRODUCTION

The gametic mating reaction of the biflagellate *Chlamydomonas reinhardi* has all the attributes of true theater. When gametes of the two mating types (mt^+ and mt^-) are mixed, they instantaneously adhere to one another so that, as one watches through the microscope, a field of randomly swimming cells is converted into a field containing large clumps of agglutinating cells. Within each clump the gametes adhere to one another by the tips of their long ($\simeq 12\ \mu m$) flagella and, thus anchored, jerk and bob about in apparent frenzy. This scene is ultimately resolved when individual pairs of mt^+ and mt^- gametes sort themselves out and fuse with one another to form single quadriflagellated zygotes. Fusion requires, first, that the gametes slip out of their walls, a feat accomplished by a lytic activity ('endolysin') released by the gametes, which digests the walls at their anterior ends. Second, fusion requires the activation of the mating structures associated with the anterior cell membrane of each gamete: each mt^+ mating structure gives rise to a long, slender fertilization tubule, filled with microfilaments, which projects towards its mt^- partner; each mt^- mating structure, meanwhile, alters its internal configuration so as to undergo fusion with the tip of the approaching fertilization tubule when it makes contact. Fusion establishes a continuous cytoplasmic bridge between pairs of gametes which quickly widens, drawing the two cells into confluence. Fusion also causes the four flagella of a zygote to lose their agglutinability. As a result, any gametes that had adhered to these flagella during the clumping stage do not continue to hang onto the zygote; instead they are freed to find new partners, and every gamete in a 1:1 mixture of the two mating types soon succeeds in becoming a zygote. The zygotes swim about for several hours, then resorb their flagella and secrete a thick zygotic cell wall.

Since flagellar agglutination, wall loss, zygotic cell fusion, and loss of flagellar adhesion can all occur within one minute, it should be clear from the foregoing description that the mating reaction is extremely efficient and involves a series of rapid signals, first from the flagellar tips to the cell body, and then from the fusing cells to their flagellar tips. Not emphasized in the foregoing description, but equally remarkable, is the extreme specificity of the reaction: an mt^+ gamete of *C. reinhardi*, for example, exhibits no flagellar interaction with other *C. reinhardi* mt^+ gametes nor with mt^- vegetative cells (cells that have not been induced, by nitrogen deprivation, to undergo gametogenesis); moreover, it fails to interact with the gametic flagella of numerous other *Chlamydomonas* species.

In this review I shall summarize studies that have sought to understand the molecular basis for the membrane recognition and membrane fusion events during *Chlamydomonas* mating. I shall focus on *C. reinhardi* but will include *C. moewusii* and *C. eugametos* where relevant, particularly since these are the species to which

Lutz and Waltraud Wiese, the pioneers in this field of research, devote most of their attention. Reviews by Wiese (1969, 1974) cover much of the early work in the field plus more general considerations on algal mating reactions, and should be consulted by all interested readers.

The usual task of the reviewer is to sort through and distill large numbers of publications so as to present a critical synopsis. My task is essentially the reciprocal: with only a few laboratories presently studying mating interactions in *Chlamydomonas*, the available information is readily summarized, leaving ample berth for both interpretation and speculation.

8.2 FLAGELLAR AGGLUTINATION

8.2.1 Morphology of the process

Vegetative and gametic cells of *Chlamydomonas* have been negatively stained, freeze-etched, or thin-sectioned and examined by transmission electron microscopy; they have also been visualized by scanning electron microscopy. From such studies the following facts and impressions have emerged.

(1) A *C. reinhardi* flagellum is covered by a membrane, continuous with the plasma membrane, which bears a conspicuous fuzzy coat, $\sim$ 20 nm wide, on its outer surface (Bergman *et al.*, 1975; Ringo, 1967; Snell, 1976a; Witman *et al.*, 1972), a coat assumed to represent surface carbohydrate in analogy with other membranes. Since this coat remains with the membranes through repeated washes (Bergman *et al.*, 1975; Snell 1976a), it apparently represents the carbohydrate moieties of intrinsic membrane glycoproteins and not a secreted 'slime'. The distal half of each flagellum also carries two longitudinal rows of hairlike mastigonemes, each $\simeq 0.9$ μm in length and 15 nm wide (Bergman *et al.*, 1975; Ringo, 1967; Snell, 1976a; Witman *et al.*, 1972), which project from the membrane surface and are believed to function in creating an effective flagellar stroke (Bouck, 1972). The flagellar surface of *C. moewusii* has a similar morphology (McLean *et al.*, 1974).

(2) No difference in flagellar membrane or mastigoneme morphology is evident when vegetative cells and gametes of *C. reinhardi* are visualized in thin section or in negative stain (Bergman *et al.*, 1975; Snell, 1976a).

(3) The interior of the *C. reinhardi* flagellar membrane bears particles when visualized by freeze-fracture, and the outer (E) face of the membrane tends to carry more particles in gametes than in vegetative cells (Bergman *et al.*, 1975; Snell, 1976a). No change in particle distribution or in the smooth flagellar surface (revealed by freeze-etching) is apparent when gametes agglutinate (Goodenough, unpublished).

(4) Agglutinating flagella usually wind loosely around one another and appear to interact most strongly at the tips (Bergman *et al.*, 1975; Snell, 1976a; Wiese, 1969); in *C. reinhardi*, one tip typically protrudes out slightly beyond the other (Bergman *et al.*, 1975).

(5) Serial sections through agglutinating flagella reveal that the flagellar membrane bilayers do not touch one another directly: contact instead appears to be made *via* the fuzzy surface coats (Fig. 8.1; from Weiss and Goodenough, unpublished studies). Since mastigonemes project almost a micron from the flagellar surface, these observations seem to rule out agglutination between mastignoeme *tips,* but since mastigonemes might wrap themselves close to the flagellar surface during agglutination, their involvement or non-involvement with the mating process cannot be decided by visual inspection alone.

8.2.2 Membrane participation in agglutination

A convincing demonstration of membrane participation in agglutination has been made by isolating and purifying flagellar membrane vesicles from gametes of one *mt* and presenting them to gametes of the opposite *mt*: the gametes adhere to the vesicles, and then to one another, so that with the light microscope it appears that a true clumping reaction is taking place (Bergman *et al.,* 1975; McLean *et al.,* 1976; Snell, 1976a). Such *isoagglutination* reactions occur only between gametes and gamete-derived membrane vesicles of opposite *mt*; specifically, isoagglutination does not occur if either the cells or the vesicles are in the vegetative state, nor if gametic vesicles are derived from the same *mt* as the gametes. Thus it seems clear that flagellar membranes represent at least one of the participants in the agglutination reaction.

Isolated mastigonemes are unable to cause isoagglutination (Bergman *et al.,* 1975; McLean *et al.,* 1976; Snell, 1976a), but even this observation fails to rule out a mastigoneme involvement in adhesion. It is conceivable, for example, that each mt^+ mastigoneme bears a single agglutination site, in which case mt^+ mastigonemes would readily bind to mt^- flagella but would lack the requisite 'multivalency' to cause the clumping together of many flagella. This question could probably be settled by binding studies using radioactively labeled mastigonemes, but interest in mastigonemes appears for the present to have waned in favor of membranes.

It should be noted that the observations concerning membrane involvement in agglutination (Bergman *et al.,* 1975; McLean *et al.,* 1974; Snell, 1976a) were made with preparations obtained by taking medium in which vegetative or gametic cells had been swimming, spinning out the cells, and then pelleting at high speed the flagellar membranes and mastigonemes which are continuously sloughed from the cells into the media. This pellet, which is equivalent to the 'gamone' preparations of Moewus (1933) and the 'isoagglutinin'* preparations of Wiese (1969), can be further

---

* The term 'isoagglutinin' has sometimes been used with the implication that its activity resides in an identifiable molecule, an implication which may have gone beyond the original intent of the term. The introduction of electron microscopic techniques has clarified the situation, showing that the activity resides in the membrane vesicle fraction, as described in the text.

fractionated into membrane or mastigoneme preparations by various modes of gradient centrifugation (Bergman et al., 1975; McLean et al., 1974; Snell, 1976a; Witman et al., 1972). A rather laborious protocol has been developed to isolate an active vesicle fraction uncontaminated by other cellular material and free of proteolytic modifications (Bergman et al., 1975) (see Fig. 8.2); for many experimental purposes, however, a cruder preparation is sufficient and can be obtained quite easily (McLean et al., 1974; Snell, 1976a; Wiese, 1969).

8.2.3 Polypeptide composition of membranes and mastigonemes

Granted that membranes and, possibly, mastigonemes participate in agglutination, it was logical to ask next whether the polypeptide composition of these components changes as flagella acquire gametic agglutinability. Witman et al. (1972) were the first to report that vegetative mastigonemes consist of repeating units of a single glycoprotein, the glycopolypeptide having an apparent molecular weight of 170 000 by sodium dodecyl sulfate (SDS) polyacrylamide gel electrophoresis (a somewhat higher apparent molecular weight is found by Bergman et al. (1975)). They also made the striking observation that vegetative flagellar membranes yield a single polypeptide band when subjected to SDS-gel electrophoresis (in contrast to the polypeptide complexity of most membranes that have been analyzed). This band migrates even more slowly than the mastigoneme component, and also contains carbohydrate.

Bergman et al. (1975) were able to detect, using gradient slab gels, very minor amounts of several additional glycopolypeptides of extremely high molecular weights in vegetative membranes. By overloading such gels, moreover, the single major band of Witman et al. (1972) appears to consist of several species of glycopolypeptide with very similar but not identical electrophoretic mobilities; whether this 'microheterogeneity' is due to differences in carbohydrate moieties and/or in amino acid sequences is not known. Thus the vegetative flagellar membrane can be described as possessing a closely related family of major glycopolypetides plus two or three minor glycopolypeptides, all migrating in gels more slowly than myosin (mol. wt. = 200 000) (Bergman et al., 1975; Bergman and Goodenough, unpublished).

Two laboratory groups went on to ask whether gametic flagellar membranes or mastigonemes of *C. reinhardi* contain any novel polypeptides, and somewhat variant results were obtained. Bergman et al. (1975) could detect no differences between vegetative and gametic isoagglutinin preparations of either mating type. Snell (1976a), on the other hand, often found an additional polypeptide of 70 000 daltons (the 'U band') in preparations of mt^+ flagellar membranes and mastigonemes; 'U' was sometimes faintly present in mt^- membrane preparations as well. While Bergman et al. occasionally encounter such a polypeptide in their gels, it is not restricted to mt^+ gametes and they believe it may be generated by proteolysis of the large glycoprotein species. Until these differences are resolved, it is probably fair to state that both research groups have isolated gametic flagellar membranes which retain full mt-specific isoagglutinability and which are electrophoretically indistinguishable from their vegetative counterparts. In other words, no novel polypeptide has as yet been definitively correlated with the acquisition of agglutinability in this very simple membrane.

The invariance in the gel-electrophoresis patterns of *C. reinhardi* flagellar membranes is consistent with at least four hypotheses:
(1) proteins play no role in the agglutination reaction;
(2) the gametic flagellar membranes possess (glyco)polypeptide agglutinins that are present in so few copies that they are not detected by gel electrophoresis;
(3) the acquisition of agglutinability entails sufficiently subtle modifications of the amino-acid sequence and/or the carbohydrate moieties of vegetative membrane glycoproteins that their electrophoretic mobilities are apparently unchanged in gametic preparations; or
(4) agglutinability is mediated *via* existing vegetative-cell glycoproteins but requires the acquisition of some additional membrane feature — e.g. a change in membrane lipids — so that existing proteins are positioned differently with respect to the flagellar surface.

8.2.4 Protein participation in *C. reinhardi* agglutination

Of the four hypotheses listed above, the first seems least likely since Wiese and colleagues (Wiese, 1965; Wiese and Hayward, 1972; Wiese and Metz, 1969) have shown that proteolytic enzymes destroy the agglutinability of gametic flagella without in any way affecting motility or cell viability. Specifically, pretreatment of mt^+ *C. reinhardi* gametes by trypsin, subtilisin, pronase, thermolysin, chymotrypsin, or papain inhibits their ability to agglutinate with mt^- gametes; the same enzymes inhibit the agglutinability of mt^- gametes. The trypsin effect is prevented if soybean trypsin inhibitor is present; moreover, if trypsinized gametes are washed or placed in soybean inhibitor, the flagella regenerate full agglutinability within 1–4 h (McLean and Brown, 1974; Snell, 1976b). Thus it seems clear that a protein (or proteins) either acts as the agglutinin, bears the agglutinin, or serves to maintain the agglutinin in an active state.

When gel electrophoresis failed to demonstrate new gametic membrane polypeptides (Section 8.2.3), several attempts were made to determine whether any of the existing flagellar polypeptides changed their electrophoretic mobility following exposure to trypsin (K. Bergman and S. Minami, unpublished). Isolated flagella or isolated flagellar membranes were trypsinized and were shown to be inactive in isoagglutination assays; they were the subjected to SDS-gel electrophoresis. None of the major family of glycopolypeptides exhibited any detectable change either in amount or in apparent molecular weight following trypsirization, a result that leaves open the possibility that trypsin removes only a few amino acids from these polypeptides. Experiments are presently in progress in my laboratory to determine whether the minor glycopolypeptide components of the membrane are similarly indifferent to trypsinization. Meanwhile, gel analysis minimally indicates that trypsin can destroy agglutinability without apparent effect on the vast majority of the flagellar membrane polypeptides. Electron microscopy confirms this conclusion: trypsinized flagellar membrane vesicles bear a fuzzy coat which is indistinguishable in appearance from that of controls (Fig. 8.3), and apparently intact mastigonemes remain associated with trypsinized flagella (Goodenough, unpublished).

Two additional approaches have been taken to explore the nature of the trypsin-sensitive sites in *C. reinhardi* gametic flagella. Jurivich (1976) tried, and failed, to prevent flagellar agglutination by preincubating gametes of one mating type with concentrated tryptic digests of gametes of the opposite mating type; he was also unable to detect any specific binding of ^{14}C-labeled tryptic digests to flagella of opposite *mt*. While negative results can be interpreted many ways, these results suggest either that trypsin may destroy the agglutinin as it is released from the flagellar surface, or else that trypsin does not release the agglutinin at all but rather destroys some surface architectural or enzymatic feature which allows the agglutinin to function.

A second clue to the nature of the trypsin target comes from the following observations. *C. reinhardi* gametes can be fixed in 2% glutaraldehyde; the resultant 'corpses' are highly cross-linked and survive sonication, detergent treatments, and freeze-thawing. They retain, however, their *mt*-specific agglutinability (Goodenough and Weiss, 1975) such that living gametes of opposite *mt* adhere strongly to them (see also Wiese and Jones, 1963). Even more strikingly, gametic corpses lose their agglutinability following brief exposure to trypsin or chymotrypsin, but are resistant to trypsin plus trypsin inhibitor (Goodenough and Jurivich, unpublished observations). Thus the agglutinin is not rendered inactive by glutaraldehyde cross-linking, nor does glutaraldehyde render the trypsin-sensitive site inaccessible to proteolysis.

8.2.5 Carbohydrate role in *C. reinhardi* agglutination

The association of the dense 'fuzzy coat' with the flagellar surface (Fig. 8.1–8.3) suggests the possible participation of carbohydrate in the agglutination reaction. We tested this possibility by subjecting gametic corpses (Section 8.2.4) to periodate oxidation, a procedure that alters the native configuration of polysaccharide chains (Rothfus and Smith, 1963), and found that a 1 h incubation in 50 mM $NaIO_4$ destroys corpse agglutinability in both mating types whereas 5 mM is without effect (similar results occur with living cells but the experiment is less convincing in that most of the gametes shed their flagella in the presence of the reagent). Since periodate is not specific for carbohydrates (it can, for example, act on certain amino acids as well (Rothfus and Smith, 1963)), these observations suggest, rather than prove, a carbohydrate role in the mating reaction.

Exposure of *C. reinhardi* gametic flagella to neuraminidase, galactose oxidase, β-galactosidase, β-glucuronidase, or α-mannosidase has no effect on their agglutinability (Wiese and Hayward, 1972; Wiese and Wiese, 1975; Goodenough, unpublished observation). Moreover, mating proceeds in the presence of a wide variety of sugars,

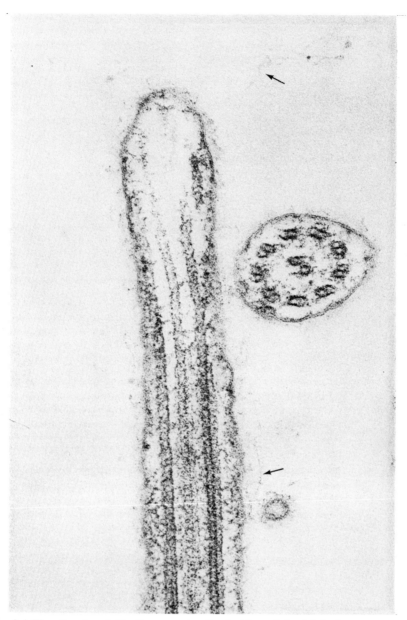

Fig. 8.1 Flagella of agglutinating gametes fixed at low glutaraldehyde concentrations (Goodenough and Weiss, 1975). The long flagellum is sectioned through its tip; a fuzzy coat is associated with its membrane, and thread-like mastigonemes (arrows) extend from the surface. The cross-sectioned flagellum is presumably making sexual contact with the long flagellum (this can never be proven for a given flagellar pair in thin section since non-specific associations occur as the sample is centrifuged). A membrane vesicle is also seen in apparent sexual contact with the long flagellum. Both contact sites appear to involve the fuzzy surface coats. Dense material also seems to accumulate in the flagellar matrix underlying the contact sites, but it is not known whether this has anything to do with agglutination. × 112 000.

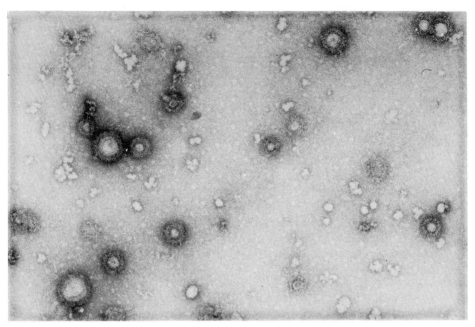

Fig. 8.2 Flagellar membrane vesicles of mt^+ gametes isolated by the method of Bergman et al. (1975), incubated with 0.2% soybean trypsin inhibitor, and negatively stained with uranyl acetate. The fuzzy coat on these vesicles is apparent. The aggregated material is presumably soybean trypsin inhibitor. x 57 000.

including a number of nucleotide sugars (Bergman and Goodenough, unpublished); thus the tested sugars do not appear capable of acting as hapten inhibitors. The role of carbohydrates in the *C. reinhardi* mating reaction, if any, remains as elusive to biochemical characterization as does the role of proteins.

8.2.6 Protein and carbohydrate participation in *C. moewusii* agglutination

The response of *C. moewusii* gametes to enzymatic digestion is significantly different from the *C. reinhardi* response. As summarized in a recent review by Wiese (1974), the agglutinability of mt^- gametes is sensitive to trypsin, pronase, and subtilisin whereas mt^+ gametes are resistant to all three enzymes. Conversely, the mt^+ gametes are sensitive to α-mannosidase while the mt^- gametes are resistant. Papain eliminates the agglutinability of both mating types, and both mating types are resistant to neuraminidase, β-galactosidase, and β-glucuronidase. Thermolysin and chymotrypsin have particularly complex effects: in syngen I of *C. moewusii*, mt^- cells (which are sensitive to trypsin, pronase, and subtilisin) are resistance to thermolysin and

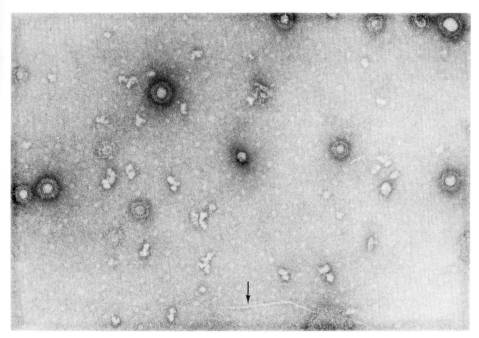

Fig. 8.3 Same sample as Fig. 8.2 but exposed to 0.1% trypsin for 2.5 h before the addition of 0.2% soybean trypsin inhibitor. The fuzzy coat is apparently unchanged, and a contaminating mastigoneme (arrow) has survived enzyme treatment. × 57 000.

chymotrypsin while mt^+ cells (resistant to trypsin, pronase, and subtilisin) are sensitive to thermolysin and chymotrypsin; in syngen II, on the other hand, both mating types are resistant to thermolysin and to chymotrypsin. *C. reinhardi*, it will be recalled, is rendered non-agglutinable by all the proteolytic enzymes listed above but is unaffected by α-mannosidase. Thus major differences exist between *C. reinhardi* and *C. moewusii,* a fact detectable by many other criteria as well (e.g. Thomas and Delcarpio, 1970; Triemer and Brown, 1975b).

Wiese and Wiese report (1965) that the products obtained by proteolytic degradation of mt^- agglutinin preparation fail to exercise any specific receptor blocking power when tested with mt^+ gametes. As noted above (Section 8.2.4), such negative results have also been obtained with *C. reinhardi.*

The sensitivity of the mt^+ gametes of *C. moewusii* to α-mannosidase represents the most direct evidence for a carbohydrate role in the agglutination reaction. Having obtained this result, Wiese reasoned that if a terminal mannose in mt^+ effects agglutination *via* a specific association with an mt^- receptor molecule, then preincubation with mannose might block the mating response, just as specific monosaccharides inhibit, for example, the agglutinability of specific lectins (Sharon and Lis, 1972). Neither D-mannose nor α-methyl-D-mannoside could be shown to have any effect, however, either on flagellar agglutinability or on the isoagglutination response to flagellar membranes (Section 8.2.2) (Wiese and Hayward, 1972; Wiese and Wiese, 1965). Therfore, the sexual interaction would appear to be more complex than a simple association between mt^+ mannosyl residues and mt^- lectin molecules. One possibility is that the postulated mt^- receptor recognizes an extended polysaccharide sequence rather than a single mannosyl residue, in which case α-methyl-mannoside would be an ineffective competitor. Alternatively, α-mannosidase may destroy a class of residues on the mt^+ flagellar surface that somehow supports or facilitates sexual agglutination but does not participate directly in the process.

8.2.7 The glycosyltransferase hypothesis

The apparent participation of both protein and carbohydrate in *C. moewusii* agglutination is compatible with Roseman's hypothesis (1970) that intercellular recognition and adhesion might be effected if surface-localized glycosyltransferases on one cell interacted with their carbohydrate substrates on a second cell. This model is particularly attractive for *Chlamydomonas* in that one can explain the loss of agglutinability of zygotic flagella by postulating that fusion somehow makes available the nucleotide-sugars required to complete the transferase reaction, whereupon the flagella separate and their surface carbohydrates no longer serve as substrates for unmated flagellar enzymes.

To test the model, McLean and Bosmann (Bosmann and McLean, 1975; McLean and Bosmann, 1975) asked whether ^{14}C-labeled nucleotide phosphate sugars were transferred to gametic flagellar preparations and whether the rate of transfer was enhanced when cells or membrane vesicles of opposite *mt* were mixed. Six distinct transferase activities were detected; all were found to increase 2–3 fold when compatible gametes were mixed; all increased 4–7 fold when compatible membrane-vesicle preparations from the two mating types were mixed; and no increase occurred when vegetative cells, vegetative membrane vesicles, or incompatible species were used.

These experiments are intriguing and may well demonstrate critical reactions in the agglutination response, but they can be criticized on several counts. First, the incubations were performed at concentrations of NaCl known to cause lysis of *Chlamydomonas* cells; since mating cells open their cell walls to form a fusion bridge (see Section 8.3.2), they may be more prone to lysis than the unmated controls, thereby exposing transferase-associated internal membranes (e.g. Golgi vesicles) to the ^{14}C-nucleotide sugars. Second, the membrane-vesicle preparation used by McLean and Bosmann is reported to be contaminated by microtubule, wall, and chloroplast fragments (McLean *et al.*, 1974) so it is not clear that flagellar membranes alone are participating in the response. Finally, it can be debated whether a 2-fold increase in enzyme activity would suffice to explain the agglutination reaction. McLean and colleagues are aware of these criticisms and are endeavoring to define more clearly the possible role of glycosyltransferases in flagellar agglutination.

To my knowledge, the glycosyltransferase hypothesis has not been tested with *C. reinhardi* except that we find (unpublished) that high exogenous concentrations of nucleoside and nucleotide sugars do not inhibit or reverse *C. reinhardi* sexual agglutination.

8.2.8 Acquisition of agglutinability

Gametogenesis in *C. reinhardi* is most readily studied in a synchronous liquid culture, under which conditions cells become capable of mating some 12 hours after they are deprived of nitrogen (Kates and Jones, 1964; Martin and Goodenough, 1975). A mitotic division usually precedes the acquisition of agglutinability (Chiang *et al.*, 1970; Martin and Goodenough, 1975), and since *C. reinhardi* cells pull in their flagella at the onset of mitosis and grow out new flagella at the conclusion of mitosis (Buffaloe, 1958; Johnson and Porter, 1968), a very different flagellar surface may be elaborated in the interval. Gametic differentiation in the absence of mitosis has been reported (Chaing *et al.*, 1970; Schmeisser *et al.*, 1973), but it is not known whether or not such cells shed their vegetative flagella and grow new gametic flagella. This question becomes important when experiments designed to follow the acquisition of agglutinability are contemplated. It would be interesting, for example, to surface-label vegetative flagella with one isotope and ask whether the accessibility of the labeled moieties changes during gametic differentiation; such approaches may, however, be precluded by mitotic flagellar shedding.

Cells of *C. reinhardi* will also differentiate into gametes when cultured on plates, following the exhaustion of nitrogen sources in the agar medium (Martin and Goodenough, 1975) and, if two mating types are cultured together on one plate, zygotes will form *in situ* (Goodenough, unpublished observation). When suspended into liquid, 'plate gametes' are usually flagellated but require from 30 min to several hours to acquire good motility and agglutinability. Schmeisser *et al.* (1973) suggest that this interval represents a differentiation period, but we find (unpublished) that the flagella of plate gametes are initially covered by an amorphous slime, possibly derived from the agar, so that the lag may simply represent the time required to shed this inhibitory coating, a lag that can be shortened considerably by washing plate gametes several times.

Gametic *C. reinhardi* flagella are longer ($\simeq$ 12 μm) than vegetative flagella ($\simeq$ 8 μm) (Randell *et al.*, 1968; Bergman, unpublished). If gametes are deflagellated (Rosenbaum *et al.*, 1969) they quickly grow out new, agglutinable flagella; if deflagellated in the presence of cycloheximide (Rosenbaum *et al.*, 1969) they produce short flagella, perhaps half-length, which are fully as agglutinable as those of normal length (Bergman, unpublished). When differentiated gametes are left starving in N-free liquid medium for several days and are then pulsed with ^{14}C-acetate to determine which polypeptides they are synthesizing, the major labeling in the resultant gels is localized to the flagellar membrane polypeptides (Minami, 1976). Thus flagellar surface components of *C. reinhardi* are being sloughed into the medium (Section 8.2.2) and being synthesized and replaced, even under intense starvation conditions. Indeed, with time, the cell bodies of gametes suspended in N-free liquid become smaller and smaller while the flagella remain of normal length, a testimony to their critical importance in attaining the next stage of the organism's life cycle.

8.2.9 Loss of agglutinability by mutation

In addition to proteolysis (Section 8.2.4) and Ca^{2+} depletion (Lewin, 1954; Wiese and Jones, 1963), flagellar agglutinability can be lost in *C. reinhardi* by gene mutation. In considering these mutant strains, it is important first to note that *C. reinhardi* strains may lose gametic agglutinability under 3 sets of circumstances.

(1) For obscure reasons wild-type stocks tend to accumulate cell lines which mate poorly (Chiang *et al.*, 1970) so that the overall mating efficiency of the strain becomes low. A strain that mates well with one wild-type stock, moreover, may mate very poorly with a second (Chiang *et al.*, 1970). This situation can usually be remedied by cloning the wild-type stocks, testing individual clones, and selecting two that mate efficiently with one another; these may, however, acquire poor mating with time so that the selection process may need to be repeated (Chiang *et al.*, 1970). We have in our collection at least two sets of wild-type strains that have not required such cloning for several years, so that it is not an invariant phenomenon. The question 'What is wild-type?' is often raised at this point, and for an organism that has been in laboratory culture for some 40 years, the question is probably unanswerable.

(2) Mutant strains selected for auxotrophy, photosynthetic disabilities, or abnormal motility are often reluctant maters (unpublished experience in many laboratories). Since these strains have general physiological handicaps, their poor mating can be assumed to be a non-specific side effect.

(3) Mutant strains can be selected that have unimpaired growth and motility but fail to develop agglutinability when starved for nitrogen, even though in other respects (e.g. in the elaboration of a mating structure) they undergo an apparently normal gametogenesis. These strains are considered in the following paragraphs. In a later section of this review, additional mutant strains are considered that have apparently normal agglutinability but fail to fuse.

Five non-agglutinable impotent (*imp*) strains have been isolated by UV-irradiation of mt^+ cells, followed by a selection procedure which enriches for cells that fail to become incorporated into the sheet of zygotes that forms after mating (Goodenough *et al.*, 1976). All but one of the non-agglutinating *imp* strains can be induced by centrifugation to fuse with mt^- gametes, albeit at very low efficiencies, and since non-agglutinating meiotic clones can be recovered from the zygotes (Goodenough *et al.*, 1976), the trait is clearly heritable (in contrast to the 'low-efficiency mating' trait of wild-type strains, described above, which is not clearly transmitted to meiotic progeny). In none of the strains is the *imp* mutation closely linked to mt^+, as evinced by the emergence of normally agglutinable mt^+ clones from many of the zygotes (Goodenough et al., 1976). Gametes of the imp-2 mt^+ strain occasionally appear to form transient contacts with mt^- gametes immediately after the two are mixed, but these cannot be seen after 1–2 minutes and no fusion results; the other non-agglutinable strains (imp-5, imp-6, imp-7, and imp-8) show no interactions with mt^- or normal mt^+ gametes. Since it has not yet been possible to mate non-agglutinating imp strains with one another, the 5 mutations may represent 5 alleles of one locus or may mark up to 5 independent loci.

Since no morphological differences distinguish vegetative and gametic cells, it is perhaps not surprising that the flagella of imp strains are indistinguishable from their wild-type counterparts (Bergman et al., 1975). While other features of the mutants will be considered in subsequent sections, it is appropriate to state here that just as the molecular basis for wild-type agglutinability is not known, neither is it known why the mutant strains fail to agglutinate.

Flagellar agglutination normally occurs over a wide temperature range (2°–37°C). Forest and Togasaki (1976, 1977) have recently developed an elegant screening procedure for isolating mutant strains conditionally defective in gametogenesis: irradiated gametes of the strain pet-10-1, which has defective photosynthetic electron transport and is thus resistant to killing by the dye methyl viologen, are allowed to mate at 35°C with gametes of opposite mt; all zygotes and all wild-type cells are then killed by exposure to methyl viologen so that only non-mating pet-10-1 mutants survive. Using this procedure, the strain gam-4 has been isolated which has normal motility, agglutinates normally at 25°C, but fails to agglutinate at 35°C. The gam-4 gene is not linked to mt and is expressed in both mt^+ and mt^- gametes (Forest and Togasaki, 1977). Nothing is yet known about the structure or biochemistry of gam-4 flagella.

8.2.10 Lectin agglutinability in *C. reinhardi*

A widely used method to probe the carbohydrate nature of a membrane surface is to measure its lectin agglutinability, lectins having specific affinities for particular sugars or classes of sugars (Sharon and Lis, 1972). Preliminary experiments with the lectins wheat-germ agglutinin and soybean agglutinin revealed no obvious effect on vegetative or gametic cells of *C. reinhardi* (Bergman, unpublished observations). Concanavalin A (Con A), on the other hand, brings about a striking flagellar-tip isoagglutination of mt^+ or mt^- gametes but not vegetative cells (Wiese and Shoemaker, 1970). Gametes but not vegetative cells also adhere by their flagellar tips to Con A-conjugated Sephadex beads (Wiese and Shoemaker, 1970).

Several lines of evidence indicate that Con A interacts with components of the gametic surface that are independent of the mt-specific agglutinins.

(1) Trypsin abolishes mt-specific agglutination but is without effect on Con A isoagglutinability in either mt (McLean and Brown, 1974).

(2) The four flagella of a zygote lose their mt-specific agglutinability but remain

Con A-agglutinable (Goodenough, unpublished).

(3) Gametes of the five non-agglutinating *imp* strains (Section 8.2.9) are all fully agglutinable with Con A (Bergman, unpublished).

(4) The acquisition of Con A agglutinability by gametic cells reportedly shows no temporal correspondence with the acquisition of *mt*-specific agglutinability (McLean and Brown, 1974).

(5) The monosacharide α-methyl-mannoside effectively reverses *C. reinhardi* gametic interactions with Con A but has no inhibitory effect on *mt*-specific interactions (McLean and Brown, 1974; Wiese and Shoemaker, 1970).

(6) Gametes engaged in Con A isoagglutination are fixed in their agglutinating configurations when the medium is made 2% glutaraldehyde; 2% glutaraldehyde, on the other hand, causes sexually agglutinating cells to come apart (Goodenough and Weiss, 1975; Goodenough unpublished).

Despite the clear differences between the two types of agglutination, the lectin remains a useful probe for exploring how vegetative and gametic flagellar surfaces might differ from one another. In a recent study, Bergman inquired whether the acquisition of Con A agglutinability by gametes could be correlated with an increased binding capacity for the lectin. Using ^{3}H-Con A and highly purified preparations of flagella (the cell wall binds the lectin and whole cells therefore cannot be used), Bergman (unpublished) finds no difference in the amount of Con A specifically bound by vegetative and gametic preparations; if anything, the vegetative flagella bind slightly more lectin than the gametic (in a typical experiment, gamete binding saturated at 6.66 μg Con A mg^{-1} flagellar protein while vegetative binding saturated at 8.94 μg mg^{-1} protein). Such a result, which has become familiar in mammalian cell studies (reviewed in Hynes, 1974), suggests some change in the state of the lectin binding sites on the gametic membranes such that binding produces membrane-to-membrane adhesions. Rutishauser and Sachs (1974), for example, have proposed that cell—cell binding induced by Con A requires short-range lateral movement of cell receptors for the lectin so as to form multi-point bridges between two cells. Possibly, therefore, the Con A receptors in the gametic flagellar membranes are freer to move about and establish such bridges at the flagellar tips.

8.2.11 Lectin agglutinability in *C. moewusii*

Gametes of *C. moewusii* are also isoagglutinable by high concentrations of Con A and will adhere to Con-A-coated Sepharose beads, both responses being readily reversed by the addition of mannose or α-methyl-mannoside (Weise and Shoemaker, 1970). In these respects, therefore, the gametic flagellar surface resembles that of *C. reinhardi* gametes (Section 8.2.9).

The apparent independence of Con A-agglutinability and *mt*-specific agglutinability in *C. reinhardi* is not, on the other hand, observed for mt^+ *C. moewusii*: Wiese and Shoemaker (1970) report that the pre-incubation of mt^+ gametes with low (non-agglutinating) concentrations of the lectin will abolish their sexual agglutinability;

a similar inhibition is not obtained when mt^- gametes are so pretreated. This result, which was questioned by McLean and Brown (1974) but reaffirmed by Wiese and Wiese (1975), is in accord with the inhibitory effect of α-mannosidase on mt^+ but not mt^- gametes (Section 8.2.6).

Wiese and Wiese (1975) did not examine the effects of α-mannosidase on Con A agglutinability or Con A inhibition of mt^+ agglutinability in *C. moewusii* (Wiese, personal communication). Such tests might shed light on the interrelationships between the low-Con A and high-Con A phenomena in this species.

8.2.12 Antibodies against the flagellar surface

Wiese prepared an antiserum against an mt^- isoagglutinin preparation from *C. eugametos*, a close relative of *C. moewusii*, and reported that the antiserum precipitated mt^- isoagglutinin and caused isoagglutination of mt^- gametes; however, it also caused a typical flagellar agglutination of mt^+ gametes and of vegetative cells (1965). No further studies of this antiserum have been reported.

Jurivich and Goodenough (unpublished) have recently launched an immunological analysis of the *C. reinhardi* flagellar surface, using antisera raised against mt^+ vegetative and gametic flagella and mt^- gametic flagella isolated by the pH shock method (Witman *et al.*, 1972). The antisera prove to have very high titers, being effective in causing cell-to-cell or corpse-to-corpse isoagglutination (Section 8.2.4) in microtitration wells to a dilution of 1:256. As Wiese found for *C. eugametos* (1965), the *C. reinhardi* antisera interact with both vegetative and gametic cells of both mating types. Similarly, the antisera interact effectively with trypsinized gametes (Section 8.2.3), with *imp* mutant strains (Section 8.2.9), and with quadriflagellate zygotes, all of which are sexually non-agglutinable. When the mt^+ gametic antiserum is fully adsorbed by mt^+ vegetative cells, moreover, the antiserum loses its ability to interact with mt^+ gametic cells as judged by microtitration-plate agglutination; similar results are obtained when gametic antisera of one mating type are exhaustively adsorbed by gametes of opposite mating type and then tested for residual self-specific agglutination activity. Finally, when the antisera are tested in Ouchterlorny plates against Triton X-100 extracts of flagellar membranes, the same major precipitin line is observed regardless of the source of antiserum or membrane. Therefore, the prominent surface antigens of vegetative and gametic flagella appear to be identical. We are now testing whether the antisera contain minor species of antibodies directed at less abundant and/or less antigenic flagellar components.

The one difference Jurivich and Goodenough have found to date is that while living gametes respond to a wide range of antiserum dilutions by isoagglutinating at their flagellar tips, living vegetative cells give quite a different response. At high antiserum titers, the flagella almost immediately fall off the vegetative cells. At titers where flagella are retained, tip agglutination is not observed; instead, the vegetative flagella make contact at random positions over their entire lengths. These observations immediately bring to mind the experiments showing that while vegetative flagella bind Con A as well as gametic flagella, they fail to agglutinate to the lectin *via* their tips (Section 8.2.10). It is as though the gametic flagellar membrane is more 'fluid' and allows a 'tipping' of surface-bound macromolecules reminiscent of the 'patching and capping' that occurs in lymphocytes (reviewed in Edelman, 1974), although this analogy may well prove to be quite superficial once these phenomena are better understood in *Chlamydomonas*.

8.2.13 The requirement for a living cell

An isoagglutination response will occur when living gametes of one *mt* are presented with flagella, flagellar membrane vesicles, or fixed corpses of opposite *mt*. Repeated attempts to observe agglutination reactions between isolated flagella, between corpses, or between membrane vesicles have, on the other hand, consistently failed (unpublished observations from this and several other laboratories). The failure of isolated flagella of opposite *mt* to interact cannot be attributed to loss of motility alone because mt^+ and mt^- gametes of *paralyzed-flagella* (*pf*) mutant strains agglutinate well with one another (Goodenough, unpublished). Since an mt^+/mt^- membrane-vesicle mixture fails to precipitate out and since the vesicles fail to neutralize each other when mixed *in vitro* (Wiese and Wiese, 1965), a simple antibody-antigen model for the initial sexual interaction of the membranes does not appear to be applicable.

The situation is quite different when non-sexual flagellar agglutinability is being probed. Gametic mt^+ corpses, for example, which show no interaction with mt^- corpses, show excellent agglutination when mixed with flagellar-surface antisera (Section 8.2.12) and exhibit a 'spreading' settling pattern in microtitration wells (Jurivich, unpublished). Similarly, isolated gametic flagella will stick to one another in the presence of Con A (Bergman, unpublished) or antibody (Jurivich, unpublished) but, as noted above, fail to interact with one another sexually. Sexual agglutination, therefore, requires that one partner be a living cell, whereas non-sexual agglutination has no such requirement.

The situation is also very different if flagella are first allowed to agglutinate and are then amputated from the cell bodies: in this case the isolated flagella remain associated with one another. Such a demonstration can be made in three ways (Goodenough, unpublished).

(1) Gametes of opposite mating type can be mixed at 2–4°C, at which temperatures they agglutinate but fail to fuse, and then treated with amphotericin (Weiss *et al.*, 1977) to cause deflagellation. Many of the detached flagella are found clumped together.

(2) Gametes of one *mt* can be isoagglutinated with flagella of opposite mating type at 22°C and then treated with amphotericin; detached flagellar clumps are again observed.

(3) Gametic corpses of one *mt* can be caused to agglutinate with living gametes of opposite *mt* and the mixture then treated with amphotericin; only the living flagella are detached, and many of these are found associated with the glutaraldehyde-fixed

flagella of the corpses, an association that persists for several days if the preparations are kept cold.

Taken together, these observations are consistent with the notion that stable sexual adhesions between *C. reinhardi* flagella can be established only if at least one of the interacting flagella is attached to a living cell; once established, the adhesions remain patent even when living cell contact is lost. The role played by the living cell in establishing sexual adhesions in *C. reinhardi* is not known.

8.2.14 Overview and speculations

The analysis of flagellar agglutination in *Chlamydomonas* that has been performed to date has perhaps revealed more about what is *not* involved than about what is involved. The acquisition of agglutinability does not involve any major change in flagellar surface morphology, in gross polypeptide composition, or in gross surface antigenicity; both carbohydrates and proteins are implicated as participants in the agglutination reaction, but individual molecular species have not been identified; and existing models for the nature of the mating reaction (e.g. surface glycosyl-transferases or antigen-antibody-type reactions) have been inadequately tested or do not readily explain the phenomena observed.

The experiments performed to date *have* established that gametogenesis confers two kinds of new properties on *Chlamydomonas* flagella: *mt*-specific changes, which allow sex-specific agglutination, and *mt*-non-specific changes, which permit tip agglutination of flagella in the presence of such exogenous proteins as concanavalin A or flagella-specific antibodies. These two kinds of changes may be inter-related to some extent in *C. moewusii* but give no evidence of overlap in *C. reinhardi*. There is clearly a lesson to be learned from *Chlamydomonas* in this regard: when studying the acquisition of surface-recognition features in other cell types, investigators should realize that the appearance of a novel trait (e.g. lectin agglutinability) may only very indirectly relate to the changes taking place which confer surface-recognition specificity.

In speculating how adhesions might be generated in *Chlamydomonas,* three general hypotheses emerge.
(1) Gametic flagellar membranes may acquire a novel class of (glyco)polypeptide, present in too few copies to be seen in gels and antigenically inert, which resides on the surface and interacts in an enzyme-substrate (Roseman, 1970) or modified antibody-antigen fashion with its opposite *mt* counterpart.
(2) The same molecular species may be present in both vegetative and gametic flagella of one *mt* but acquire agglutinative properties due to a change in the intramembranous environment (perhaps this same change permits flagellar 'tipping'). In this regard, a lipid analysis of flagellar membranes in the two states of differentiation would be of great value.
(3) Agglutinative membrane components may somehow be masked from the cell surface (and hence perhaps unavailable to the rabbit immune system) until such

time as *mt*-specific contacts are made, at which point they undergo some conformational change and/or enzymatic modification, become exposed, and develop adhesive interactions. More generally, it seems important to keep in mind that agglutination may involve two discrete stages, *mt*-specific *recognition* followed by membrane-to-membrane *adhesions*. There is at present no reason to assume that the two are effected in the same manner or even to assume that they both involve the same surface molecules.

8.3 CELL FUSION

8.3.1 The fusion apparati

The two mating types of *C. reinhardi* were long considered isogamous (i.e. identical in appearance), but this is clearly not the case at the fine-structural level. In thin section (Cavalier-Smith, 1975; Friedmann *et al.*, 1968; Goodenough and Weiss, 1975; Triemer and Brown, 1975b), unmated mt^+ gametes are found to possess a mating structure adjacent to their anterior cell membrane which consists of a thin *membrane zone* adherent to the membrane and an underlying *doublet zone*; mt^- gametes possess a broader membrane zone and no doublet zone. Of direct relevance to this review is the fact that the circular areas of membrane overlying these mating structures are also found to be differentiated in *mt*-specific fashions when viewed by freeze-cleave electron microscopy (Weiss *et al.*, 1977). Both membrane areas are essentially cleared of the usual intramembranous particles found in the surrounding plasmalemma; the few particles present within the relatively large mt^+ membrane area localize to one side of center, whereas the few particles in the smaller mt^- membrane area are centrally localized (Weiss *et al.*, 1977). The functional roles of these membrane differences will become apparent in a later section.

In *C. moewusii* the anterior papillum above the basal bodies acquires, in gametic cells, a dense material which is not reported to differ in morphology when the two mating types are compared (Triemer and Brown, 1975a).

8.3.2 Cell wall lysis

Sex-specific agglutination at the flagellar tips triggers cell wall lysis in *C. reinhardi*: lysis begins at the anterior end of the wall, the gametes slip out of the walls, and the walls floating about in the medium continue to disintegrate until they are no longer visible. Wall digestion is effected by a lytic activity, referred to as 'endolysin' or 'autolysin' (Claes, 1971, 1975; Goodenough and Weiss, 1975; Schlosser *et al.*, 1976), which is released from some unidentified compartment of the cell and which can be recovered from the medium in which cells have mated. The active component can be precipitated from the medium with ammonium sulfate (Weeks, unpublished) but little else is known about its biochemical properties. For present purposes, endolysin release is of interest inasmuch as it is one of several cellular events triggered by flagellar agglutination.

The study of endolysin release in *C. reinhardi* is complicated by the fact that unmated gametes often release the activity on their own, stimulated by as yet unknown signals. As a result, gametes of one or both mating types may be wall-less prior to mixing the cells together (e.g. Friedmann *et al.*, 1968). When walled gametes are mixed, however, it can be shown that effective flagellar agglutinations must occur if rapid wall lysis is to take place: no rapid wall loss occurs if mt^- gametes are mixed with non-agglutinating *imp* mt^+ gametes (Section 8.2.9) but it occurs when they are mixed with the agglutinating but non-fusing mt^+ *imp-1* gametes (Section 8.3.5) (Goodenough and Weiss, 1975) or with isolated gametic flagella. Isoagglutination with high Con A or antibody concentrations is not followed by cell wall loss (Claes, 1975; Goodenough and Weiss, 1975), a result that is probably due to the fact that both Con A and antibody bind to cell walls (Bergman and Jurivich, unpublished): isoagglutination with low antibody concentrations produces quantitative wall lysis.

In *C. moewusii*, cell wall lysis is confined to the wall directly overlying the anterior papilla of the two mating types; the cells otherwise remain within their walls during the subsequent events of plasmogamy and karyogamy (Triemer and Brown, 1975a).

8.3.3 Mating structure activation

In *C. reinhardi*, cell wall lysis is probably the rate-limiting step in zygote formation: when gametes of both mating types are pre-treated with endolysin to remove their walls and are then mixed together, quadriflagellated zygotes appear within 30 s; if walls are present on mt^+ but not mt^- cells, quadriflagellates do not appear until 90 s after mixing (Spath and Hwang, unpublished).

During the 30 seconds between flagellar agglutination and cell fusion, dramatic changes occur in the mating-structure regions of both mating types. The mt^+ gamete sends microfilaments from its doublet zone into a long thin fertilization tubule (Cavalier-Smith, 1975; Friedmann *et al.*, 1975; Goodenough and Weiss, 1975; Triemer and Brown, 1975b). Viewing these events by freeze-cleave microscopy (Weiss *et al.*, 1977) it can be seen that the few asymmetrically localized particles in the membrane area overlying the mt^+ mating structure (Section 8.3.1) appear to travel with the growing tip of the fertilization tubule. The 'extra' membrane which forms the sides of the fertilization tubule is found to be essentially particle-free, as if composed of lipid but essentially no protein, and appears to derive from contractile vacuole membrane (Goodenough, unpublished). The mt^- gamete exhibits a very different response: its membrane-zone material moves centripetally, leaving a central dome of membrane which, in freeze-fracture replicas, is found to be densely populated with intramembranous particles. No microfilaments appear to participate in mt^- activation, nor is 'extra' membrane generated.

Mating-structure activation, then, consists of fertilization tubule formation in

mt^+ cells and particle clustering in mt^- cells. Both events can be elicited without attendant cell fusion by presenting gametes of one mt with flagella or corpses of the opposite mt. Of considerable interest is the fact that while mt^+ mating structures will activate at 3°C to form full-length fertilization tubules, fusion does not occur at 4°C or less, is very slow at 6°C, and does not occur at normal rates until 14°C (Weiss et al., 1977; Hwang, Spath, and Goodenough unpublished). Freeze-cleave electron microscopy reveals that mt^- gametes maintained in the cold do not bring about a doming of their mating structures and a concomitant clustering of central particles when mixed with mt^+ cells (Spath, Goodenough, and Goodenough, in preparation). A possible interpretation of these observations, which we are testing, is that mt^- mating-structure activation requires a cold-sensitive change in the physical state of mating-structure-region membrane lipids (see, for example, Luzzati, 1968).

Mating-structure activation, like cell wall lysis, occurs only in response to 'sexual' flagellar contacts. The mating structure of non-agglutinating *imp* strains, although normal in morphology, are never observed to activate when *imp* mt^+ and mt^- gametes are mixed, nor does Con A isoagglutination elicit activation (Goodenough and Weiss, 1975).

In *C. moewusii*, fibrillar material is present in the bridge connecting mating gametes but its microfilamentous nature is less apparent than in *C. reinhardi* and its origin is not apparent from published micrographs (Triemer and Brown, 1975a). A freeze-fracture study of *C. moewusii* mating has not been reported.

8.3.4 Zygotic cell fusion

Zygotic cell fusion in *C. reinhardi* presumably occurs within moments after the tip of the fertilization tubule meets the activated mt^- mating structure. Freeze-fracture studies indicate that fusion occurs within the particle-rich dome of the mt^- membrane and not within the particle-free region that surrounds it (Weiss et al., 1977). As noted earlier, the tip of the fertilization tubule also bears particles. Assuming that the particles represent proteins, therefore, it seems likely that proteins — be they mt^+, mt^-, or both — participate in the membrane-fusion event, a supposition supported by the report that fusion in *C. moewusii* is sensitive to trypsin, β-mercaptoethanol, and SDS (Wiese and Jones, 1963; Wiese and Shoemaker, 1970). The mt^- particles are of particular interest in that they 'partition' (Satir and Satir, 1974) to the ectoplasmic (E) face of the membrane during freeze facture, as though they are closely associated with the surface at which membrane fusion is to take place (Weiss et al., 1977). The few particles associated with the tip of the fertilization tubule membrane, on the other hand, are found on both membrane faces (Weiss et al., 1977).

Assuming a participation of proteins in the membrane-fusion process does not necessarily imply that such proteins function in an mt-specific fashion. In other words, it is possible that fusion requires that specific mt^+ proteins make contact with specific mt^- proteins; conversely, the fact that mt^+ gametes fuse only with mt^- gametes may essentially be a foregone conclusion once mt-specific associations are established between the outspread flagella of a mating pair and the mt^+ fertilization tubule extends towards its mt^- 'target'. Also unknown is the mechanism of the fusion reaction itself, e.g. whether it is an enzymatic reaction (Poole et al., 1970), a protein-mediated destabilization of lipid bilayers (Van der Bosch and McConnell, 1975), or some other process. Ignorance of membrane-fusion mechanisms, I should note, is essentially complete for all organisms (see Lucy, 1975 and Weiss et al., 1977 for reviews).

8.3.5 Mutations affecting zygotic cell fusion

Two mutations in *C. reinhardi* – *imp-1* and *imp-1-15* – interfere with zygotic cell fusion without affecting flagellar agglutination. Gametes in an *imp-1* x mt^- mating agglutinate for 24–48 h and only rarely manage to fuse to form zygotes for genetic analysis. The mt^- mating structures activate in such crosses, but no mt^+ mating structure activation is observed. Many *imp-1* gametes, moreover, bear abnormal-looking mating structures (Goodenough and Weiss, 1975). The *imp-1* mutation has not been observed to recombine with mt^+ in numerous crosses (Goodenough et al., 1976), suggesting that it may affect a gene within the mt^+ locus.

The mutation carried by the *imp-1-15* strain appears to affect the membrane fusion process more specifically. The *imp-1-15* strain, a clone isolated from the *imp-1* stock, is capable of generating a microfilament-filled fertilization tubule from its doublet zone but nonetheless fails to fuse with mt^- gametes (Goodenough, unpublished). The existence of such a phenotype suggests that the fertilization-tubule tip is indeed an active participant in the fusion process.

8.4 THE FLAGELLUM-TO-GAMETE AND ZYGOTE-TO-FLAGELLUM SIGNALS

8.4.1 The flagellum-to-gamete signal

The previous section described at least two phenomena—cell wall lysis and mating-structure activation – which were triggered by *mt*-specific flagellar membrane agglutination. The flagellum-to-gamete signal might travel down the flagellar membrane and out the plasma membrane to the mating-structure region which lies about 1μm from one of the flagellar bases. Alternatively, it might travel down the axoneme microtubules, into the basal bodies, and out to the mating structure *via* microtubules and a striated fiber that extends from one basal body directly to one edge of the mating structure (Goodenough and Weiss, manuscript in preparation). Finally, it might be transmitted *via* matrix components of the flagella. The inability to generate or to 'receive' such a signal would presumably result in a non-fusing phenotype, although this does not appear to be the defect in the *imp-1-15* strain (Section 8.3.5).

Poor signal transmission may, on the other hand, account for the *imp-3* and *imp-4* phenotypes: cells carrying either mutation agglutinate well but fuse with low (but variable) efficiency (a phenotype distinct from the 'physiological poor maters' described in Section 8.2.9, which usually agglutinate poorly). The morphology of the mating structure is normal in *imp-3* and *imp-4* cells; the mutations are unlinked to *mt* and to each other, and can be expressed in either mating type (Goodenough *et al.*, 1976).

Poor signal transmission appears to be the defect in the temperature-sensitive mutant *gam-1*. Forest and Togasaki isolated this strain (Forest and Togasaki, 1975) and established that it could agglutinate, but not fuse, at 35°C and acquired the ability to fuse if shifted to 25°C for a one-hour period. Tetrad analysis (Forest and Togasaki, 1976) revealed that the *gam-1* mutation is unlinked to *mt* but is expressed only in mt^- strains: it can be carried by mt^+ cells without affecting their fusibility, and transmitted through them to mt^- zygotic products which once again express the trait (Forest and Togasaki, 1976). Structural studies of the *gam-1* mutant (Forest, Goodenough, and Goodenough, in preparation) reveal that at 35°C, agglutination activates neither the *gam-1* mt^- mating structure *nor* wild-type mt^+ mating structures. When such agglutinating cells are shifted to 25°C for an hour, activation of both mating structures occurs and fusion ensues. The agglutination reaction generated by *gam-1* mt^- cells at restrictive temperature, therefore, is apparently devoid of signalling activity.

8.4.2 The zygote-to-flagellum signal

The zygote-to-flagellum signal, transmitted from a fusing zygote, abolishes the *mt*-specific agglutinability of all four zygote flagella (Lewin, 1954) without affecting either the Con A or antibody agglutinability of the flagellar surfaces (Bergman, Goodenough, and Jurivich, unpublished). This abolition of sexual agglutinability is somehow conferred upon the flagellar-membrane 'precursor pool' as well, for if zygotes are deflagellated, they grow back 4 flagella that are as fully non-agglutinable as the ones they replace (Bergman, unpublished). Nothing is known about the nature of this signal.

The loss of flagellar agglutinability by *Chlamydomonas* zygotes may have two selective advantages. First, quadriflagellates are thereby able to swim about seeking an appropriate environment for encystment (they become negatively phototactic at this stage (Adams, 1975)) without being encumbered by extraneous gametes hanging onto their flagella. Second, gametes that participate in an initial large 'clumping' reaction are released from any cells that form zygotes and are free to seek new partners, resulting in 100% mating efficiencies for large cell populations. As noted at the outset of this article, such efficiency, coupled with the apparent simplicity of the interacting membranes and the availability of mutant strains, makes the mating interactions of *Chlamydomonas* a fertile area of study.

REFERENCES

Adams, M. (1975), Effects of sunlight on inheritance of chloroplast genes in *Chlamydomonas reinhardi. Genetics,* **80**, 58.

Bergman, K., Goodenough, U.W., Goodenough, D.A., Jawitz, J. and Martin, H. (1975), Gametic differentiation in *Chlamydomonas reinhardtii.* II. Flagellar membranes and the agglutination reaction. *J. Cell Biol.,* **67**, 606–622.

Bosmann, H.B. and McLean, R.J. (1975), Gametic recognition: lack of enhanced glycosyltransferase ectoenzyme system activity on nonsexual cells and sexually incompatible gametes of *Chlamydomonas. Biochem. biophys. Res. Commun.* **63**, 323–327.

Bouck, G.B. (1972), Architecture and assembly of mastigonemes. In: *Advances in Cell and Molecular Biology,* Vol. 2. (Dupraw, E.J., ed.), New York, Academic Press, pp. 237–271.

Buffaloe, N.D. (1958), A comparative cytological study of four species of *Chlamydomonas. Bull. Torrey Bot. Club,* **85**, 151–163.

Cavalier-Smith, T. (1975), Electron and light microscopy and gametogenesis and gamete fusion in *Chlamydomonas reinhardii. Protoplasma,* **86**, 1–18.

Chiang, K.S., Kates, J.R., Jones, R.F. and Sueoka, N. (1970), On the formation of homogeneous zygotic populations in *Chlamydomonas reinhardtii. Dev. Biol.,* **22**, 655–669.

Claes, H. (1971), Autolyse der Zellwand bei den Gameten von *Chlamydomonas reinhardii. Arch. Mikrobiol.,* **78**, 180–188.

Claes, H. (1975), Influence of concanavalin A on autolysis of gametes from *Chlamydomonas reinhardii. Arch. Mikrobiol.,* **103**, 225–230.

Edelman, G.M. (1974), Surface alterations and mitogenesis in lymphocytes. In: *Control of Proliferation in Animal Cells.* (Clarkson, B. and Baserga, R. eds.), Cold Spring Harbor Laboratory, pp. 357–378.

Forest, C.L. (1976), A genetic analysis of gametogenesis in *Chlamydomonas reinhardi.* Ph. D. Thesis, Indiana University, Bloomington, Indiana.

Forest, C.L. and Togasaki, R.K. (1975), Selection for conditional gametogenesis in *Chlamydomonas reinhardi. Proc. natn. Acad. Sci. U.S.A.,* **72**, 3652–3655.

Forest, C. and Togasaki, R.K. (1976), Conditional gamete formation in *Chlamydomonas reinhardi. J. Cell Biol.,* **70**, 189a.

Forest, C.L. and Togasaki, R.K. (1977), A selection procedure for obtaining conditional gametogenic mutants using a photosynthetically incompetent strain of *Chlamydomonas reinhardi. Molec. gen. Genet.* (in press).

Goodenough, U.W., Hwang, C. and Martin, H. (1976), Isolation and genetic analysis of mutant strains of *Chlamydomonas reinhardi* defective in gametic differentiation, *Genetics,* **82**, 169–186.

Goodenough, U.W. and Weiss, R.L. (1975), Gametic differentiation in *Chlamydomonas reinhardi.* III. Cell wall lysis and microfilament-associated mating structure activation in wild-type and mutant strains. *J. Cell Biol.,* **67**, 623–637.

Hynes, R.O. (1974), Role of surface alterations in cell transformation; the importance of proteases and surface proteins. *Cell,* **1**, 147–156.

Johnson, U.G. and Porter, K.R. (1968), Fine structure of cell division in *Chlamydomonas reinhardi.* Basal bodies and microtubules. *J. Cell Biol.,* **38**, 403–425.

Jurivich, D.A. (1976), Agglutination in *Chlamydomonas reinhardtii* gametes. Undergraduate honors thesis, Dept. of Biology, Harvard University.

Kates, J.R. and Jones, R.F. (1964), The control of gametic differentiation in liquid cultures of *Chlamydomonas*. *J. Cell comp. Physiol.*, **63**, 157–164.

Lewin, R.A. (1954), Sex in unicellular organims. In: *Sex in Micro-organisms*, (Wenrich, D.H., ed), *Am. Assoc. Adv. Sci.*, Washington, pp. 100–133.

Lucy, J.A. (1975), Aspects of the fusion of cells *in vitro* without viruses. *J. Reprod. Fert.*, **44**, 193–205.

Luzzati, V. (1968), X-Ray diffraction studies of lipid-water systems In: *Biological Membranes. Physical Fact and Function*, (Chapman, D. ed.), Academic Press, New York, pp. 71–124.

Martin, N.C. and Goodenough, U.W. (1975), Gametic differentiation in *Chlamydomonas reinhardtii*. I. Production of gametes and their fine structure. *J. Cell Biol.*, **67**, 587–605.

McLean, R.J. and Bosmann, H.B. (1975), Cell–cell interactions: Enhancement of glycosyltransferase ectoenzyme systems during *Chlamydomonas* gametic contact. *Proc. natn. Acad. Sci., U.S.A.*, **72**, 310–313.

McLean, R.J. and Brown, R.M. (1974), Cell surface differentiation of *Chlamydomonas* during gametogenesis. I. Mating and concanavalin A agglutinability. *Dev. Biol.*, **36**, 279–285.

McLean, R.J., Laurendi, C.J. and Brown, R.M. Jr. (1974), The relationship of gamone to the mating reaction in *Chlamydomonas moewusii*. *Proc. natn. Acad. Sci. U.S.A.*, **71**, 2610–2613.

Minami, S. (1976), Protein synthesis during differentiation of *Chlamydomonas reinhardi*. Ph. D. thesis, Harvard University.

Moewus, F. (1933), Untersuchungen uber die sexualitat und entwicklung von *Chlorophyceae*. *Arch. Prot.*, **80**, 469–526.

Poole, A.R., Howell, J.I. and Lucy, J.A. (1970), Lysolecithin and cell fusion. *Nature*, **227**, 810–813.

Randall, J., Cavalier-Smith, T., McVittie, A., Warr, J.R. and Hopkins, J.M. (1968), Developmental and control processes in the basal bodies and flagella of *Chlamydomonas reinhardi*. *Symp. Soc. Dev. Biol.*, **26**, 43–83.

Ringo, D.L. (1967), Flagellar motion and fine structure of the flagellar apparatus in *Chlamydomonas*. *J. Cell Biol.*, **33**, 543–571.

Roseman, S. (1970), The synthesis of complex carbohydrates by multi-glycosyltransferase systems and their potential function in intercellular adhesion. *Chem. Phys. Lipids*, **5**, 270–297.

Rosenbaum, J.L., Moulder, J.E. and Ringo, D.L. (1969), Flagellar elongation and shortening in *Chlamydomonas* – The use of cycloheximide and colchicine to study the synthesis and assembly of flagellar proteins. *J. Cell Biol.*, **41**, 600–619.

Rothfus, J.A. and Smith, E.L. (1963), Glycopeptides IV. The periodate oxidation of glycopeptides from human γ-globulin. *J. biol. Chem.*, **238**, 1402–1410.

Rutishauser, U. and Sachs, L. (1974), Receptor mobility and the mechanism of cell–cell binding induced by concanavalin A. *Proc. natn. Acad. Sci. U.S.A.*, **71**, 2456–2460.

Satir, P. and Satir, B. (1974), Design and function of site-specific particle arrays in the cell membrane. In: *Control and Proliferation in Animal Cells.* (Clarkson, B. and Baserga, R. eds.), Cold Spring Harbor Laboratory, pp. 233–249.

Schlosser, U.G., Sachs, H. and Robinson, D.G. (1976), Isolation of protoplasts by means of a 'species-specific' autolysine in *Chlamydomonas. Protoplasma,* **88**, 51–64.

Schmeisser, E.T., Baumgartel, D.M. and Howell, S.H. (1973), Gametic differentiation in *Chlamydomonas reinhardi*: Cell cycle dependency and rates in attainment of mating competency. *Dev. Biol.,* **31**, 31–37.

Sharon, N. and Lis, H. (1972), Lectins: Cell-agglutinating and sugar-specific proteins. *Science,* **177**, 949–959.

Snell, W.J. (1976a), Mating in *Chlamydomonas*: A system for the study of specific cell adhesion. I. Ultrastructural and electrophoretic analysis of flagellar surface components involved in adhesion. *J. Cell Biol.,* **68**, 48–69.

Snell, W.J. (1976b), Mating in *Chlamydomonas*: a system for the study of specific cell adhesion. II. A radioactive flagella-binding assay for quantitation of adhesion. *J. Cell Biol.,* **68**, 70–79.

Thomas, D.L. and Delcarpio, J.B. (1970), Isozymes of mating types of *Chlamydomonas. J. Phycol.,* **6**, 8–9.

Triemer, R.E. and Brown, R.M. Jr. (1975a), The ultrastructure of fertilization in *Chlamydomonas moewusii. Protoplasma,* **84**, 315–326.

Triemer, R.E. and Brown, R.M. Jr. (1975b), Fertilization in *Chlamydomonas reinhardi,* with special reference to the structure, development, and fate of the choanoid body. *Protoplasma,* **95**, 99–108.

Tsai, C.M., Huang, C.C. and Canellabis, E.S. (1973), Iodination of cell membranes I. optimal conditions for the iodination of exposed membrane components. *Biochim. biophys. Acta,* **332**, 47–58.

van der Bosch, J. and McConnell, H. (1975), Fusion of dipalmitoylphosphatidyl choline vesicle membranes induced by concanavalin A. *Proc. natn. Acad. Sei. U.S.A.,* **72**, 4409–4413.

Weiss, R.L., Goodenough, D.A. and Goodenough, U.W. (1977), Membrane differentiations at sites specialized for cell fusion. *J. Cell Biol.,* **72**, 144–160.

Wiese, L. (1965), On sexual agglutination and mating-type substances (gamones) in isogamous heterothallic chlamydomonads. I. Evidence of the identity of the gamones with the surface components responsible for sexual flagellar contact. *J. Phycol.,* **1**, 46–54.

Wiese, L. (1969), Algae. In: *Fertilization: Comparative Morphology, Biochemistry, and Immunology.* Vol. 2. (Metz, C.B. and Monroy, A. eds), Academic Press, New York, pp. 135–188.

Wiese, L. (1974), Nature of sex-specific glycoprotein agglutinins in *Chlamydomonas. Ann. N.Y. Acad. Sci.,* **234**, 383–394.

Wiese, L. and Hayward, P.C. (1972), On sexual agglutination and mating-type substances in isogamous dioecious *Chlyamydomonads.* III. The sensitivity of sex cell contact to various enzymes. *Am. J. Bot.,* **59**, 530–536.

Wiese, L. and Jones, R.F. (1963), Studies on gametic copulation in heterothallic Chlamydomonads. *J. cell comp. Physiol.,* **61**, 265–274.

Wiese, L. and Metz, C.B. (1969), On the trypsin sensitivity of fertilization as studied with living gametes in *Chlamydomonas*. *Biol. Bull.*, **136**, 483–493.

Wiese, L. and Shoemaker, D.W. (1970), On sexual agglutination and mating type substance (gamones) in isogamous heterothallic chlamydomonads. II. The effect of concanavalin A upon the mating-type reaction. *Biol. Bull.* **138**, 88–95.

Wiese, L. and Wiese, W. (1965), On sexual agglutination and mating type substances in *Chlamydomonas*. *Am. J. Bot.*, **52**, 621.

Wiese, L. and Wiese, W. (1975), On sexual agglutination and mating type substances in isogamous dioecious Chlamydomonads. IV. Unilateral inactivation of the sex contact capacity in compatible and incompatible taxa by α-mannosidase and snake venom protease. *Dev. Biol.*, **43**, 264–276.

Witman, G.B., Carlson, K., Berliner, J. and Rosenbaum, J.L. (1972), *Chlamydomonas* flagella. I. Isolation and electrophoretic analysis of microtubules, matrix, membranes, and mastigonemes. *J. Cell Biol.*, **54**, 507–539.

9 Cell – Cell Interactions in Ciliates: Evolutionary and Genetic Constraints

D. L. NANNEY

9.1	Introduction to the Ciliates	page	353
9.2	General requirements and constraints on cellular unions		355
	9.2.1 Nutritional state		356
	9.2.2 Genetic distance		357
	9.2.3 Stage of clonal life cycle		357
	9.2.4 Stage of cell cycle		359
	9.2.5 Other environmental circumstances		360
9.3	Mating in *Paramecium*		361
	9.3.1 Mating types, named species and sibling species complexes		361
	9.3.2 The mating act, chemical induction and *in vitro* analysis		362
	9.3.3 Mating-type determination and inheritance		365
	9.3.4 Evolutionary significance of compatibility controls		368
	9.3.5 Life history and developmental considerations		370
	(a) *Mating type determination, 370,* (b) *Mating immaturity, 370,* (c) *Adolescence, 371,* (d) *Maturation, 372*		
9.4	Mating in *Tetrahymena*		373
	9.4.1 Mating types and mating-type determination		373
	9.4.2 Analysis of the mating sequence		375
	9.4.3 Cortico-nuclear interactions in conjugation		378
9.5	Mating in *Euplotes*		381
	9.5.1 Mating behavior and mating-type inheritance		381
	9.5.2 Soluble gamones		382
9.6	Mating in *Blepharisma*		384
9.7	Summary and Conclusions		388
	References		390

Acknowledgements

The author is pleased to acknowledge the helpful criticisms provided by Sally Allen, Joseph Frankel, Akio Miyake, Dennis Nyberg, Eduardo Orias, Ellen Simon, Tracy Sonneborn and our editor, Jose Reissig. These respondents must not, however, be held responsible for residual omissions or errors of fact or interpretation.

Microbial Interactions
(*Receptors and Recognition,* Series B, Volume 3)
Edited by J.L. Reissig
Published in 1977 by Chapman and Hall, 11, New Fetter Lane, London EC4P 4EE
© Chapman and Hall

9.1 INTRODUCTION TO THE CILIATES

The chief cellular interactions in the ciliated protozoa are those associated with conjugation. In considering cell—cell interactions in ciliates, therefore, we are concerned with all the mechanisms that bring together cells of appropriate kinds for mating and/or prevent the union of inappropriate cells. We are also interested in the mechanisms that regulate and co-ordinate the complex nuclear and cortical modifications associated with conjugation, so that the processes occur in synchrony and reach a proper conclusion. Although descriptions of phenomena associated with conjugation are abundant (Grell, 1967; Raikov, 1972), and distributed over a wide variety of forms, analytical studies are much more limited, and even the best of these, though exciting in their promise, fall short of providing a complete mechanistic understanding.

The earlier work of importance was summarized capably by Hiwatashi (1969a) and by Miyake (1974); Crandall (1977) has just published a more general review of microbial mating and Miyake (1977) has prepared a critical evaluation of the molecular basis for ciliate conjugation. These reviews relieve the burden of a complete literature survey and permit a more general analysis.

One central problem underlying all attempts to synthesize the scattered studies is the relationship among the organisms, and hence, the generality of conclusions established on any one species. Although fossil records of ciliates are unavailable, we may assume that ciliates are of very ancient origin; they probably date to a time shortly after the first known eucaryotes originated and were almost certainly well-established by a billion years ago. Although some diversification has certainly occurred more recently, some of the organisms being compared are very different, and must have had an enormous time in which to modify their originally similar structures and functions. Unfortunately, no *a priori* judgment is certain concerning the mechanisms likely to be invariant, as opposed to those susceptible to marked evolutionary alterations. Some structures and some mechanisms, the form and function of cilia – or of ribosomes – for example, are notoriously stable, presumable because cilia and ribosomes represent early biological inventions of great utility which are sustained in an extremely stable form by their isolation on an adaptive peak. The necessary function is performed so well that it can scarcely be improved upon without a complete disassembly and redesign; and that radical alteration is prevented by the usual necessity for stepwise modifications in evolutionary processes.

The comparative problem within the ciliates is, of course, the same problem posed by this entire volume on a larger scale. However diversified the ciliates may be, they are still more similar among themselves than they are to bacteria or algae or fungi. Our hope, of course, is to identify primordial and pervasive principles governing cellular interactions, and mechanisms fundamentally similar in all organisms (Reissig, 1974). We may not assume in advance, however, that all the organisms surveyed — even all of those in one taxonomic basket — utilize the same means to achieve apparently similar ends. That is a question to be asked rather than a conclusion to be accepted. Some important new principles, for example, must have been included in the design of eucaryotes.

The fact that ciliates are classified together indicates that they have striking similarities. The common features are their elaborately patterned and ciliated cortex and a dimorphic nuclear system composed characteristically of one or more diploid germinal nuclei (micronuclei) and one or more compound somatic nuclei (macronuclei). I will not attempt to survey the enormous variety of cortical patterns and morphogenetic mechanisms which distinguish the ciliates (Hanson, 1967), because the organismic diversity in these respects is obvious. It is worth mentioning, however, that recent research indicates an unexpected diversity within the general constraints of 'nuclear dimorphism'.

The micronuclei are relatively inert nuclei in all ciliates, although they perform essential but unknown vegetative functions in some species (Nanney and McCoy, 1976). They are the reserve nuclei called into action only when sexual episodes occur. At these times the macronucleus is discarded via fragmentation and/or dissolution, and a new somatic nucleus is developed. This new macronucleus then becomes the metabolic control center, synthesizing RNA and manifesting its presence by specific genetically controlled phenotypes.

Nuclear dimorphism represents a solution to a particular set of biological 'needs', perhaps a need to become larger while maintaining protoplasmic continuity, a need to equip progeny with a substantial protoplasmic investment while giving them genetic independence, a need to insulate the germ line from metabolic accidents, etc. Although the details whereby ciliates bridge the generation gap differ somewhat, the general patterns of nuclear behavior and transformation are very much the same (Raikov, 1972). Following fertilization, a zygote nucleus, characteristically diploid, divides one or a few times and the identical products are distributed by controlled spindle orientations to distinctive cytoplasmic regions that determine their developmental fates — to continue as quiescent micronuclei or to undergo a period of rapid DNA synthesis and transformation into a functional somatic nucleus.

One would expect that these functional somatic nuclei would be essentially similar in their organization, but this is apparently not so (Raikov, 1976). *Paramecium* and *Tetrahymena* certainly differ greatly in their total compoundness (Allen and Gibson, 1972; Gorovsky, 1973), and they may differ in the way the chromosomal materials are organized (Allen and Gibson, 1973; Orias and Flacks, 1975; Nyberg, 1976; Preer, 1976). Nevertheless, they are similar in another important respect: the DNA of the micronucleus and that in the macronucleus are very similar, and probably differ only for a few sequences — such as that of ribosomal DNA — which are amplified in the macronucleus (Gall, 1974; Engberg *et al.*, 1974). This DNA conservatism in development is apparently not found in some other large groups of ciliates.

Nucleic acid hybridization analyses (Prescott *et al.*, 1973) indicate that only a small fraction of the micronuclear DNA sequences occur in the completed macronucleus in *Stylonychia*. Chromatin elimination of a drastic sort occurs during macronuclear development in these forms (Ammermann, 1965; Kloetzel, 1970).

An acceptable reconciliation of all the genetic, cytological and biochemical observations on the ciliate macronucleus is not yet available. What is certain is that some care is required in applying conclusions to one ciliate species which have been derived from another. Yet extrapolation is necessary for a comprehensive understanding. The diverse ciliates provide different experimental opportunities and have been studied in different ways. *Paramecium* (Sonneborn, 1975b; van Wagtendonk, 1974) has been studied intensively for the longest period of time and has the best understood genetic systems, but it does not produce soluble gamones. *Tetrahymena* (Sonneborn, 1975a; Elliott, 1973) has come under more recent extensive study, and its genetic properties and ease of cultivation in axenic media make it ideal for certain kinds of biochemical manipulation, but it does not produce soluble mating substances either. Ironically, the organism providing the best opportunity for studying certain aspects of the molecular basis of conjugation, *Blepharisma*, is very poorly understood in its genetics, nutrition and physiology except in a few special areas (Giese, 1973).

Because of the hazards of extrapolation, I have not been comfortable with a casual amalgamation of information from diverse organisms, and have segregated the observations from different genera. Because of the 'underclassification' of the ciliates, and their ancient origins, we should be wary even of comparisons within these genera. Before approaching the organismic groups, however, a few general comments concerning the circumstances of conjugation may be helpful. Then we will consider comparative genetic and molecular data that may enable us better to evaluate the comparability of mechanisms governing cellular interactions in these organisms.

9.2 GENERAL REQUIREMENTS AND CONSTRAINTS ON CELLULAR UNIONS

For all these organisms, as with most unicells, the mating act is the chief and in some cases the sole social event in the life cycle. It is an important and complicated maneuver that requires precision and co-ordination to be completed successfully. Potential mates must first be brought within mating distance. This physical association may in some instances be accomplished by random movement or by congregation in response to a common stimulus; in other cases, specific chemoattractants may be involved. Once propinquity has been achieved, cellular recognition factors must come into operation so that appropriate cell unions are promoted and inappropriate unions are discouraged. These factors are characteristically associated with the cell membranes, especially the membranes surrounding the cilia which cover much of the cell surface. Cell recognition leads to the elaboration of surface components capable of cementing cells together. Once cellular fusion has occurred, the complex cortical and nuclear changes required for meiosis, reciprocal fertilization and somatic nuclear replacement must be initiated and co-ordinated.

Unfortunately, we know very little about the molecules involved in these interactions. Because the cells that unite in conjugation usually are of different compatibility types, we sometimes loosely refer to the molecules controlling the mating specificities as 'the mating-type substances'. This usage, however, can lead to unwarranted assumptions. Some ciliates produce chemo-attractants; some produce soluble specific products that stimulate mating readiness in cells of complementary type; all ciliates have membrane-bound components responsible for specific cellular agglutination. Such specific substances which lead to the formation of conjugant pairs have all been called, by an extension of the original definition, *gamones* (Miyake, 1974). In at least one instance an identifiable soluble gamone has both a chemo-attractant role and a stimulatory role, but the generality of this role multiplicity is unknown. Perhaps the same molecules are used in different phases of the mating process, differing only in their localizations, associations (free or embedded in membranes) or in their quantities (perhaps in low quantity during recognition and in high quantity during agglutination). But these possible relations need not be correct, and the distinctions between chemo-attractants, agglutinins and other mating substances should not be blurred. The difderent phases of the mating act may be achieved by entirely distinct molecular mechanisms.

Regardless of the mechanisms, the act of conjugation is complicated and hazardous to the participants. The difficulties attendant upon conjugation may be illustrated by comparing the probabilities of death following conjugation and fission. Viability losses which can be reasonably assigned to 'errors' in completing cell division are usually far less than 1% in most ciliates. In contrast, only rarely will a cross between ciliate cultures yield 99% viable progeny; 90% is a more commonly observed value, and crosses in some species have never given more than 10% survival, even when many combinations of strains have been used. With such selection against cells that mate, it is understandable that mating is a carefully regulated activity occurring only under special circumstances which maximize its biological benefits. It is not even surprising that some ciliates have abandoned sexuality entirely.

9.2.1 Nutritional state

One constraint on mating that is rationalized in this context is the common requirement for prior starvation. Under laboratory conditions *Tetrahymena* conjugation usually requires some 12 hours between cell union and cell separation, and an additional interval of perhaps 6–8 h before the first post-zygotic cell division. During this interval a non-mating cell provided with nutrient can divide once every three hours and produce a total of 64 or more cells, in contrast to the two produced from each conjugant. Thus, the selective advantage of non-mating over mating, based entirely on differential reproduction in the presence of food, is of the order of 32:1. The restriction of mating to conditions of limited nutrient is therefore easily understood.

9.2.2 Genetic distance

Another constraint on mating involves the choice of a mating partner. Whatever the benefits of conjugation, they are related in part to the genetic properties of the individuals involved. Organisms of very distant genetic relationship may be able to conjugate even though the hybrid nuclear constitution is incapable of maintaining life. Conjugation between *Paramecium tetraurelia* and *P. octaurelia* provides an example of sterile interactions in ciliates (Haggard, 1974). The F_1 progeny are vigorous, but no F_2 or backcross progeny survive. Under natural conditions, such matings and others giving reduced hybrid survival would be strongly selected against. At the other extreme, matings which occur among cells of the same clone may have a much smaller probability of producing useful genetic combinations than do cells of somewhat more distant origin. One would therefore expect natural selection to control the genetic relationships of the mating cells, in response to the organisms' requirements for recombinational and mutational variety. Such considerations undoubtedly account for the origin of the complex systems of cellular complementarity associated with conjugation (Sonneborn, 1957; Nyberg, 1974), but are beyond our special concerns here.

9.2.3 Stage of clonal life cycle

A related constraint concerns the requirement for sexual maturity. Ciliates generally have a complete life cycle (Maupas, 1889; Jennings, 1944; Sonneborn, 1954a; Nanney, 1974). Beginning at fertilization it commonly contains an interval of sexual immaturity, a time of maturity, and finally a stage of senescence and death if a sexual event does not interrupt the progression and re-initiate the cycle. Some ciliates have short periods of immaturity and others have very long periods — lasting under some conditions for several years. The mechanisms for the changes occurring at maturation are not understood; presumably the molecular components that potentiate compatibility are not synthesized until this time, or are not synthesized in sufficient quantity, or are not located at effective external sites. Alternatively the mating molecules may be present but blocked by some additional immaturity materials.

The adaptive significance of the length of the immaturity period, and hence its regulability by genetic factors, is supported by the following considerations (Sonneborn, 1957). The length of the immaturity period is a measure of the physical distance a ciliate, when it achieves maturity, has travelled from its point of origin, and also a measure of the genetic distance of potential mates. At its origin a cell is associated with cells of its parental clones and with siblings produced in the same cross. After some weeks or months of reproduction and dispersal it is more likely to be associated with unrelated cells. Species whose evolutionary strategies involve inbreeding are expected on this interpretation to have short periods of immaturity, while ciliates seeking recombinational variety with strangers may inhibit inbreeding by long immaturity periods. Sonneborn (1957) found the length of the immaturity period to be one of the best general indicators of the position of a ciliate on an inbreeding-outbreeding scale, and Nyberg (1974) showed a strong correlation between immaturity intervals and other evidence of outbreeding genetic economies.

Another indication of the adaptive role of the immaturity period comes from efforts to modify it. Most clones of *T. thermophila* require 50 or more cell divisions to achieve full maturity, but mutations frequently occur to shorten the immaturity period, to about 12–15 cell divisions. The Em (early mature) mutants are distributed over a number of genetic loci, but they manifest one common characteristic: they all show slow and irregular growth in the interval immediately following conjugation (Bleyman, 1971). This observation was rationalized recently when it was discovered that various kinds of mistreatment applied to conjugating pairs during macronuclear development gave rise to early mature phenocopies, i.e. lines which showed early maturity, but unlike the Em mutants did not transmit this property to their progeny (Nanney and Meyer, 1977). The common factor revealed in these studies was a genotype-environment incompatibility manifested during or immediately after the development of a new macronucleus. The explanation suggested for both kinds of early maturity is the activation of a 'maturity shunt', which bypasses the usual long-term temporal control system, or the removal of an immaturity block. In either case, a ciliate that discovers a hostile environment at its origin, has recourse to a rapid return to the genetic lottery, which is ordinarily forbidden by the ecological strategy of its species. How a cell might sense an incompatibility, or how it might use such signals to shorten the temporal program, cannot be understood until we are better informed in general about the mechanisms of temporal programming.

One other evidence of the adaptive significance of mating in the clonal life cycle concerns the phenomenon sometimes referred to as 'senile selfing'. Selfing is a general term covering conjugation within a clonal culture, ordinarily between cells of the same genotype. Some species seem to contain only selfing clones, i.e. no barriers to intraclonal mating are apparent though interspecific conjugation is forbidden. In other species, some clones are of pure mating type while others are selfers. In these cases the mating type restrictions on cell union are not necessarily violated in selfing clones; at least in many cases, selfing is due to the heterogeneity of the clone with respect to mating type; two or more different mating types are expressed within the same clone, and the cells uniting in conjugation are of different type. Clonal heterogeneity of mating types is often related to the clonal life cycle. Cells of *T. thermophila*, for example, when young often have the ability to produce two or more mating types (Nanney and Caughey, 1955), but older cells lose this capability. This capacity is of little practical consequence because it is usually lost well before the time of sexual maturity, so that it does not commonly potentiate the conjugation of cells separated by only a few cell divisions.

This 'juvenile selfing', however, is different from senile selfing – which occurs only as the organisms approach the limits of their life cycles without encountering appropriate mates. Jennings (1942, 1944) first reported senile selfing in *Paramecium bursaria,* and more recently Heckmann (1967) has analyzed an interesting case in

Euplotes crassus. Only clones of certain genotypes manifest selfing, particularly clones heterozygous for the mating-type alleles which determine mating types in this species. Ordinarily one of the alleles is fully dominant over the others, but old heterozygous clones — 400 or more cell divisions from their origin — simultaneously begin manifesting their recessive mating types in some of their cells. Just as genotype—environment incompatability seems to trigger a normally forbidden early conjugation in *Tetrahymena,* so an imminent decline into senility and death in *Euplotes* may potentiate an ordinarily forbidden conjugation among close relatives. The adaptive significance of such an alternative, particularly for a colonizing species, is apparent.

9.2.4 Stage of cell cycle

Another general constraint on mating should also be noted. Conjugation is a complex series of alterations affecting eventually all parts of the organism; from the tips of the cilia to the micronuclear histones. These changes in the synthetic patterns and nuclear behavior of the two cells must be precisely co-ordinated so that at a particular moment in time — some hours after the initiation of the processes — pronuclei are simultaneously ready for co-ordinated transfer. Some of this integration of activities is achieved by a synchronization of the cells with respect to their cell cycles prior to the initiation of mating (Pieri, 1965; Pieri *et al.,* 1968; Woodard *et al.,* 1966; Golicova, 1974; Doerder and Debault, 1976). Cells mating under natural conditions have usually shared a common environment for at least the recent past; they are therefore likely to be in similar chronotypic and physiological states. They are also usually not in an active growth state, as mentioned above, and hence, have come to rest at a characteristic stage of the cell cycle — usually in the G1 stage. The significance of this common environmental experience is shown by the consequences of mating cells in the laboratory which have had different recent environmental histories. Although cell fusion may occur, the cells have only a limited capacity to adjust to their partners' physiological rhythms. *Tetrahymena* conjugation, for example, between strains adapted in advance to different temperature conditions, is seldom successfully completed (Nanney, 1963; see also Sonneborn, 1954b). Although cellular adhesion may occur, the persistent differences in cellular tempos prevent a normal synchronization of events. The fact that ciliate conjugation ordinarily occurs in the G1 stage of the macronuclear cell cycle is probably not to be understood as signifying some special feature of G1, common to all processes of 'cellular differentiation'. Rather it probably indicates only that conjugating cells must be co-ordinated, and that G1 is a somewhat easier stage within which to achieve synchrony than others. In any case, ciliates are not uniform in the stage of the cell cycle at which conjugation is initiated (Luporini and Dini, 1975).

The impression should not be given that mating co-ordination is achieved entirely by common environmental entrainment. In nearly all ciliates an interval of interaction precedes actual cell union. In some cases this interaction is manifested by a preliminary ciliary agglutination, or by a stereotyped 'mating dance'. In some species it involves soluble gamones; in other species it requires surface interactions achieved by repeated cell contacts. Although the details are various, they all attest to the need for preparation for the mating act, and one of the necessities for a successful act, is that the partners be fine-tuned with respect to their temporal pattern. This fine-tuning is probably achieved by the preparatory cellular interactions. The possibility of continuing co-ordination during conjugation is also suggested by certain observations.

9.2.5 Other environmental circumstances

From the preceding account, it should be apparent that the circumstances under which mating occurs in ciliates, and the choice of the participants in it, are carefully regulated so as to optimize its contribution to the genetic economy of the species. Because of the diverse ecological niches and evolutionary strategies within the phylum Ciliata, one would expect a great diversity of special regulatory systems to be superimposed upon any possible common mechanisms of cell fusion and co-ordination. Just as some species restrict mating to older organisms, or to relatives at a specified genetic distance, so other species arrange conjugation to occur in co-ordination with various environmental conditions. The availability of food is one constraining environmental variable already mentioned, but by no means the only one. Most ciliates, for example, are required to survive a considerable interval of time without abundant food, but mating is limited even within this time of nutritional deprivation by other factors. Strains of *Paramecium bursaria,* for example, have a diurnal mating reactivity which continues in the absence of environmental cues for several days (Ehret, 1953; Cohen, 1964, 1965; see also Sonneborn, 1938 and Karakashian, 1968). The persistent circadian rhythm of mating reactivity would suggest that a particular time of day may be the optimal time for this species to initiate a successful conjugation, except that different clones have their mating activated at different times in the diurnal cycle.

Another diurnal periodicity associated with mating is that reported for *Paramecium multimicronucleatum* (Barnett, 1966; Clark, 1972). In this case, the temporally patterned condition is not mating reactivity but mating type itself. At a particular time of day certain clones regularly shift from one of the two mating types of the species to the other. The adaptive significance of this tactic is obscure, but at certain times of day a clone is converted from a pure mating type into a kind of selfer, distinct from the juvenile and senile selfers mentioned earlier. At the transition times each day, as a culture transforms in only approximate synchrony from one mating type to the other, it contains two mating types which may engage in conjugation. It is hard to imagine that inbreeding (or outbreeding) has selective advantages at sunrise and sunset which are overruled at noon and midnight. But one may doubt that *some* of these auxiliary regulatory systems are adaptive without doubting that most of them contribute to the strategy for survival.

We may finally mention in passing that many special conditions of temperature,

9.3 MATING IN *PARAMECIUM*

9.3.1 Mating types, named species and sibling species complexes

The union of two ciliates in the act of conjugation was examined and properly interpreted as a sexual event almost as soon as improved microscopic techniques made the necessary observations possible (Bütschli, 1876). Indeed, some of the earliest reports of fusion of nuclei at fertilization came from the study of conjugation in *Paramecium*. Although conjugation was studied in *Paramecium* by Hertwig (1889) in the 1880's, and on a broad comparative basis by his Algerian contemporary, E. Maupas (1889), the ability to induce mating and hence to study conjugation under controlled conditions was delayed until T.M. Sonneborn's (1937) discovery of mating types in *Paramecium*.

Sonneborn's basic observation was that cultures of *Paramecium aurelia* from diverse sources often underwent conjugation when mixed, even though mating pairs were rare or absent from unmixed controls. Because conjugation is a reciprocal process in which the two participants play equivalent roles, the cell types in a union may not be designated as female or male; both cells are both contributors and recipients of pronuclei. Hence, the strain differences, clearly perceived by the paramecia but difficult for a human observer to recognize, were called *mating-type* differences.

The discovery of mating types led quickly to the discovery that *Paramecium aurelia* is a complex of sibling species, each containing two mating types. These mating types were assigned sequential Roman numerals while their genetic species were designated by Arabic numbers. Genetic species 1 possessed mating types I and II; genetic species 2 had types III and IV, etc. These species were initially called 'varieties' (Sonneborn, 1947), later 'syngens' (Sonneborn, 1957), and have eventually been assigned Latin binomials (Sonneborn, 1975c). Thus variety 1 was transformed into syngen 1 and finally into *Paramecium primaurelia*. Its mating types are still referred to as I and II, while the mating types of *P. tetraurelia*, for example, are still called types VII and VIII.

Although the mating types of sibling species are clearly different, they are nevertheless demonstrably homologous (Sonneborn, 1947, 1957). Gene flow between sibling species has not been found even when serious efforts were made to detect it (Haggard, 1974). Mating sometimes occurs between species, but it always leads to death in either the first or the second generation. When such mating occurs it may involve only one mating type of one species with one mating type of the other, or it may involve both mating types of both species. Never, however, has one mating type of one species conjugated with both mating types of another species. Sufficient numbers of interspecific matings have been observed to identify the homologous types in most of the species. These homologous classes are often referred to as 'odd' and 'even', in reference to the original enumeration. Each species thus has two complementary classes of cells, homologous to but not identical with the corresponding classes in the sibling species.

The *P. aurelia* complex is only one of several species complexes in the genus, though it is the most extensively studied group (Vivier, 1974) and the only one in which the species have been assigned Latin binomials (Sonneborn, 1975c). Experimental work on mating has also been extensive among the species of the *P. caudatum* group (Hiwatashi, 1968, 1969b; Takahashi and Hiwatashi, 1974; Myohara and Hiwatashi, 1975). These species are considered closely related to those of the *P. aurelia* complex, though they differ from them in being larger and in having distinctive micronuclei. They resemble them, however, in having dual mating systems, as do also the sibling species of the *P. multimicronucleatum* group.

The only other species complex providing a significant body of experimental studies is that of *P. bursaria*. These species appear more distantly related and are notable for their symbiotic algae, which give them their green color, as well as for their readily detectable form difference. For present purposes, however, we are more interested in the number of mating types per species. Shortly after Sonneborn's initial discovery of mating types in *P. aurelia*, H. S. Jennings (1939) reported a comparable situation in *P. bursaria*, except that the number of classes of mating type per species was more than two. More recent work (Bomford, 1966) indicates that some species contain four mating types and others eight mating types. As in the case of the *P. aurelia* species, some interspecific mating reactions have been reported, but only between the 8-type species. No mating interactions between named species have been observed, except for non-specific interactions observed after chemical induction; thus, no mating-type homologies between the complexes can be assigned. Direct one-to-one homologies could not be expected in any case if the number of types is not the same.

9.3.2 The mating act, chemical induction and *in vitro* analysis

In all the species of *Paramecium*, (Hiwatashi, 1969a; Miyake, 1974; Vivier, 1974) the mating act is initiated by an immediate agglutination of cells of diverse type into clumps in which the cells are associated by heterotypic ciliary adhesions. After approximately an hour, paired cells emerge from the clumps. These pairs are characteristically heterotypic, but homotypic pairs in low frequency are formed in some species. The attachments are initially in various positions, but they tend to involve the ventral and anterior surfaces to the greatest extent. Later, pairs are formed by a ventral homopolar alignment of two cells. The cortical structures in the areas of contact, cilia and trichocysts, undergo loss or dedifferentiation and intimate cortical union is achieved.

The first area of firm contact in conjugating pairs is the anterior ventral surface referred to as the 'holdfast' region. The loss of the ciliary reaction follows this union.

Later a second region of fusion develops in the vicinity of the mouth, located in *Paramecium* just posterior to the midregion of the cell. This region includes the 'paroral cone', the structure through which pronuclei are transferred to the mate, and no nuclear transfer occurs unless a firm paroral junction is established. A third but facultative attachment area has been observed at some time in *P. aurelia* (Sonneborn, 1955.) It is important to note that the homotypic pairs which form in some species usually do not achieve paroral unions while heterotypic pairs usually do have paroral unions. Nevertheless, nuclear reorganization beginning with meiosis, occurs in all pairs with holdfast attachments. Migratory pronuclei are produced and move to the paroral region, but cannot be transferred to the mate in homotypic pairs. Instead, the migratory nucleus returns to fertilize the stationary nucleus in the same cell and establish a uniparental and homozygous offspring. This process may also occur in heterotypic pairs when firm paroral unions fail to occur. It is called *cytogamy* to distinguish it from *conjugation* which requires mutual cross-fertilization and from *autogamy,* which is a similar process which may occur in unpaired cells at certain times in the life histories of some species. The cortical events associated with mating thus trigger a complex nuclear response, but this same response may be induced by a very different kind of signal which integrates nutritional and life cycle information.

Mating does not always require mating-type differences, even for the initial stimulation which results in homopolar pairs. As Miyake (1958) first showed for *P. caudatum,* and as was found later in a number of species (Miyake, 1968, 1974; Hiwatashi, 1969a) intraclonal conjugation may be induced by certain environmental manipulations. Although the precise conditions for optimal mating vary from organism to organism, a number of general features were discovered to be necessary of important.

(1) Non-specific mating occurs chiefly under Ca-poor conditions; it is greatly enhanced by EDTA in *P. caudatum* (Hiwatashi, 1959).
(2) It is inhibited by unidentified components of the culture medium, which can be removed by washing in salt solutions.
(3) It is enhanced by a number of ions, including K, Mg, Rb, Cs, and Mn.
(4) It is facilitated by certain organic compounds, including urea, biuret, and acetamide. Heparin and acriflavine may also be helpful; acriflavin or some other flavine is essential in the *P. aurelia* species.

Induced mating differs from normal mating in several significance respects. Not only do cells of the same mating type unite, but cells belonging to different species or even species complexes may become associated; two major classes of incompatability are therefore bypassed by the inducing treatments. The induced matings, moreover, are not preceded by the usual agglutination of the cilia, but the same time interval occurs between the start of the stimulation and the formation of the cell union. The cellular attachments with chemical induction are not as regular and stereotyped as are mating type-mediated unions. Cells are usually attached to each other by their holdfast regions, but a holdfast region may be attached to some other part of the ventral surface of the mate, as in the paroral cone region. Multiple associations and chains occur. Consistent with these irregularities is the fact that nuclear transfer often or, in some strain combinations, always fails, even though nuclear reorganization (via cytogamy) proceeds to completion.

One might think that induced non-specific matings are mechanistically distinct from mating type mediated matings, but certain observations show that this is not so. Manipulations which easily induce conjugation in mature cells are entirely ineffective in immature cells and even mature cells can be induced only when they are highly reactive. Mating type-specific recognition factors, however, may not be required for induced mating. Cells prevented from mating by treatment with trypsin are still able to be induced by chemical means.

Although studies on chemically induced matings are intriguing, they can only supplement knowledge concerning the more usual pathway of induction and particularly the role of mating type-specific substances. Because all these components are membrane-bound in *Paramecium*, and because membrane biochemistry is only now being well developed, our knowledge of these components is minimal. The protcinaceous nature of these substances is suggested by the effect of enzyme treatments (Metz and Butterfield, 1951), but until the substances can be isolated and assayed, such indirect evidence must be treated with caution.

Immunological approaches were attempted many years ago (Metz and Fusco, 1948), but the studies with *P. aurelia* showed no evidence for mating-type specificity in the action of antiserum. Antibodies may block mating, but the blocking is correlated with antigenic characteristics rather than the mating types of the strains. Hiwatashi and Takahashi (1967) have reported more recently that interference with mating can be achieved without ciliary immobilization in *P. caudatum* (see also Sasaki *et al.*, 1972); if repeatable such observations would offer some hope for immunological analysis.

Generally the attempts to develop *in vitro* systems of analysis have proceeded very slowly. Metz (1948) showed that formalin-killed cells of *P. aurelia* still have mating reactivity, can clump with live (but not with killed cells) of complementary mating type and induce homotypic holdfast attachments among live cells. Either mating type of the species of *P. aurelia* used by Metz retained activity when formalin-killed, but Hiwatashi (1949), using strains of *P. caudatum*, found that only one of the two mating types in each species retained activity when treated this way. These observations suggest chemical differences in agglutination substances and consistent differences in the homologous substances in the various species, but the action of formalin is too poorly understood to permit a molecular interpretation.

A similar differential effect on mating types in *P. calkinsi* by nitrous acid (Metz, 1954) is also difficult to interpret, particularly since in this case only one pair of strains was used. The differential response may have involved strain characteristics only coincidentally associated with mating types. The early studies on the effects of various treatments on the reactivity of killed cells are reviewed by Metz (1954) and Miyake (1968).

To escape the complications of working with whole cells, efforts have been made to fractionate cells into components with mating type-specific properties. Following observations of Metz on *P. aurelia*, Cohen and Siegel (1963) found that cilia removed from *P. bursaria* in mating condition were able to adhere to the ventral surfaces of mature reactive cells of complementary mating type. They could distinguish the cilia derived from the 4 different mating types of species 1 by their heat inactivation curves. Bomford (1967) has similar evidence for differential heat sensitivity of the three classes of mating type substances in an eight mating type system in another of the *P. bursaria* species. Miyake (1964) has reported mating type-specific properties of cilia from strains of *P. multimicronucleatum* also. Takahashi *et al.* (1974) have finally been able to obtain ciliary agglutination between cilia from complementary mating types without involving live cells in the reaction. Membranous vesicles from ciliary preparations may be even more advantageous (Kitamura and Hiwatashi, 1976). Such preparations from *P. caudatum* induce agglutination, homotypic pairing and nuclear reorganization.

With the better means of resolving molecular components of membranes now available, several laboratories are attempting to identify membrane components associated with mating type or maturity in ciliates. Preliminary reports of some of these studies have been circulated, but their significance is still problematical. The chief problem is a reliable assay for the isolated substances, and preferably one which does not require living cells. Recently Miwa *et al.* (1975) have used a bioassay system involving injection of materials into other live cells. They find that cellular fractions from immature cells of *P. caudatum*, when injected into mature cells, may impose immaturity on the treated cells for a period up to 15 or more cell divisions. They characterize this component as a heat-labile macromolecule, but it has not been further analyzed.

9.3.3 Mating-type determination and inheritance

The essential similarities in the mating act and in the subsequent nuclear reorganizations in the various *Paramecium* species support an interpretation of uniformity of mechanisms, and of probable homology. Homology implies a genetic commonality of some sort, and an interpretation of homology requires a consideration of genetic information where available.

In spite of the similarities in the events of conjugation, the various species show a remarkable diversity in the genetic mechanisms controlling the circumstances under which conjugation occurs. Because these mechanisms are important in evaluating homologies, not only among the paramecia, but also with other ciliates, they need to be briefly reviewed. The most thoroughly studied genetic controls are those of the *P. aurelia* group (Sonneborn, 1947; 1975b; Butzel, 1974). These two-type or dual species fall into two major sets of approximately equal size (Fig. 9.1). The first of these is often referred to as 'Group A', and the pattern of mating-type determination and inheritance is said to be 'caryonidal'. The term caryonide refers to a clone

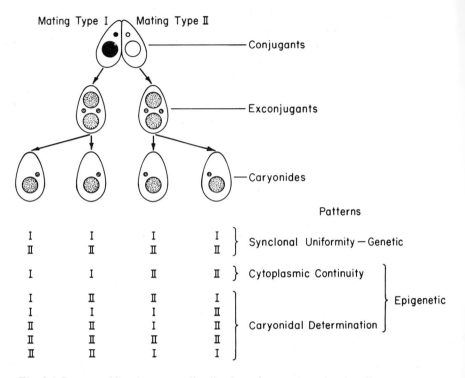

Fig. 9.1 Patterns of mating-type distribution after conjugation in ciliates. Conjugating pairs are usually heterotypic (composed of cells of different mating type). The exconjugants from a single pair are characteristically alike in their genotypes. Hence, if the exconjugants of single pairs regularly have the same mating types, the mating types are directly determined by the genotype. In other species, each pair regularly produces one exconjugant clone of each of the parental mating types; they show cytoplasmic continuity of mating types of the 'Group B' pattern. In still other species, the subclones (caryonides) with different new macronuclei in each pair are independently determined as to mating type. Uniform synclones or uniform clones may be produced, but only in the frequencies expected by chance. This is the caryonidal or 'Group A' pattern.

composed of the vegetative products sharing derivatives of one original macronucleus. Each mating cell (in the *P. aurelia* complex) usually produces two new macronuclei and, hence, two caryonides; each conjugating pair produces four caryonides. The essential feature of the Group A pattern is that caryonides are commonly pure for a mating type while sister caryonides often have different mating types. Indeed sister caryonides are independent in mating type, and no correlation can be found with the cytoplasmic parent. These observations are interpreted as showing that mating type determination is a developmental event, occurring during macronuclear development and maintained thereafter throughout the life of the clone.

The other major set of *P. aurelia* species is called 'Group B'. In spite of the fact that mating reactions may occur between A and B species, the presumably homologous specificities appear to be controlled in a very different way. The mating type of an exconjugant clone (including two caryonides) is usually uniform, but the two exconjugant clones from the same pair − proved to be genetically equivalent − are characteristically of different mating type. The mating type of the progeny generally corresponds to the mating type of the cytoplasmic parent, and this pattern is sometimes referred to as showing 'cytoplasmic continuity'.

Although these two patterns appear to be very different, more thorough analysis (Sonneborn, 1954b) reveals their common basis. In both groups, mating-type determination is a developmental event localized in the macronucleus. The difference lies in the induction of mating-type differentiation in the B species by the physiological system established in the parental macronuclei. This inductive influence is transmitted through the cytoplasm but does not require an autonomous cytoplasmic component.

Both A and B species may contain genetic variables which condition mating-type determination and expression. Often a single mating-type locus possesses alleles which regulate whether both or only one of the mating types is expressed. Because in such instances mating type is usually restricted to the Odd mating type, a persistent suggestion has arisen to the effect that the Odd mating type may in some way be precursor to the Even (Butzel, 1974). Beyond this observation, the scattered mutational analyses (see Butzel, 1955; Taub, 1963; Barnett, 1966; Siegel and Cole, 1967; Byrne, 1973; Cronkite, 1975; for examples of mutational analyses in several *Paramecium* species) available are difficult to summarise, and are outside the scope of this review. The important point for present purposes is this: genetic analysis can reveal some basic similarities in apparently diverse control systems (Group A versus Group B), but mutational analysis has not revealed in detail how the systems work.

One exceptional species in the *P. aurelia* complex serves as a bridge to other paramecia and indeed to other ciliates: the species formerly referred to as syngen 13 (Sonneborn, 1966), shows synclonal uniformity, i.e. all the progeny of an individual pair are of the same mating type. The mating type is determined not by a developmental event, but by the genotype established at fertilization. Although the 'group' is composed of a single species in *P. aurelia* complex, synclonal uniformity is sometimes said to characterize the 'Group C' pattern; genetic determination is the essential feature.

The *P. caudatum* species complex has not been so well studied genetically as the *P. aurelia* complex, but the available information suggests many common features (Hiwatashi, 1968; Myohara and Hiwatashi, 1975). The dual systems which have been examined have a basic genetic determination of mating type, overridden at times by epigenetic controls which permit mating type diversities to be expressed in genetically uniform clones. The *P. multimicronucleatum* species which have been studied (Barnett, 1966) on the other hand have epigenetic controls, regularly producing diverse sister caryonides, but with genes affecting the circumstances under which these diversities are expressed. Although the information is not as extensive as would be desired, the complexes of *P. aurelia, P. caudatum* and *P. multimicronucleatum* appear to have permutations of the same few patterns of regulatory systems.

The comparative task becomes more difficult when the *P. bursaria* complex is included, primarily because of the change in numbers of mating types from two to either 4 or 8. As mentioned earlier, no one-to-one homologies of mating types can be expected even between the 4 and 8-type species of *P. bursaria,* much less between these mating types and those of the dual systems previous considered, Yet, one might hope that the essential features of the compatibility system might be derived from a common source. Indeed, the fact that *Paramecium* species are restricted to 2 or 8-types suggested to Metz (1948; 1954) long ago that higher numbered species were derived from lower numbered species by a process of compounding elementary binary compatibility elements. A two-type system involves a molecular complementarity which might be symbolized by elements A and a. A four-type system could be generated by a genetic duplication followed by a differentiation — so that A generates B, and a gives rise to b. This duplicated system could then yield four classes of cells: AB, Ab, aB and ab. Any combination of different classes would be compatible for either one or both sets of components. An additional duplication would be expected to yield 8 mutually compatible classes: ABC, ABc, AbC, Abc, aBC, aBc, abC and abc.

Supporting this interpretation is the evidence from breeding analysis in *P. bursaria* (Siegel and Larison, 1960; Bomford, 1966). In a four-type species, mating types are determined simply and directly by the genotype at two loci, with two-alleles per locus with dominance. The eight-type species probably have three loci involved. How the A allele and the a allele, or the B allele and the b allele, influence surface complementaries is not known. The genetic loci identified may be regulatory loci affecting a number of related cellular properties, rather than structural loci for mating-type substances.

The plausibility of the compounding hypothesis does not constitute proof of its correctness. Nevertheless, when it is combined with other evidences of important similarities in the organisms involved, and with the lack of some otherwise more acceptable explanation, we should perhaps accept provisionally this interpretation involving a fundamental similarity and homology of the compatibility systems within the genus *Paramecium.*

9.3.4 Evolutionary significance of compatibility controls

With respect to the diversities noted, however, we should perhaps raise the teleonomic question: why should such similar organisms regulate their mating specificities in such different ways? Sonneborn (1957) has pointed out that most of the otherwise unexplained differences in the life histories of these two groups can be explained on a simple and consistent hypothesis: the *P. aurelia* species are all more or less committed to an inbreeding genetic economy while the *P. bursaria* species are generally outbreeders. This interpretation is related to our present concerns in the following way. Epigenetic mating-type determination, whether of the Group A or the Group B type, provides among the progeny of most individual pairs (synclones) sublines of different mating type and hence the capacity to mate with other cells of the same genotype. In contrast, genetic mating-type determination results in synclonal uniformity and prevents matings among cells of this close relationship. The mode of mating type determination becomes in this context a secondary adaptation to a way of life, not necessarily an intrinsic feature of the compatibility system *per se*. The regulation of functions by genic, as opposed to epigenetic, mechanisms need not reflect a basic difference in the actions of the genes or their products.

Similar considerations rationalize the numbers of mating types in the two groups. Again Sonneborn (1957) has noted that the probability that a chance encounter between strangers will result in conjugation depends upon the numbers of mating types in a system. In a dual mating type system, with equal frequencies of the types, half of the encounters between 'strangers' are potentially sexual. If four mating types occur in equal frequency only 1/4 of chance encounters are of like-type cells and 3/4 are potentially sexual. In a similar manner, an eight-type system raises the probability of conjugation with a stranger to 7/8. Thus, a selective advantage for a compound system can be proposed for an organism committed to an outbreeding program.

Other features of these groups make this interpretation plausible and consistent. All species of the *P. aurelia* complex, if denied conjugation, will undergo autogamy, a substitute sexual process leading to homozygosis; no species of *P. bursaria* undergoes autogamy. Generally the species of the *P. aurelia* complex are able to mate again shortly after undergoing conjugation; the *P. bursaria* species have very long periods of sexual immaturity after conjugation. As mentioned earlier, the time after a sexual event for such organisms is a measure of physical distance from their origins, and hence a measure of the probability that they will be among strangers when they acquire the capacity to mate.

Although the evolutionary rationalization of changes in the control systems is possible, the acceptability of the hypothesis of homology requires also a brief consideration of the ease of interconversion of the diverse methods of mating-type determination in the two species arrays. Since a complete understanding has not been achieved of any of the mechanisms, a final answer cannot be given. We can mention, however, that the simple genetic interpretation of the *P. bursaria* system does not account for all observations. Particularly, some older clones of *P. bursaria* are 'selfers'; cells in the same clone, presumably of the same genotype, are sometimes capable of mating (Jennings, 1942). In such clones cells of diverse mating types can be demonstrated. The appearance of intraclonal diversities can be rationalized on an evolutionary basis: old clones which have failed to meet a stranger are advised to self rather than to terminate without any progeny. The phenomenon requires an epigenetic mechanism superimposed on a genetic mechanism. In a converse way gene mutations have been found which can convert an epigenetic regulatory system in the

P. aurelia species to a genetic system, and one of the *P. aurelia* species in fact has a simple genetic system (Sonneborn, 1966).

Our parsimonious hypothesis, therefore, is that a common elementary system of surface complementation underlies all the mating type systems of the *P. aurelia* and *P. bursaria* species complexes (as well as those of *P. caudatum, P. multimicronucleatum,* and the other 'species' of the genus). Changes in the specificities of the reactants provide species barriers of sufficient magnitude to prevent sure assignment of homologies by investigators as well as fruitless sexual adventures by the organisms. Changes in the regulatory systems account for the apparent difference in fundamentally similar compatibility mechanisms, which differ primarily in their degree of compoundness.

9.3.5 Life history and developmental considerations

The various *Paramecium* species are not equally well studied, and some show features not found in others. Nevertheless, some value may be gained from a 'composite' picture of the control systems involved in cell union. In some cases the 'unusual' mechanism may be truly unique, but in other cases it is certainly only cryptic or compressed in the species in which it is not known. The synthetic sequence (Fig. 9.2) has the following general features, beginning with fertilization.

(a) *Mating-type determination*

Mating type determination characteristically occurs at or shortly after fertilization. In some paramecia the zygote genotype is determinative. In other paramecia the new macronuclei become fixed epigenetically in their mating capabilities before the first macronuclear division. This fixation of capacities may involve a restriction to one of the two mating-type specificities; in other cases it involves a fixation of a circadian pattern of expression of two mating types or a progressive pattern of changing expression during the clonal life cycle. In any case, the determination of the pattern of expression occurs very early and prior to any manifestation of the cell specificities themselves.

(b) *Mating immaturity*

Characteristically, but not invariably, paramecia manifest a refractory period immediately following fertilization. This inability to mate may be due to the presence of an inhibitor, to the absence of mating-type agglutinins, to the inappropriate location of those substances, or to their insufficient quantity. A decision as to these or other alternatives must await the identification, localization, and quantification of the hypothetical substances. The simplest hypothesis at the moment is that mating-type agglutinins exist as discrete molecular species (as opposed to spatial patterns of organization of identical molecules in the membranes of cells of different mating type) and that they are not synthesized during the period of immaturity. The mechanism preventing the synthesis of the molecules for a fixed and often long period of time are unknown.

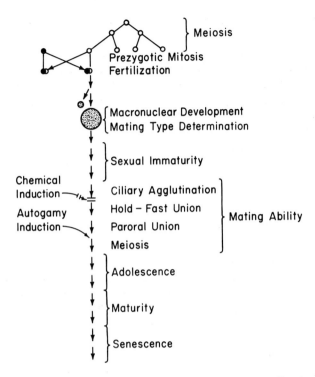

Fig. 9.2 The clonal life history of a generalized paramecium. Following meiosis a pre-zygotic mitosis produces identical haploid nuclei in each conjugating pair member; one pronucleus is exchanged with the mate. The fertilization nucleus gives rise to a new somatic apparatus (macronucleus) as well as a new germ line (micronucleus). The mating type is usually determined during macronuclear development unless it is fixed genetically at the time of fertilization. An interval of sexual immaturity, characteristic of the species, follows. Mating ability is usually associated with several capabilities acquired simultaneously, but these may sometimes be disjoined either developmentally or by genetic blocks ||. Chemical induction of mating bypasses ciliary agglutination but permits holdfast union. Natural autogamy bypasses all cellular interactions, and the ability to undergo autogamy (in species with autogamy) comes later than the ability to initiate meiosis via conjugation. In some species a stage of 'adolescence', during which the mating specificities are not fully developed, precedes full maturity. The final stage of senescence includes a slower growth rate, a lowered ability to survive a nuclear reorganization, and in some species the capacity to change mating type.

(c) *Adolescence*

The term 'adolescence' has been used in *Paramecium* in a restricted sense. It has been applied only to multi-locus systems in which two or more loci become active asynchronously. Clones sometimes first acquire the ability to mate with certain of the mating types to which they will eventually become compatible, but not with all of them. In the interval between the release of inhibition for the first and last genetic locus, the cells are said to be adolescent. The simplest interpretation for adolescent cells is that some of the structural genes necessary for the formation of mating-type agglutinins are blocked from transcription, but that others are already capable of being induced under appropriate circumstances. Since many changes in the synthetic capacities of the cells are involved in converting immature into mature cells, restricting the term adolescence to this one class of cellular alterations may be confusing. Perhaps the term will be broadened as our understanding of maturation develops.

(d) *Maturation*

The appropriate circumstances for the induction of mating differ in different species. In all species these include a cell cycle co-ordination, because the cells must be synchronized. This cell cycle co-ordination is usually achieved in part by restricting mating to a particular portion of the growth cycle. In some species mating capacity is also regulated by light or by circadian chronometers, or by other environmental conditions. Generally we are ignorant of the ways environmental signals trigger mating activity, but our simple scheme suggests that they cause the transcription and/or the translation of molecular messages for the synthesis of mating-type substances.

We should not think of the mating act as an all-or-none phenomenon. Takagi (1971) for example, has found for *P. multimicronucleatum* that during the maturation process, several stages can be identified. Cells first acquire the ability to undergo *ciliary agglutination* between the special cilia on the ventral surfaces of complementary types. Later they acquire the ability to form more intimate unions in the *hold-fast* region on the anterior ventral surface. Later still they become capable of fusing in the *paroral* region where the pronuclei are exchanged. This sequence of maturation could be considered a characterization of the period of adolescence, except that the term has a special meaning in the ciliates; fully 'mature' cells, in the sense of capacity to complete the sequence, may be 'adolescent' with respect to their specificity. This maturation sequence may have a fundamentally quantitative basis, perhaps a gradually increasing amount of mating substances whose reactions trigger the other effects through threshold reactions, but the discreteness of the other effects provides a useful frame of reference for considering the various influences on the mating act.

Maturation may not, however, involve a simple increase in a single mating type-specific substance or complex of substances. We do not know whether the molecules that stimulate the synthesis of agglutinins are themselves involved in either ciliary or cellular adhesions. Certainly the recognition factors may be mating type-specific even when the hold-fast agglutinins seem to be non-specific. Moreover, we have mentioned examples in which hold-fast agglutination is significantly less specific than paroral agglutination. Maturation may require a sequential synthesis of distinctive complementary complexes with little, or only distance evolutionary homology, and with different degrees of specificity. The compounding of homologous systems for more precise regulation during developmental sequences could parallel the compounding of homologous systems proposed in the comparison of species groups.

Long ago Metz and Foley (1949) described a mutant isolated by Sonneborn in *P. primaurelia* which was able to agglutinate, but not able to form hold-fast unions. The strain was apparently blocked between these two developmental stages. Because this CM (can't mate) strain retained the ability to clump indefinitely, the hold-fast union (or some later event) must be responsible through a feed-back reaction for the loss of the ciliary response which normally occurs later in conjugation. This step (or these steps) between ciliary agglutination and hold-fast union is also the point at which chemical induction enters the sequence. As we have noted, in chemical induction, no ciliary interaction is observed, but hold-fast union occurs at the time at which such unions normally occur with ciliary stimulation. Chemical induction does not involve exactly the same steps as does ciliary induction, because a gene mutation can block the capacity to respond to chemical induction without the loss of the ciliary route (Cronkite, 1974, 1975). The pathways are also differentially sensitive to Ca and to trypsin (see Miyake, 1974).

According to Takagi, maturing cells may form hold-fast unions before acquiring the capacity to form paroral unions. Apparently, meiosis and macronuclear breakdown do not occur in these pairs. Meiosis, macronuclear breakdown and paroral cone formation have been difficult to separate. Certainly meiosis and macronuclear breakdown may occur without paroral union under some circumstances. When this is triggered by hold-fast unions, it is called *cytogamy*; when it is triggered by a combination of environmental and intrinsic 'calendric' controls it is called *autogamy*. The acquisition of the autogamic capacity is a developmental event coming later than the ability to complete the mating reaction in paramecia that undergo autogamy, i.e., the species of the *P. aurelia* complex.

Taken all-in-all, the evidence from mutational, physiological and developmental analysis of the mating sequence in *Paramecium* is perhaps more supportive of a sequential synthesis of different molecular complementarity systems rather than of a system of quantitative and spatial regulation of a single substance.

9.4 MATING IN *TETRAHYMENA*

9.4.1 Mating types and mating-type determination

The tetrahymenas are smaller in size, simpler in design, and less demanding in their nutrition than are the paramecia (see Elliott, 1973). They are assigned to a number of named species (Corliss, 1973), but those known to undergo conjugation have all been assigned to the *T. pyriformis* complex (Nanney and McCoy, 1976). Not all the strains in this group are capable of mating, however, and even those that can mate comprise a set of sibling species (see Sonneborn, 1975a). We may diregard from the outset the strain that have not been observed to conjugate. Most of these are amicronucleate, and they are not closely related to the breeding strains, as judged by enzyme.

mobilities (Borden *et al.*, 1973). They appear to consist of several distinctive groups that have had separate evolutionary histories of considerable duration.

The breeding strains of the *T. pyriformis* complex have been assigned to ten species, after a period of being regarded as 'varieties' or 'syngens'. The most extensively studied strains are those of *T. thermophila*, formerly syngen 1, but some significant information is available about others, particularly *T. borealis, T. canadensis, T. americanis, T. pigmentosa,* and *T. hyperangularis.* All these species have more than two mating types, the numbers respectively being 7, 7, 5, 9, 3, and 5 (Nyberg, unpublished). Four other species for which we have relatively less information have 2, 3, 4, and 5 known mating types.

Generally, therefore, the mating systems of the tetrahymenas are multiple systems (not dual) but not highly multiple. Moreover, the numbers of mating types do not fall into a simple 2^n series. This fact should not be taken too seriously at this point. *T. thermophila,* for example, has only been collected a half dozen times and three of its seven known mating types were derived solely from laboratory crosses. The most commonly collected species, *T. americanis,* also has the most known mating types, but little breeding analysis has been carried out with this species. Judgement as to the significance of the numbers of mating types in the *Tetrahymena* species must await a more complete and careful inventory. I am inclined to maintain hope of homology between the mating systems of *Paramecium* and *Tetrahymena,* at least until stronger evidence against such homology is provided.

Perhaps the strongest reason for homologizing the compatability systems in the two genera is the pattern of mating-type inheritance. Three species, *T. thermophila, T. borealis,* and *T. canadensis,* manifest caryonidal determination of mating types. This pattern is essentially identical to that in *P. primaurelia* except for its compoundness, and includes a high sensitivity to temperature during the critical time of macronuclear development (Nanney, 1956; Phillips, 1969).

Not all the species of the complex show epigenetic determination, however. *T. americanis, T. pigmentosa,* and *T. hyperangularis* show synclonal uniformity, the hallmark of direct genetic determination of mating types. More detailed analysis shows, however, that the genetic mechanism employed to determine mating types in these species is different from that in *P. bursaria,* and instead similar to that in *Euplotes* and in other hypotrichs to be discussed later. The multiple mating types in the *P. bursaria* system are achieved by increasing the number of loci while retaining two alleles per locus. The multiple mating types in these *Tetrahymena* species are achieved by a one-locus system with multiple alleles. Of course, if the locus in the one-locus species is compound, this apparent conflict may be resolved. The alleles manifest serial or 'peck-order' dominance; a heterozygote manifests the mating type of the allele with the highest 'rank' in the series.

These mating-type genes reveal their likely regulatory nature, as to those in *P. bursaria,* by certain manifestations of instability. Selfing clones are common and alleles may undergo 'mutation' at rates as high as several percent per generation (Orias, 1963). The mode of determination of mating type may provide little information concerning the mechanism whereby cellular complementarities are controlled. Genetic and epigenetic systems, caryonidal and cytoplasmic, one-locus and multiple-locus systems can be associated by plausible evolutionary inferences with homologous specificities. These diverse tactics achieve the same ends: the establishment of complementary classes of cells in proportions and genetic relationships appropriate to that organism's genetic economy.

The multiple mating types without a 2^n progression raise questions about the hypothesis of binary complementarity. An alternative proposal is that each mating type in a serial dominance system produces a unique mating-type substance, or recognition factor (A, or B, or C ...) which permits it to interact with any cell bearing any alternative factor. A large array of mutually complementary but not self-complementary molecules is hard to design with stereochemical models. One would prefer to think, for the moment, that the genetic evidence is misleading, and that the compatibility systems of *Paramecium* and *Tetrahymena* are similar, and fundamentally binary in their elementary units. The multiple *Tetrahymena* systems may be 2^n systems with a certain amount of degeneracy.

9.4.2 Analysis of the mating sequence

Considerable recent effort (Bruns and Brussard, 1974; Wolfe, 1973) has been expended in the analysis of the early stages of conjugation in *Tetrahymena*, and particularly in *T. thermophila* (formerly syngen 1 of *T. pyriformis*). The most immediate difference between *Paramecium* and *Tetrahymena* is the lag observed between the mixing of diverse cultures, and the onset of mating. Although a starved condition is required for mating in *Tetrahymena*, starvation is not a sufficient condition. Cells may be starved under some circumstances without acquiring the ability to mate. Thus, cells allowed to starve by depleting the nutrients in a peptone medium do not become competent; both exponential cells and 'starved' cells require about two hours after removal to a salt solution before being able to mate. Even the concentration of the salt solution is important, because cells do not become competent in 50 mM Tris or phosphate buffer, even though they do so in buffers below 40 mM (Bruns and Brussard, 1974).

This first phase of the mating reaction, therefore, is not simply a general starvation process, but a specific alteration in the cells' metabolic state. It is referred to as 'initiation' and the cells which have passed through the process are said to be 'initiated'. Initiation is not a process requiring the presence of complementary types, but can occur in unmixed cultures. It is the autonomous process which prepares the cells to send and to receive signals from other cells of complementary type. Initiation can be prevented in many ways. We have already mentioned the effect of exhausted proteose peptone medium, and of high salt concentrations. Large amounts of amino acids, at 10 times or more the concentrations in defined media, also prevent initiation, but individual components of the defined media in the usual concentrations do not interfere. Many antibiotics (actinomycin D, cycloheximide, caffeine, colchicine, and dithiothreitol) prevent initiation, but the significance of such studies is difficult to evaluate (Allewell et al., 1976).

In any case, initiation does not lead directly to cell union, but only to the capacity to respond to mating-type signals if complementary cells are present (McCoy, 1972; Bruns and Palestine, 1975). This second stage also requires an extended period of time — lasting usually one to two hours, and complementary types must be physically present (see Takahashi, 1973). Cell filtrates have not been found efficacious, nor do cells separated by filters manifest signs of stimulation. Ionic substitutes for mating types have not yet been identified. Even mixed cultures which are kept agitated by swirling do not acquire the capacity to mate, even though cell—cell contacts must be occurring in such cultures. The effective contacts seem to be associations of greater duration than afforded in an agitated culture. These contacts can, however, be distributed in time by a succession of periods of quiescence in periodically agitated culture. Bruns and Brussard (1974) use a swirling schedule of one-half minute on and three-fourths minute off. Cells do not pair under this regimen, but after about two hours, massive mating occurs when the swirling is stopped. This incremental second stage of the process, requiring the presence of complementary mating types and the opportunity for fairly extended cellular contacts is called 'co-stimulation' (Fig. 9.3). Co-stimulation results in the capacity to form conjugating pairs, all of which are heterotypic.

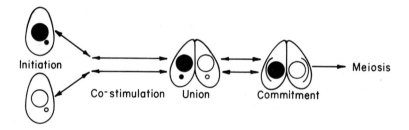

Fig. 9.3 Summary of early events in *Tetrahymena* conjugation. The process of 'initiation' requires the starvation of mature cells, but it may occur in unmixed cultures. Initiated cells become capable of cell union only through a process of 'co-stimulation' which requires the physical presence of cells of a complementary mating type. Cell union may be reversed by the addition of nutrient, but only up to the time at which the micronuclei become committed to meiosis. Each of these processes requires about 1—2 hours, depending on the conditions.

Co-stimulation may be distinguished from initiation not only by the requirement for complementary types, but also by the fact that co-stimulation can occur under some conditions that prevent initiation. Mixed cultures may be co-stimulated at high salt concentrations, for example, provided they have previously become initiated at lower salt concentrations.

Co-stimulation leads to the third stage of the conjugation sequence — cell union. This stage, like the previous two stages, is reversible. If, within an hour of union, conjugating pairs are returned to nutrient media, the component cells separate and re-enter a vegetative cell cycle. After about an hour the union is irreversible; the micronuclei have passed the preliminary swelling stages and have begun to elongate into the characteristic 'crescent' stage of the first meiotic division. These operationally distinctive stages of conjugation have not apparently been named. I will refer to them as the interval of 'reversible union', and the stage of 'commitment'. Allewell et al. (1976) suggest that commitment marks the time at which the oral apparatus ceases to function, but the capacity of *Tetrahymena* to absorb most nutrients even when its oral apparatus is non-functional (Rasmussen and Orias, 1975) makes this interpretation doubtful. Meiosis is probably irreversible after a certain stage, and selection would probably act to prevent the establishmen of a haploid or aneuploid micronucleus which would be subject to severe disadvantage at subsequent matings. Allewell et al. also note that the concentrations of actinomycin D which prevent initiation and co-stimulation do not interrupt the stage of reversible union. Cycloheximide, on the other hand, can block the union and result in the separation of pair members. Perhaps by this stage the essential messenger RNA's for continuation into the commitment stage have been formed, but continued protein synthesis is required. Ofer et al. (1976) report on the effects of polylysine, concanavalin A and bivalent metals.

Two other observations concerning these early conjugal events require separate consideration. Wolfe (1974) reports that when late exponential cells are harvested and mixed with complementary cells, the cells not only pass through initiation and co-stimulation to the formation of pairs — as described above — but a synchronous wave of cell division also occurs, accounting for about 70% of the original cells, at about the time pair formation begins. This 'wave' of division occurs only when exponential phase cells are mated, and not when infradian mode cells are used. It is not, therefore, an essential feature of the conjugation sequence, and may possibly be unrelated to the mating events *per se*.

The second observations are those of Phillips (1971) who reports a special kind of cellular communication associated with initiation in *Tetrahymena canadensis* (formerly syngen 7). Phillips found that small aliquots of mixed cultures of complementary mating types often showed an all-or-none mating reaction. Large samples, in contrast, showed uniformly strong reactions. This phenomenon was discovered not to be mating type-specific, but a special aspect of initiation. In large unmixed mass cultures all the cells become competent; in small aliquots, only some samples become competent. This is interpreted in terms of a 'spreading effect' from rare initial cells which spontaneously undergo initiation (or some related changes) and which are able to induce at high efficiency a similar initiation of associated cells which have not spontaneously undergone that change. The spreading effect does not require cell–cell contact but can be mediated by supernatants from starved cultures. Because supernatants boiled and stored for 2 days at 4°C were still active, Phillips concludes that the active agent is a heat-stable low molecular weight compound. It has not been further characterized.

9.4.3 Cortico-nuclear interactions in conjugation

We noted that the surface interactions in *Paramecium* may be temporally and regionally separated; in *P. bursaria,* at least, the ventral cilia are involved, then the hold-fast region and finally the paroral regions. We have also seen that the specificities of these interactions with respect to mating types are not consistent in different species. The ciliary reaction is always specific, but the hold-fast union can sometimes be homotypic, even when paroral union requires mating-type differences. In *Tetrahymena* one cannot separate these several functions spatially. A ciliary interaction is probably responsible for co-stimulation, but this has not been directly demonstrated. It is inferred from the distribution of the cilia over the entire cell surface; contact of surface components other than cilia would be difficult to arrange. In order for co-stimulation to occur the cilia must possess recognition factors after initiation, even though ciliary agglutination is not seen. The signal function of these mating substances may then be separated from the agglutination function. This distinction does not require that the mating type-specific substances of the cilia and the mating surface be different in kind; the mating surface might be a region of high accumulation of the same materials on the ciliary membranes which is provoked by the signal transfer occurring during co-stimulation.

Cell union occurs in the region just anterior to the mouth and pronuclear transfer occurs in this same region. Although spatially separated mating surfaces are not known in *Tetrahymena,* evidence of functional diversification among contiguous regions is at hand. Particularly, the attachment of pronuclei during pregamic divisions and prior to pronuclear exchange is highly specific. The migratory pronuclei, for example, distinguish between the right and left sides of their own membranes and invariably attach on the right side. This dextrosinistral differentiation provides for an orderly movement of traffic between cells at the time of nuclear exchange. Interestingly enough, Maupas noted long ago (1889) that among the ciliates, pronuclear transfer is always to the right, thus establishing the primordial correctness of this traffic pattern, at least for this phylum. The uniformity in even such a simple matter suggests in any case that we have here an example of an early biological 'invention' of great stability because it can scarcely be improved upon even though it may be entirely arbitrary in certain respects.

Incidentally, the internal differentiation of the fusion membranes perceived by the pronuclei may also have an external correlate. If the fusion membranes were uniformly 'sticky', adhesions between cells of complementary mating type could occur with any orientation – including a heteropolar orientation. The forward motion of both mates might tend to disrupt heteropolar unions and allow only homopolar unions to persist. Alternatively, the fusion membranes might themselves be asymmetric externally as well as internally. A complementarity may exist only between the left surface of one cell and the right surface of a complementary cell. This complementary asymmetry would provide a more precise 'docking' of the mating pairs, assuring a better regulation of position in anticipation of nuclear transfer. Clear evidence of such left-right differentiation of mating surfaces has recently been reported by Honda and Miyake (1976) for *Blepharisma*.

Thus far we have no understanding of the nature of this right-left asymmetry in the fusion area. Its existence, however, is clear and it allows the suggestion that a similar anterior-posterior distinction among adjacent surface areas may also exist as homologues to the hold-fast and paroral regions in *Paramecium*. Such differentiations could be thought of either as gradients or as patchworks, but they again raise the possibility that the agglutinins are complexes of diverse molecules rather than single substances.

Homotypic cellular unions have only been reasonably well established in *Tetrahymena* in the instance of certain permanently selfing strains (Nanney, 1953). Some suggestive evidence for homotypic unions comes from the juvenile selfers of *T. thermophila*. Allen (unpublished) found that single cells from strongly selfing cultures isolated in capillary tubes with minimal food divided either once or twice. Under such circumstances pairing regularly occurred between the two or four present. These selfers have been studied in great detail (Allen and Nanney, 1958) and are interpreted as lineages with macronuclei mosaic for mating type determinants. Because the number of 'determinants' is fairly high (90 before cell division), sister cells usually differ from each other only quantitatively. Because mating regularly occurs between the products of a single division, even these quantitative differences are not likely to be important in the mating. Rather, one must consider that the cells mating are alike, not in the sense of having the same mating type, but in the sense of both being 'mosaic', not only in the macronucleus but also in their surface membranes.

In contrast to these observations, all the evidence from lines of pure mating-type indicates invariably heterotypic cell unions. Recent studies on conjugation among three cells emphasize the specificity of these *Tetrahymena* unions (Preparata and Nanney, 1977). When three mating types are mixed, most of the triplets formed consist of three different mating types, as can be demonstrated by genetic markers. These associations are symmetrical and 'tripodal', and fertilization is tripolar, i.e. each cell contributes a migratory pronucleus to one cell and receives a migratory pronucleus from the other. Normal fertilization and nuclear reorganization occurs in all cells.

When only two mating types are mixed, the frequency of triplet conjugation is much lower, and each triplet must consist of two cells of one mating type and one of the other. Thus two cells of the same mating type are held in close association by their mutual attachment to a limited area of a cell of another mating type. Yet, even under these conditions the cells of the same type apparently do not unite. Certainly they do not ordinarily transfer pronuclei to each other. Instead, both transfer pronuclei to the single cell of different mating type, and only one of them receives a pronucleus in exchange. As a consequence, most of the triplets studied finish conjugation with one haploid, one diploid and one triploid exconjugant. Very rarely mixtures of two mating types may yield a triplet with all diploid exconjugants. These cases could indicate a rare homotypic pronuclear transfer, but are more likely to be caused by cytogamy. The genetic tests to distinguish these possibilities have not yet been undertaken.

Certainly cytogamy can occur in *Tetrahymena,* even though it is usually forbidden in these generally outbreeding forms. Orias and Hamilton (unpublished) have recently found that over 30% of conjugating pairs may be induced to undergo cytogamy if they are transferred from the salt solution in which they are mated to nutrient medium just before the time of nuclear transfer. The information now available does not permit a distinction between an effect of the treatment on the internal attachment of a nucleus at the mating surface prior to the formation of gametic nuclei, and an effect on pronuclear transfer through the membrane. In either case the customary interaction between the nuclear membrane and the surface membranes is modified and subsequent nuclear behavior is altered. The controlling role of the mating surface is suggested by the failure of homotypic nuclear transfer through these surfaces, in contrast to the ease of homotypic nuclear fusion when transfer is prevented. Homotypic nuclear fusion is perhaps not surprising. Indeed, *Tetrahymena rostrata,* and inbreeding relative of the outbreeding *T. pyriformis* complex, habitually undergoes autogamy and is not known to mate (Corliss, 1965).

Little is known of the mechanism whereby surface fusion triggers the onset of meiosis and regulates nuclear transfer. Ultrastructural studies (Elliott and Tremor, 1958) show extensive changes in the cortical organelles in the region of fusion, but these changes give little illumination. The triplet conjugants, however, provide an indication that the signal associated with the membranes has a quantitative aspect similar to that of the ciliary reaction. When two cells of one type share a fusion surface with one cell of another type, only a fraction of their surfaces can be involved. Indeed the two like-type members often share unequally in this limited space. A characteristic feature of these two-type triplets is a variablility in the time of onset of meiosis, strikingly different from the synchrony in normal pairs and three-type triplets. Generally the unique type cell is somewhat advanced with reference to its two mates, and the two mates are sometimes asynchronous, particularly when their regions of attachment are of different extent. These observations suggest that the time of initiation of meiosis is proportional to the area of association between different mating types. The signal is probably not a simple trigger, but rather a sustained and quantifiable message, with perhaps a threshold required before subsequent events can be provoked.

The communication between the cortex and the nucleus in ciliate morphogenesis has been recently documented and discussed by deTerra (1974, 1975) with special reference to *Stentor,* but ciliate conjugation provides some of the most striking examples (Sonneborn, 1954b; Nanney, 1953) of co-ordinated nucleocytoplasmic interactions. One of the most bizzare of these concerns the relationship between the cortex in the area of fusion and the macronucleus. Ordinarily the *Tetrahymena* micronuclei go through a stereotyped sequence of divisions and migrations which seems to be correlated with this anterior cortical region. They seem to move away from it at some times; they must move toward it at other times; they undergo different kinds of differentiation depending on their distance from it. During all these events the macronuclei manifest no obvious reactions. But when conjugating cells are deprived of a micronucleus, at the time in the sequence at which the migratory pronucleus ordinarily moves to the membranes, the macronucleus extends a finger-like projection to the surface, touching the membrane at the precise position on the cell's right side where the pronucleus is ordinarily attached (Nanney and Nagel, 1964). These observations strongly suggest structural, as opposed to molecular, signals which associate and co-ordinate cellular organelles, but which are invisible with nuclear stains (see Ruiz et al., 1976).

9.5 MATING IN *EUPLOTES*

9.5.1 Mating behavior and mating-type inheritance

The only other genus of ciliates with a substantial body of experimental and genetic studies is *Euplotes*. Sonneborn (1946) once suggested that *Paramecium* and *Euplotes* might be as distantly related as bats and gorillas. At least as measured by the time from their common origins, and in terms of their accumulated molecular diversities, this estimate is probably far too conservative (Nanney, 1977). Yet one might still hope for homologies of structures and functions which are apparent after such long separation.

The first obvious connection between the compatability systems in *Euplotes* and those in the organisms previously discussed is the method of mating-type determination. Just as *Tetrahymena* and *Paramecium* are linked by the occurrence of caryonidal determination, so *Tetrahymena* and *Euplotes* are linked by one-locus control systems. Most *Euplotes* species have serial dominance but *Euplotes* is, like *Paramecium* and *Tetrahymena*, of very ancient origin and its modern representatives are heterogeneous in the details of their regulatory systems.

Generally, the cells of *Euplotes* (Heckmann, 1963, 1964; Nobili, 1966, Kosaka, 1973), and of the other hypotrichs such as *Stylonychia* (Downs, 1956, 1959; Ammermann, 1965) and *Oxytricha* (Siegel, 1956; Ricci et al., 1975), enter a period of activation when complementary types are mixed. This activation period is not manifested in ciliary adhesion, as in *Paramecium*, but it is not as visibly asymptomatic as in *Tetrahymena*. Instead the cells manifest a stereotyped and ritualistic 'courtship dance', which ends in cell pairing after an interval of an hour or more. Probably in all these cases, the cells of complementary type trigger in their mates the cellular changes necessary for conjugation (Heckmann and Siegel, 1964). The mating act requires new molecules, as is indicated by the effects of cycloheximide (Dini and Miyake, 1976), but the stimuli for these syntheses have not been identified.

Although most of the systems of mating-type determination in *Euplotes* appear superficially to resemble the *Tetrahymena* serial dominance patterns, one possibly important question concerning them should be raised. We noted that the *Tetrahymena* systems are multiple systems, but low-multiple; the number of mating types thus far found in a species is less than 10, and some species have been repeatedly collected in only a few mating types. In some organisms, particularly among the fungi (see Raper, 1966), essentially open-ended multiple compatibility systems have been described, in which literally hundreds of alleles are found at each locus, in either one-locus or multiple-locus systems. To impose a 2^n series on a system with 150 multiple alleles, is perhaps not reasonable, while a 2^n series with 3–16 mating types might reasonably be constrained into a partially degenerate binary system. Yet even with the fungi, the multiple 'alleles' are derived by compounding loci, and once an efficient system of compounding has been achieved, perhaps a 3-element locus (and 8 alleles) is no more easily achieved or maintained than an 8-element locus (and 128 alleles). Whether the *Euplotes* systems in general are low-multiple systems or high-multiple systems is not well established. Relatively few mating types per species are reported, but this may reflect only the lack of effort in collecting new strains. Certainly the experimental manipulation of the strains becomes more burdensome as the number of alternative types increases. One species of *Stylonychia* has been reported (Downs, 1959) to have 15 mating types, but Ammermann (1965) observed 48 in *S. mytilus*.

9.5.2 Soluble gamones

Some species of *Euplotes* depart from the usual requirement of direct cell–cell interaction for activation. The first of these, *E. patella*, was studied long ago by Kimball (1939, 1942) and Powers (1943), and recent confirmations and extensions are not available. Nevertheless these studies are of considerable interest. In *E. patella* clones of cells do not mate when unmixed, but conjugation occurs when certain clones are brought together. As in other ciliate examples, clones are considered of like type if no mating occurs in mixtures, and of different types if conjugation does occur. However, in this particular example, the cells which mate may not have come from different clones, and in some examples all the pairs in a mixture are homotypic and derive from one of the original clones.

This situation was elucidated when cell-free filtates of some clones were found to induce mating among cells of other clones. In some cases these interactions were reciprocal, and in other cases they were unilateral. In unilateral reactions, whether induced by whole cultures or filtrates, the pairs were all homotypic. In bilateral reactions, both kinds of cells were induced to mate, but the pairs were formed at random; some were homotypic of one type, some were homotypic of the other type and some were biparental. This situation is somewhat similar to that mentioned in *P. caudatum*, except that in *E. patella* the various pair combinations are formed simply in proportion to the frequencies of the reactants. Moreover, unlike the usual situation in *P. caudatum*, the remainder of conjugation proceeds normally.

As mentioned earlier, most *Euplotes* mating control systems are one-locus multiple allele systems. But unlike most *Euplotes* systems — which show serial dominance — the *E. patella* system manifests co-dominance; the heterozygote is distinctive from both homozygotes. In Sonneborn's (1947) rationalization of this system, each allele at the *mt* locus is responsible for the production of a specific activating substance, and this substance, or gamone, is effective on any cell of the same species which does not produce it. An allele then has a dual effect, to cause the production of a specific substance and to cause the producer to be immune to its activity. This interpretation is confirmed by the special properties of heterocaryotic doublets, containing various combinations of the known alleles. These doublets manifest mating types consistent with their allelic composition, regardless of the location of the alleles; they have the ability to activate any of the other types not bearing an allele they carry and lack the capacity to respond to any mating type not carrying an allele different from their own. These doublets could be resolved into singlets with a single macronucleus, and these singlets manifested the component mating types.

In the case of *E. patella*, then, activation is a mating type-specific process — occurring only in the presence of a substance produced by (certain) genetically different cells. Once activated, however, homotypic matings may be consummated. In terms of the breeding economy, this mechanism suggests an organism tolerant to inbreeding, even though selfing does not generally occur unless 'strangers' are in the vicinity.

Another 'atypical' breeding system with somewhat similar ecogenetic consequences has been described by Katashima (1959, 1961) for *E. eurystomus*. Katashima identified 5 genetic species in this species complex, each containing only two mating types. The mating types in this case are characterized by their ability to inhibit mating among cells of the complementary type. Unmixed clones will self when allowed to starve. This selfing is reduced in the presence of the complementary type or of cell-free fluid from cultures of the complementary type. The ascertainment of mating patterns requires morphologically marked strains, because all starving cultures manifest mating. Some homotypic pairing occurs even when complementary types are present, but in some instances over 90% of the pairs are heterotypic. This pattern of behavior suggests an organism even more committed to an inbreeding way of life than in the case of *E. patella,* but an organism able to acquire some variety through intercourse with strangers when opportunities are available.

Unfortunately, the cell-free mating type-specific compound in *E. patella* and *E. eurystomus* have not been isolated or characterized. Later attempts by Katashima (1961) to find strains of *E. patella* in Japan with filtrate activity were not successful. Two genetic species were studied, one with six mating types and the other with three, but culture filtrates were ineffective in either case. The Japanese species appear to be more generally committed to outbreeding than the American species studied by Kimball. The Japanese strains have an immaturity period lasting months and an extended 'adolescence' stage following that. Selfing is limited to a few clones and to 'senescent' cultures. The behavior of adolescent clones is reminiscent of that of strains of *P. bursaria* studied by Jennings (1942) and later by Siegel (1965). The ability to mate with the several complementary types is not acquired simultaneously, but step-wise. In the Japanese syngen (species) 1 of *E. patella*, a clone which will eventually become type I, for example, may first acquire the ability to mate with types II and IV and V, and later become able to mate with types III and VI. Mating-type differentiation is viewed by Katashima as a sequential process: first clones differentiate into type A (I, III, VI) or type B (II, IV, V), and later complete the process. Such observations are of course compatible with a multiple binary system of differentiation in which the 8 possible types have degenerated to 6. Unfortunately, genetic studies on this species are not yet available. One would particularly like to know the number of loci involved in determining mating type.

The most recent report of gamone activity in the hypotrichs (Esposito *et al.*, 1976) concerns *Oxytricha bifaria*. Esposito and his colleagues show in this multiple mating system that mating type-specific filtrates are able to shorten the 'courtship' stage preliminary to cell union, and to permit immediate pairing when cells of different mating type are mixed. Although the gamone is mating type-specific, acting only on cells complementary to those producing it, the response is generalized in the sense that the activated cell is able to mate with *any* complementary cell and not just the kind of cell producing the gamone. Thus, the system must contain as many gamones as it contains mating types. Only one response to gamone action, however, appears to occur. One is tempted to suppose that similar though less stable gamones may occur in all the 'dancing' hypotrichs, and these gamones might be responsible for a chemoattraction which holds the dancing partners together in the prenuptial activities.

9.6 MATING IN *BLEPHARISMA*

Studies on mating in *Blepharisma* (see Miyake, 1974) are in some ways the most advanced among the ciliates. They are also the best known and most recently reviewed in detail. For this reason, we provide here an abbreviated account incommensurate with their importance.

Most notably, soluble gamones in *Blepharisma* have been demonstrated, isolated and partially characterized and one has even been synthesized (Tokoroyama *et al.*, 1973). In other respects, however, studies on *Blepharisma* are still primitive and a firm connection with comparable phenomena in other ciliates has not yet been established.

The classification and domestication of *Blepharisma* has lagged, partly because of its breeding systems; several different species have been described, but the genetic relations are hard to examine. Nearly all strains undergo selfing in clonal culture, and the compatibility between strains can usually be assessed only when morphological markers are available. Fortunately, some strains have been found to 'differentiate' into types which do not mate when separated but which do conjugate when mixed (Miyake and Beyer, 1973). These strains have recently been reassigned (Hirschfield et al., 1973) to *B. japonicum* var. *intermedium*, although they were previously designated as *B. intermedium*. Among these strains only two mating types have been found and these are designated as I and II. As in the case of *Euplotes patella* mentioned earlier, culture fluids may effect the mating of complementary types.

A careful analysis of the activities of the fluids leads to the following (Fig. 9.4) scheme (Miyake, 1974). Mating type I secretes a mating type-specific substance, called gamone I, which activates cells of type II, rendering them capable of mating

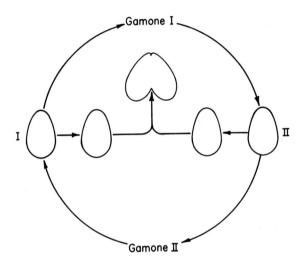

Fig. 9.4 Diagrammatic representation of cell interactions in early conjugation in *Blepharisma japonicum* (from Miyake and Beyer, 1973). Gamone I, produced by mating type I activates type II cells to become reactive and also increase the production of gamone II which has a symmetrical effect on mating type I. Gamone II is also a chemo-attractant for cells of type I (Honda and Miyake, 1975).

after an interval of about two hours. This activation involves increased protein synthesis and is blocked by cycloheximide (Miyake and Honda, 1976). Similarly, mating type II cells secrete gamone II, which renders type I cells capable of mating after a comparable period of time. Either type, after activation but in the absence of cells of the complementary type, will unite in homotypic pairs. When cells of both types are present and activated, both heterotypic and homotypic pairs develop. Although united, homotypic pairs do not continue into meiosis and subsequent re-organization, but may persist as pairs for long periods of time, so long as gamone concentrations are maintained at high levels. Heterotypic pairs, in contrast, undergo meiosis, fertilization and nuclear re-organization.

Miyake and Beyer (1974) show that both gamones not only activate the complementary cells, but also induce those cells to secrete larger amounts of their own gamones in a mutual positive feedback system. Gamone II is also a chemo-attractant for type I cells (Honda and Miyake, 1975).

One of the more interesting features of this system is the contrast between the behavioral symmetry of the mating-type responses which contrasts with the chemical distinctiveness of the gamones. Gamone I (blepharmone) is a glycoprotein of about 20 000 daltons (Miyake and Beyer, 1974; Braun and Miyake, 1975). It has an unusually high content of tyrosine, estimated at 13 residues among the total of 175 amino acids (excluding tryptophan). Glutamic acid in contrast is low and estimated at 7 residues per molecule. Aspartic acid, threonine and serine are high; they are estimated to compose 26, 17 and 19 residues respectively. These high proportions are common in glycoproteins because these amino acids serve as attachment sites for carbohydrate units. In the case of blepharmone, the sugars consist of hexosamine and mannose, each represented by three residues per polypeptide chain. The sugar content is approximately 5% of the total molecular weight of 20 000 excluding tryptophan. The polypeptide chain has a high content of polar amino acids (46%) and is easily soluble in water. For this reason Braun and Miyake speculate that the interaction between the gamone and its target membrane is by means of the sugar residues rather than by hydrophobic groups of the protein. Gamone II (blepharismone) is a small molecule related to tryptophan (Kubota *et al.*, 1973): calcium-3-(2′ formyl-amino-5′-hydrozybenzol)-lactate (Fig. 9.5).

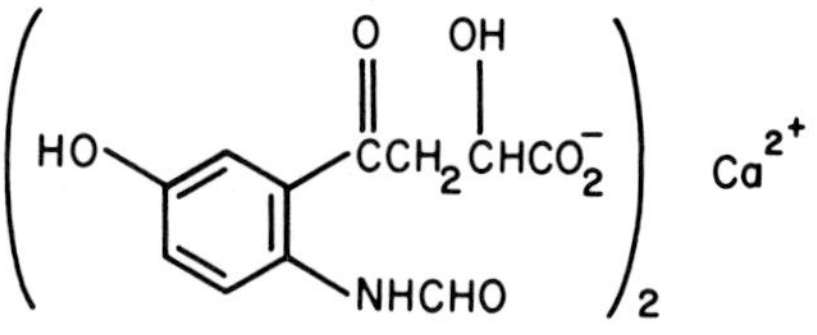

$$\left(HO - \underset{NHCHO}{\underset{|}{\bigcirc}} - \overset{O}{\overset{\|}{C}}CH_2 \overset{OH}{\overset{|}{C}}HCO_2^- \right)_2 Ca^{2+}$$

Fig. 9.5 The chemical structure of blepharismone, Gamone II. (Miyake and Beyer, 1974).

Miyake and Bleyman (1976) have used the gamones as a means of studying the relationships among the blepharismas. Gamone II from *B. japonicum var intermedium* is widely effective in inducing mating in strains of other named species. Gamone I, in contrast, is highly specific, inducing mating only within the same species. Culture filtrates from other strains have effects on the *B. japonicum var intermedium* strains which are consistent with these results. The small molecule of gamone II may in fact be a common gamone shared by several species, while the large gamone I molecule undergoes variation and provides the basis for species specificities. Even though mating occurs in an interspecific mixture of types I and II, the pairs formed are usually homotypic pairs of type I; genetic exchange between species has not been clearly demonstrated.

The studies on *Blepharisma* can be explained by either of two different mechanisms of interaction at the molecular level. The first, and superficially the simplest, requires a direct interaction between the gamones, with the small key (Gamone II) interacting with the large lock (Gamone I) in an allosterically significant combination capable of provoking cell union, meiosis and subsequent conjugal changes. Attempts to detect this direct interaction between the gamones have been unsuccessful; the gamones combined *in vitro* do not antagonize each other. However, the molecules might very well interact when one is embedded in a membrane, even though no interaction of free molecules can be found.

The comparative studies would suggest that the lock varies while the key remains constant. The alternative interpretation is that the mating type I activation and the type II activation are initiated by separate molecular signals. Gamone I activates mating type II cells by combining with its own specific membrane-bound receptor, I–R; gamone II in a similar manner activates type I by combining with its specific receptor, II–R. The gamone I–IR complex might be similar to the gamone II–IIR complex, i.e. IR might be similar to gamone II and IIR might be similar to gamone I. If this were so, the apparent molecular asymmetry of the reactions would be greatly minimized. On the other hand, only aesthetic principles support the idea of commonality. In any case the gamone I–IR complex appears to be relatively more labile in an evolutionary sense than the gamone II–IIR complex, which is understandable on a lock and key basis, but not predicted in the case of two independent mechanisms.

The *Blepharisma* work is also beginning to provide some insight into the mechanisms whereby meiosis and nuclear re-organization are induced. As mentioned earlier, homotypic unions do not lead to meiosis. Miyake (1975) constructed long chains of homotypic cells by treating doublet cells with gamones. The doublets have two ventral surfaces capable of union with two different cells, and are thus able to form a series of connected cells of the same mating type. An introduced cell of the other mating type may then be incorporated at the end of such a chain if the chain has not closed into a circle. This union is heterotypic and the two cells directly involved are stimulated to initiate meiosis. At a later time the next cell in the series begins meiosis, and the onset of meiosis moves progressively through the chain (Miyake *et al.*, 1977).

The passage of the meiotic stimulus can be measured by a surgical separation of the attached cells at various times before the visible manifestations of meiosis. Miyake *et al.* interpret the results as due to the transfer of a substance, rather than to the passage of a stimulus. Because the terminal cells on chains of 5 or more cells are not stimulated, the investigators propose that a finite amount of this substance diffuses from a fixed source, and that the amount is only sufficient to induce about four cells. Longer chains in fact have fewer induced cells than shorter chains. This result may be explained by a dilution of the substance below a threshold of activity in the longer chains. The investigators hope to extract the meiotic stimulant, to purify and characterize it. Injection procedures might provide a practicable assay.

9.7 SUMMARY AND CONCLUSIONS

A question that should be answered prior to serious comparative efforts with non-ciliates, is whether the mechanisms promoting and restricting the union of cells in different ciliate species are fundamentally alike. This question has no satisfactory answer at the present time, even though the tempo of discovery is increasing. A firm answer requires an understanding of the mechanisms in at least two distantly related organisms, and at the moment we do not possess a fully satisfactory explanation for even one organism.

The comparative survey among the ciliates shows large differences in the details of the compatability systems. These details presumably reflect two diverse kinds of influence on the organisms: the molecular devices which govern their short-term activities and the population forces which govern their long-term ecological and evolutionary strategies. The primordial molecular mechanisms which must underlie the apparent diversity are largely obscured in an organismic survey. Most of the differences observed can be assigned a plausible teleonomic rationale, but thus far they resist satisfactory causal analysis.

Among the sibling species of the *P. aurelia* complex we can be confident that the mechanisms co-ordinating cellular intercourse are very similar, in spite of the superficially large differences in detail — particularly with respect to mating type determination and inheritance. This confidence arises from several factors:

(1) the occasional interspecific matings which assure the similarity of the intercellular signals,
(2) the general similarities of mating behavior in the several species, including the similar modes of cellular attachment, and
(3) the common capability for chemical induction by similar ionic manipulations.

With respect to the complexes designated as *P. multimicronucleatum* and *P. caudatum,* we may be similarly confident insofar as available information permits, except that natural interspecific matings between complexes have not been reported. The taxonomic distinctions among these complexes may indeed be so superficial that some species combinations within a complex are more distantly removed from a common origin than are some species combinations in different complexes. This suspicion is especially directed to the *P. multimicronucleatum* in comparison with the *P. aurelia* complex.

The extension of the *P. aurelia-caudatum-multimicronucleatum* mating system to the *P. bursaria* species requires a larger imagination. Although the organisms are similar in major respects, the *P. bursarias* have significant differences in design — quite aside from their symbiotic algae. I am not aware that the technics of chemical induction so generally effective in the other *Paramecium* species are successful with *P. bursaria.* Of greater importance, however, is the shift from a dual mating-type system to a multiple, 2^n, mating-type system. This discordance may be understood, as explained earlier, by compounding the elementary binary compatibility system of the other paramecia. The shift from an epigenetic mechanism of mating-type determination to a genetic mechanism need not be disturbing, because such a shift has been observed even within the *aurelia* complex. The hypothesis of a compounding of the compatibility system need not suggest a compounding of the entire genetic apparatus; gene duplication may occur without chromosomal or genomic duplication. Nevertheless, the size, the DNA content and the chromosome number of the micronucleus in *P. bursaria* is notably larger than that in the other species discussed. Moreover, the general genetic conservatism of polyploids is consistent with the outbreeding genetic economy of *P. bursaria,* which contrasts so strongly with the inbreeding character of most of the *P. aurelias.*

The next question concerns the relationship between the two genera of hymenostome holotrichs, *Tetrahymena* and the *Paramecium.* Are there available significant evidences of homology in the compatibility system? All paramecia agglutinate immediately when brought into contact with appropriately prepared mates; tetrahymenas, in contrast, always require a significant interval of time between recognition and agglutination. *Paramecium* conjugation involves two isolated surface areas with distinctive properties; the tetrahymenas have only one identified mating area. The *Paramecium* mating systems are dual or plausibly compound dual systems; the *Tetrahymena* mating systems do not appear to form a 2^n series, and the multiple systems do not seem to involve multiple locus control.

Some of these contrasts between paramecia and tetrahymenas can be explained away. The *Tetrahymena* mating surfaces, for example, may be composed of adjacent functional areas which are separated in *Paramecium.* The *Tetrahymena* mating system may have become compound by tandem duplication instead of by polyploidization; its original binary nature may be obscured in part by incomplete sampling and in part by evolutionary degeneracy. Quite aside from such rationalizations, however, I find the detailed similarities between the phenomena of caryonidal determination of mating-type in the two genera to be persuasive. Although one should be as careful to avoid misleading similarities as in discounting superficial divergences, I am inclined to consider the mating mechanisms of *Paramecium* and *Tetrahymena* to be homologous and fundamentally similar in their present manifestations.

Assurance of homology fades as one moves from the holotrich genera to the hypotrich spirotrichs. The mating dance of *Euplotes* and its relatives sets them apart from *Paramecium* and *Tetrahymena.* Its choreography may be directed via short-range soluble gamones which are difficult to detect under most circumstances. Mating type-specific gamones with remarkably diverse roles have been reported among the hypotrichs, while all specific mating substances are firmly bound in the holotrichs. The chief connection between the hypotrichs and the holotrichs is through their one-locus multiple mating-type system. One-locus serial dominance mating-type determination occurs in some of the tetrahymenas and many of the hypotrichs. We rationalized the *Tetrahymena* device as being the basic binary complementarity system of the holotrichs, compounded by a local duplication of the genetic elements to provide multiple mating types. The hypotrich *mt* locus might have a similar explanation. But the reason for retaining a belief in such an homology is a hope for simplicity, rather than a convincing recital of similarities.

When one considers *Blepharisma,* a genus of spirotrichs in the heterotrich suborder, the connections become even more tenuous, because here we have not even the questionable association through a common genetic control, and very little comparative information. Now that unprecedented advances have been achieved in the isolation of the gamones of *Blepharisma,* we may hope for an acceleration of research on this organism and eventually a firmer basis of comparison. For the present, however, in associating *Blepharisma* and *Paramecium* we must rely on the same *a priori* considerations that bring together in this book ciliates and fungi, bacteria and algae.

Just as we have difficulty in attributing common mechanisms to different ciliates, because of our lack of understanding of those mechanisms, we are not likely to find sure comparisons with organisms in other phyla. We are more likely to find in comparative analyses important suggestions for renewed study on the ciliates. Perhaps the most important contribution that ciliates can provide in such exercises is a reminder that the phenomena observed at the cellular level are not necessarily derived directly and simply from the properties of the molecular mechanisms, but that these phenomena have an evolutionary history and may reflect ecological strategies and genetic economies as well as physiological imperatives.

REFERENCES

Allen, S.L. and Gibson, I. (1972), Genome amplification and gene expression in the ciliate macronucleus, *Biochem. Genet.,* **6**, 293–313.

Allen, S.L. and Gibson, I. (1973), Genetics of Tetrahymena. In: *Biology of Tetrahymena,* (Elliott, A.M. ed.), Dowden, Hutchinson and Ross, Stroudsburg, Pa, pp. 307–373.

Allen, S.L. and Nanney, D.L. (1958), An analysis of nuclear differentiation in the selfers of *Tetrahymena. Am. Nat.,* **92**, 139–160.

Allewell, N.M., Oles, J. and Wolfe, J. (1976), A physiochemical analysis of conjugation in *Tetrahymena pyriformis. Exp. Cell Res.,* **97**, 394–405.

Ammermann, D. (1965), Cytologische und genetische Untersuchungen an dem Ciliaten *Stylonychia mytilus* Ehrenberg. *Arch. Protistenk.,* **108**, 109–152.

Barnett, A. (1966), A circadian rhythm of mating type reversals in *Paramecium multimicronucleatum,* syngen 2, and its genetic control. *J. cell. comp. Physiol.,* **67**, 239–270.

Bleyman, L.K. (1971), Temporal patterns in the ciliated protozoa. In: *Developmental Aspects of the Cell Cycle,* (Cameron, I.L., Padilla, G.M. and Zimmerman, A.M., eds.), Academic Press, New York and London, pp. 67–91.

Bomford, R. (1966), The syngens of *Paramecium bursaria*: New mating types and intersyngenic mating reactions. *J. Protozool.,* **13**, 497–501.

Bomford, R. (1967), Stable changes of mating type after abortive conjugation in *Paramecium bursaria. Exp. Cell Res.,* **47**, 30–41.

Borden, D., Whitt, G.S. and Nanney, D.L. (1973), Electrophoretic characterization of classical *Tetrahymena pyriformis* strains. *J. Protozool.*, **20**, 693–700.

Braun, V. and Miyake, A. (1975), Composition of blepharmone, a conjugation-inducing glycoprotein of the ciliate *Blepharisma. FEBS Letters*, **53**, 131–134.

Bruns, P.J. and Brussard, T.B. (1974), Pair formation in *Tetrahymena pyriformis*, an inducible developmental system. *J. exp. Zool.*, **188**, 337–344.

Bruns, P.J. and Palestine, R.F. (1975), Co-stimulation in *Tetrahymena pyriformis*: a developmental interaction between specially prepared cells. *Dev. Biol.*, **42**, 75–83.

Bütschli, O. (1876), Studien über die erster Enwicklungsvorgange des Einzelle, die Zellteilung und der Konjugation der Infusorien. *Abhandl. Senckenb. Naturforsch. Ges.*, **10**, 1–150.

Butzel, H.M. (1955), Mating-type mutations in variety 1 of *Paramecium aurelia* and their bearing upon the problem of mating-type determination. *Genetics*, **40**, 321–330.

Butzel, H.M., Jr. (1974), Mating-type determination and development in *Paramecium aurelia*. In: *Paramecium: A Current Survey* (Van Wagtendonk, W.J., ed.), Elsevier, Amsterdam, London, New York, pp. 91–130.

Byrne, B.C. (1973), Mutational analysis of mating-type inheritance in syngen 4 of *Paramecium aurelia. Genetics*, **74**, 63–80.

Clark, M.A. (1972), Control of mating-type expression in *Paramecium multimicronucleatum*, syngen 2. *J. cell Physiol.*, **79**, 1–14.

Cohen, L. (1964), Diurnal intracellular differentiation in *Paramecium bursaria. Exp. Cell Res.*, **36**, 398–406.

Cohen, L.W. (1965), The basis for the circadian rhythm of mating in *Paramecium bursaria. Exp. Cell Res.*, **37**, 360–367.

Cohen, L.W. and Siegel, R.W. (1963), The mating-type substances of *Paramecium bursaria. Genet. Res.*, **4**, 143–150.

Corliss, J.O. (1965), Láutogamie et la sénescence du cilié hyménostome *Tetrahymena rostrata* (Kahl.) *Ann. Biol.*, **4**, 49–69.

Corliss, J.O. (1973), History, ecology and evolution of species of *Tethrahymena*. In: *Biology of Tetrahymena* (Elliott, A.M., ed.), Dowden, Hutchinson and Ross, Stroudsburg, Pa. pp. 1–55.

Crandall, M. (1977). Mating-type interactions in micro-organisms. In: *Receptors and Recognition* (Cuatrecasas, P. and Greaves, M.F., eds.), Series A, Vol. 3, Chapman and Hall, pp. 45–100.

Cronkite, D.L. (1974), Genetics of chemical induction of conjugation in *Paramecium aurelia. Genetics*, **76**, 703–714.

Cronkite, D.L. (1975), A supressor gene involved in chemical induction of conjugation in *Paramecium aurelia. Genetics*, **80**, 13–21.

De Terra, N. (1974), Cortical control of cell division. *Science*, **184**, 530–537.

De Terra, N. (1975), Evidence for cell surface control of macronuclear DNA synthesis in *Stentor. Nature*, **258**, 300–303.

Dini, F. and Miyake, A. (1977), Preconjugant interaction in *Euplotes crassus*. *J. Protozool.* (in press).

Doerder, F.P. and Debault, L.E. (1976), Cytofluorometric analysis of nuclear DNA during meiosis, fertilization and macronuclear development in the ciliate, *Tetrahymena pyriformis*, syngen 1. *J. Cell Sci.*, **17**, 471–93.

Downs, L.E. (1956), The breeding system in *Stylonychia putrina. Proc. Soc. exp. Biol. Med.*, **93**, 586–587.

Downs, L.E. (1959), Mating-types and their determination in *Stylonychia putrina, J. Protozool.*, **6**, 285–292.

Ehret, C.F. (1953), An analysis of the role of electromagnetic radiation in the mating reaction of *Paramecium bursaria. Physiol. Zool.*, **26**, 274–300.

Elliott, A.M. (1973), *Biology of Tetrahymena.* Dowden, Hutchinson and Ross, Stroudsburg, Pa.

Elliott, A.M. and Tremor, J.W. (1958), The fine structure of the pellicle in the contact area of conjugating *Tetrahymena pyriformis. J. Biophys. biochem. Cytol.*, **4**, 839–840.

Engberg, J., Christiansen, G. and Leick, J.V. (1974), Autonomous rDNA molecules containing single copies of the ribosomal RNA genes in the macronucleus of *Tetrahymena pyriformis. Biochem. biophys. Res. Comm.*, **59**, 1356–1265.

Esposito, F., Ricci, N. and Nobili, R. (1976), Mating-type specific soluble factors (gamones) in cell interaction of conjugation in the ciliate *Oxytricha bifaria. J. exp. Zool.*, **197**, 275–282.

Gall, J.C. (1974), Free ribosomal genes in the macronucleus of *Tetrahymena. Proc. natn. Acad. Sci.*, **71**, 3078–3081.

Giese, A.C. (1973), *Blepharisma:* The biology of a light-sensitive protozoan. Stanford University Press, Stanford, California.

Golicova, M.N. (1974), Cytophotometrical study of DNA amount in micronucleus of *Paramecium bursaria. Acta Protozool.*, **13**, 109–18.

Gorovsky, M. (1973), Macro- and micronuclei of *Tetrahymena pyriformis:* A model system for studying the structure and function of eucaryotic nuclei. *J. Protozool.*, **20**, 19–25.

Grell, K.G. (1967), Sexual reproduction in protozoa. In: *Research in Protozoology II* (Chen, T.T. ed.), Pergamon, Oxford-New York.

Haggard, B. (1974), Interspecies crosses in *Paramecium aurelia* (syngen 4 by syngen 8). *J. Protozool.*, **21**, 152–159.

Hanson, E.D. (1967), Protozoan development. In: *Chemical Zoology I Protozoa* (Florkin, M. *et al.*), Academic Press, New York.

Heckmann, K. (1963), Paarungssysteim und genabhängige Parrungstypdifferenzierung bei dem hypotrichen Ciliaten *Euplotes vannus.* O.F. Müller. *Arch. Protistenk., Arch. Protistenk.*, **106**, 393–421.

Heckmann, K. (1964), Experimentelle Untersuchungen an *Euplotes crassus* I Paarungssystem, Konjugation und Determination der Paarungstypen. *Z. Verebungsl.*, **95**, 114–124.

Heckmann, K. (1967), Age-dependent intraclonal conjugation in *Euplotes crassus. J. exp. Zool.*, **165**, 269–278.

Heckmann, K. and Siegel, R.W. (1964), Evidence for the induction of mating-type substances by cell to cell contacts. *Exp. Cell Res.*, **36**, 688–691.

Hertwig, R. (1889), Über die Konjugation der Infusorien. *Abh. Bayr. Akad. Wiss.*, **17**, 150–233.

Hirshfield, H.I., Isquith, I.R. and Di Lorenzo, A.M. (1973), Classification, distribution and evolution. In: *Blepharisma* (Giese, A.C. ed.,) 304–322. Stanford University Press.

Hiwatashi, K. (1949), Studies on the conjugation of *Paramecium caudatum*. II. Induction of pseudoselfing pairs by formalin-killed animals. *Sci. Rep. Tôhoku Univer., Biol.*, **18**, 141–143.

Hiwatashi, K. (1959), Induction of conjugation by ethylene diamine tetraacetic acid (EDTA) in *Paramecium aurelia. Sci. Rep. Tôhoku Univer., Biol.*, **25**, 81–90.

Hiwatashi, K. (1968), Determination and inheritance of mating-type in *Paramecium caudatum. Genetics*, **58**, 373–386.

Hiwatashi, K. (1969a), *Paramecium*. In: *Fertilization II* (Metz, C.B. and Monroy, A. eds.), Academic Press, New York.

Hiwatashi, K. (1969b), Genetic and epigenetic control of mating-type in *Paramecium caudatum. Jap. J. Genet.*, **44** (suppl. 1), 383–387.

Hiwatashi, I. and Takahashi, M. (1967), Inhibition of mating reaction by antisera without immobilization in *Paramecium. Sci. Rep. Tôhoku Univer., Biol.*, **33**, 281–290.

Honda, H. and Miyake, A. (1975), Taxis to a conjugation-inducing substance in the ciliate *Blepharisma. Nature* **257**, 678–680.

Honda, H. and Miyake, A. (1976), Cell-to-cell contact by locally differentiated surfaces in conjugation in *Blepharisma. Dev. Biol.*, **52**, 221–230.

Jennings, H.S. (1939), Genetics of *Paramecium bursaria*. I. Mating types and groups, their interrelations and distribution; mating behavior and self sterility. *Genetics*, **24**, 202–233.

Jennings, H.S. (1942), Genetics of *Paramecium bursaria*. III. Inheritance of mating type, in cross and in clonal self-fertilization. *Genetics*, **27**, 193–211.

Jennings, H.S. (1944), *Paramecium bursaria:* Life history. I. Immaturity, maturity and age. *Biol. Bull.*, **86**, 131–145.

Karakashian, M.W. (1968), The rhythm of mating in *Paramecium aurelia* syngen 3. *J. Cell Physiol.*, **71**, 197–209.

Katashima, R. (1959), Mating types in *Euplotes eurystomus. J. Protozool.*, **6**, 75–83.

Katashima, R. (1961), Breeding system of *Euplotes patella* in Japan. *Jap. J. Zool.*, **13**, 39–61.

Kimball, R.F. (1939), Mating types in *Euplotes. Am. Nat.*, **73**, 451–456.

Kimball, R.F. (1942), The nature and inheritance of mating types in *Euplotes patella. Genetics*, **27**, 269–285.

Kitamura, A. and Hiwatashi, K. (1976), Mating-reactive membrane vesicles from cilia of *Paramecium caudatum. J. Cell Biol.*, **69**, 736–740.

Kloetzel, J.A. (1970), Compartmentalization of the developing macronucleus following conjugation in *Stylonychia* and *Euplotes. J. Cell. Biol.*,

Kosaka, T. (1973), Mating types of marine stocks of *Euplotes woodruffi* (Ciliata) in Japan. *J. Sci. Hiroshima Univ. Ser. B.*, **24**, 135–144.

Kubota, T., Tokoryama, T., Tsukuda, Y., Koyama, H. and Miyake, A. (1973), Isolation and structure determination of blepharismin, a conjugation-initiating gamone in the ciliate *Blepharisma. Science*, **179**, 400–402.

Luporini, P. and Dini, R. (1975), Relationships between cell cycle and conjugation in 3 hypotrichs. *J. Protozool.*, **22**, 541–544.

Maupas, E. (1889), Le rajeunissement karyogamique chez les ciliés. *Arch. Zool. exp. gén. 2e Sér.*, **7**, 149–517.

McCoy, J.W. (1972), Kinetic studies on the mating reaction in *Tetrahymena pyriformis*, syngen 1. *J. exp. Zool.*, **180**, 271–278.

Metz, C.B. (1948), The nature and mode of action of the mating-type substances. *Am. Nat.*, **82**, 85–95.

Metz, C.B. (1954), Mating substances and the physiology of fertilization in ciliates. In: *Sex in Micro-organisms* (Wenrich, D.H. ed.), 284–334. Am. Ass. Adv. Sci., Washington.

Metz, C.B. and Butterfield, W. (1951), Action of various enzymes on the mating-type substances of *Paramecium calkinsi*. *Biol. Bull.*, **101**, 99–105.

Metz, C.B. and Foley, M.T. (1949), Fertilization studies in *Paramecium aurelia:* an experimental analysis of a non-conjugating stock. *J. exp. Zool.*, 505–528.

Metz, C.B. and Fusco, E.M. (1948), Inhibition of the mating reaction in *Paramecium aurelia* with antiserum. *Anat. Rec.*, **101**, 654–655.

Miwa, I., Nobuyuki, H. and Hiwatashi, K. (1975), Immaturity substances: material basis for immaturity in *Paramecium*. *J. Cell Sci.*, **19**, 369–378.

Miyake, A. (1958), Induction of conjugation by chemical agents in *Paramecium caudatum*. *J. Inst. Polytech. Osaka City Univ. Ser. D.*, **9**, 251–296.

Miyake, A. (1964), Induction of conjugation by cell-free preparations in *Paramecium multimicronucleatum*. *Science*, **146**, 1583–1585.

Miyake, A. (1968), Induction of conjugation by chemical agents in *Paramecium*. *J. exp. Zool.*, **167**, 359–380.

Miyake, A. (1974), Cell interaction in conjugation of ciliates. *Curr. Top. Microbiol. Immunol.*, **64**, 49–77.

Miyake, A. (1975), Control factor of nuclear cycles in ciliate conjugation: cell-to-cell transfer in multicellular complexes. *Science*, **189**, 53–55.

Miyake, A. (1977), Cell communication, cell union and initiation of meiosis in ciliate conjugation. *Curr. Topics in Develop. Biol.*, **12** (in press).

Miyake, A. and Beyer, J. (1973), Cell interaction by means of soluble factors (gamones) in conjugation of *Blepharisma intermedium*. *Exp. Cell Res.*, **76**, 15–24.

Miyake, A. and Beyer, J. (1974), Blepharmone: A conjugation-inducing glycoprotein in the ciliate *Blepharisma*. *Science*, **185**, 621–623.

Miyake, A. and Bleyman, L.K. (1976), Gamones and mating types in the genus *Blepharisma* and their possible taxonomic application. *Genet. Res.*, **27**, 267–275.

Miyake, A. and Honda, H. (1976), Cell union and protein synthesis in conjugation in *Blepharisma*. *Exp. Cell Res.*, **100**, 31–40.

Miyake, A., Maffei, M. and Nobili, R. (1977), Nature of meiosis-initiating factor deduced from the mode of cell-to-cell propagation of nuclear activation in multicellular complexes of *Blepharisma*. *Exp. Cell Res.*, (in press).

Myohara, K. and Hiwatashi, K. (1975), Temporal patterns in the appearance of mating-type instability in *Paramecium caudatum*. *Jap. J. Genet.*, **50**, 133–139.

Nanney, D.L. (1953), Nucleocytoplasmic interaction during conjugation in *Tetrahymena*. *Biol. Bull.*, **105**, 133–148.

Nanney, D.L. (1956), Caryonidal inheritance and nuclear differentiation. *Am. Nat.*, **90**, 291–307.

Nanney, D.L. (1963), The inheritance of H-L serotype differences at conjugation in *Tetrahymena*. *J. Protozool.*, **10**, 152–155.

Nanney, D.L. (1964), Macronuclear differentiation and subnuclear assortment in ciliates. In: *The Role of Chromosomes in Development* (Locke, M. ed.), 253–273. 23rd Symp. Soc. Study Dev. Growth, Academic Press, New York.

Nanney, D.L. (1974), Aging and long-term temporal regulation in ciliated protozoa. A critical review. *Mech. Ageing Dev.*, **3**, 81–105.

Nanney, D.L. (1977), Molecules and morphologies: the perpetuation of pattern in the ciliated protozoa. *J. Protozool.*, **24**, 27–35.

Nanney, D.L. and Caughey, P.A. (1955), An unstable nuclear condition in *Tetrahymena pyriformis. Genetics*, **40**, 388–398.

Nanney, D.L. and McCoy, J.W. (1976), Characterization of the species of the *Tetrahymena pyriformis* complex *Trans. Am. Microsc. Soc.*, **95**, 664–682.

Nanney, D.L. and Meyer, E.B. (1977), Traumatic induction of early maturity in *Tetrahymena. Genetics*, **86**, 103–112.

Nanney, D.L. and Nagel, M.J. (1964), Nuclear misbehavior in an aberrant inbred *Tetrahymena. J. Protozool.*, **11**, 465–473.

Nobili, R. (1966), Mating types and mating-types inheritance in *Euplotes minuta* Yocum (Ciliata, Hypotrichida). *J. Protozool.*, **13**, 38–41.

Nyberg, D. (1974), Breeding system and resistance to environmental stress in ciliates. *Evolution*, **28**, 367–380.

Nyberg, D. (1976), Are macronuclear subunits in *Paramecium* functionally diploid? *Genet. Res.*, **27**, 239–248.

Ofer, L., Leukovitz, H. and Loyter, A. (1976), Conjugation in *Tetrahymena pyriformis*. The effect of polylysine, concanavalin A and bivalent metals on the conjugation process. *J. Cell Biol.*, **70**, 287–293.

Orias, E. (1963), Mating-type determination in variety 8, *Tetrahymena pyriformis Genetics*, **48**, 1509–1518.

Orias, E. and Flacks, M. (1975), Macronuclear genetics of *Tetrahymena* I. Random distribution of macronuclear gene copies in *T. pyriformis*, syngen 1. *Genetics*, **79**, 187–206.

Phillips, R.B. (1969), Mating-type inheritance in syngen 7 of *Tetrahymena pyriformis:* Infra- and interallelic interactions. *Genetics*, **63**, 349–359.

Phillips, R.B. (1971), Induction of competence for mating in *Tetrahymena* by cell-free fluids. *J. Protozool.*, **18**, 163–165.

Pieri, J. (1965), Interprétation cytophotométrique des phénomènes nucléaires au cours de la conjugasion chez *Stylonchia pustulata. C.R. Acad. Sci. Paris*, **261**, 2742–2744.

Pieri, J., Vaugien, C. and Touillier, M. (1968), Interprétation cytophotométrique des phénomènes micronucléaires au cours de la division binaire et des divisions prégamiques chez *Paramecium trichium. J. Cell. Biol.*, **36**, 664–668.

Powers, E.L. (1943), The mating type of double animals in *Euplotes patella. Am. Midland Nat.*, **30**, 175–195.

Preer, J.R. Jr. (1976), Quantitative predictions of random segregation models of the ciliate macronucleus. *Genet. Res.*, **27**, 227–238.

Preparata, R.M. and Nanney, D.L. (1977), Cytogenetics of triplet conjugation in *Tetrahymena:* Origin of haploid and triploid clones. *Chromosoma*, **60**, 49–57.

Prescott, D.J., Murti, K.G. and Bostock, C.J. (1973), Genetic apparatus of *Stylonychia sp. Nature*, **242**, 576 and 297–600.

Raikov, I.B. (1972), Nuclear phenomena during conjugation and autogamy in ciliates. In: *Research in Protozoology IV* (Chen, T.T. ed.), Pergamon, Oxford, New York.

Raikov, I.B. (1976), Evolution of macronuclear organization. *Ann. Rev. Genet.,* **10**, 413–440.

Raper, J.R. (1966), *Genetics of Sexuality in Higher Fungi.* Ronald Press, New York.

Rasmussen, L. and Orias, E. (1975), *Tetrahymena:* Growth without phagocytosis. *Science,* **190**, 464–465.

Reissig, J.L. (1974), Decoding of regulatory signals at the microbial surface. *Curr. Top. Microbiol. Immunol.,* **67**, 43–96.

Ricci, N., Esposito, F. and Nobili, R. (1975), Conjugation in *Oxytricha bifaria:* Cell interaction. *J. exp. Zool.,* **192**, 343–348.

Ruiz, F., Adoutte, A., Rossignol, M. and Beisson, J. (1976), Genetic analysis of morphogenetic processes in *Paramecium I.* A mutation affecting trichocyst formation and nuclear division. *Genet. Res.,* **27**, 109–122.

Sasaki, S., Ito, A., Kitamura, A. and Hiwatashi, K. (1972), Mating-type specific antigen in *Paramecium caudatum. Genetics,* **71**, s55.

Siegel, R.W. (1956), Mating types in *Oxytricha* and the significance of mating-type systems in ciliates. *Biol. Bull.,* **110**, 352–357.

Siegel, R.W. (1965), Hereditary factors controlling development in *Paramecium. Brookhaven Symp. Biol.,* **18**, 55–65.

Siegel, R.W. and Cole, J. (1967), The nature and origin of mutations which block a temporal sequence for genic expression in *Paramecium. Genetics,* **55**, 606–617.

Siegel, R.W. and Larison, L.L. (1960), The genetic control of mating-types in *Paramecium bursaria. Proc. natn. Acad. Sci. U.S.A.,* **46**, 344–349.

Sonneborn, T.M. (1937), Sex, sex inheritance and sex determination in *Paramecium aurelia. Proc. natn. Acad. Sci. U.S.A.,* **23**, 378–385.

Sonneborn, T.M. (1938), Mating-types in *Paramecium aurelia:* diverse conditions for mating in different stocks; occurrence, number and interrelations of the types. *Proc. Am. Phil. Soc.,* **79**, 411–434.

Sonneborn, T.M. (1947), Recent advances in the genetics of *Paramecium* and *Euplotes. Adv. Genet.,* **1**, 263–358.

Sonneborn, T.M. (1954a), The relation of autogamy to senescence and rejuvenescence in *Paramecium aurelia. J. Protozool.,* **1**, 38–53.

Sonneborn, T.M. (1954b), Patterns of nucleocytoplasmic integration in *Paramecium.* **6**, (suppl.), 307–325.

Sonneborn, T.M. (1955), A third point of attachment between conjugants in *Paramecium aurelia* and its significance. *J. Protozool.,* **2**, (suppl.), 13.

Sonneborn, T.M. (1957), Breeding systems, reproductive methods and species problems in Protozoa. In: *The Species Problem* (Mayr, E. ed,), pp. 155–324. *Am. Ass. Adv. Sci.,* Washington, D.C.

Sonneborn, T.M. (1966), A non-conformist genetic system in *Paramecium aurelia. Am. Zool.,* **6**, 589.

Sonneborn, T.M. (1975a), *Tetrahymena pyriformis* In: *Handbook of Genetics,* (King, R.C. ed), **2**, pp. 433–467. Plenum Press, New York, N.Y.

Sonneborn, T.M. (1975b), *Paramecium aurelia* In: *Handbook of Genetics,* (King, R.C. ed.), **2**, pp. 469–594. Plenum Press, New York, N.Y.

Sonneborn, T.M. (1975), The *Paramecium aurelia* complex of 14 siblings species. *Trans. Am. Microbl. Soc.,* **94**, 155–178.

Takagi, Y. (1971), Sequential expression of sex-traits in the clonal development of *Paramecium multimicronucleatum*. *Jap. J. Genet.*, **46**, 83–91.

Takahashi, M. (1973), Does fluid have any function for mating in *Tetrahymena?* *Sci. Rep. Tôhoku Univ. Biol.*, **36**, 223–229.

Takahashi, M. and Hiwatashi, K. (1974), Potassium: A factor necessary for the expression of mating reactivity in *Paramecium caudatum*. *Exp. Cell Res.*, **85**, 23–30.

Takahashi, M., Takeuchi, N. and Hiwatashi, K. (1974), Mating agglutination of cilia detached from complementary mating types of *Paramecium*. *Exp. Cell Res.*, **87**, 415–417.

Taub, S.R. (1963), The genetic control of mating-type differentiation in *Paramecium*. *Genetics*, **48**, 815–834.

Tokoroyama, T., Hori, S. and Kubota, T. (1973), Synthesis of blepharismone, a conjugation-inducing gamone in ciliate *Blepharisma*. *Proc. Jap. Acad.*, **49**, 461–463.

Van Wagtendonk, W.J. (1974), *Paramecium: A Current Survey*. Elsevier Scientific Publishing Co., Amsterdam and New York.

Vivier, E. (1974), Morphology, taxonomy and general biology of the genus *Paramecium*, In: *Paramecium: A Current Survey* (Van Wagtendonk, W.J. ed.), pp. 1–89.

Wolfe, J. (1973), Conjugation in *Tetrahymena:* The relationship between the division cycle and cell pairing. *Dev. Biol.*, **35**, 221–231.

Wolfe, J. (1974), Reciprocal induction of cell division by cells of complementary mating-types in *Tetrahymena*. *Exp. Cell Res.*, **87**, 39–46.

Woodard, J., Woodard, M., Gelber, B. and Swift, H. (1966), Cytochemical studies of conjugation in *Paramecium aurelia*. *Exp. Cell Res.*, **41**, 55–63.

10 An Overview

JOSE L. REISSIG

10.1	Introduction	*page*	401
10.2	Long-range interactions		402
10.3	Short-range interactions		405
10.4	Docking		406
10.5	Modulation and second messengers		408
10.6	Specificity		410
	10.6.1 Components and wiring		410
	10.6.2 Mating types		411
10.7	Mix		412
	References		414

Acknowledgements

I am thankful to the contributors to this volume for their unfailing co-operation, and to them, to Rollin D. Hotchkiss and to Mike Shodell for the critical reading of this manuscript. The secretarial assistance of Amanda Jensen and Gail Allan is gratefully acknowledge. Richard R. Griffith introduced me to the Lamia and other beasts.

Microbial Interactions
(*Receptors and Recognition,* Series B, Volume 3)
Edited by J.L. Reissig
Published in 1977 by Chapman and Hall, 11 New Fetter Lane, London EC4P 4EE
© Chapman and Hall

10.1 INTRODUCTION

Having confidently stated in the Preface that the search for unity in this work belongs in the best traditions of microbiology, I am now faced with the task of delivering the unitary interacting microbe. The endeavour conjures visions nurtured by another tradition, far older and seemingly grotesque. I am refering to the tradition represented by the composite beasts of the primitive bestiaries, like the Lamia of Fig. 10.1.

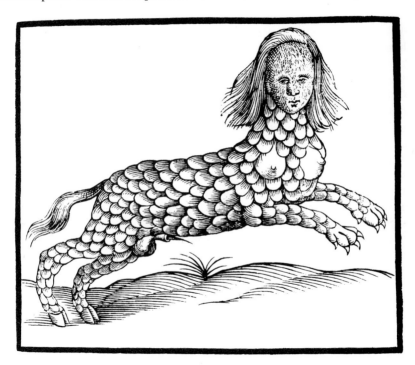

Fig. 10.1 A Lamia: an embellishment of the legendary Lamia, Queen of Libya, loved by Zeus. It is variously said to have had hind feet of a cow or goat, front feet of a bear or cat, scales of a fish, a snake or a dragon, an unattributed penis, and other attributes of a woman. From *The History of Four-footed Beasts* by Edward Topsell (1658), E. Cotes, London. By courtesy of the Wellcome Trustees.

Whichever attribution of cultural heritage should be more appropriate, the fact remains that a unitary view of the mechanisms of microbial interaction is not easily achieved at this time. In order to facilitate the task, the major observations reported in the volume have been organized conceptually in what is called the Thesaurus (Appendix). The reader may choose to use the Thesaurus as an aid to put together a do-it-yourself overview. Alternatively, the Thesaurus may be used to retrieve the information on which this summarizing chapter is based.

An earlier volume of this series contains a review on mating interactions in micro-organisms (Crandall, 1977). Crandall's review is a comprehensive one. The present volume has concentrated, instead, on a number of selected systems, both sexual and otherwise. The two approaches are therefore complementary.

10.2 LONG-RANGE INTERACTIONS

Long-range interactions between micro-organisms are mediated by chemical messengers. Such messengers are often called pheromones, following the proposal of Karlson and Lüscher (1959) who had primarily in mind substances which mediate communication between animals of the same species.

The chemical nature of microbial pheromones is quite varied. Like the hormones and pheromones of higher organisms they range from cAMP and other low molecular weight substances, to large proteins or glycoproteins. The ubiquity of cAMP warrants some comments. Ordinarily cAMP functions as a 'second messenger' (Sutherland, 1972); i.e. its concentration inside the cell is regulated by the levels of hormones ('first messengers') in the outside. However, in *Dictyostelium discoideum* cAMP acts as a 'first messenger'. One cannot but wonder why cells had to come up with the same chemical for such different jobs. Does this molecular conservatism reflect a stringent interdependence of elements put together in the course of evolution, compelling cells to adopt the system as a package ? This is surely not a comprehensive explanation because another slime mold (*Polysphondylium violaceum*) uses a peptide instead of cAMP as chemotactic messenger, while retaining cAMP for other intercellular communications. Hence opportunism, rather than conservatism, would seem to be at the root of the choice of pheromone in the slime molds, at least for chemotaxis.

Let us now consider the informational content of the messages exchanged. Three different types of commands are encoded, to wit:

1 'Come close !'
2 'Get ready !'
3 'Tell others !'

Chemotactic messages are typical examples of the first category. Get-ready messages encompass several subgroups: 'get ready for development !', 'get ready for mating !' and 'get ready for DNA uptake !'. The third category (tell-others) refers to instances in which the cells relay and amplify the messages they receive, as when a pheromone induces the production of more of the same or of a complementary pheromone.

Slime mold amoebae exchange all three kinds of messages. As discussed by Newell, the object of their complex developmental exercise is to evade unfavorable environments by ultimately producing spores capable of dispersion. The first step in this exercise is aggregation, and the usefulness of come-close messages as a prelude to it is self-evident. The territorial reach of this call is amplified by a relay system (tell-others messages). The contribution of the get-ready signals to genetic fitness becomes apparent when one considers that they permit cells to initiate preparations for aggregation and development back at the beginning of their journey towards the point of congregation. In *D. discoideum* all three messages are coded as cAMP, but this does not imply that they are sensed by the same receptors. In fact, there is evidence indicating that the tell-others signals are sensed by a specific class of receptors.

Among the ciliate case histories discussed by Nanney, the one which best illuminates the nature of long-range interactions is that of *Blepharisma*. The object of their exercise is mating; hence communication occurs primarily between cells of different mating types. Type I cells produce gamone I (a glycoprotein) which is only sensed by type II cells, while type II cells produce gamone II (a low molecular weight compound) which is only sensed by cells of the complementary type. All three kinds of messages are exchanged: come-close, get-ready (for mating), and tell-others. Come-close messages are unidirectional, being encoded by gamone II only. This chemotactic asymmetry can be easily understood in terms of the intended result. Indeed, once the attraction of one mating type by the other has been elicited, there seems to be little point in securing the reciprocal movement (Current grooming practices among human males notwithstanding). On the other hand, the reciprocity of the get-ready messages is clearly a must: getting ready means to become cohesive, and only cells so instructed can agglutinate, and hence mate, with each other. The tell-others messages are also reciprocal; that is to say, gamone I induces type II cells to make gamone II, and vice versa.

The mating system of *Saccharomyces* reviewed by Manney and Meade, is characterized by two mating types, *a* and α, each producing specific pheromone(s), called *a* and α factors respectively. There are no chemotactic responses because yeast is not motile. Insofar as is known, the communication between the two mating types is symmetrical, but the nature of the messages emanating from α cells is better understood. Type α cells produce α-factor, a family of four closely related oligopeptides. This factor encodes some or all of the get-ready commands, which induce in type *a* cells the following effects: agglutinability with α cells, alterations in the structure of the cell wall — with provisions for the elongation of cells into 'shmoos' — and transient arrest of the cell cycle in G1. All these commands prepare the cells for fusion and fertilization.

The interactions facilitating genetic transformation in bacteria, surveyed by Lacks, show distinct parallelism with the ones considered above. No exchange of chemotactic signals has been reported in this system; an omission which should come as no surprise since here the cells do not beckon their partners, but ship away their DNA (Borenstein and Ephrati-Elizur, 1969). The get-ready and tell-others messages are conveyed by a small protein, called activator, in the better known systems.

Get-ready commands include all the necessary provisions for the cells to be able to take up DNA: availability of a DNA-binding protein, induction of autolysin activity, etc. The tell-others messages consist of the induction of activator by activator itself, but apparently only by activator approaching the cells from the outside. Accordingly, an autocatalytic increase in the basal (subthreshold) level of activator in the medium occurs when cells reach a density high enough for effective communication, which must also mean a density high enough to warrant setting the transformation machinery in motion. This is cybernetically analogous to the situation in *Saccharomyces* and *Blepharisma,* where complementary cells induce each other into competence using signals which are reinforced by the proximity of the complementary types.

The get-ready commands discussed in the above paragraphs cannot become effective in the absence of protein synthesis in the target cells (Sections 1.2.6, 5.3.3, 5.3.4, 5.4.3, 7.4.3 and 9.6).

Although the sensory system used by bacteria in foraging for an optimal environment is not ordinarily utilized for cell–cell interaction, the chapter on bacterial chemotaxis by Hazelbauer and Parkinson is most germane to this volume because it provides a penetrating insight into the operation of sensory mechanisms. Bacteria detect attractants and repellents using peripheral membrane proteins as sensory receptors. When a ligand (e.g. an attractant) with affinity for some of those stereospecific receptors becomes available, the cell senses this as a signal, probably quantified in terms of the extent of binding which occurred. The information is processed by a series of components, and the extent of site occupancy at that moment is compared with that prevailing at an immediately preceeding moment. If occupancy is on the increase, the tumble generator, which normally reverses the direction of flagellar rotation at regular intervals and thus causes tumbling, is restrained: bacteria continue to swin in the same direction – up the gradient. The analysis of this system, using an adroit combination of genetic and biochemical methods, has provided some fascinating glimpses into its workings, whetting our appetites for further revelations and for eventual comparative studies. We would like to compare the strategy used by bacteria for the detection of chemical gradients with that of other micro-organisms. The former, faced with the problem of detecting gradients which vary only insignificantly along their minute bodies, have circumvented it by effecting time comparisons of the concentrations encountered in the course of their swimming. Slime mold amoebae, although themselves relatively quiescent before stimulation, are exposed to moving cAMP gradients as a consequence of pulsation at the source. Further opportunities for temporal comparisons are provided in this system by the ability of the amoebae to scan surrounding areas by protruding and contracting pseudopodia and filopodia (Cooke *et al.,* 1976), but the question of whether slime molds and other non-bacterial micro-organisms use memory-based comparisons for chemotactic orientation remains open.

10.3 SHORT-RANGE INTERACTIONS

The systems considered in the previous section are characterized by overt long-range interactions. It is quite possible that these cells also conduct additional informational transactions after they are in close proximity (cf. Section 7.4.5), but whatever 'whispering' — to use Newell's expression — takes place among them is overshadowed by the more obvious 'shouting'. It is easier to analyze this whispering in systems which display it conspicuously and exclusively, but this should not lead us to believe that the two modes of interaction — long and short-range — are mutually exclusive. *Chlamydomonas moewusii* var *rotunda* for instance (Wiese, 1969) uses, as a prelude for sexual union, both chemotaxis and a short-range ritual analogous to that observed in other varieties. A similar duality has been reported for some ciliates (Section 9.2).

The *Chlamydomonas* mating interactions, described by Goodenough, are choreographed in such a way that the major steps are easily discernable. Agglutination of 'plus' and 'minus' *Chlamydomonas* cells via the flagellar tips — possibly mediated by their fuzzy coats — induces wall lysis and activation of complementary mating structures, steps which are a prelude for cell fusion and fertilization. Flagella are directly implicated in this induction, as shown by the fact that isolated flagella, when added to gametes of the opposite mating type, also elicit wall lysis and mating structure activation; while mutational blockage of the agglutination step abolishes these two effects.

An unidentified early step in that sequence activates a control circuit which causes the flagella to lose agglutinability, and disengage. With the availability of an increasing number of 'impotent' mutants, one can expect that a more discerning analysis of the *Chlamydomonas* mating foreplay should soon be possible.

Mating rituals in paramecia show distinct homologies with the above system. As described by Nanney, the initial get-ready-to-mate signals can be traced back to a ciliary foreplay stage, which is then followed by a firmer cellular union — at the holdfast region first, then at the paroral region. The cilia lose their agglutinability in the course of this sequence. Mating interactions requiring cell—cell contact, rather than diffusible messengers, have also been reported for other ciliates. They go under the name of co-stimulation in *Tetrahymena*, or courtship dance in the hypotrichs, but their details are still obscure.

The smaller size of *Escherichia coli* makes it more difficult to eavesdrop on its mating ritual. Nevertheless, as is made clear in Achtman and Skurray's account, the wealth of genetic tools available in this organism makes up for the skimpiness of visual detail. It appears that cells are first held together by pili, then by unstable wall-wall contacts, and finally by dilution-stable links. An event occurring at an undetermined stage in this sequence triggers the get-ready signal which initiates conjugal replication of DNA, i.e. replication initiated at the *oriT* site. The signals for actual DNA transfer are apparently not given until mating aggregates become stabilized.

Get-ready messages are also exchanged between phages and bacteria as a prelude to infection. Attachment of the phage to its receptor on the outer membrane of enteric bacteria, as reviewed by Braun and Hantke, triggers the release of the nucleic acid from the capsid and prepares the bacterial cell to receive it. The details of what passes between guest and host — often also victim — are not known, but the simplicity of the system recommends it for more exhaustive study.

A comprehensive and comparative understanding of the systems mentioned in this section must await a more detailed elucidation of the proceedings, but it may be appropriate to formulate here some of the unanswered questions. While, as pointed out by Jones (Section 4.6.2), electrostatic charge relationships may enjoin cells to approach each other via appendages before attempting a closer docking, it is clear that the short-range interactions described encode get-ready-for-mating messages equivalent to those transmitted by pheromones. What is not clear, however, is whether short-range and long-range interactions are analogous in their molecular mechanisms (cf. p. 358). It is tempting to postulate such an analogy on the premise of simplicity. One may assume that short-range interactions are mediated by pheromone-like effectors, the only difference between the two situations being that in one case the effectors are freely diffusible, while in the other they are tightly bound to the cell surface. This view is implicit in a number of working models, yet it remains unsubstantiated in spite of some sustained attempts to verify it. In *E. coli* for instance, two surface proteins, required respectively by male and female cells for interaction during the early mating stages, have been identified. Yet it turns out, as pointed out by Achtman and Skurray, that these proteins — F-pilin and the *ompA* gene product — are active at different stages, and do not interact with each other. The step mediated by F-pili is not specific for male-female pairs, but leads to the formation of homotypic pairs as well. It is therefore not surprising that the search for F-pilus receptor molecules on the female surface — as distinct from the male surface — should have been fruitless. Models to explain the mode of action of F-pili must also take into account the observation that they do not attach to the surface of *Pseudomonads*, for instance (Section 6.3.2(g)). The possibility that their ability to form intercellular bridges is based on hydrophobic affinity for certain membrane systems, rather than affinity towards stereo-specific receptors, has not been ruled out.

In summary, it may be fitting to keep an open mind about the way in which short-range interactions are transacted. Various alternative mechanisms of microbial interaction are discussed in Section 10.7.

10.4 DOCKING

One would expect short-range interactions to include not only informationally meaningful exchanges, like the get-ready signals discussed above, but also docking maneuvers, i.e. maneuvers simply directed at holding the cells together. It is nevertheless unclear whether these conceptually different tasks are performed by the same or different molecules. Even if different molecules are used, which appears plausible, the problem remains of how to distinguish operationally between the two types of interaction. In an attempt to provide a congruent overview, I am implying that the early stages of short-range interactions are primarily informational, while the latter stages only provide mechanical stability. One redeeming feature of this premature attempt at classification is that it invites improvement.

The better understood microbial docking systems involve complementary molecules. Complementary mannanproteins (Crandall, 1977) hold together *Hansenula wingeii* cells of complementary mating type. The role of pallidin and its complement in the cellular docking of *P. pallidum* appears well-established; while the part played by discoidin, contact sites A, and their respective complements in *D. discoideum* is less clear. Far more tentative, as pointed out by Jones, is the role and nature of the adhesins which bridge the attachment of bacteria to the cellular surfaces of their animal hosts. Equally tentative, yet no less fascinating, are the recent studies on the attachment of nitrogen-fixing bacteria to the roots of leguminous plants. It has been shown, for instance, that there is a specific matching between the O-antigens of *Rhizobia* and the lectins produced by their legume hosts: *in vitro* binding occurs only between those antigen-lectin pairs derived from natural symbionts (Wolpert and Albersheim, 1976; Bhunvaneswari *et al.*, 1977). Further detail is provided by the work of Dazzo and Brill (1977), suggesting that the multivalent plant lectin molecules cross-bridge the surfaces of the root hairs and of the symbiotic bacteria by attaching to immunologically similar sites located on both surfaces. Studies on the molecules which bind parasites to their plant hosts are still in their infancy, but there are indications that matching molecular specificities may also explain the selectivity of these associations (DeVay and Adler, 1976).

Some comments on the possible ways of distinguishing between informational exchanges at close range, and mere docking maneuvers, are pertinent here. One approach is to try to dissect the elements of a system using the genetic scalpel, but this may not be easy since a docking failure may interfere with informational exchange simply by destabilization of the cellular union. Furthermore, mutations causing cell surface changes often engender pleiotropic effects on the membrane (Section 10.7). Another approach is to treat cells with isolated surface components which are presumed to be part of the interacting system: the responses elicited by these components may provide a clue to their physiological role. Thus, the addition of phage proteins (ghosts) to bacteria, and their consequent attachment to the corresponding receptors, leads to cellular alterations akin to — though far more drastic than — those obtained with the intact phage particle (Section 3.1).

Even if it should turn out to be possible to draw a clear distinction between informational and mechanical contacts at the molecular level, the distinction between these two categories remains blurred at other levels since the mere joining of cells together is bound to have indirect but far-reaching consequences on their physiological and ecological fate (Section 4.2.2(e) and 4.7.1).

10.5 MODULATION AND SECOND MESSENGERS

One example of the modulation of microbial interactions is the 'undocking' of cells; i.e., the loss of agglutinability by cilia and flagella in the course of mating in *Paramecium* and *Chlamydomonas,* and the disaggregation which follows the termination of DNA transfer in *E. coli* mating. All three are active processes, requiring the completion of a specified program before taking place. The difference in the nature of the events triggering disaggregation in the three systems — cellular union in the protozoa, introduction of the last genes in *E. coli* — is as would be expected from the different requirements of the mating programs. The molecular mechanism of undocking is unknown.

A number of other ways in which microbial interactions can be modulated is listed in the Thesaurus (Section II). Of particular interest is the listing (II.B.1) of the requirements of divalent cations (Ca^{2+} and/or Mg^{2+}) for aggregation of *Chlamydomonas* flagella and *Dictyostelium* amoebae, for cAMP pulsing in the latter cells, for DNA entry during transformation, and for straight swimming and chemotaxis in bacteria. Further down (II.B.3) there is reference to the observation that high K^+/Ca^{2+} ratios, as well as other ionic manipulations, can induce non-specific mating in *Paramecium*. It is unclear whether the effects of divalent cations on aggregation are due to the formation of intercellular ionic bridges, or to any other direct or indirect effect on surface properties. The more explicitly functional effects — on cAMP production, DNA entry, chemotaxis and mating — may result from the well known capacity of divalent cations to act as co-factors for a variety of enzymes and contractile proteins; but to leave things at that, in the context of current biological thinking, would amount to an unwarranted dismissal of an important viewpoint. Rather, one should try to put these observations into perspective by asking whether, under physiological conditions, the ions in question regulate reactions which are rate-limiting for the system under consideration; and whether the concentration of these ions is in turn regulated by the signalling network. If this is so, then a strong case can be made for the hypothesis that the ion in question is an intracellular messenger ('second messenger') in that system; a hypothesis which enjoys currency in many non-microbial situations (Berridge, 1975). A role for Ca^{2+} as intracellular messenger has been substantiated in a microbial system which, although not reviewed in this volume, is closely related to others which have been discussed. I am referring to the work of Naitoh and Eckert (1974; Eckert, 1972) on the swimming behaviour of *Paramecium*. These authors have shown that the 'avoiding reaction' — i.e. the transient reversal in swimming direction which occurs when an animal collides with an object — results from the following chain of events:

(i) transient depolarization of the membrane, due to mechanical stimulation at the anterior end;

(ii) increase in the Ca^{2+} conductance of the membrane caused by depolarization;
(iii) influx of Ca^{2+} into the cell, down the electrochemical gradient, through the newly open Ca^{2+}-gates;
(iv) reorientation of the cilia, altering the direction of effective stroke so as to cause backward swimming, as a consequence of the higher (10^{-6} M) levels of intracellular Ca^{2+}.

This avoiding reaction can also be elicited by chemical stimulation, e.g. by transfering the organisms to media high in K^+ relative to Ca^{2+}. While the detailed interpretation of this ionic effect is somewhat problematic, in all likelihood it is also mediated by an increment in the intracellular concentration of free Ca^{2+} (Naitoh and Eckert, 1974). The evidence for all this will not be reviewed here, except to point out that mutants exist which are defective in the voltage-sensitive Ca^{2+}-gating mechanism, and which are concomitantly incapable of performing the avoiding reaction (Kung et al., 1975). Such mutants are called 'pawns' since they, too, can only move forward. Interestingly enough, pawns cannot be induced to mate non-specifically by chemical treatments (e.g. high K^+, low Ca^{2+} conditions) which induce such mating in the wild type (Cronkite, 1976). Therefore, although the steps in the ionic induction of mating are still obscure, it appears likely that the levels of intracellular Ca^{2+} play a critical role. Whether or not Ca^{2+} acts as intracellular messenger in the course of mating type-mediated mating is far less clear. The observation that the same pawn mutants which cannot be induced to mate by ionic manipulations can nevetheless mate heterotypically without difficulty (Cronkite, 1976) argues against the likelihood that the same ionic mechanism is used as a signalling intermediate in both situations. Nonetheless, a relationship between the Ca^{2+} gate and heterotypic mating is suggested by the report that some of the pawn mutants are sterile (Chang et al., 1974), but it remains uncertain whether this phenotypic concomitance is due to pleiotropic effects of one mutation, or to simulatenous mutation at more than one locus (cf. Cronkite, 1977).

The possibility that the bacterial chemotactic signals are mediated by ionic fluxes was discussed in Section 2.3.4.

Finally, some comments on the relationship of cAMP and Ca^{2+} appear in order. As argued by Berridge (1975) both these compounds are often cast in the role of 'second messengers', some times in the same system. In such cases, Ca^{2+} is frequently active on the target function — thus being a 'primary second messenger' in Berridge's terminology — while cAMP plays the secondary role of modulating the Ca^{2+} concentration by stimulating its outflow. A parallel situation may exist in *Dictyostelium*, where Chi and Francis (1971) found that cAMP (acting there as 'first messenger') enhances the outflow of Ca^{2+} (a potential 'second messenger').

10.6 SPECIFICITY

10.6.1 Components and wiring

The specificity of a regulatory circuit may be considered at two different levels, one concerned with the nature of the component elements, and the other with the configuration of the circuit — i.e. with how these elements are wired together. In the case of allosteric enzymes, for instance, the basic elements are complementary molecular moieties: allosteric site, allosteric effector, active site and substrate. Some of the wiring connecting these elements is the results of stereospecific interaction between molecular moeieties. Additional wiring results from the cohabitation of two particular sites on the same protein molecule, and from a complex functional interaction between them. Our understanding of what determines the specificity of microbial interactions at the two levels is patently more primitive than for allosteric enzymes.

The receptors for cAMP in slime molds; and for attractants, repellents, and transformation activator in bacteria, probably interact stereospecifically with their effectors. This statement can be made more forceful if restricted to the better known chemoreceptors. On the other hand, the receptors for phages, colicins and substrates of intermediate molecular weight located on the outer membrane of *E. coli*, appear unorthodox in their specificity since the same molecule can recognize ligands in these three structurally heterogeneous categories while exhibiting remarkable specificity within each category. This paradox may be resolved by assuming that the same protein contains various combining sites, although — as indicated by Braun and Hantke (Section 3.7.1) — such an explanation is not without complications.

Some interesting features of the wiring have been uncovered in the bacterial chemotaxis system. Inputs from a large number of chemoreceptors are processed through components which decrease in diversity at successive steps (Section 2.5). The final target of the commands must be a component of the flagellar motor. A tendency for input pathways to converge is also noticed in the processing of signals originating at the phage-colicin-etc. receptors on the outer membrane of *E. coli* (Section 3.3.2(c)). It would be interesting to know how general the stratagem of stepwise convergence turns out to be, since recourse to it may illustrate the molecular difficulties of an alternative system of wiring, namely the direct wiring of many input pathways onto a single molecular species.

The fundamental questions about the nature of much of the wiring remain unanswered. Is molecular contact between stereospecific proteins required for informational processing? Current work on mammalian hormone receptors (Cuatrecasas and Hollenberg, 1976) has led to the hypothesis that signalling components located on the membrane can contact each other, and hence exchange information in the form of conformational changes, by virtue of their mobility within the fluid membrane bilayer. The alternative of long-distance interactions mediated by phase transitions in the overall organization of the membrane may also be considered (Section 10.7).

Evidence has also been presented for the existence of divergent pathways. For instance the chemoreceptors of *E. coli* are the point of entry both for chemotactic information, which follows one pathway, and for molecular transport, which follows another (Section 2.4.2(e)). Similar bifurcations, informational and vectorial, occur in the systems reviewed by Braun and Hantke (Section 3.8). The situation is reminiscent of the bifurcated pathways of intermediate metabolism. Extensive research in this latter area has shown that, in the course of evolution, cells have been able to choose between two major options: to utilize a single enzyme for each of the shared steps, often incorporating special features so as to provide the necessary regulatory versatility; or to utilize multiple enzymes, thus in fact establishing parallel pathways (Brown *et al.*, 1975). Are these two options equally available for membrane transduction systems of the kind discussed in this volume, or does the supramolecular economy of the membrane favor the first option by putting too valuable a premium on parsimony?

10.6.2 Mating types

Mating systems are usually designed so that some or all of the prescribed interactions only take place between cells of different mating type, i.e. heterotypic cells. Systems in which all the interacting components are specific for mating type (i.e. insure heterotypic interactions) would have the necessary redundancy to effectively prevent any homotypic outcome or wasteful preliminaries. Yet the more extensively studied systems reviewed in this volume include both specific and unspecific steps in their mating routines. The extent of this equivoque can be appreciated by consulting the Thesaurus (Appendix) where the superscript e denotes interactions restricted to cells of different mating type, while the superscript o denotes the absence of such a restriction. For instance, by consulting **Processes** under I.A.2 it can be seen that the development of competence for mating is heterotypically restricted in *Saccharomyces* and *Blepharisma*; but further consideration of mating interactions in these organisms under **Specific responses** reveals that, once competence is developed, no restrictions are imposed on the type of cells which can agglutinate with each other. Similarly, as can be seen in Section I.B.1 of the Thesaurus, only some of the mating stages are heterotypically restricted in *E. coli* or *Paramecium*. There are also cases in which environmental modulation permits cells to bypass heterotypic specificity (Section II.B.3).

Anthropomorphic bigotry notwithstanding, microbial sexual ambiguity should not be taken as necessarily aberrant or unplanned. As discussed by Nanney (Sections 9.2, 9.3 and 9.7) evolutionary strategies among the ciliates often involve options in the direction of an inbreeding economy, and this can be achieved either by permitting homotypic unions, or by the choice of genetic method of sex determination. The complexity of the factors weighing on that choice is well illustrated by the ciliates (Sections 9.3.4, 9.4.1 and 9.5.1), and it is interesting to note that other systems ordinarily viewed as sexually 'straight' also toy with homotypic alternatives.

For instance, the repressed sex factors of *E. coli* (Section 6.4.2) provide this organism with an escape from heterosexual confinement; and some strains of yeast have developed a library of sex gene 'cassettes' in order to keep the homothallic option open (Hicks *et al.*, 1977).

10.7 MIX

One salient feature of microbial interactions is the involvement of the cell surface as a prime substation for the reception and transduction of signals encoding regulatory commands. It is convenient to refer to regulatory processes so mediated as 'perisemic' — from the Greek: περι, all around; and σημα, signal (Reissig, 1974). The interactions of mammalian cells with each other, with hormones, with antigens, etc., are also generally perisemic; i.e., mediated by transducers located on the cell surface (Greaves, 1976). Here we are concerned with the nature of the microbial perisemic mechanisms: What can we learn from the systems reviewed?

The involvement of receptor proteins at the input end of the sensory chain, and the likelihood that the signals received are transduced and transmitted by the stepwise propagation of conformational changes in the receptor and other protein components of the signalling pathway, was briefly recapped in Section 10.6.1. In the present section (for that reason labeled 'Mix') I wish to argue that we should be prepared that there be not one, but several basically different mechanisms for perisemic interaction. This doctrine of perisemic plurality finds support both in theoretical considerations and in experimental observations. The former can be presented most simply by saying that membranes are anisotropic structures, characterized not only by the molecules they contain, but also by the arrangement of these components; and, since (a) the different topographic displays contain informational capabilities, and (b) evolution is opportunistic; it is unlikely that this versatility would remain unexploited. A couple of examples of models covered by this pluralistic view may help to clarify the way in which it differs from the more restrictive models which only take into account protein—protein interaction. Consider the following possibilities:

(1) The binding of a ligand to a surface receptor causes a conformational change in the latter; this in turn affects the structure of the lipid matix, and the effects of this change propagate to diverse functional elements associated with the membrane. This model has been discussed by Perkins (1973), among others, and is given credence by the observation that co-operative transitions do in fact occur in biological membranes (Verma and Wallach, 1976).

(2) The receptor is a transmembrane protein, and ligand binding causes several subunits to congregate in such a way as to interrupt the hydrophobic domain of the bilayer with a hydrophilic channel. Pinto da Silva and Nicolson (1974) have proposed that such structures be called permeaphores. Permeaphores represent topological rearrangements of protein sub-units, and need not involve conformational changes of the polypeptide chain. It is important to keep in mind — in the context of this and

other alternative models — that, although membranes are characterized by a fair degree of fluidity, the mobility of the constitutive elements is modulated by a variety of restraints (Nicolson, 1976).

In order to assess the role of supramolecular structure in perisemic signalling among the systems reviewed in this volume, the experimental observations which might support such a role will be listed below.

(1) Evidence for the inhomogeneous distribution of functional elements in the plane of the membrane is provided by the fact that sexual agglutination is localized at the tips of *Chlamydomonas* flagella (Section 8.1) and *Paramecium* cilia (Hiwatashi, 1969), and that additional regional differentiation in agglutinability is observed among the ciliates (Section 9.4.3).

(2) The requirement that the surface of at least one of the partners be unfixed, for agglutination to occur in *Chlamydomonas* (Section 8.2.13) or *Paramecium* (Section 9.3.2), indicates that receptor mobility is required in that process. In *Chlamydomonas*, various lines of evidence suggest that the development of mating competence is accompanied by increased mobility of membrane elements (Sections 8.2.10, 8.2.12 and 8.2.14). In *E. coli* the formation of pili results from the aggregation of molecules from a precursor pool in the membrane bilayer (Section 6.3.1(a)), and intercellular membrane rearrangements occur during mating (Section 6.3.2(b)).

(3) Changes in the topographic display of membrane elements occur during development in slime molds (Section 1.6.6; see also 1.3 and 1.6.2) and in the development of the mating structures of *Chlamydomonas* (Section 8.3.3).

(4) The attachment of phage particles to their receptors on the bacterial surface leads to drastic membrane changes, as monitored with fluorescent probes (Section 3.8). Functional interactions between the inner and outer membranes of *E. coli* are indicated by the energization requirement for the irreversible binding of phages (Section 3.1).

(5) Understanding of bacterial motility along with its chemotactic modulation requires the assumption that the reversals of flagellar rotation be synchronized over the entire surface of the cell (Section 2.3.4). Such co-ordination is best understood as a result of transitions in the properties of the membrane behaving as a supramolecular unity.

(6) Temperature requirements for transformation in pneumococci (Section 5.3.3) or for mating structure activation in *Chlamydomonas* (Section 8.3.3) could be the result of thermotropic membrane transitions; although other explanations, including explanations based on the effects of temperature on protein conformation, have not been excluded.

(7) Mutations affecting cell surface functions are often pleiotropic, as in the case of α factor-deficient mutants of *Saccharomyces* (Section 7.4.4, 7.5) and of mutations affecting proteins on the outer membrane of *E. coli* (Section 6.3.1(b)). Extensive pleiotropism is expected to occur in a system composed of elements showing tight interdependency.

(8) The possibility that supramolecular integration may put a premium on parsimony

was discussed in Section 10.6.1. The jamming in chemotactic signals emanating from one receptor by signals arriving from another chemoreceptor (Section 2.5.1) might also have a similar explanation.

(9) One or possibly more cases in which divalent cations function as 'second messengers' were discussed in Section 10.5. Signalling is achieved simply by transient alterations in membrane permeability, capable of generating the necessary ionic fluxes. Membrane depolarization may result from electrical or mechanical stimuli, from topological rearrangements forming permeaphores, etc.

By itself, none of the above observations can be taken as proof that the perisemic idiom includes supramolecular information; but, as a whole, they carry a strong suggestion that in Nature's bestiary there may be more to recognition than just receptors, and more to perisemic transduction than the helpless collision of stereospecific membrane proteins.

REFERENCES

Berridge, M.J. (1975), The interaction of cyclic nucleotides and calcium in the control of cellular activity. *Adv. Cyclic Nucleotide Res.*, **6**, 1–98.

Bhuvaneswari, T.V., Pueppke, S.G. and Bauer, W.D. (1977), The role of lectins in plant microorganism interactions. I. Binding of soybean lectin to rhizobia *Plant Physiol.*, (in press).

Borenstein, S. and Ephrati-Elizur, E. (1969), Spontaneous release of DNA in sequential genetic order by *Bacillus subtilis, J. mol. Biol.*, **45**, 137–152.

Brown, C.S., Kline, E.L. and Umbarger, H.E. (1975), Single reactions with multiple functions: multiple enzymes as one of three patterns in microorganisms, In: *Isozymes II: Physiological Function* (Markert, C.L., ed.), pp. 249–273, Academic Press, New York, San Francisco, London.

Chang, S.Y., Van Houten, J., Robles, L.J. Lui, S.S. and Kung, C. (1974), An extensive behavioural and genetic analysis of the Pawn mutants in *Paramecium aurelia. Genet. Res.*, **23**, 165–173.

Chi, Y.Y. and Francis, D. (1971), Cyclic AMP and calcium exchange in a cellular slime mold. *J. cell Physiol.*, **77**, 169–174.

Cooke, R., Clarke, M., von Wedel, R.J. and Spudich, J.A. (1976), Supramolecular forms of *Dictyostelium* actin. In: *Cell Motility* (Goldman, R., Pollard, T. and Rosenbaum, J., eds), book B, pp. 575–587, Cold Spring Harbor Laboratory, Cold Spring Harbor.

Crandall, M. (1977), Mating-type interactions in microorganisms. In: *Receptors and Recognition* (Cuatrecasas, P. and Greaves, M.F., eds.) series A, Vol. 3, pp. 45–100, Chapman and Hall, London.

Cronkite, D.L. (1976), A role of calcium ions in chemical induction of mating in *Paramecium tetraurelia, J. Protozool.*, **23**, 431–433.

Cronkite, D.L. (1977), An analysis of the mechanism of rapid micronuclear migration in *Paramecium caudatum, Proc. 5th Int. Cong. Protozool.* (in press).

Cutrecasas, P. and Hollenberg, M.D. (1976), Membrane receptors and hormone action. *Adv. Protein Chem.*, **30**, 251–451.

Dazzo, F.B. and Brill, W.J. (1977), Receptor site on clover and alfalfa roots for
   Rhizobium. App. environ. Microbiol., **33**, 132–136.
DeVay, J.E. and Adler, H.E. (1976), Antigens common to hosts and parasites.
   A. Rev. Microbiol., **30**, 147–168.
Eckert, R. (1972), Bioelectric control of ciliary activity. *Science*, **176**, 473–481.
Greaves, M.F. (1976), Cell surface receptors: a biological perspective. In: *Receptors
   and Recognition* (Cuatrecasas, P. and Greaves, M.F., eds.), series A, Vol. 1,
   pp. 1–32, Chapman and Hall, London.
Hicks, J.B., Strathern, J.N. and Herskowitz, I. (1977), The cassette model of mating-
   type interconversion. In: *DNA Insertion Elements, Plasmids and Episomes*
   (Bukhari, A.I., Shapiro, J. and Adhya, S., eds), pp. 457–462.
   Cold Spring Harbor Laboratory, Cold Spring Harbor.
Hiwatashi, K. (1969), *Paramecium.* In: *Fertilization. Comparative Morphology,
   Biochemistry and Immunology* (Metz, C.B. and Monroy, A., eds.), Vol. 2,
   pp. 255–293. Academic Press, New York.
Karlson, P. and Lüscher, M. (1959), 'Pheromones': a new term for a class of biologically
   active substances. *Nature*, **183**, 55–56.
Kung, C., Chang, S.Y., Satow, Y., Van Houten, J. and Hansma, H. (1975), Genetic
   dissection of behavior in *Paramecium, Science*, **188**, 898–904.
Naitoh, Y. and Eckert, R. (1974), The control of ciliary activity in Protozoa. In:
   Cilia and Flagella (Sleigh, M.A., ed.), pp. 305–352, Academic Press, London-
   New York.
Nicolson, G.L. (1976), Transmembrane control of the receptors on normal and
   tumor cells. I. Cytoplasmic influence over cell surface components.
   Biochim. biophys. Acta, **457**, 57–108.
Perkins, J.P. (1973), Adenyl cyclase. *Adv. Cyclic Nucleotide Res.*, **3**, 1–64.
Pinto da Silva, P. and Nicolson, G.L. (1974), Freeze-etch localization of concanavalin A
   receptors to the membrane-intercalated particles of human erythrocyte ghost
   membranes. *Biochim. biophys. Acta*, **363**, 311–319.
Reissig, J.L. (1974), Decoding of regulatory signals at the microbial surface.
   Curr. Top. Microbiol. Immunol., **67**, 43–96.
Sutherland, E.W. (1972), Studies on the mechanism of hormone action. *Science*,
   177, 401–408.
Verma, S.P. and Wallach, D.F.H. (1976), Erythrocyte membranes undergo cooperative,
   pH-sensitive state transitions in the physiological temperature range: evidence
   from Raman spectroscopy. *Proc. natn. Acad. Sci., U.S.A.*, **73**, 3558–3561.
Wiese, L. (1969), Algae. In: *Fertilization. Comparative Morphology, Biochemistry and
   Immunology* (Metz, C.B. and Monroy, A. eds), Vol. 2, pp. 135–188.
   Academic Press, New York.
Wolpert, J.S. and Albersheim, P. (1976), Host-symbiont interactions. I. The lectins
   of legumes interact with the O-antigen-containing lipopolysaccharides of their
   symbiont *Rhizobia. Biochem. biophys. Res. Commun.*, **70**, 729–737.

Thesaurus of Microbial Interactions

Superscripts e and o refer to specificity in mating systems: e indicates an interaction restricted to heterotypic cells or sources, while o indicates (some) lack of discrimination between heterotypic and homotypic interaction. Some subunit molecular weights are indicated in multiples of 1000 units (K).

I. PRIME PATHWAYS

A. LONG-RANGE COMMUNICATION

A.1. COME-CLOSE MESSAGES

(Summons for development, mating or migration)

Processes

- chemotaxis [*D. discoideum*, 1.1, 1.2.1–1.2.5; *P. violaceum*, 1.1]
  - chemotaxis [bacteria, 2]
    - chemotaxis [bacteria, 4.7.1]
      - chemotaxis [*C. moewusii* var *rotunda*[e], 10.3]
      - chemotaxis [*B. intermedium*[e], 9.6]

Agents

- acrasin = cAMP, pulsed [*D. discoideum*, 1.1, 1.2.1/.2]; acrasin = 1K peptide [*P. violaceum*, 1.1, Table 1.1]
  - attractants or repellents [bacteria, 2.2.2/.4]
    - gamone II = blepharismone = 2′-formylamino-5′-hydroxybenzoyl lactate [*B. intermedium*[e], 9.6]

Receptors

- cAMP receptors on membrane [*D. discoideum*, 1.2.2, 1.3]
  - chemoreceptors = peripheral membrane proteins [*E. coli*, *S. typhimurium*, 2.4.2, Tables 2.1–2.4]

Signal processing

- signalling components, including Methyl-accepting-Chemotactic Protein (MCP) [*E. coli*, 2.5]

Specific responses

- extension/contraction of pseudopodia [*D. discoideum*, 1.2.5]
  - modulation of the frequency of reversal of flagellar rotation, hence tumbling [*E. coli*, *S. typhimurium*, 2.3.2–2.3.4, 2.5]

A.2. GET-READY MESSAGES

(For development, mating or DNA uptake)

Processes

- development of cohesiveness, initiation of differentiation [*D. discoideum*, 1.1, 1.2.6, 1.6]
  - development of competence for transformation [bacteria, 5.3.3, 5.4.3]
    - development of competence for mating [*S. cerevisiae*^{e}, 7.2]
      - development of competence for mating
        [*B. intermedium*^{e}, 9.6; *T. canadensis*^{o}, 9.4.2;
        E. patella, E. eurystomus, O. bifaria^e, 9.5.2]

Agents

- cAMP, pulsed [*D. discoideum*, 1.1, 1.2.6; *P. violaceum*, 1.1]
  - activators = 5 to 10K proteins [*S. sp.*, 5.3.3, 5.4.3c]
    - *a* factor(s) = (glyco)protein(s) (etc.)/α factor
      (comprising α_1 through α_4) = oligopeptides
      [*S. cerevisiae*^{e}, 7.3.3–7.3.6, 7.4.1/.2, Table 7.2]
      - gamone I = blepharmone = 20K glycoprotein/
        gamone II (see I.A.1) [*B. intermedium*^{e}, 9.6];
        heat-stable factor [*T. canadensis*^{o}, 9.4.2]

Receptors

- cAMP receptors on membrane [*D. discoideum*, 1.2.2, 1.3]
  - activator receptors on membrane [*S. pneumoniae*, 5.3.3]

Specific responses

- protrusion of filopodia, development of relay competence, formation of contact sites A, of lectins and of extracellular phosphodiesterase; increased affinity of lectin receptors, differentiation of stalk cells [*D. discoideum*, 1.1, 1.2.4/.6/.7, 1.4.1/.4]
  - induction of autolysin and of DNA-binding protein; agglutinability at low pH; binding of DNA; entry of DNA, single-strand degradation [*S. pneumoniae*, 5.3, 5.4, 6.3.1d; bacteria, 5.4]
    - development^e of competence for aggregation^o, wall alterations^e, shmoo formation^e, reversible arrest in G1^e; fertilization^e [*S. cerevisiae*, 7.2.3/.5, 7.4.3e, Table 7.3]
      - development^e of competence for aggregation^o; meiosis^e and fertilization^e [*B. intermedium*, 9.6]

Thesaurus of Microbial Interactions 421

A.3. TELL-OTHERS MESSAGES

(Relay of come-close and get-ready messages)

Processes

- signal relay [*D. discoideum*, 1.2.4]
    - autoinduction of activator [*S. sp.*, 5.3.3]
        - induction of *a* factor by α factor [*S. cerevisiae*[e], 7.2.2]
            - induction of gamones [*B. intermedium*[e], 9.6];
              spreading effect [*T. canadensis*[o], 9.4.2]

Agents (Same as in I.A.1/.2)

Receptors

- cAMP receptors on membrane for signal relay differ from receptors for other messages [*D. discoideum*, 1.3]

Specific responses

- induction of adenylate cyclase [*D. discoideum*, 1.2.4]

B. SHORT-RANGE COMMUNICATION

B1. GET-READY MESSAGES

(For mating or phage infection)

Processes

- phage infection [enterobacteria plus phages, 3.1/.8]
    - conjugation [*E. coli*, 6.3]
        - mating [*C. reinhardi*, 8.1]
            - mating [ciliates, 9.3.2, 9.4.2, 9.5.1]

Signalling and early stages

- reversible and irreversible attachment; membrane alterations, superinfection exclusion; DNA entry [*E. coli* plus phages, 3.1/.8]; entry may be subdivided in stages requiring gene expression in order to proceed [*E. coli* plus T5, 3.3.1]
    - F-pili binding[o], wall-wall contact, stabilization of mating aggregates[e]; conjugal DNA replication (involving endonuclease cut at *oriT*), wall alterations; DNA transfer *E. coli*, 6.3.1a/c, 6.3.2b/c/e, 6.4, Fig. 6.9]

Signalling and early stages (continued)

- flagellar agglutination^e; wall lysis, activation of mating structures; zygotic cell fusion [*C. reinhardi*, 8.1, 8.2.1, 8.3.2/.3/.4, 8.4.1]
  - ciliary adhesion^e, holdfast union^o, paroral union^e, nuclear reorganization^o, and fertilization^o [paramecia, 9.3.2]; co-stimulation^e, reversible union^e, commitment, and meiosis [*T. thermophila*, 9.4.2]; courtship dance^e [hypotrichs, 9.5.1]

Molecular basis

- phage proteins(s)/phage receptors on bacterial membrane = pilin, flagellin, matrix protein; *ompA, tonA, tonB, bfe, lamB, tsx* gene products [*E. coli* plus various phages, 3.2–3.6, 5.7.2, 6.3.1a/d]
  - male: F-pili containing F-pilin = 11K protein/female: pOmpA = 28K protein [F$^+$ *E. coli*, 6.3.1a/b, 6.3.2g]; male: I-pili/female: LPS [I$^+$ *E.coli*, 6.3.1b]

C. DOCKING

Agents

- pallidin = 25K + 18K protein/complementary oligosaccharide-containing molecule [*P. pallidum*, 1.4]; discoidins I (= 26K protein) + II (= 25K protein)/complementary molecule [*D. discoideum*, 1.4]; contact sites A/complementary molecule [*D. discoideum*, 1.5]
  - bacterial adhesins/complementary molecules (bacteria plus animal cells, 4.2/.3/.5]. K88 adhesin = 2.3K protein [porcine *E. coli*, 4.3.1a/b]
    - complementary mannanproteins^e = type 5 factor/type 21 factor [*H. wingeii*, 7.2.5c]
      - O-antigen-containing LPS/plant lectin [rhizobia plus legumes, 10.4]

II. MODULATION

A. INITIATION

A.1. TRIGGERED BY STARVATION

Processes

- cAMP pulsing [*D. discoideum*, 1.1, 1.2.6]
  - gametogenesis [*C. reinhardi*, 8.2.8]
    - initiation [*T. thermophila*, 9.4.2]; mating reactivity [ciliates, 9.2.1]

A.2. TRIGGERED OTHERWISE

Processes

- cAMP pulsing periodicity [*D. discoideum*, 1.2.1]
  - development of competence [bacteria, 5.4.3a]
    - adolescence, maturity and senile selfing [paramecia, 9.3.5]; circadian mating rythms [*P. bursaria, P. multimicronucleatum*, 9.2.5]

B. MODIFICATION

B.1. DUE TO DIVALENT CATIONS

Processes

- cAMP pulsing; aggregation; myosin-ATPase [*D. discoideum*, 1.2.5, 1.6.1/.6]
  - straight swimming and chemotaxis [*B. subtilis*, 2.3.4]
    - DNA binding and entry [*S. pneumoniae*, 5.3.1/.2/.4, 5.4.2a–5.4.2d, *B. subtilis*, 5.4.2a/d]
      - gametic agglutination [*C. reinhardi*, 8.2.9]
        - backward swimming [paramecia, 10.5]

B.2. DUE TO ENDOGENOUS ANTAGONISM

Signal-destroying agents

- phosphodiesterase [*D. discoideum*, 1.2.3]
  - barrier factor = α factor antagonist [*S. cerevisiae*, 7.2.5b, 7.3.5b]

424   *Microbial Interactions*

Blocking agents

> • products of surface exclusion genes *traS*, *traR* and *traT*, including 24K membrane protein [*E. coli*, 6.2.3, 6.3.1d, 6.3.2e]
>> • selfing inhibition [*E. eurystomus*, 9.5.2]

Refractory periods

> • for cAMP relay [*D. discoideum*, 1.2.4]; for movement [*D. discoideum*, 1.2.5]
>> • implicit in temporal comparisons for chemotaxis [bacteria, 2.4.1]

B.3. DUE TO A BYPASS

Agents and processes

> • uncouplers of oxidative phosphorylation acting as repellents in chemotaxis [*B. subtilis*, 2.4.2e]
>> • *trt* gene product bypasses activator requirement for competence [*S. pneumoniae*, 5.3.3]
>>> • starvation induces selfing° [*E. eurystomus*, 9.5.2]; ionic conditions (e.g. high K^+/Ca^{2+} ratio) can bypass° the ciliary adhesion step in mating [paramecia, 9.3.2, Fig. 9.2, 10.5]

C. TERMINATION

C.1 DISENGAGEMENT

Processes

> • temporariness of adhesivity [bacteria plus animal cells, 4.7.1]
>> • disaggregation after mating [*E. coli*, 6.3.2d]
>>> • loss of flagellar agglutinability [*C. reinhardi*, 8.1]
>>>> • loss of ciliary adhesivity [paramecia, 9.3.2]; undocking after mating [ciliates]

Cues

> • gene(s) on transferred sex factor [*E. coli*, 6.3.2d]
>> • unidentified contacts [*C. reinhardi*, 8.4.2]
>>> • holdfast union [*P. primaurelia*, 9.3.5d]

C.2. FAILED OUTCOME

Processes

- lethal effects of phage ghosts [*E. coli* plus phages, 3.1]
- lethal zygosis [*E. coli*, 6.3.2f]
- mating errors [ciliates, 9.2]

III. SUMMARY

- microbes interact perisemically [microbes, 10.7]

Index

a-factor, 285, 289, 290, 294, 296–298, 300–303, 319, 403
α-factor, 283, 285, 288–295, 298–303, 318, 319, 403
   induction, 285, 293, 404
Acrasin, 5, 7, 402
Actin, 22
Activator, 188–193, 212, 213, 403, 404
   receptors, 190–193, 410
Adenylate cyclase, 11, 20, 21, 79
Adhesins, bacterial, 141, 147, 151, 167, 168, 407
   receptors, 142, 155–161, 166, 167
   See also Fibrillae; Fimbriae; K88-antigen; K99-antigen; Lipopolysaccharide; Lipoteichoic acid; O-antigen
Adhesion of bacteria to
   animal cells, 141–168
   crop, 148, 155
   epithelia, 142, 144–148, 153–155, 160, 167
   erythrocytes, 142, 144–149, 152–161, 167
   eukaryotic cells, ecological advantages of, 166, 167, 407
   glass, 149, 163–165
   intestine, 145, 155, 158–160
   roots, 407
   teeth, 147, 160
   tongue, 147, 148
Adolescence, See Maturity stages
Agglutination, See Aggregation
Agglutination factors, See Agglutinins
Agglutinins, 406, 407
   ciliates, 356, 364, 370, 372, 378, 389
   yeasts, 285, 292, 298, 299
   See also Adhesins; Conjugation bridge;

Agglutinins, (continued)
   Contact sites A;
   Discoidin; Haemagglutinins;
   Isoagglutinins; Lectins;
   ompA protein; Pallidin; Pili
Aggregation, 406, 407
   in development, 3–7, 23–25, 29, 30, 38–48, 403, 408
   acquisition of competence to aggregate, 23, 24, 31, 34, 35
   aggregation-stimulating factor, 24
   in mating mixtures, 286–288, 290, 299 304–310, 325–334, 341–346, 359, 360, 362, 363, 371, 372, 389, 405, 408
   acquisition of competence to aggregate, 288, 290, 293, 294, 298, 299, 301, 309, 318, 335, 336, 341, 403
   homotypic vs. heterotypic, 362–365, 376, 378, 380, 382, 383, 385, 387, 406, 411
   loss of competence to aggregate, 325, 334, 337, 346, 362, 373, 405, 408
   stages, 288, 310, 316
   See also Mating aggregates, E. coli.
   in ordinary cultures, 143, 144, 152, 158
   adding antibodies, 339–341, 343, 346
   adding lectins, 337–340, 343, 344, 346
   with subcellular fractions or corpses, 33, 330, 340, 341, 364, 365, 405, 413
   of transformation-competent cells at low pH, 189, 190

* The assistance of Alison Holtzschue in the preparation of the Index is gratefully acknowledged.

Aggregation (*continued*)
    See also Agglutinins; Charge on cell surface; Mutants, aggregation
Agrobacterium tumefaciens, 215
Albomycin, 110
    See also Mutants, insensitive, albomycin
Anesthetics, 85
ATP, 28, 67, 85, 86
ATPase, 22, 67, 85, 86
ATP-pyrophosphodydrolase, II
Autogamy, 363, 369, 371, 373
Autolysins, *See* Cell wall, autolysis
Avoiding reaction, 408, 409

Bacillus, 215, 267
Bacillus subtilis, 61, 65, 67, 68, 72, 75, 84, 85, 87, 89, 179, 181, 183, 184, 194–196, 198-214, 257
Bacteriophages, *See* Phages
Bacterium photometricum, 72
Barrier factor, 294, 297, 298
Blepharisma intermedium, *See Blepharisma japonicum*
Blepharisma japonicum, 355, 379, 384–387, 390, 403, 404, 411
Blepharismone, 385–387, 403
    receptors, 387
Blepharmone, 385–387, 403
    receptors, 387
Bordetella, 147, 150–152, 167

Ca^{2+}, 22, 38, 44, 48, 67, 85, 86, 165, 187, 194, 202, 204, 205, 208, 215, 216, 336, 363, 373, 408, 409
    receptors, 38
Cell–cell interactions
    long-range, 402–404
        See also Chemotaxis; Cyclic AMP, signal relay; Pheromones
    short-range, 319, 356, 359, 360, 405–407
        See also Aggregation; Mating
Cell division
    arrest in G1, 283, 285, 286, 290, 293–295, 297–301, 359, 403
    synchronization, 285, 308, 359, 372, 377
Cell expansion, 294
Cell fusion, 283, 285, 287, 288, 318, 325,

Cell fusion (*continued*)
    342–346, 354, 355, 380, 405
    reversible union vs. irreversible commitment, 337, 376, 378
Cell wall
    alterations during mating, 288, 290, 292, 293, 318, 403
    autolysis, 183, 190, 212, 325, 342–345, 404, 405
    susceptibility to lysis, 290, 301
    See also Peptidoglycan
Charge on cell surface, 153, 157, 161–166, 406
Chemoreceptors, 61, 70, 71, 73, 74, 76, 79–86, 90, 404, 410, 411, 414
    for amino acids, 84, 85
    fluorescence quenching, 81–83
    for ions, 85, 86
    occupancy, 74, 75
    for repellents, 78, 85
    for sugars, 70, 71, 74, 76, 80–84, 86, 87, 90, 117, 118, 130
Chemosensors, 70–88, 90, 404, 410, 414
Chemotaxis, 402
    bacteria, 61–91, 146, 167, 404, 408–411, 413
        attractants, 355, 356, 385
        repellents, 62, 78, 84, 85
    Chlamydomonas, 405
    ciliates, 355, 356, 385, 403
    slime molds, 3–5, 7–23, 29, 30, 46–48, 403, 404
        refractory period for movement, 18, 22, 23
Chlamydomonas eugametos, 325, 339
Chlamydomonas moewusii, 325, 326, 332–334, 338, 339, 341–344, 405
Chlamydomonas reinhardi, 325–346, 405, 408, 413
Choline, 190
Cilia, 353, 355, 362, 413
Ciliary agglutination, *See* Aggregation, in mating mixtures
Ciliates, 353–390, 411
Circadian rhythms of mating reactivity, 360, 370

Clostridium welchii, 167
Colicins
  receptors, 101–137, 410
  See also Mutants, insensitive, colicins
Competence, *See* Transformation
Competence factors, 189, 212
  See also Activator
Concanavalin A, 24, 29, 154, 292, 301, 302, 338–340, 344, 377
  glycoproteins binding to 38–41, 48
Conjugation, *See* Mating
Conjugation bridge, 249, 269
Contact sites A, 23–25, 36–38, 48, 407
Co-operativity, 35, 44, 412
  See also Kinetics, non-linear
Corynebacteria, 148, 152
Co-strimulation, 376–378, 405
Courtship, 285, 286, 316–318
  dance, 381, 384, 405
Cyclic AMP, 3–30, 44–48, 212, 213, 402, 403, 409
  analogues, 12
  pulses, 6–12, 14, 18–24, 46–48, 404, 408
  receptors, 3, 12–14, 25–30, 46, 48, 403, 410
  signal relay, 5, 17–22, 29, 30, 48
    refractory period, 5, 12, 18, 23
Cyclic GMP, 13, 26
Cytochromes, 10, 11
Cytogamy, 363, 364, 373, 380

Development, 6, 23, 24, 28, 29, 36, 39–41, 43–46, 403, 413
Dictyostelium discoideum, 3–48, 402, 403, 407–409, 413
Diplococcus pneumoniae, *See Streptococcus pneumoniae*
Disaggregation after mating, 235, 257, 261, 264, 266–269, 408
Discoidin, 24, 31–35, 37, 48, 407
  receptors, 34, 35, 407
Divalent cations, 61, 85, 86, 183–185, 194, 200–202, 205, 363, 377, 408, 409, 414
  receptors, 38
  See also individual ions

DNA
  binding to bacteria, 179, 181, 183–185, 190–203, 205–207, 210
    reversible vs. irreversible, 181, 196–199, 201, 202, 219
  cohesive ends, 214, 215
  conversion to single strands
    in conjugation, 255, 269
    in transformation, 185–187, 192–194, 198, 257
  entry, 103, 179–181, 185–188, 190–193, 196, 199–211, 404
  fragmentation
    in conjugation, 217
    in encapsidation, 218
    in transformation, 184–187, 191–194, 197–201, 205–209, 219, 257
    See also Endonuclease, major pneumococcal
  integration, 180, 186, 187, 194, 203, 206, 207, 209–211, 216
  proteins binding to, 204, 206, 212, 217–219, 257, 258 (for molecules on the cell surface which bind to DNA, *See* DNA, receptors)
  DNA, *See* DNA, receptors)
  receptors, 191–194, 198–202, 208, 209, 212, 404
  recombination, 186, 187, 192, 193, 204, 207, 218
  release (of transforming DNA), 403
  replication, 211, 241, 354
    conjugational, 217, 255–257, 265, 268, 405
    rolling circle mechanism, 235, 256, 257
    inhibition of, 283, 285, 290, 293
  specificity in transformation, 194, 195, 204
  transfer
    conjugational, 217, 235, 247, 249, 255–258, 263–266, 268, 405
    phage, 102, 103, 105–108, 128, 217, 218, 244, 258, 406
    See also DNA, entry; Pili, F and

DNA, transfer (*continued*)
　　DNA transfer; Transfection DNases, 181, 183, 201–205, 207–209, 216, 257
　　　See also Endonuclease, major pneumococcal
Docking, *See* Aggregation

Electron transport, 67
　　See also Cytochromes
Endonuclease
　　major pneumococcal, 182, 183, 187, 188, 190–194, 205, 207, 208, 219
　　　See also Mutants, nuclease
　　sex-factor coded, 256, 269
Energy requirements, 67, 80, 104, 108, 184, 187, 191, 194, 196, 200–203, 205, 218, 219
　　See also Membrane-energized state
Enterobacteria, 143, 215
Enterochelin, 113, 116, 120, 121, 124
　　See also Mutants, transport
Escherichia coli, 61–91, 101–137, 143, 145, 146, 150, 152, 153, 157, 161, 163, 164, 167, 195, 215–217, 235–269, 405, 408, 410–413
Euplotes, 359, 381–385, 389

F factor, 237–239, 242
　　Hfr, 236, 241, 261, 263, 268
　　origin of transfer (oriT), 240, 243, 256, 269, 405
　　origin of vegetative replication (*oriV*), 240, 241
　　See also Mutants
F pili, *See* Pili
F pilin, *See* Pilin
Ferrichrome, 109–111, 120, 121, 123, 124
　　See also Mutants, transport
Fertilization, 354, 356, 371, 379, 385, 405
　　tubule, *See* Mating structures
Fibrillae, 141, 145, 148, 153, 154
Filopodia, 14, 24, 25, 404
Fimbriae (i.e. non-sex pili), 141, 143, 144, 146, 147, 150–152, 157, 158, 161, 163, 167, 245, 250, 251

Flagella
　　bacterial, 63–70, 90, 146, 167, 245, 404, 410, 413
　　Chlamydomonas, 325–327, 331, 335, 340, 346, 405, 413
　　fuzzy coat, 326, 327, 330–333, 405
　　See also Mutants, flagella
Flagellin, 65, 71, 104
Flavins, 69
Folic acid, 46, 47

Gametogenesis (including comparison of gametes with vegetative cells), 325–329, 335, 337–339, 341, 413
Gamones, *See* Pheromones
Glucan, 290, 292
Glycocalyx, 155, 156
Glycoproteins
　　Blepharisma, 386, 402, 403
　　Hansenula, *See* Agglutinins, yeasts
　　salivary, 160
　　See also a-factor; Membrane, glycoproteins; Concanavalin A
Glycosphingolipids, 41, 42, 48
Glycosyltransferases, 334, 335, 341

Haemagglutinins, *See* Adhesion of bacteria to, erythrocytes
Haemophilus, 195, 215, 257, 258
Haemophilus influenzae, 179, 181, 195, 196, 199–207, 211–214
Hansenula wingei, 292
Holdfast attachment, 362–364, 371–373, 378, 405
Hormones, 402, 410, 412
　　See also Pheromones

Inbreeding vs. outbreeding genetic economies, 357, 358, 361, 362, 368, 369, 383, 388, 389, 411, 412
Initiation, 375, 376
Interactions, *See* Cell–cell interactions
Isoagglutinins, 327

Isoagglutinins (*continued*)
   See also Membrane, vesicles, flagellar

K^+, 187, 194, 202, 204, 205, 363, 408, 409
K88 antigen, 145, 150–152, 157, 159, 160, 167
K99 antigen, 145
Kinetics
   of DNA binding to cells, 195–197
   non-linear, 17, 29
Klebsiella, 143, 144, 161
Klinokinesis, 63

Lactobacillus, 148, 153, 158
Lectins, 35, 36, 157, 407
   See also Concanavalin A; Discoidin; Pallidin
Lethal zygosis, 242, 266–268
   See also Mutants, lethal zygosis
Life cycles
   Paramecium, 371
      See also Maturity stages
   yeast, 284
Lipids, See Membrane, lipids
Lipopolysaccharide, 103–105, 154, 161, 267
   See also Mutants, lipopolysaccharide
Lipoprotein, 104, 258
Lipoteichoic acid, 153, 154, 161, 168

Macronucleus, 354, 355, 366, 367, 370, 371, 379, 381
Mannan and mannanproteins, 290, 292, 407
Mastigonemes, 326–328, 331, 333
Mating
   Chlamydomonas, 325–346, 405
   ciliates, 353–390, 403, 405, 409
      chemical induction, 362–364, 371, 373, 388, 408, 409
      hazards of, 356
   E. coli
      F-factor-coded, 235–269, 405
      I-factor-coded, 238, 239, 255
   yeasts, 283–319, 403
Mating aggregates
   E. coli, 235, 243, 258–265, 268

Mating aggregates, *E. coli* (*continued*)
   stabilization, 143, 235, 254, 262–264, 267–269, 405
   wall–wall contacts, 235, 249, 263–265, 268, 405
   See also Disaggregation after mating
   yeast, See Aggregation, in mating mixtures
Mating bridge, See Conjugation bridge
Mating dance, 359, 389
Mating pairs, See Mating aggregates
Mating signal, 256, 405
Mating structures, *Chlamydomonas*, 342, 344, 345
   activation, 325, 343–345, 405, 413
Mating substances
   diffusible, See Pheromones, mating
   surface-bound, See Agglutinins
Mating-type determination
   ciliates, 358, 365–371, 373–375, 380–384, 389, 411
      See also Circadian rhythms of mating reactivity; Inbreeding vs. outbreeding genetic economies
   yeasts, 284, 412
Maturity stages
   adolescence, 371, 372, 383, 384
   immaturity, 357, 365, 369–371, 383
   maturity shunt, 357, 358, 371–373
   senescence, 357, 371, 383
Meiosis, 356, 363, 371, 373, 376, 377, 380, 385, 387
Membrane, 103–105, 155, 156, 191–193, 201, 219, 326–329, 339, 345, 346, 355, 364
   adhesion zones, 129, 258
   channels, 104, 124, 125, 129, 130
   energized state, 67, 103, 129, 413
   fluidity, See Membrane, topological rearrangements
   fusion, 258, 263, 344, 345, 413
   glycoproteins, 155–157, 159, 160, 326, 328, 329, 341
   lipids, 155, 160, 161, 341, 344, 345
   particles, 44–46, 48, 326, 342–344

Membrane (continued)
  potential, 68, 69, 88, 408
  proteins, 42–46, 80, 104, 127, 155, 212, 414
    matrix, 104, 127, 130
    Methyl-accepting-Chemotactic Protein, 89, 90
    periplasmic, 76–84, 90, 404
    See also Chemoreceptors; Chemosensors; Mutants; Transport; individual proteins
  sialic acids, 157, 159, 167, 181
  structural transitions, 129, 410, 412, 413
  topological rearrangements, 29, 41, 44–46, 48, 338, 340, 341, 344, 410, 413, 414
  vesicles, 365
    flagellar, 327, 328, 331–334, 340
Memory, See Sensory mechanisms, temporal comparisons
Mesosomes, 199
Methionine, 75, 89, 90, 113
Mg^{2+}, 67, 68, 85, 86, 187, 188, 194, 202, 206, 209, 363, 408
Micrococcus radiodurans, 215
Microfilaments, 325, 343, 345
Micronuclei, 354, 355, 371, 376, 377, 380, 381, 389
Microtubules, 345
Minicells, 255, 256, 265
Mn^{2+}, 38, 48, 187, 188, 202, 208, 209, 363
Mucus, 145, 146, 150, 166
Murein, See Peptidoglycan
Mutants
  α-factor, yeast, 289, 303, 310–316, 413
  activator-independent, pneumococcus (*trt*), 190
  aggregation
    Chlamydomonas (*imp-2, imp-5, imp-6, imp-7, imp-8, gam-4*), 336–339, 343, 344, 405
    slime mold, 23, 37, 42, 44, 47, 48
    yeast, 310, 315
    See also Mutants, mating; Mutants, DNA transfer
  avoiding reaction, paramecium (pawn), 409

Mutants (continued)
  carbohydrate utilization, bacteria (*car, chr*), 109, 111
  cell division, yeast, 288
  cell surface, 407
  chemical induction of mating, paramecium, 373
  chemotaxis
    E. coli (for mutant symbols see p.71), 62, 63, 66, 69, 73–79, 86–89
    slime mold, See Mutants, aggregation
  conjugation, See Mutants, mating
  DNA replication, *E. coli* (*dnaE, dnaG, frp*), 239–240, 243, 255, 256, 263, 264
  DNA transfer, *E. coli* (*traA-traM, traO, traS, traT*), 240–243, 247, 256, 263–269
  F-factor replication, See Mutants, DNA replication
  F-factor repression, *E. coli* (*finO, finP*), 240, 242, 243
  flagella
    bacterial, 65, 66, 69, 71, 89
    Chlamydomonas (*pf*), 340
  fusion, *Chlamydomonas* (*imp-1, imp-1-15*), 345
  fusion response, *Chlamydomonas* (*imp-3, imp-4, gam-1*), 346
  incompatability, *E. coli* (*inc*), 239–241, 243
  insensitive
    α-factor, yeast, 289
    albomycin, *E. coli* (*sid, tonA*), 109–112, 117, 121, 130
    colicins, *E. coli* (*bfe, cbr, cbt, cir, exb, feuA, feuB, tolG, tonA, tonB, tsx*), 103, 110–117, 119–121, 124, 127, 128, 130, 249
    phage, *E. coli* (*bfe, con, lamB, ompA, pel, pif, tfrA, tonA, tonB, tra, tut, tsx*), 103, 108–124, 127–130, 239–241, 243, 247, 249, 252–254

Mutants (*continued*)
  lethal zygosis, *E. coli* (*ilzA, ilzB*), 240, 242, 266, 267
  lipopolysaccharide, *E. coli* (rough, *tfrA*), 104, 105, 253–255
  mannanprotein, yeast, 292
  mating
    deficient as donors in conjugation, See Mutants, DNA transfer
    deficient as recipients in conjugation, *E. coli* (*con, ompA, tfrA, tolG, tut*), 102, 104, 249, 252, 255, 266, 267
    early mature, paramecium, 358
    sterile, *Chlamydomonas* (*imp*), See Mutants, aggregation; Mutants, fusion; Mutants, fusion response
    sterile, *Paramecium* (CM), 373
    sterile, yeast, 288–290, 298, 310, 311
  morphological, yeast, 312, 316, 318
  nuclease, pneumococcus (*end, noz*), 182–185, 188, 196–198, 204, 206–209, 214, 216
  phosphodiesterase, slime mold, 26
  pili, See Mutants, DNA transfer
  recombination, *E. coli* (*rec-2, recB, recC, sbcA, sbcB*), 204, 206, 214, 216
  relay, slime mold, 29, 30
  resistant, See Mutants, insensitive
  sensitive to antibiotics, etc. *E. coli* (*ompA*, etc.), 249, 252, 254
  sporulation, *B. subtilis*, 211
  transformation, pneumococcus (KB, *ntr*), 182, 185, 190, 201, 204, 218
  See also Mutants, nuclease
  transport, bacteria
    ferric citrate (*cit*), 115–117, 127
    ferric enterochelin (*cbr, cbt, exb, fep, feuB, tonB*), 113–117, 125–130
    ferrichrome (*sid, tonA*), 111, 114, 117, 121, 124, 127–130
    nucleoside (*tsx*), 120, 121, 127
    sugars (*mal, mgl, lamB*, etc.), 71, 75, 77–79, 84, 116–118, 122, 123, 127, 128, 130

Mutants (*continued*)
  vitamin B_{12} (*btu, bfe*) 119, 124, 127–130
  tumbly, *E. coli*, 73, 89
Mycoplasma, 149, 150, 158, 164
Myoblasts, 36
Myosin, 22
Myotubes, 36

Neisseria, 146, 147, 150, 152, 153, 163, 211, 215
Nuclear reorganization, 363–365, 371, 373, 379, 385, 387
Nuclear transfer, 363, 364, 371, 378–380
Nucleo-cortical interactions, 380, 381

O-antigens, 407
Octanoic acid, 294
ompA protein, 102, 104, 243, 249, 252, 254, 255, 266–269, 406
Orgy, See Courtship
Oxytricha, 381, 384

Pallidin, 24, 31–35, 48, 407
  receptors, 35, 407
Paramecium, 354, 355, 361–373, 375, 378, 379, 381, 388–390, 405, 408, 411, 413
Paramecium aurelia complex, 357, 361–370, 373, 374, 388, 389
Paramecium bursaria, 358, 360, 362, 365, 368–370, 374, 378, 383, 388, 389
Paramecium caudatum, 362–367, 368, 370, 382, 388
Paramecium multimicronucleatum, 360, 362, 365, 367, 368, 370, 372, 388
Paroral attachment, 363, 371–373, 378, 405
Parsimony, 411, 413
Pathogenicity
  animal, 143–145, 147, 149, 150, 167
  plant, 407
Peptidoglycan, 103, 104

Peptidoglycan (*continued*)
  changes during conjugation, 265, 268
Periplasmic proteins, See Membrane, proteins, periplasmic
Perisemic, 412–414
Permaphores, 412, 414
Phages, 84, 195, 210, 214–218, 239, 241, 244, 246, 247, 258, 405, 406
  DNA transfer, See DNA transfer, phage
  encapsidation, 218
  ghosts, 101, 407
  receptors, 84, 101–137, 217, 258, 406, 407, 410, 413
  See also Mutants, insensitive, phage
Pheromones, 190, 191, 212, 213, 402–404, 406
  mating
    inhibitory, 383
    stimulatory, 283, 284, 288, 289, 293, 294, 297, 298, 319, 356, 360, 382–385, 389, 390
    See also a-factor; α-factor; Blepharismone; Blepharmone
  other, See Acrasin; Activator; Competence factors; Cyclic AMP
Phosphodiesterase, 3, 13, 15–17, 20, 23, 24, 26–28, 40, 41
Pili, sex, 102, 141, 405, 413
  F, 235, 241, 243–251, 258, 262–269
    binding to bacteria, 246, 250, 251, 261, 267, 268, 406
    and DNA transfer, 235, 249, 262, 263
    retraction, 244, 248, 249
  F-like, 238, 239, 244
  I-like, 238, 239, 244
  other, 238, 239, 246–248
Pilin, 104, 241, 244, 246, 406
Pilot protein, See DNA, proteins binding to
Plants, 407
Plasmids
  general and non-sex, 200, 216, 217, 235–237, 256
  sex, 236–239, 242, 266, 269
    incompatibility, 237–239
    repressed, 242, 269, 412
    See also F factor; Mutants, incompatibility

Pleiotropism, 114, 252–254, 298, 311, 316, 319, 407, 413
Polylysine, 377
Polysphondylium pallidum, 7, 31–35, 407
Polysphondylium violaceum, 5, 7, 15, 402
Prostaglandins, 46, 47
Protein kinases, 24
Protein synthesis requirements, 404
  for the development of aggregation competence, 288, 309, 310
  for the development of mating competence, 375, 377, 381, 385
  for the development of transformation competence, 189–191, 199, 212
  for disaggregation, 264
  for DNA transfer, 105, 257
  implicit in slime mold interactions, 20, 21, 23, 24, 33, 34, 36–40, 43, 44, 48
Proteus, 143, 144
Pseudomonas, 152, 157, 237, 267, 406
Pseudopodia, 404

Receptors, 101–103, 412
  induction, 118, 122, 125–128
  See also Membrane, adhesion zones; Membrane, proteins; Specificity of molecular interactions; Transport, receptor-dependent uptake systems; individual ligands.
Refractory period for cyclic AMP signal relay, See Cyclic AMP
Refractory period for movement, See Chemotaxis, slime molds
Rhizobium, 215, 407
RNA
  synthesis requirement for DNA transfer, 257
  transfer, phage, 244

Saccharomyces cerevisiae, 283–319, 403, 404, 411–413

Salmonella, 143, 144, 167
Salmonella typhimurium, 61–91, 109–113
Selfing, 359, 369, 379, 383, 384
   juvenile, 358, 379
   senile, 358
   See also Aggregation, in mating mixtures, homotypic vs. heterotypic
Senescence, See Maturity stages
Sensory mechanisms, 404
   adaptation, 72–75, 89–91
   jamming, 62, 75, 85–87, 414
   generation, See Cyclic AMP, pulses
   processing, See Chemosensors
   relay, 5, 17–22, 29, 30, 48, 189–191, 285, 377, 385, 402–404
   signal destruction, 15–17, 290, 297, 298
   signal transduction, 86–88, 90, 101, 380, 387
      cilia-to-cell, 378
      flagellum-to-gamete, 345, 346
      zygote-to-flagellum, 346
      See also Perisemic; Receptors
   temporal comparisons, 14, 72–74, 91, 404
Serratia, 143, 144
Sex-factors, See Pheromones, mating; Plasmids, sex
Shigella, 143, 167
Shmooing, 283, 289, 290, 292–302, 403
Slime molds, See *Dictyostelium discoideum*; *Polysphondylium* spp.
Specificity of molecular interactions, 120–122, 410, 411
Spreading effect, 377
Staphylococcus, 148, 161, 215
Starvation requirements for the development of mating competence, 335, 336, 356, 360, 375
Stentor, 380
Streptococcus, 147, 148, 153, 154, 158, 160, 161, 166–168, 183, 184, 189, 195, 203, 207, 211–213, 215, 236
Streptococcus pneumoniae, 179–219, 257, 403, 404, 413

Streptomyces, 236
Stylonychia, 355, 381, 382
Sulfhydryl requirement for transformation, 184
Superinfection exclusion, 128
Surface exclusion, 242, 256, 258, 261, 264–269
   See also Mutants, DNA transfer (*traS* and *traT*)

Teichoic acids, 190
Tetrahymena, 354–356, 359, 373–382, 389, 405
Tetrahymena canadensis, 374, 377
Tetrahymena pyriformis, 373–375, 380
Tetrahymena thermophila, 358, 374, 375, 379
Transduction, 218
Transfection, 180, 213–216
Transformation, 179–219
   competence, 179, 180, 182, 188–193, 199, 209, 211–213, 218
      See also Activator
   donor marker eclipse, 186, 203, 204
   non-physiological, 215, 216
      See also Transfection
   phenotypic expression, 180
   and protoplasts, 212
   and surface exclusion, 258, 265
Transport, 77–80, 90
   chemiosmotic hypothesis, 67
   gates or channels, 69, 129, 130, 258, 409
   phosphotransferase system, 76, 79, 80
   receptor-dependent uptake systems, 109–130, 410
Trichocysts, 362
Tumble generator, 68–71, 74, 87–90, 404
Tumbling, bacteria, 63, 66, 72
   See also, Mutants, tumbly; Tumble generator

Uncouplers of oxidative phosphorylation, 67, 85

Vibrio, 145, 146, 157, 158, 166
Vitamin B$_{12}$, 119, 121, 122, 124
   See also Mutants, transport

Zn^{2+}, 248, 264
Zygote formation, *See* Cell fusion; Fertilization